AF863983

RILEM State-of-the-Art Reports

RILEM STATE-OF-THE-ART REPORTS
Volume 31

RILEM, The International Union of Laboratories and Experts in Construction Materials, Systems and Structures, founded in 1947, is a non-governmental scientific association whose goal is to contribute to progress in the construction sciences, techniques and industries, essentially by means of the communication it fosters between research and practice. RILEM's focus is on construction materials and their use in building and civil engineering structures, covering all phases of the building process from manufacture to use and recycling of materials. More information on RILEM and its previous publications can be found on www.RILEM.net.

The RILEM State-of-the-Art Reports (STAR) are produced by the Technical Committees. They represent one of the most important outputs that RILEM generates – high level scientific and engineering reports that provide cutting edge knowledge in a given field. The work of the TCs is one of RILEM's key functions.

Members of a TC are experts in their field and give their time freely to share their expertise. As a result, the broader scientific community benefits greatly from RILEM's activities.

RILEM's stated objective is to disseminate this information as widely as possible to the scientific community. RILEM therefore considers the STAR reports of its TCs as of highest importance, and encourages their publication whenever possible.

The information in this and similar reports is mostly pre-normative in the sense that it provides the underlying scientific fundamentals on which standards and codes of practice are based. Without such a solid scientific basis, construction practice will be less than efficient or economical.

It is RILEM's hope that this information will be of wide use to the scientific community.

Indexed in SCOPUS, Google Scholar and SpringerLink.

More information about this series at http://www.springer.com/series/8780

Dr. Alexis Courtois, EDF Direction Ingénierie et Projets Nouveau Nucléaire—Direction Technique, 19 rue Pierre Bourdeix, 69007 Lyon, France.
e-mail: alexis.courtois@edf.fr
Dr. João Custódio, LNEC, Lisbona, Portugal. e-mail: jcustodio@lnec.pt
Dr. Steven Feldman, NIST, Materials and Structural Systems Division, Gaithersburg, MD, USA. e-mail: steven.feldman@nist.gov
Dr. Eric R. Giannini, RJ Lee Group, La Crosse, WI, USA.
e-mail: egiannini@rjleegroup.com
Nathalie Ishak, Hydro-Quebec, Canada.. e-mail: Ishak.Nathalie@hydro.qc.ca
Prof. Laurence Jacobs, Georgia Institute of Technology, Atlanta, USA.
e-mail: laurence.jacobs@coe.gatech.edu
Dr. Tetsuya Katayama, Taiheiyo Consultant Co Ltd, Japan.
e-mail: katayama@catv296.ne.jp
Dr. Yuichiro Kawabata, Â Port and Airport Research Institute, Japan.
e-mail: kawabata-y@pari.go.jp
Dr. Yann Le Pape, Oak Ridge National Laboratory, TN, USA.
e-mail: lepapeym@ornl.gov
Dr. Andreas Leemann, Empa, Swiss Federal Laboratories for Materials Testing and Research, Laboratory for Concrete and Construction Chemistry, Switzerland.
e-mail: andreas.leemann@empa.ch
Dr. Renaud-Pierre Martin, Université Paris Est, Materials and Structures Department, Urban and Civil Engineering Testing and Modeling Laboratory (EMGCU), IFSTTAR, French institute of science and technology for transport, development and networks, F-77447 Marne la Vallée, France.
e-mail: renaud-pierre.martin@ifsttar.fr
Dr. Esperanza Menendez Mendez, Institute of Construction Science Eduardo Torroja, Madrid, Spain. e-mail: emm@ietcc.csic.es
Dr. Christine Merz, Merz Ingenieurberatung gmbh, Möriken, Switzerland.
e-mail: merz@merz-ingenieurberatung.ch
Dr. Stéphane Multon, Civil Engineering, Paul Sabatier University, and Construction and Durability Materials Laboratory of Toulouse (France).
e-mail: stephane.multon@insa-toulouse.fr
Dr. David Rothstein, DRP, A Twining Company, Boulder, CO, USA.
e-mail: petro@drpcinc.com
Prof. Leandro Sanchez, University of Ottawa, Canada. Leandro.
e-mail: Sanchez@uottawa.ca
Prof. Victor E. Saouma, University of Colorado, Boulder, USA.
e-mail: saouma@colorado.edu
Prof. Alain Sellier, Civil Engineering, Paul Sabatier University, and Construction and Durability Materials Laboratory of Toulouse (France).
e-mail: alain.sellier@insa-toulouse.fr
Dr. Henrik Erndahl Sorensen, Teknologisk Institut, Denmark.
e-mail: hks@teknologisk.dk
Dr. Yuya Takahashi, University of Tokyo, Japan.
e-mail: takahashi@concrete.t.u-tokyo.ac.jp

Victor E. Saouma
Editor

Diagnosis & Prognosis of AAR Affected Structures

State-of-the-Art Report of the RILEM Technical Committee 259-ISR

Springer

RILEM Technical Committee 259-ISR

Prognosis of deterioration and loss of serviceability in structures affected by alkali–silica reactions

Chairman

Victor Saouma

Deputy Chairmen

Leandro Sanchez
Yann LePape

Active Committee Members

This book is authored and edited under the auspices of the RILEM Technical Committee TC 259-ISR. It presents technical contributions from RILEM members and international partners that provide an outline of the state-of-the-art in diagnosis and prognosis of concrete structures affected by alkali–silica reaction (ASR).

Francesco Amberg, Lombardi Consulting, Switzerland.
e-mail: francesco.amberg@lombardi.ch
Prof. Ignacio Carol, Polytechnic of Catalunya, Barcelona, Spain.
e-mail: ignacio.carol@upc.edu
Dr. Derek Cong, WJE, Austin TX, USA. e-mail: dcong@wje.com

Editor
Victor E. Saouma
Department of Civil Engineering
University of Colorado
Boulder, CO, USA

ISSN 2213-204X ISSN 2213-2031 (electronic)
RILEM State-of-the-Art Reports
ISBN 978-3-030-44013-8 ISBN 978-3-030-44014-5 (eBook)
https://doi.org/10.1007/978-3-030-44014-5

© RILEM 2021
No part of this work may be reproduced, stored in a retrieval system, or transmitted in any form or by any means, electronic, mechanical, photocopying, microfilming, recording or otherwise, without written permission from the Publisher, with the exception of any material supplied specifically for the purpose of being entered and executed on a computer system, for exclusive use by the purchaser of the work. Permission for use must always be obtained from the owner of the copyright: RILEM.
The use of general descriptive names, registered names, trademarks, service marks, etc. in this publication does not imply, even in the absence of a specific statement, that such names are exempt from the relevant protective laws and regulations and therefore free for general use.
The publisher, the authors and the editors are safe to assume that the advice and information in this book are believed to be true and accurate at the date of publication. Neither the publisher nor the authors or the editors give a warranty, expressed or implied, with respect to the material contained herein or for any errors or omissions that may have been made. The publisher remains neutral with regard to jurisdictional claims in published maps and institutional affiliations.

This Springer imprint is published by the registered company Springer Nature Switzerland AG
The registered company address is: Gewerbestrasse 11, 6330 Cham, Switzerland

Preface

With the aging of our infrastructure, ASR will become increasing prevalent, more so than we have suspected so far. Whereas this particular aging process is unsightly and mildly disruptive for buildings, pavements, or railroad ties, ASR is a major societal problem when it affects dams, large bridges, or even nuclear structures. In those situations, one must decide whether to maintain the structure under continuous auscultation or simply decommission it. The associated costs are enormous.

Given the increasing prevalence of ASR, this book is timely, pertinent, and necessary. While tools for assessing this particularly complex and confounding phenomenon have long existed, until now they have not been assembled into a single authoritative source. Unfortunately, in my own observation, the lack of organized information has already allowed inadequate assessments and poorly informed decisions about some critical ASR-infected infrastructures. Recently, for example, I evaluated the technical basis for a license amendment request for the life extension of a major nuclear reactor suffering from ASR. The regulating agency evaluation was shockingly simplistic and ill-informed, and yet the license for continued operation for an additional 30 years was approved. Thus, it is my hope, as a scientist and concerned citizen, that this book will strengthen government oversight of the risks posed by ASR in critical infrastructures by providing them with a comprehensive assessment methodology.

This volume represents four years of work by the RILEM Technical Committee 259-ISR Prognosis of Deterioration and Loss of Serviceability in Structures Affected by Alkali–Silica Reactions. I had the honor and pleasure to chair it and closely work with some of the world's best experts in the field.

As is often the case, the addressed topics do not represent what I anticipated during our first meeting in 2014. Our original focus was narrow and possibly ill defined, but after many, at times spirited, discussions I realized that the scope should be broadened. At times, this was based on committee discussions, and at others, I allowed myself to unilaterally seek additional participation. This led to substantial reshaping and enrichment of our original ideas.

In the process of editing, I have avoided two major pitfalls. First, I realized the folly of trying to build a consensus on a proposed method of diagnosis/prognosis emanating from our committee, but have rather let "a hundred flowers blossom" for the benefit of allowing many perspectives. Secondly, I have refrained from constraining the content of solid contributions by some of the participants who wanted to dwell in more details than others. Likewise, I have at times curtailed some coverage, or simply not included others whose work was not yet sufficiently mature for inclusion.

There are twenty-six chapters and four major appendices broken into four sections. Again, no chapter length was imposed, and one should not assume that those who embrace brevity are deficient in quality.

It is my hope that this book will be accessible as a mine for both engineers consulting it as a starting point of an investigation and for researchers starting with a literature survey.

I have reviewed each contribution, at times questioned the authors, and redone some of the figures. In so doing, I have converted most of the submitted Word files into the more esthetically pleasing format provided by LaTeX. Likewise, each of the 380 citations found an entry in the BibTeX database. Finally, I have tried, albeit imperfectly, to index the book but am certain to have missed some key entries. In the end, I am pleased to provide our publisher with a copy-ready manuscript that would only require minimal editing before publication.

Finally, and on a personal level, whereas I am both indebted and humbled by the knowledge of all the contributors, my greatest satisfaction was not in assiduously editing this book, but in meeting so many interesting colleagues and getting to know them on a personal level. They have my deep appreciation.

Boulder, CO, USA
June 2020

Victor E. Saouma

Acknowledgements

The editor would like to thank Dr. Manouchehr Hassanzadeh who was an early supporter of this endeavor and who has written the template followed by all contributors in the Prognostic part.

In addition, he would like to acknowledge the financial support of the US Nuclear Regulatory Commission through Grant No. NRC-HQ-60-14-G-0010, and the US Bureau of Reclamation through Assistance Agreement R18AC00055. Their support has greatly contributed to make this report a reality.

RILEM Publications

The following list is presenting the global offer of RILEM Publications, sorted by series. Each publication is available in printed version and/or in online version.

RILEM Proceedings (PRO)

PRO 1: Durability of High Performance Concrete (ISBN: 2-912143-03-9; e-ISBN: 2-351580-12-5; e-ISBN: 2351580125); *Ed. H. Sommer*

PRO 2: Chloride Penetration into Concrete (ISBN: 2-912143-00-04; e-ISBN: 2912143454); *Eds. L.-O. Nilsson and J.-P. Ollivier*

PRO 3: Evaluation and Strengthening of Existing Masonry Structures (ISBN: 2-912143-02-0; e-ISBN: 2351580141); *Eds. L. Binda and C. Modena*

PRO 4: Concrete: From Material to Structure (ISBN: 2-912143-04-7; e-ISBN: 2351580206); *Eds. J.-P. Bournazel and Y. Malier*

PRO 5: The Role of Admixtures in High Performance Concrete (ISBN: 2-912143-05-5; e-ISBN: 2351580214); *Eds. J. G. Cabrera and R. Rivera-Villarreal*

PRO 6: High Performance Fiber Reinforced Cement Composites - HPFRCC 3 (ISBN: 2-912143-06-3; e-ISBN: 2351580222); *Eds. H. W. Reinhardt and A. E. Naaman*

PRO 7: 1st International RILEM Symposium on Self-Compacting Concrete (ISBN: 2-912143-09-8; e-ISBN: 2912143721); *Eds. Å. Skarendahl and Ö. Petersson*

PRO 8: International RILEM Symposium on Timber Engineering (ISBN: 2-912143-10-1; e-ISBN: 2351580230); *Ed. L. Boström*

PRO 9: 2nd International RILEM Symposium on Adhesion between Polymers and Concrete ISAP '99 (ISBN: 2-912143-11-X; e-ISBN: 2351580249); *Eds. Y. Ohama and M. Puterman*

PRO 10: 3rd International RILEM Symposium on Durability of Building and Construction Sealants (ISBN: 2-912143-13-6; e-ISBN: 2351580257); *Eds. A. T. Wolf*

PRO 11: 4th International RILEM Conference on Reflective Cracking in Pavements (ISBN: 2-912143-14-4; e-ISBN: 2351580265); *Eds. A. O. Abd El Halim, D. A. Taylor and El H. H. Mohamed*

PRO 12: International RILEM Workshop on Historic Mortars: Characteristics and Tests (ISBN: 2-912143-15-2; e-ISBN: 2351580273); *Eds. P. Bartos, C. Groot and J. J. Hughes*

PRO 13: 2nd International RILEM Symposium on Hydration and Setting (ISBN: 2-912143-16-0; e-ISBN: 2351580281); *Ed. A. Nonat*

PRO 14: Integrated Life-Cycle Design of Materials and Structures - ILCDES 2000 (ISBN: 951-758-408-3; e-ISBN: 235158029X); (ISSN: 0356-9403); *Ed. S. Sarja*

PRO 15: Fifth RILEM Symposium on Fibre-Reinforced Concretes (FRC) - BEFIB'2000 (ISBN: 2-912143-18-7; e-ISBN: 291214373X); *Eds. P. Rossi and G. Chanvillard*

PRO 16: Life Prediction and Management of Concrete Structures (ISBN: 2-912143-19-5; e-ISBN: 2351580303); *Ed. D. Naus*

PRO 17: Shrinkage of Concrete – Shrinkage 2000 (ISBN: 2-912143-20-9; e-ISBN: 2351580311); *Eds. V. Baroghel-Bouny and P.-C. Aïtcin*

PRO 18: Measurement and Interpretation of the On-Site Corrosion Rate (ISBN: 2-912143-21-7; e-ISBN: 235158032X); *Eds. C. Andrade, C. Alonso, J. Fullea, J. Polimon and J. Rodriguez*

PRO 19: Testing and Modelling the Chloride Ingress into Concrete (ISBN: 2-912143-22-5; e-ISBN: 2351580338); *Eds. C. Andrade and J. Kropp*

PRO 20: 1st International RILEM Workshop on Microbial Impacts on Building Materials (CD 02) (e-ISBN 978-2-35158-013-4); *Ed. M. Ribas Silva*

PRO 21: International RILEM Symposium on Connections between Steel and Concrete (ISBN: 2-912143-25-X; e-ISBN: 2351580346); *Ed. R. Eligehausen*

PRO 22: International RILEM Symposium on Joints in Timber Structures (ISBN: 2-912143-28-4; e-ISBN: 2351580354); *Eds. S. Aicher and H.-W. Reinhardt*

PRO 23: International RILEM Conference on Early Age Cracking in Cementitious Systems (ISBN: 2-912143-29-2; e-ISBN: 2351580362); *Eds. K. Kovler and A. Bentur*

PRO 24: 2nd International RILEM Workshop on Frost Resistance of Concrete (ISBN: 2-912143-30-6; e-ISBN: 2351580370); *Eds. M. J. Setzer, R. Auberg and H.-J. Keck*

PRO 25: International RILEM Workshop on Frost Damage in Concrete (ISBN: 2-912143-31-4; e-ISBN: 2351580389); *Eds. D. J. Janssen, M. J. Setzer and M. B. Snyder*

PRO 26: International RILEM Workshop on On-Site Control and Evaluation of Masonry Structures (ISBN: 2-912143-34-9; e-ISBN: 2351580141); *Eds. L. Binda and R. C. de Vekey*

PRO 27: International RILEM Symposium on Building Joint Sealants (CD03; e-ISBN: 235158015X); *Ed. A. T. Wolf*

PRO 28: 6th International RILEM Symposium on Performance Testing and Evaluation of Bituminous Materials - PTEBM'03 (ISBN: 2-912143-35-7; e-ISBN: 978-2-912143-77-8); *Ed. M. N. Partl*

PRO 29: 2nd International RILEM Workshop on Life Prediction and Ageing Management of Concrete Structures (ISBN: 2-912143-36-5; e-ISBN: 2912143780); *Ed. D. J. Naus*

PRO 30: 4th International RILEM Workshop on High Performance Fiber Reinforced Cement Composites - HPFRCC 4 (ISBN: 2-912143-37-3; e-ISBN: 2912143799); *Eds. A. E. Naaman and H. W. Reinhardt*

PRO 31: International RILEM Workshop on Test and Design Methods for Steel Fibre Reinforced Concrete: Background and Experiences (ISBN: 2-912143-38-1; e-ISBN: 2351580168); *Eds. B. Schnütgen and L. Vandewalle*

PRO 32: International Conference on Advances in Concrete and Structures 2 vol. (ISBN (set): 2-912143-41-1; e-ISBN: 2351580176); *Eds. Ying-shu Yuan, Surendra P. Shah and Heng-lin Lü*

PRO 33: 3rd International Symposium on Self-Compacting Concrete (ISBN: 2-912143-42-X; e-ISBN: 2912143713); *Eds. Ó. Wallevik and I. Níelsson*

PRO 34: International RILEM Conference on Microbial Impact on Building Materials (ISBN: 2-912143-43-8; e-ISBN: 2351580184); *Ed. M. Ribas Silva*

PRO 35: International RILEM TC 186-ISA on Internal Sulfate Attack and Delayed Ettringite Formation (ISBN: 2-912143-44-6; e-ISBN: 2912143802); *Eds. K. Scrivener and J. Skalny*

PRO 36: International RILEM Symposium on Concrete Science and Engineering – A Tribute to Arnon Bentur (ISBN: 2-912143-46-2; e-ISBN: 2912143586); *Eds. K. Kovler, J. Marchand, S. Mindess and J. Weiss*

PRO 37: 5th International RILEM Conference on Cracking in Pavements – Mitigation, Risk Assessment and Prevention (ISBN: 2-912143-47-0; e-ISBN: 2912143764); *Eds. C. Petit, I. Al-Qadi and A. Millien*

PRO 38: 3rd International RILEM Workshop on Testing and Modelling the Chloride Ingress into Concrete (ISBN: 2-912143-48-9; e-ISBN: 2912143578); *Eds. C. Andrade and J. Kropp*

PRO 39: 6th International RILEM Symposium on Fibre-Reinforced Concretes - BEFIB 2004 (ISBN: 2-912143-51-9; e-ISBN: 2912143748); *Eds. M. Di Prisco, R. Felicetti and G. A. Plizzari*

PRO 40: International RILEM Conference on the Use of Recycled Materials in Buildings and Structures (ISBN: 2-912143-52-7; e-ISBN: 2912143756); *Eds. E. Vázquez, Ch. F. Hendriks and G. M. T. Janssen*

PRO 41: RILEM International Symposium on Environment-Conscious Materials and Systems for Sustainable Development (ISBN: 2-912143-55-1; e-ISBN: 2912143640); *Eds. N. Kashino and Y. Ohama*

PRO 42: SCC'2005 - China: 1st International Symposium on Design, Performance and Use of Self-Consolidating Concrete (ISBN: 2-912143-61-6; e-ISBN: 2912143624); *Eds. Zhiwu Yu, Caijun Shi, Kamal Henri Khayat and Youjun Xie*

PRO 43: International RILEM Workshop on Bonded Concrete Overlays (e-ISBN: 2-912143-83-7); *Eds. J. L. Granju and J. Silfwerbrand*

PRO 44: 2nd International RILEM Workshop on Microbial Impacts on Building Materials (CD11) (e-ISBN: 2-912143-84-5); *Ed. M. Ribas Silva*

PRO 45: 2nd International Symposium on Nanotechnology in Construction, Bilbao (ISBN: 2-912143-87-X; e-ISBN: 2912143888); *Eds. Peter J. M. Bartos, Yolanda de Miguel and Antonio Porro*

PRO 46: ConcreteLife'06 - International RILEM-JCI Seminar on Concrete Durability and Service Life Planning: Curing, Crack Control, Performance in Harsh Environments (ISBN: 2-912143-89-6; e-ISBN: 291214390X); *Ed. K. Kovler*

PRO 47: International RILEM Workshop on Performance Based Evaluation and Indicators for Concrete Durability (ISBN: 978-2-912143-95-2; e-ISBN: 9782912143969); *Eds. V. Baroghel-Bouny, C. Andrade, R. Torrent and K. Scrivener*

PRO 48: 1st International RILEM Symposium on Advances in Concrete through Science and Engineering (e-ISBN: 2-912143-92-6); *Eds. J. Weiss, K. Kovler, J. Marchand, and S. Mindess*

PRO 49: International RILEM Workshop on High Performance Fiber Reinforced Cementitious Composites in Structural Applications (ISBN: 2-912143-93-4; e-ISBN: 2912143942); *Eds. G. Fischer and V.C. Li*

PRO 50: 1st International RILEM Symposium on Textile Reinforced Concrete (ISBN: 2-912143-97-7; e-ISBN: 2351580087); *Eds. Josef Hegger, Wolfgang Brameshuber and Norbert Will*

PRO 51: 2nd International Symposium on Advances in Concrete through Science and Engineering (ISBN: 2-35158-003-6; e-ISBN: 2-35158-002-8); *Eds. J. Marchand, B. Bissonnette, R. Gagné, M. Jolin and F. Paradis*

PRO 52: Volume Changes of Hardening Concrete: Testing and Mitigation (ISBN: 2-35158-004-4; e-ISBN: 2-35158-005-2); *Eds. O. M. Jensen, P. Lura and K. Kovler*

PRO 53: High Performance Fiber Reinforced Cement Composites - HPFRCC5 (ISBN: 978-2-35158-046-2; e-ISBN: 978-2-35158-089-9); *Eds. H. W. Reinhardt and A. E. Naaman*

PRO 54: 5th International RILEM Symposium on Self-Compacting Concrete (ISBN: 978-2-35158-047-9; e-ISBN: 978-2-35158-088-2); *Eds. G. De Schutter and V. Boel*

PRO 55: International RILEM Symposium Photocatalysis, Environment and Construction Materials (ISBN: 978-2-35158-056-1; e-ISBN: 978-2-35158-057-8); *Eds. P. Baglioni and L. Cassar*

PRO 56: International RILEM Workshop on Integral Service Life Modelling of Concrete Structures (ISBN 978-2-35158-058-5; e-ISBN: 978-2-35158-090-5); *Eds. R. M. Ferreira, J. Gulikers and C. Andrade*

PRO 57: RILEM Workshop on Performance of cement-based materials in aggressive aqueous environments (e-ISBN: 978-2-35158-059-2); *Ed. N. De Belie*

PRO 58: International RILEM Symposium on Concrete Modelling - CONMOD'08 (ISBN: 978-2-35158-060-8; e-ISBN: 978-2-35158-076-9); *Eds. E. Schlangen and G. De Schutter*

PRO 59: International RILEM Conference on On Site Assessment of Concrete, Masonry and Timber Structures - SACoMaTiS 2008 (ISBN set: 978-2-35158-061-5; e-ISBN: 978-2-35158-075-2); *Eds. L. Binda, M. di Prisco and R. Felicetti*

PRO 60: Seventh RILEM International Symposium on Fibre Reinforced Concrete: Design and Applications - BEFIB 2008 (ISBN: 978-2-35158-064-6; e-ISBN: 978-2-35158-086-8); *Ed. R. Gettu*

PRO 61: 1st International Conference on Microstructure Related Durability of Cementitious Composites 2 vol., (ISBN: 978-2-35158-065-3; e-ISBN: 978-2-35158-084-4); *Eds. W. Sun, K. van Breugel, C. Miao, G. Ye and H. Chen*

PRO 62: NSF/ RILEM Workshop: In-situ Evaluation of Historic Wood and Masonry Structures (e-ISBN: 978-2-35158-068-4); *Eds. B. Kasal, R. Anthony and M. Drdácký*

PRO 63: Concrete in Aggressive Aqueous Environments: Performance, Testing and Modelling, 2 vol., (ISBN: 978-2-35158-071-4; e-ISBN: 978-2-35158-082-0); *Eds. M. G. Alexander and A. Bertron*

PRO 64: Long Term Performance of Cementitious Barriers and Reinforced Concrete in Nuclear Power Plants and Waste Management - NUCPERF 2009 (ISBN: 978-2-35158-072-1; e-ISBN: 978-2-35158-087-5); *Eds. V. L'Hostis, R. Gens, C. Gallé*

PRO 65: Design Performance and Use of Self-consolidating Concrete - SCC'2009 (ISBN: 978-2-35158-073-8; e-ISBN: 978-2-35158-093-6); *Eds. C. Shi, Z. Yu, K. H. Khayat and P. Yan*

PRO 66: 2nd International RILEM Workshop on Concrete Durability and Service Life Planning - ConcreteLife'09 (ISBN: 978-2-35158-074-5; ISBN: 978-2-35158-074-5); *Ed. K. Kovler*

PRO 67: Repairs Mortars for Historic Masonry (e-ISBN: 978-2-35158-083-7); *Ed. C. Groot*

PRO 68: Proceedings of the 3rd International RILEM Symposium on 'Rheology of Cement Suspensions such as Fresh Concrete (ISBN 978-2-35158-091-2; e-ISBN: 978-2-35158-092-9); *Eds. O. H. Wallevik, S. Kubens and S. Oesterheld*

PRO 69: 3rd International PhD Student Workshop on 'Modelling the Durability of Reinforced Concrete (ISBN: 978-2-35158-095-0); *Eds. R. M. Ferreira, J. Gulikers and C. Andrade*

PRO 70: 2nd International Conference on 'Service Life Design for Infrastructure' (ISBN set: 978-2-35158-096-7, e-ISBN: 978-2-35158-097-4); *Ed. K. van Breugel, G. Ye and Y. Yuan*

PRO 71: Advances in Civil Engineering Materials - The 50-year Teaching Anniversary of Prof. Sun Wei' (ISBN: 978-2-35158-098-1; e-ISBN: 978-2-35158-099-8); *Eds. C. Miao, G. Ye, and H. Chen*

PRO 72: First International Conference on 'Advances in Chemically-Activated Materials – CAM'2010' (2010), 264 pp, ISBN: 978-2-35158-101-8; e-ISBN: 978-2-35158-115-5, *Eds. Caijun Shi and Xiaodong Shen*

PRO 73: 2nd International Conference on 'Waste Engineering and Management - ICWEM 2010' (2010), 894 pp, ISBN: 978-2-35158-102-5; e-ISBN: 978-2-35158-103-2, *Eds. J. Zh. Xiao, Y. Zhang, M. S. Cheung and R. Chu*

PRO 74: International RILEM Conference on 'Use of Superabsorsorbent Polymers and Other New Addditives in Concrete' (2010) 374 pp., ISBN: 978-2-35158-104-9; e-ISBN: 978-2-35158-105-6; *Eds. O.M. Jensen, M.T. Hasholt, and S. Laustsen*

PRO 75: International Conference on 'Material Science - 2nd ICTRC - Textile Reinforced Concrete - Theme 1' (2010) 436 pp., ISBN: 978-2-35158-106-3; e-ISBN: 978-2-35158-107-0;
Ed. W. Brameshuber

PRO 76: International Conference on 'Material Science - HetMat - Modelling of Heterogeneous Materials - Theme 2' (2010) 255 pp., ISBN: 978-2-35158-108-7; e-ISBN: 978-2-35158-109-4; *Ed. W. Brameshuber*

PRO 77: International Conference on 'Material Science - AdIPoC - Additions Improving Properties of Concrete - Theme 3' (2010) 459 pp., ISBN: 978-2-35158-110-0; e-ISBN: 978-2-35158-111-7; *Ed. W. Brameshuber*

PRO 78: 2nd Historic Mortars Conference and RILEM TC 203-RHM Final Workshop – HMC2010 (2010) 1416 pp., e-ISBN: 978-2-35158-112-4; *Eds J. Válek, C. Groot, and J. J. Hughes*

PRO 79: International RILEM Conference on Advances in Construction Materials Through Science and Engineering (2011) 213 pp., ISBN: 978-2-35158-116-2, e-ISBN: 978-2-35158-117-9; *Eds Christopher Leung and K.T. Wan*

PRO 80: 2nd International RILEM Conference on Concrete Spalling due to Fire Exposure (2011) 453 pp., ISBN: 978-2-35158-118-6, e-ISBN: 978-2-35158-119-3; *Eds E.A.B. Koenders and F. Dehn*

PRO 81: 2nd International RILEM Conference on Strain Hardening Cementitious Composites (SHCC2-Rio) (2011) 451 pp., ISBN: 978-2-35158-120-9, e-ISBN: 978-2-35158-121-6; *Eds R.D. Toledo Filho, F.A. Silva, E.A.B. Koenders and E.M.R. Fairbairn*

PRO 82: 2nd International RILEM Conference on Progress of Recycling in the Built Environment (2011) 507 pp., e-ISBN: 978-2-35158-122-3; *Eds V.M. John, E. Vazquez, S.C. Angulo and C. Ulsen*

PRO 83: 2nd International Conference on Microstructural-related Durability of Cementitious Composites (2012) 250 pp., ISBN: 978-2-35158-129-2; e-ISBN: 978-2-35158-123-0; *Eds G. Ye, K. van Breugel, W. Sun and C. Miao*

PRO 84: CONSEC13 - Seventh International Conference on Concrete under Severe Conditions – Environment and Loading (2013) 1930 pp., ISBN: 978-2-35158-124-7; e-ISBN: 978-2- 35158-134-6; *Eds Z.J. Li, W. Sun, C.W. Miao, K. Sakai, O.E. Gjorv & N. Banthia*

PRO 85: RILEM-JCI International Workshop on Crack Control of Mass Concrete and Related issues concerning Early-Age of Concrete Structures – ConCrack 3 – Control of Cracking in Concrete Structures 3 (2012) 237 pp., ISBN: 978-2-35158-125-4; e-ISBN: 978-2-35158-126-1; *Eds F. Toutlemonde and J.-M. Torrenti*

PRO 86: International Symposium on Life Cycle Assessment and Construction (2012) 414 pp., ISBN: 978-2-35158-127-8, e-ISBN: 978-2-35158-128-5; *Eds A. Ventura and C. de la Roche*

PRO 87: UHPFRC 2013 – RILEM-fib-AFGC International Symposium on Ultra-High Performance Fibre-Reinforced Concrete (2013), ISBN: 978-2-35158-130-8, e-ISBN: 978-2-35158-131-5; *Eds F. Toutlemonde*

PRO 88: 8th RILEM International Symposium on Fibre Reinforced Concrete (2012) 344 pp., ISBN: 978-2-35158-132-2, e-ISBN: 978-2-35158-133-9; *Eds Joaquim A.O. Barros*

PRO 89: RILEM International workshop on performance-based specification and control of concrete durability (2014) 678 pp, ISBN: 978-2-35158-135-3, e-ISBN: 978-2-35158-136-0; *Eds. D. Bjegović, H. Beushausen and M. Serdar*

PRO 90: 7th RILEM International Conference on Self-Compacting Concrete and of the 1st RILEM International Conference on Rheology and Processing of Construction Materials (2013) 396 pp, ISBN: 978-2-35158-137-7, e-ISBN: 978-2-35158-138-4; *Eds. Nicolas Roussel and Hela Bessaies-Bey*

PRO 91: CONMOD 2014 - RILEM International Symposium on Concrete Modelling (2014), ISBN: 978-2-35158-139-1; e-ISBN: 978-2-35158-140-7; *Eds. Kefei Li, Peiyu Yan and Rongwei Yang*

PRO 92: CAM 2014 - 2nd International Conference on advances in chemically-activated materials (2014) 392 pp., ISBN: 978-2-35158-141-4; e-ISBN: 978-2-35158-142-1; *Eds. Caijun Shi and Xiadong Shen*

PRO 93: SCC 2014 - 3rd International Symposium on Design, Performance and Use of Self-Consolidating Concrete (2014) 438 pp., ISBN: 978-2-35158-143-8; e-ISBN: 978-2-35158-144-5; *Eds. Caijun Shi, Zhihua Ou, Kamal H. Khayat*

PRO 94 (online version): HPFRCC-7 - 7th RILEM conference on High performance fiber reinforced cement composites (2015), e-ISBN: 978-2-35158-146-9; *Eds. H.W. Reinhardt, G.J. Parra-Montesinos, H. Garrecht*

PRO 95: International RILEM Conference on Application of superabsorbent polymers and other new admixtures in concrete construction (2014), ISBN: 978-2-35158-147-6; e-ISBN: 978-2-35158-148-3; *Eds. Viktor Mechtcherine, Christof Schroefl*

PRO 96 (online version): XIII DBMC: XIII International Conference on Durability of Building Materials and Components(2015), e-ISBN: 978-2-35158-149-0; *Eds. M. Quattrone, V.M. John*

PRO 97: SHCC3 – 3rd International RILEM Conference on Strain Hardening Cementitious Composites (2014), ISBN: 978-2-35158-150-6; e-ISBN: 978-2-35158-151-3; *Eds. E. Schlangen, M.G. Sierra Beltran, M. Lukovic, G. Ye*

PRO 98: FERRO-11 – 11th International Symposium on Ferrocement and 3rd ICTRC - International Conference on Textile Reinforced Concrete (2015), ISBN: 978-2-35158-152-0; e-ISBN: 978-2-35158-153-7; *Ed. W. Brameshuber*

PRO 99 (online version): ICBBM 2015 - 1st International Conference on Bio-Based Building Materials (2015), e-ISBN: 978-2-35158-154-4; *Eds. S. Amziane, M. Sonebi*

PRO 100: SCC16 - RILEM Self-Consolidating Concrete Conference (2016), ISBN: 978-2-35158-156-8; e-ISBN: 978-2-35158-157-5; *Ed. Kamal H. Kayat*

PRO 101 (online version): III Progress of Recycling in the Built Environment (2015), e-ISBN: 978-2-35158-158-2; *Eds I. Martins, C. Ulsen and S. C. Angulo*

PRO 102 (online version): RILEM Conference on Microorganisms-Cementitious Materials Interactions (2016), e-ISBN: 978-2-35158-160-5; *Eds. Alexandra Bertron, Henk Jonkers, Virginie Wiktor*

PRO 103 (online version): ACESC'16 - Advances in Civil Engineering and Sustainable Construction (2016), e-ISBN: 978-2-35158-161-2; *Eds. T.Ch. Madhavi, G. Prabhakar, Santhosh Ram and P.M. Rameshwaran*

PRO 104 (online version): SSCS'2015 - Numerical Modeling - Strategies for Sustainable Concrete Structures (2015), e-ISBN: 978-2-35158-162-9

PRO 105: 1st International Conference on UHPC Materials and Structures (2016), ISBN: 978-2-35158-164-3, e-ISBN: 978-2-35158-165-0

PRO 106: AFGC-ACI-fib-RILEM International Conference on Ultra-High-Performance Fibre-Reinforced Concrete – UHPFRC 2017 (2017), ISBN: 978-2-35158-166-7, e-ISBN: 978-2-35158-167-4; *Eds. François Toutlemonde & Jacques Resplendino*

PRO 107 (online version): XIV DBMC – 14th International Conference on Durability of Building Materials and Components (2017), e-ISBN: 978-2-35158-159-9; *Eds. Geert De Schutter, Nele De Belie, Arnold Janssens, Nathan Van Den Bossche*

PRO 108: MSSCE 2016 - Innovation of Teaching in Materials and Structures (2016), ISBN: 978-2-35158-178-0, e-ISBN: 978-2-35158-179-7; *Ed. Per Goltermann*

PRO 109 (2 volumes): MSSCE 2016 - Service Life of Cement-Based Materials and Struc-tures (2016), ISBN Vol. 1: 978-2-35158-170-4, Vol. 2: 978-2-35158-171-4, Set Vol. 1&2: 978-2-35158-172-8, e-ISBN : 978-2-35158-173-5; *Eds. Miguel Azenha, Ivan Gabrijel, Dirk Schlicke, Terje Kanstad and Ole Mejlhede Jensen*

PRO 110: MSSCE 2016 - Historical Masonry (2016), ISBN: 978-2-35158-178-0, e-ISBN: 978-2-35158-179-7; *Eds. Inge Rörig-Dalgaard and Ioannis Ioannou*

PRO 111: MSSCE 2016 - Electrochemistry in Civil Engineering (2016), ISBN: 978-2-35158-176-6, e-ISBN: 978-2-35158-177-3; *Ed. Lisbeth M. Ottosen*

PRO 112: MSSCE 2016 - Moisture in Materials and Structures (2016), ISBN: 978-2-35158-178-0, e-ISBN: 978-2-35158-179-7; *Eds. Kurt Kielsgaard Hansen, Carsten Rode and Lars-Olof Nilsson*

PRO 113: MSSCE 2016 - Concrete with Supplementary Cementitious Materials (2016), ISBN: 978-2-35158-178-0, e-ISBN: 978-2-35158-179-7; *Eds. Ole Mejlhede Jensen, Konstantin Kovler and Nele De Belie*

PRO 114: MSSCE 2016 - Frost Action in Concrete (2016), ISBN: 978-2-35158-182-7, e-ISBN: 978-2-35158-183-4; *Eds. Marianne Tange Hasholt, Katja Fridh and R. Doug Hooton*

PRO 115: MSSCE 2016 - Fresh Concrete (2016), ISBN: 978-2-35158-184-1, e-ISBN: 978-2-35158-185-8; *Eds. Lars N. Thrane, Claus Pade, Oldrich Svec and Nicolas Roussel*

PRO 116: BEFIB 2016 – 9th RILEM International Symposium on Fiber Reinforced Concrete (2016), ISBN: 978-2-35158-187-2, e-ISBN: 978-2-35158-186-5; *Eds. N. Banthia, M. di Prisco and S. Soleimani-Dashtaki*

PRO 117: 3rd International RILEM Conference on Microstructure Related Durability of Cementitious Composites (2016), ISBN: 978-2-35158-188-9, e-ISBN: 978-2-35158-189-6; *Eds. Changwen Miao, Wei Sun, Jiaping Liu, Huisu Chen, Guang Ye and Klaas van Breugel*

PRO 118 (4 volumes): International Conference on Advances in Construction Materials and Systems (2017), ISBN Set: 978-2-35158-190-2, Vol. 1: 978-2-35158-193-3, Vol. 2: 978-2-35158-194-0, Vol. 3: ISBN:978-2-35158-195-7, Vol. 4: ISBN:978-2-35158-196-4, e-ISBN: 978-2-35158-191-9; *Eds. Manu Santhanam, Ravindra Gettu, Radhakrishna G. Pillai and Sunitha K. Nayar*

PRO 119 (online version): ICBBM 2017 - Second International RILEM Conference on Bio-based Building Materials, (2017), e-ISBN: 978-2-35158-192-6; *Eds. Sofiane Amziane, Moham-med Sonebi*

PRO 120 (2 volumes): EAC-02 - 2nd International RILEM/COST Conference on Early Age Cracking and Serviceability in Cement-based Materials and Structures, (2017), Vol. 1: 978-2-35158-199-5, Vol. 2: 978-2-35158-200-8, Set: 978-2-35158-197-1, e-ISBN: 978-2-35158-198-8; *Eds. Stéphanie Staquet and Dimitrios Aggelis*

PRO 121 (2 volumes): SynerCrete18: Interdisciplinary Approaches for Cement-based Materials and Structural Concrete: Synergizing Expertise and Bridging Scales of Space and Time, (2018), Set: 978-2-35158-202-2, Vol.1: 978-2-35158-211-4, Vol.2: 978-2-35158-212-1, e-ISBN: 978-2-35158-203-9; *Eds. Miguel Azenha, Dirk Schlicke, Farid Benboudjema, Agnieszka Knoppik*

PRO 122: SCC'2018 China - Fourth International Symposium on Design, Performance and Use of Self-Consolidating Concrete, (2018), ISBN: 978-2-35158-204-6, e-ISBN: 978-2-35158-205-3; *Eds. C. Shi, Z. Zhang, K. H. Khayat*

PRO 123: Final Conference of RILEM TC 253-MCI: Microorganisms-Cementitious Materials Interactions (2018), Set: 978-2-35158-207-7, Vol.1: 978-2-35158-209-1, Vol.2: 978-2-35158-210-7, e-ISBN: 978-2-35158-206-0; *Ed. Alexandra Bertron*

PRO 124 (online version): Fourth International Conference Progress of Recycling in the Built Environment (2018), e-ISBN: 978-2-35158-208-4; *Eds. Isabel M. Martins, Carina Ulsen, Yury Villagran*

PRO 125 (online version): SLD4 - 4th International Conference on Service Life Design for In-frastructures (2018), e-ISBN: 978-2-35158-213-8; *Eds. Guang Ye, Yong Yuan, Claudia Romero Rodriguez, Hongzhi Zhang, Branko Savija*

PRO 126: Workshop on Concrete Modelling and Material Behaviour in honor of Professor Klaas van Breugel (2018), ISBN: 978-2-35158-214-5, e-ISBN: 978-2-35158-215-2; *Ed. Guang Ye*

PRO 127 (online version): CONMOD2018 - Symposium on Concrete Modelling (2018), e-ISBN: 978-2-35158-216-9; *Eds. Erik Schlangen, Geert de Schutter, Branko Savija, Hongzhi Zhang, Claudia Romero Rodriguez*

PRO 128: SMSS2019 - International Conference on Sustainable Materials, Systems and Struc-tures (2019), ISBN: 978-2-35158-217-6, e-ISBN: 978-2-35158-218-3

PRO 129: 2nd International Conference on UHPC Materials and Structures (UHPC2018-China), ISBN: 978-2-35158-219-0, e-ISBN: 978-2-35158-220-6;

PRO 130: 5th Historic Mortars Conference (2019), ISBN: 978-2-35158-221-3, e-ISBN: 978-2-35158-222-0; *Eds. José Ignacio Álvarez, José María Fernández, Íñigo Navarro, Adrián Durán, Rafael Sirera*

PRO 131 (online version): 3rd International Conference on Bio-Based Building Materials (ICBBM2019), e-ISBN: 978-2-35158-229-9; *Eds. Mohammed Sonebi, Sofiane Amziane, Jona-than Page*

PRO 132: IRWRMC'18 - International RILEM Workshop on Rheological Measurements of Ce-ment-based Materials (2018), ISBN: 978-2-35158-230-5, e-ISBN: 978-2-35158-231-2; *Eds. Chafika Djelal, Yannick Vanhove*

PRO 133 (online version): CO2STO2019 - International Workshop CO2 Storage in Concrete (2019), e-ISBN: 978-2-35158-232-9; *Eds. Assia Djerbi, Othman Omikrine-Metalssi, Teddy Fen-Chong*

RILEM Reports (REP)

Report 19: Considerations for Use in Managing the Aging of Nuclear Power Plant Concrete Structures (ISBN: 2-912143-07-1); *Ed. D. J. Naus*

Report 20: Engineering and Transport Properties of the Interfacial Transition Zone in Cementitious Composites (ISBN: 2-912143-08-X); *Eds. M. G. Alexander, G. Arliguie, G. Ballivy, A. Bentur and J. Marchand*

Report 21: Durability of Building Sealants (ISBN: 2-912143-12-8); *Ed. A. T. Wolf*

Report 22: Sustainable Raw Materials - Construction and Demolition Waste (ISBN: 2-912143-17-9); *Eds. C. F. Hendriks and H. S. Pietersen*

Report 23: Self-Compacting Concrete state-of-the-art report (ISBN: 2-912143-23-3); *Eds. Å. Skarendahl and Ö. Petersson*

Report 24: Workability and Rheology of Fresh Concrete: Compendium of Tests (ISBN: 2-912143-32-2); *Eds. P. J. M. Bartos, M. Sonebi and A. K. Tamimi*

Report 25: Early Age Cracking in Cementitious Systems (ISBN: 2-912143-33-0); *Ed. A. Bentur*

Report 26: Towards Sustainable Roofing (Joint Committee CIB/RILEM) (CD 07) (e-ISBN 978-2-912143-65-5); *Eds. Thomas W. Hutchinson and Keith Roberts*

Report 27: Condition Assessment of Roofs (Joint Committee CIB/RILEM) (CD 08) (e-ISBN 978-2-912143-66-2); *Ed. CIB W 83/RILEM TC166-RMS*

Report 28: Final report of RILEM TC 167-COM 'Characterisation of Old Mortars with Respect to Their Repair (ISBN: 978-2-912143-56-3); *Eds. C. Groot, G. Ashall and J. Hughes*

Report 29: Pavement Performance Prediction and Evaluation (PPPE): Interlaboratory Tests (e-ISBN: 2-912143-68-3); *Eds. M. Partl and H. Piber*

Report 30: Final Report of RILEM TC 198-URM 'Use of Recycled Materials' (ISBN: 2-912143-82-9; e-ISBN: 2-912143-69-1); *Eds. Ch. F. Hendriks, G. M. T. Janssen and E. Vázquez*

Report 31: Final Report of RILEM TC 185-ATC 'Advanced testing of cement-based materials during setting and hardening' (ISBN: 2-912143-81-0; e-ISBN: 2-912143-70-5); *Eds. H. W. Reinhardt and C. U. Grosse*

Report 32: Probabilistic Assessment of Existing Structures. A JCSS publication (ISBN 2-912143-24-1); *Ed. D. Diamantidis*

Report 33: State-of-the-Art Report of RILEM Technical Committee TC 184-IFE 'Industrial Floors' (ISBN 2-35158-006-0); *Ed. P. Seidler*

Report 34: Report of RILEM Technical Committee TC 147-FMB 'Fracture mechanics applications to anchorage and bond' Tension of Reinforced Concrete Prisms – Round Robin Analysis and Tests on Bond (e-ISBN 2-912143-91-8); *Eds. L. Elfgren and K. Noghabai*

Report 35: Final Report of RILEM Technical Committee TC 188-CSC 'Casting of Self Compacting Concrete' (ISBN 2-35158-001-X; e-ISBN: 2-912143-98-5); *Eds. Å. Skarendahl and P. Billberg*

Report 36: State-of-the-Art Report of RILEM Technical Committee TC 201-TRC 'Textile Reinforced Concrete' (ISBN 2-912143-99-3); *Ed. W. Brameshuber*

Report 37: State-of-the-Art Report of RILEM Technical Committee TC 192-ECM 'Environment-conscious construction materials and systems' (ISBN: 978-2-35158-053-0); *Eds. N. Kashino, D. Van Gemert and K. Imamoto*

Report 38: State-of-the-Art Report of RILEM Technical Committee TC 205-DSC 'Durability of Self-Compacting Concrete' (ISBN: 978-2-35158-048-6); *Eds. G. De Schutter and K. Audenaert*

Report 39: Final Report of RILEM Technical Committee TC 187-SOC 'Experimental determination of the stress-crack opening curve for concrete in tension' (ISBN 978-2-35158-049-3); *Ed. J. Planas*

Report 40: State-of-the-Art Report of RILEM Technical Committee TC 189-NEC 'Non-Destructive Evaluation of the Penetrability and Thickness of the Concrete Cover' (ISBN 978-2-35158-054-7); *Eds. R. Torrent and L. Fernández Luco*

Report 41: State-of-the-Art Report of RILEM Technical Committee TC 196-ICC 'Internal Curing of Concrete' (ISBN 978-2-35158-009-7); *Eds. K. Kovler and O. M. Jensen*

Report 42: 'Acoustic Emission and Related Non-destructive Evaluation Techniques for Crack Detection and Damage Evaluation in Concrete' - Final Report of RILEM Technical Committee 212-ACD (e-ISBN: 978-2-35158-100-1); *Ed. M. Ohtsu*

Report 45: Repair Mortars for Historic Masonry - State-of-the-Art Report of RILEM Technical Committee TC 203-RHM (e-ISBN: 978-2-35158-163-6); *Eds. Paul Maurenbrecher and Caspar Groot*

Report 46: Surface delamination of concrete industrial floors and other durability related aspects guide - Report of RILEM Technical Committee TC 268-SIF (e-ISBN: 978-2-35158-201-5); *Ed. Valerie Pollet*

Contents

Contributors

Ammar Abd-elssamd Department of Civil and Environmental Engineering, University of Tennessee, Knoxville, TN, USA

Alexandre Boule EDF - Direction Industrielle - Département Techniques de réalisation et Expertise en Géosciences et Génie civil, Aix-en-Provence Cedex 02, France

Alexis Courtois EDF Direction Ingénierie et Projets Nouveau Nucléaire - Direction Technique, Lyon, France

Stéphane Cuchet Holcim Suisse, Laboratoire des matériaux, Eclépens, Switzerland

João Custódio LNEC - National Laboratory for Civil Engineering, Lisbon, Portugal

Diego Jesus de Souza University of Ottawa, Ottawa, Canada

William Dressel Waterways & Concrete Dams, US Bureau of Reclamation, Denver, CO, USA

Benoît Fournier Universié Laval, Montreal, Canada

Eric R. Giannini RJ Lee Group, La Crosse, WI, USA

Bruno Godart Materials and Structures Department, IFSTTAR, French Institute of Science and Technology for Transport, Development and Networks, Université Paris Est, Marne la Vall'ee, France

Etienne Grimal Electricité de France, Centre d'Ingénierie Hydraulique, EDF-CIH Technolac, Le Bourget du Lac Cedex, France

M. Amin Hariri-Ardebili Department of Civil Engineering, University of Colorado, Boulder, CO, USA

Jean-Marie Hénault EDF - R&D, Chatou Cedex, France

Laurence Jacobs Georgia Institute of Technology, Atlanta, USA

Z. John Ma Department of Civil and Environmental Engineering, University of Tennessee, Knoxville, TN, USA

Samuel Johnson Electric Power Research Institute, Charlotte, NC, USA

Tetsuya Katayama Taiheiyo Consultant Co., Ltd., Sakura, Japan

Philippe Kolmayer Electricité de France, Centre d'Ingénierie Hydraulique, EDF-CIH Technolac, Le Bourget du Lac Cedex, France

Sihem Le Pape Department of Civil and Environmental Engineering, University of Tennessee, Knoxville, TN, USA

Yann Le Pape Oak Ridge National Laboratory, Oak-Ridge, TN, USA

Andreas Leemann Empa, Swiss Federal Laboratories for Materials Testing and Research, Laboratory for Concrete and Construction Chemistry, Dübendorf, Switzerland

Daniel Liechty Waterways & Concrete Dams, US Bureau of Reclamation, Denver, CO, USA

Renaud-Pierre Martin Urban and Civil Engineering Testing and Modeling Laboratory (EMGCU), Materials and Structures Department, IFSTTAR, French Institute of Science and Technology for Transport, Development and Networks, Université Paris Est, Marne la Vall'ee, France

Benoit Masson EDF Direction Ingénierie et Projets Nouveau Nucléaire - Direction Technique, Lyon, France

Esperanza Menéndez Institute of Construction Science Eduardo Torroja, Madrid, Spain

Christine Merz Merz Ingenieurberatung, Möriken, Switzerland

Pierre Morenon Civil Engineering, Paul Sabatier University, and Construction and Durability Materials Laboratory of Toulouse, Toulouse, France

Kozo Mukai Taiheiyo Consultant Co., Ltd., Sakura, Japan

Stéphane Multon Civil Engineering, Paul Sabatier University, and Construction and Durability Materials Laboratory of Toulouse (France), Toulouse, France; Université de Toulouse, Toulouse, France

Boumediene Nedjar Materials and Structures Department, Urban and Civil Engineering Testing and Modeling Laboratory (EMGCU), IFSTTAR, French institute of science and technology for transport, development and networks, Université Paris Est, Marne la Vall'ee, France

Pierre Nicot Transfer Division, Paul Sabatier University, and Construction and Durability Materials Laboratory of Toulouse (France), Toulouse, France

Chunyu Qiao DRP, A Twining Company, Boulder, CO, US

Patrice Rivard Department of Civil Engineering and Building Engineering, Sherbrooke, Canada

Claude Rospars Materials and Structures Department, Urban and Civil Engineering Testing and Modeling Laboratory (EMGCU), IFSTTAR, French Institute of Science and Technology for Transport, Development and Networks, Université Paris Est, Marne la Vall'ee, France

Simon-Nicolas Roth Hydro-Québec, Direction Expertise Barrages et Infrastructures, Vice-présidence Planification, Stratégies et Expertises, Montréal (Québec), Canada

David Rothstein DRP, A Twining Company, Boulder, CO, US

Jerzy Salamon Waterways & Concrete Dams, US Bureau of Reclamation, Denver, CO, USA

Leandro Sanchez University of Ottawa, Ottawa, Canada

Victor Saouma Department of Civil Engineering, University of Colorado, Boulder, CO, USA

Tomomi Sato Taiheiyo Consultant Co., Ltd., Sakura, Japan

Jerǒme Sausse EDF Hydro - Direction Technique Générale - Département Surveillance, Brive la Gaillarde, France

Alain Sellier Civil Engineering, Paul Sabatier University, and Construction and Durability Materials Laboratory of Toulouse (France), Toulouse, France;
Université de Toulouse, Toulouse, France

Yuya Takahashi Department of Civil Engineering, School of Engineering, The University of Tokyo, Bunkyo-ku, Tokyo, Japan

François Toutlemonde Materials and Structures Department, IFSTTAR, French Institute of Science and Technology for Transport, Development and Networks, Université Paris Est, Marne la Vall'ee, France

Denis Vautrin EDF - R&D, Chatou Cedex, France

Jonathan Wood Structural Studies Design, Chiddingfold, UK

Part I
Diagnosis

Chapter 1
Introduction to Diagnosis

Andreas Leemann and Tetsuya Katayama

1.1 Introduction

ASR in concrete typically follows a certain sequence. On the level of the structure, there is an initial period after construction where no expansion occurs. Then the concrete starts to expand and cracks develop, often in a linear way, and can eventually damage the structure (NB slight cracking and signs of AAR are usually insufficient to "damage" the serviceability or safety of a structure).

The duration of the induction period and the subsequent rate of expansion are dependent on the alkali in the concrete mix design, the aggregates used and the environment (temperature and availability of water or water vapour (relative humidity). In constant conditions the initial rate of expansion may decrease with time and expansion may even cease following a sigmoidal expansion curve [1]. Such a leveling of expansion in steady moisture conditions can have two different reasons. Firstly, the amount of alkalis in the concrete falls below a critical level due to leaching and to chemical binding of alkalis in the ASR products. Secondly, the reactive minerals providing silicon for the reaction are depleted. In structures with a small section, a decrease and even halt of expansion is due partly to a decrease of available alkalis, and partly to drying of concrete. However, in many structures containing late-expansive aggregate, the stage of a decreasing expansion rate is not reached. Dams and bridges for example often show a linear expansion rate for decades [2] where monitoring started long after construction. By contrast, expansion of several dams

A. Leemann (✉)
Empa, Swiss Federal Laboratories for Materials Testing and Research, Laboratory for Concrete and Construction Chemistry, Überlandstrasse 129, 8600 Dübendorf, Switzerland
e-mail: andreas.leemann@empa.ch

T. Katayama
Taiheiyo Consultant Co. Ltd., Sakura, Japan
e-mail: katayama@catv296.ne.jp

© RILEM 2021

V. E. Saouma (ed.), *Diagnosis & Prognosis of AAR Affected Structures*,
RILEM State-of-the-Art Reports 31,
https://doi.org/10.1007/978-3-030-44014-5_1

and bridges containing early-expansive volcanic aggregates ceased within 40 years after construction following a sigmoidal expansion curve [3, 4].

On a microstructural level ASR is showing a sequence as well [5–7]. The first ASR products are formed in the interface to the cement paste creating a rim in the aggregate. Supersaturation is reached there in an early stage of the reaction, as thermodynamic modelling shows [8]. More reaction products are formed along minerals grain boundaries further into aggregate. The increasing pressure induced by the ASR products leads to the cracking of the aggregate. Typically these cracks run from the aggregates into the cement paste eventually connecting with cracks caused by the expansion of other aggregates. First these cracks are empty. But as reaction products are continuously formed, the cracks in aggregates are slowly filled. Later, reaction products are extruding into the cracked cement paste [5, 9, 10]. The start of linear expansion in the structure goes together with crack formation in the concrete on a microstructural level (see paragraph 3).

For the owner of a structure it is of great importance to know, how the damages due to ASR will develop in the future and how the expansion will affect the structure. However, before a prognosis of ASR-induced deterioration and eventual loss of serviceability of a structure can be made, its current stage has to be investigated. Such a diagnosis involves several steps. In the RILEM State-of-the-Art-Report "Guide to diagnosis and Appraisal of AAR damage to concrete structures" [11] the required procedures to investigate a structure are described in detail. Following the above mentioned publication they can be divided as follows:

1. Symptoms of an expansive reaction during routine inspection.
2. Confirmation investigation.
3. Severity investigation.
4. Reporting.

Coring and material analysis are included in the second step, while the expansion tests covered in WP-1 in this report are part of the third step. On a microstructural level, the RILEM report concentrates on a general description of the applicable methods and of the ASR phenomena. In addition to this, the microstructural features observed by polarizing microscopy (PM) and scanning electron microscopy (SEM) allow to assess the stage of ASR and with it the point in the sequence of ASR development. In the following a short outline of the required information is specified and the principle is shown, how such an assessment can be made.

1.2 Petrographic Examination

1.2.1 General Background

ASR can be roughly classified into two types: early-expansive ASR and late-expansive ASR. Late-expansive ASR was formerly called "slow/late-expanding alkali-silicate/silica-reaction" in the CSA standard in the 1980s and 1990s (e.g.

CAN/CSA-A23.1-M90). The deleterious reactivity of such aggregates cannot always be judged reliably by the conventional expansion tests (e.g. [12]) without using concrete prism test (eg. CSA-23.2-a4A [13]). For convenience, this type of ASR was renamed as "late-expansive ASR" by Katayama [14], Table 1.1.

There are principal differences between ASR in concrete structures and in the concrete prism tests performed at a constant temperature and humidity without stress. The most problematic rock types that cause ASR in large dams and power plants in the world contain extremely late-expansive rocks (quartzite, gneiss, granite) and late-expansive rocks (greywacke, siltstone, quartz sandstone). It is generally difficult to obtain reliable expansion data for these rock types even in the concrete prism test, because their expansion does not come to an end at 38 °C within one year as specified in the test. Expansion tests are designed to make a minimal engineering judgement whether the aggregate is potentially deleterious or not, and terminate within an acceptable time limit. Furthermore, there is evidence that the potential expansion is underestimated due to leaching of alkali during the test [15].

1.2.2 Identification of the Type of ASR and Reactive Rocks

Stereomicroscopy can provide an overview of the distribution of ASR in a core and can facilitate the selection of areas for thin section preparation.

Thin section petrography using a PM is performed to confirm the type of rocks that are reactive and identify alkali-reactive minerals within these rocks. These observations can determine whether the aggregate in concrete is producing early-expansive ASR, late-expansive ASR or intermediate of these (Table 1.1). It is important to note that the aggregate even with the same rock name could present a different reactivity, depending on the geologic history (e.g. shale with or without opal; andesite with or without cristobalite or volcanic glass) as a result of recrystallization of metastable silica in younger age into stable form in older age [16].

1.2.3 The Stage of ASR in Concrete

The stages of ASR in concrete can be classified based on microscopic textures and microstructural features such as those listed in Table 1.2. Figure 1.1 illustrates these features in concrete with late expansive aggregate. Note that it is important to examine each rock type in the coarse and fine aggregate because the reactivity may likely differ.

In order to identify the stage of ASR as described above, the following information has to be considered:

- Reaction rims may not be readily detected on the cut or polished surface of concrete, particularly with particles of the late-expansive aggregates (e.g. quartzite, schist, gneiss, granite). They may be more easily detected on the fracture sur-

Table 1.1 Reaction rates of ASR, rock types and associated reactive minerals (adapted from Grattan-Bellew [13], Katayama [14] expansion compiled from Fig. D.3)

Reactivity	Concrete prism tests to reach 0.04%*	Rock type	Reactive mineral
Early-expansive	<50 days	Andesite, rhyolite, opaline chert opalineshale	Opal, cristobalite, tridymite, rhyolitic glass
Intermediate-expansive	50–90 days	Chert, devitrified rhyolite, silicious limestone	Some volcanic glass, chalcedony, cryptocrystalline quartz
Late-expansive	90–180 days	Greywacke, siltstone, shale, metamorphic rocks (schist, quartzite)	Cryptocrystalline quartz, microcrystalline quartz
Extremely late expansive	> 180 days	Quartzite, gneiss, granite	Microcrystalline quartz

*CSA A23.2-14A, ASTM C1293, RILEM AAR-3 and their equivalents (38 °C, Na_2O_{eq} 5.1–5.5 kg/m³)

Table 1.2 Stages of ASR in concrete [4, 9, 17]

Stage	Petrographic features
i	Formation of reaction rim, no cracks
ii	Rimming of ASR sol/gel (halo) in cement paste around the reacted aggregate; deposits of gel within voids near the aggregates, no cracks
iii	Cracking of aggregate filled with ASR gel
iv	Propagation of gel-filled cracks from the reacted aggregate into surrounding cement paste
v	Increase of the crack width and precipitation of ASR gel along cracks into air voids distant from the reacted aggregate
vi	Formation of network of gell-filled cracks connecting the reacted aggregate

face. Detailed examination in thin sections or SEM on sections may confirm the formation of reaction rims due to ASR (as opposed to surface weathering).

- Exudation of ASR sol/gel into the surrounding cement paste is characterized by darkening of the contacting paste (rimming, halo) under reflected or transmitted light. At this stage, capillary pores within aggregate and surrounding paste may be filled with gel. Note that this type of exudation may be less evident in the late-expansive rock types because the amount of reactive silica is scarce.
- After the appearance of cracks within cement paste, i.e. stages (iv), (v), and (vi), a variety of methods may be used to assess the severityof damage. These include

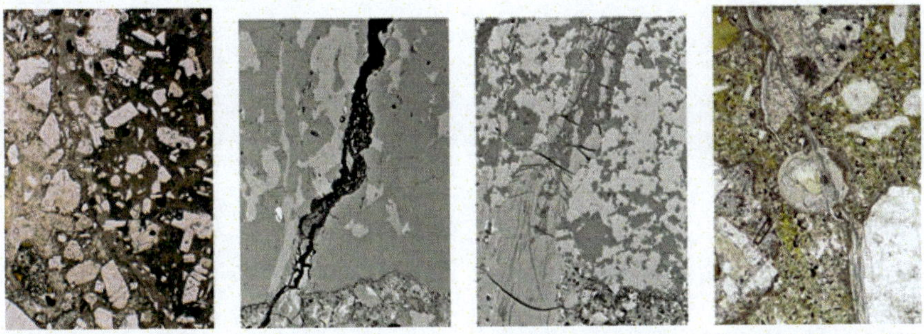

(a) stage i; reaction rim

(b) stage iii; crack within reacted aggregate (SEM)

(c) stage iv; gel-filled crack extending from the aggregate into cement paste (SEM)

(d) stagev; void-filling gel along the crack distant from the aggregate (PM)

Fig. 1.1 Sequence of textural development in ASR in field concrete

(i) crack width or the crack index (total crack width (mm/m) as described by Leemann and Griffa [18] and its percentage (%) by Katayama [4], (ii) damage levels as described by Menéndez, García, and Prendes [19], (iii) damage rating index [20–22].

1.2.4 Identification of Alkali-Reactive Minerals

After completing observation under the polarizing microscope, SEM observation on the same polished thin section is done to confirm the very point of reaction of the reactive minerals within reacted aggregate (Fig. 1.2). It is important to identify the species of the reacting mineral in concrete, as well as to assess whether the type of the reaction is early-expansive or late-expansive. This can be done by identifying early-expansive silica minerals, because they are highly alkali-reactive having small pessimum proportions. They can produce ASR even if they are present in a small proportion at a lower alkali level of concrete (i.e. $Na_2Oeq < 3.0\,kg/m^3$).

1.2.5 Composition of ASR Products

Compositions of amorphous and crystalline ASR products in the aggregates and in the cement paste can be determined by energy-dispersive X-ray spectroscopy (EDS) analysis in the SEM. It can be used as further confirmation for ASR as the mechanism for expansion. ASR products within aggregates, where the expansion is generated,

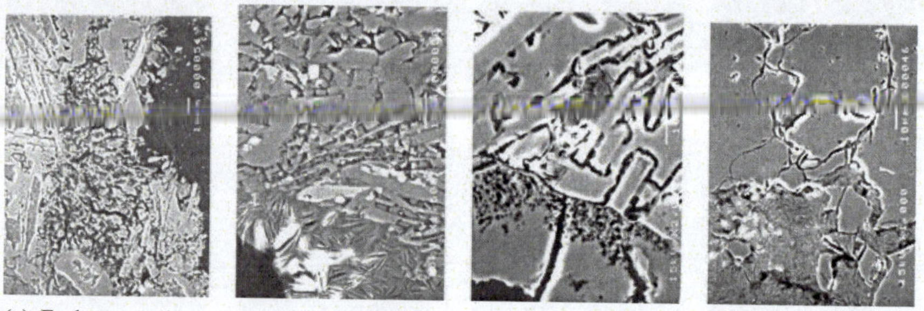

(a) Early-expansive cristobalite in andesite

(b) Early-expansive tridymite in andesite

(c) Early-expansive rhyolitic glass in andesite

(d) Late-expansive microcrystalline quartz in gneiss

Fig. 1.2 Reacting silica minerals and volcanic glass in the ASR-affected field concretes (partly from [5]

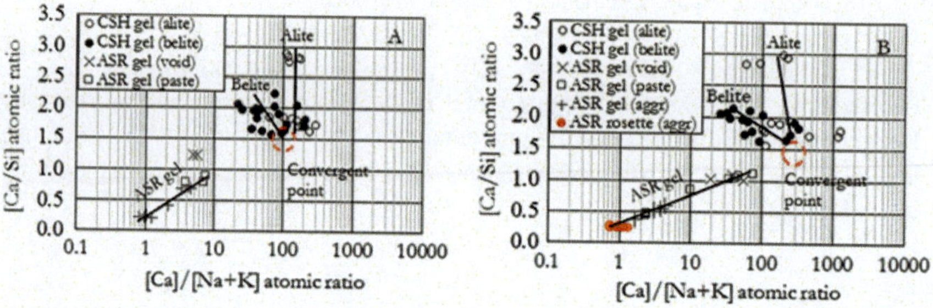

Fig. 1.3 Type I evolutionary changes of compositions of ASR gel and CSH gels in field concretes. (**a**) Early stage of ASR (externally sound), (**b**) middle stage of ASR (moderately deteriorated) [24]

display a typical composition rich in alkalis and a change of this composition, when they extrude into the cement paste [5, 10, 17, 23].

The composition of the ASR products may evolve with the stage of reaction and typically show a partial replacement of alkalis by calcium from cement paste, finally approaching an end-point of reaction, i.e. convergent point, at which ASR gel has a composition corresponding to non-expansive CSH gel [5]. Figure 1.3 gives a general compositional trend of ASR in concrete. In the early stage (A), compositional line of ASR gel is short. In the middle stage of ASR (B), compositional range of ASR gel extends and expansive alkali-rich ASR gel in the aggregate changes into rosette crystals. In aged field concretes, the range and trend of compositional lines can vary widely, depending on the stage of ASR and the mode of weathering, whereas concrete prism tests are terminated generally at the middle stage of ASR (Table 1.3).

Table 1.3 Compositional ranges of crack-filling ASR gel in aggregate and paste, reflecting the stage of ASR and the mode of weathering (e.g. Katayama 2008, 2012, Katayama et al. 2008)

	Weathering	Composition	Type I				Type II	Type III
			Single line				Parallel lines	Deflected line
			Early stage	Middle stage	Late stage	End point*		
Concrete	None	[Ca/Si]	<0.8	<1.5		1.7-1.8		
		[Ca]/[Na+K]	<10	<100		200–400		
Prism	Additional	[Ca/Si]		<1.4		1.5		
	freeze-thaw	[Ca]/[Na+K]		<100		200		
Field	Negligible	[Ca/Si]	<1.0	<1.5	<1.5	1.3-1.6		
		[Ca]/[Na+K]	<10	<100	<200	60–300		
	Carbonation along	[Ca/Si]					<1.3	
	Cracks	[Ca]/[Na+K]					<200<1000	
Structure	Intense leaching	[Ca/Si]			<1.3	1.3		<1.3
	and freeze-thaw	[Ca]/[Na+K]			<1000	1000		<1000

*Convergent point in equilibrium with CSH gel

1.3 Preparation and Conclusions

The sequence of ASR on the level of the structure can be correlated with the sequence of ASR on the level of the micro-structure, Fig. 1.4. Petrography is essential for diagnosis of AAR, the root cause of expansion and micro and macrocracking, as well as estimation of the present [25] and past expansion [4]. The caution of using microstructure and petrography in isolation is that it is a time consuming procedure for examining small 2-D sections in a heterogeneous volume. The structure concrete at macro and micro level is combined with the reinforcement and the stress fields which modify the cracking and expansion in 3D. This is the configuration of the "structure" for which "prognosis" as well as diagnosis is required. It needs to be considered as part of a complementary range of tests which each partially answer specific questions and in combination provide a better but still much less than 100% prognosis.

For this reason, preparation should be made using core samples extracted with a field survey, confirming the information, such as location, environmental factors which affects ASR (access to water, moisture, deicing salt), direction relative to rebars, structural aspects, etc. On the other hand, monitoring crack width and large scale displacement of a structure is easy to perform, being possibly automatized compared to petrographic examination. However, this may miss the heterogeneous expansion within a large section of the structure. For instance, large dam (e.g. height 60m) has a distinctive gradient of vertically decreasing temperature and vertically increasing pressure to the depth, which results in different petrographic stages of ASR and past expansion by depth [4]. This aspect cannot be verified by the bulk structural monitoring, unless petrographic examination is performed.

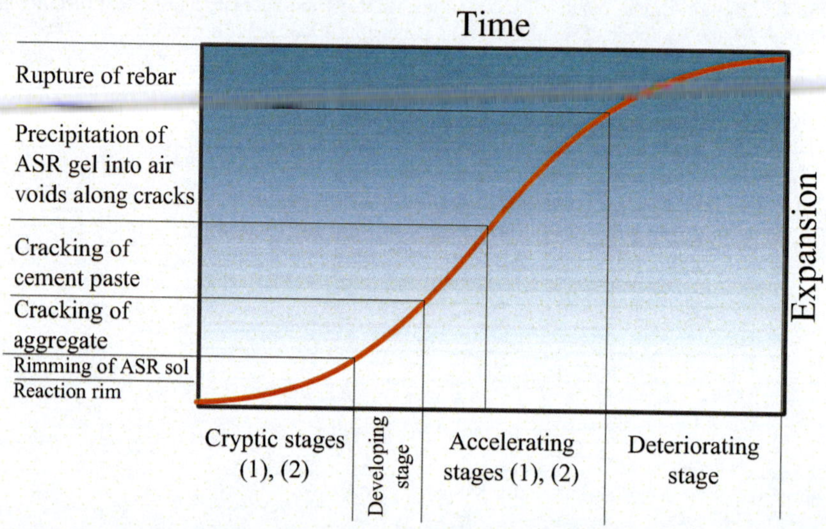

Fig. 1.4 Petrographic stages of ASR and expansion curve of concrete (late-expansive ASR) [26]

1. Induction stage: This is a dormant period before expansion starts. The concrete surface looks sound. However, ASR is taking place internally, forming reaction rims and sol/gel migrates through the concrete micro-structure. Gel may form deposits in voids. If the alkali level of concrete is sufficient for a particular rock type, ASR will proceed, especially if further alkalis are contributed from the aggregate.
2. Development stage: this ranges from crack formation in aggregates to the initial stage of cracking of cement paste that extends from the cracked aggregate.
3. Acceleration stage: Expansion and cracking of concrete proceed steadily at a high rate, increasing the width of gel-filled cracks that extend from the reacted aggregate. Precipitation of ASR gel along cracks distant from the reacted aggregate becomes pronounced. Lamination of gel deposits in fractures are characteristic of this stage.
4. Deterioration stage: Expansion rate decreases due to equilibrium of ASR in concrete, but displacement and crack width of concrete reach a maximum. In RC structures built with highly reactive aggregate (e.g. andesite) and a less-ductile steel, rebar fracture may occur at the stirrup hoops when small bending radius was applied resulting in initial defects before use.

Table 1.4 and Fig. 1.4 further illustrate the correlation. For detail, see Appendix.

It should be noted that, where non-homogeneous expansion is generated by the aggregate like a gravel, strain data (expansion in mm/m or in %) could differ considerably depending on the measuring length. In such a case, when the total displacement is averaged for the entire height or length of the structure, strain could be much smaller than is measured in a small span of concrete (e.g. 10 cm) around a particular

Table 1.4 General correlation between the progress of ASR and petrographic stages in field concretes

Item	Process	Evolution of ASR time →			
		Early Middle Late			
		Induction*	Development	Acceleration	Deterioration
Reaction	Dissolution of silica	x	x	x	x
	Formation of ASR sol/gel		x	x	x
	Crystallization of ASR gel			x	x
	ASR gel altering to CSH gel			x	Max
Expansion	Generation of pressure		x	Max	x
	crack width		x	x	Max
	Lamination and displacement			x	Max
Petrographic stages	i) Reaction rim	x	x	x	x
	ii) Sol/gel around aggregate		x	x	x
	iii) Cracking of aggregate		x	x	x
	iv) Cracking of cement paste			x	Max
	v) Gel filling in distant voids			x	Max

*Latent period in pathology, or cryptic stage in algology With early-expansive ASR, the process towards the acceleration stage is short

crack for monitoring. This aspect and a correlation between the petrographic stages of ASR and the total past expansion estimated by the fitting to the Larive's sigmoidal curve have recently been clarified [4].

As such, the petrographic investigation on the level of the microstructure is an important tool for not only diagnosing ASR, but additionally for assessing the stage of ASR and for correlating it with the stage of ASR in the concrete structure. This provides a base for tests trying to assess the residual expansion potential of concrete extracted from ASR-damaged structures.

References

1. Larive, C.: Apports Combinés de l'Expérimentation et de la Modélisation à la Compréhension del'Alcali-Réaction et de ses Effets Mécaniques. Ph.D. thesis. Paris: Laboratoire Central des Ponts et Chaussées, 1998. https://hal.inria.fr/docs/00/52/06/76/PDF/1997TH_LARIVE_C_NS20683.pdf
2. Wood, J.: When does AAR stop? In: Proceedings of the 12th ICAAR, Beijing, China, pp. 1016-1024 (2004)
3. Tuthill, L.H.: Alkali-silica reaction-40 years later. What Happened? Where are we? What now? Concrete Int. **4**(04), 32–36 (1982)
4. Katayama, T.: An attempt to estimate past expansion of concrete based on petrographic stage of alkali-silica reaction. In: Proceedings of the 39th International Conference on Cement Microscopy. Toronto, Canada, pp. 217–236 (2017)
5. Katayama, T.: Petrographic study of alkali-aggregate reactions in Concrete. Ph.D. thesis. University of Tokyo (2012)
6. Saouma, V., Martin, R., Hariri-Ardebili, M.A., Katayama, T.: A mathematical model for the kinetics of the Alkali-Silica chemical reaction. Cement Concrete Res. **68**, 184–195 (2015)
7. Leemann, A., Katayama, T., Fernandes, I., Broekmans, M.: Types of alkali-aggregate reactions and the products formed. Proc. Inst. Civil Eng. Construct. Mater. **169**, 128–135 (2016)
8. Guthrie, G., Carey, J.: A thermodynamic and kinetic model for paste-aggregate interactions and the alkali-silica reaction. Cement Concrete Res. **76**, 107–120 (2015)
9. Katayama, T., Tagami, M., Sarai, Y., Izumi, S., Hira, T.: Alkali-aggregate reaction under the influence of deicing salts in the Hokuriku district. Japan. Mater. Char. **53**, 105–122 (2004)
10. Leemann, A., Lura, P.: E-modulus of the alkali-silica reaction product determined by micro-indentation. Construct. Build. Mater. **44** (2013)
11. Guide to diagnosis and appraisal of AAR damage to concrete in structures: part 1 diagnosis (AAR 6.1). In: Godart, B., de Rooij, M., Wood, J. (eds.) vol. 12. Springer Science & Business Media (2013)
12. ASTM C227-10: Standard test method for potential Alkali reactivity of cement-aggregate combinations (Mortar-Bar Method). Annual Book of ASTM Standards (2010). https://doi.org/10.1520/C0227-10
13. Grattan-Bellew, P.: A review of test methods for alkali-expansivity of concrete aggregates. In: Proceedings of the 5th International Conference on Alkali-Aggregate Reaction (ICAAR). Cape Town, South Africa (1981)
14. Katayama, T.: Petrography of alkali-aggregate reactions in concrete-reactive minerals and reaction products. In: East Asia Alkali-Aggregate Reaction Seminar, Tottori. Tottori University, Supplementary paper. Tokyo, A45–A59 (1997)
15. Lindgård, J., Thomas, M., Sellevold, E., Pedersen, B., Andiç-çakir, Ö., Justnes, H., Rønning, T.: Alkali-silica reaction (ASR)-performance testing: influence of specimen pre-treatment, exposure conditions and prism size on alkali leaching and prism expansion. Cement Concrete Res. **53**, 68–90 (2013)
16. Katayama, T., Kaneshige, Y.: Diagenetic changes in potential alkaliaggregate reactivity of volcanic rocks in Japan-a geological interpretation. In: P.E. Grattan-Bellew (ed.) Proceedings of the 7th International Conference on Alkali-Aggregate Reaction. Ottawa, Canada, pp. 489–493 (1987)
17. Katayama, T., Oshiro, T., Sarai, Y., Zaha,K., Yamato, T.: Late-expansive ASR due to imported sand and local aggregates in Okinawa Island, southwestern Japan. In: Proceedings of the 13th International Conference on Alkali-Aggregate Reaction (ICAAR). Trondheim, Norway, pp. 862–873 (2008)
18. Leemann, A., Griffa, M.: Diagnosis of alkali-aggregate reaction in dams. Tech. rep. Publication 290934. SFOE-Project SI/500863-01. EMPA, Laboratory for Concrete/Construction Chemestry (2013). http://www.bfe.admin.ch/dokumentation/energieforschung/index.html?lang=de&publication=11174

19. Menéndez, E., García, R., Prendes, N.: Damage assessment of dams affected by expansive reactions using stereomicroscopy and electron microscopy. Relationship Between The Damage Levels DRI And DL. Tech. rep. Madrid. Instituto Eduardo Torroja, Spain (2015)
20. Grattan-Bellew, P.E., Danay, A.: Comparison of laboratory and field evaluation of AAR in large dams. In: Proceedings of the International Conference on Concrete Alkali-Aggregate Reactions in Hydroelectric Plants and Dams, Fredericton, NB. 23 p. 1992
21. Rivard, P., Ballivy, G.: Assessment of the expansion related to alkalisilica reaction by the Damage Rating Index method. Construct. Build. Mater. **19** (2005)
22. Sanchez, L., Fournier, B., Jolin, M., Bastien, J.: Evaluation of the Stiffness Damage Test (SDT) as a tool for assessing damage in concrete due to alkalisilica reaction (ASR): Input parameters and variability of the test responses. Construct. Build. Mater. **77**, 20–32 (2015)
23. Thaulow, N., Jakobson, U., Clark, B.: Composition of alkali silica gel and ettringite in concrete railroad ties: SEM-EDX and X-ray diffraction analysis. Cement Concrete Res. **26** (1996)
24. Katayama, T., Sarai, Y., Sawaguchi, H.: Diagnosis of ASR in airport pavements in Japan; Early-expansive sand aggregates missed by conventional tests. In: Proceedings of the 15th International Conference on Alkali-Aggregate Reaction (ICAAR). Sao Paolo, Brazil, pp. 1219–1228 (2016)
25. Menéndez, E.: Evaluation and gradation of simultaneous damage in concrete affected by Alkali-Silica reaction and sulfatea attack. Int. Rev. Civil Eng. **2.2** (2011)
26. Katayama, T.: Introduction to alkali-silica reaction, Part 3. How to diagnose ASR. Concrete J. (Japan Concrete Institute, JCI) **52**(12) (2014) (in Japanese)

Chapter 2
Assessment of Damage and Expansion

Andreas Leemann, Esperanza Menéndez, and Leandro Sanchez

2.1 Crack Indices

Andreas Leemann

2.1.1 Macroscopic Scale

An indication about the expansion of the concrete can be derived from the crack-index [1]. The crack-index is determined by measuring the crack width along pre-drawn lines, Fig. 2.1 and is expressed as crack-width per measured length. However, it has to be kept in mind that the formation of the cracks may not be attributable solely to AAR. Still, the crack index indicates a concrete expansion at the studied location in mm/m.

A. Leemann (✉)
Empa, Swiss Federal Laboratories for Materials Testing and Research, Laboratory for Concrete and Construction Chemistry, Überlandstrasse 129, 8600 Dübendorf, Switzerland
e-mail: andreas.leemann@empa.ch

E. Menéndez
Institute of Construction Science Eduardo Torroja, Madrid, Spain
e-mail: emm@ietcc.csic.es

L. Sanchez
University of Ottawa , Ottawa, Canada
e-mail: Leandro.Sanchez@uottawa.ca

© RILEM 2021

V. E. Saouma (ed.), *Diagnosis & Prognosis of AAR Affected Structures*,
RILEM State-of-the-Art Reports 31,
https://doi.org/10.1007/978-3-030-44014-5_2

Fig. 2.1 Determination of the crack-index along the lines A-B, A-C, A-D and B-C. Side of the square = 1 m, Leemann and Griffa [2]

2.1.2 *Microscopic Scale*

Due to the problems related to sample preparation without impregnation especially in the case of microcracks with a width below 20 μm, it seems mandatory to work with impregnated samples. As thin sections are an important tool for diagnosing AAR, it is obvious to use them for assessing the extent of the reaction as well. Similar to the determination of a crack index on the surface of concrete (see previous paragraph), the crack width can be determined along defined lines (Fig. 2.2). Here, several points are important:

1. The crack index has to be determined at different concrete depth, as ASR is not homogeneously distributed in concrete structures.
2. Thin sections from the same concrete depth have to be prepared perpendicular and parallel to the length axis of the concrete core to account for case of crack orientation.
3. It is recommended to use a fair number of thin sections for reasons of representativeness (e.g. covered area of 100 cm^2 per concrete depth depending as well on maximum aggregate size).

The thin section is moved at a magnification of 50× or 100× in the fluorescent light mode along the lines shown in Fig. 2.2. The width of the AAR-induced cracks intersecting with these lines is measured. The crack index can be calculated as crack width per measured length. The determination of the crack index indicates the past expansion of the concrete at a given location and concrete depth in mm/m.

An important confirmation between cracking and expansion is given by Lindgard et al. [3]. The surface of cut concrete prisms (70 × 280 mm^2) was impregnated with epoxy resin containing a fluorescent dye and photographed in UV-light. Crack segmentation was done with image analysis. A good relation between expansion in the concrete prism test (0.04–0.43%) and the crack area was established. As the linear measurements done in the determination of the crack index are equal to the percentage of crack area [4], the crack index as a way to assess past expansion can be confirmed.

Fig. 2.2 Thin section (52 x 85 mm^2) with lines along which the crack index is determined [2]

Additionally, the results of the crack index can be combined with a determination of the number of reacted aggregates [2, 5]. This approach further improves the assessment of the extent of ASR and the comparison of different parts of an investigated structure or different structures.

2.2 Damage Rating Index

Leandro Sanchez

2.2.1 Introduction

The following is adapted from Sanchez et al. [6] Sanchez et al. [7], and Sanchez, Fournier, Jolin, Mitchell, and Bastien [8]

The DRI is a microscopic test procedure performed with the use of a stereomicroscope (15–16x magnification) where damage features generally associated with alkali-aggregate reaction (AAR) are counted through a 1 cm^2 grid (i.e. 10 x 10 mm units) drawn on the surface of polished concrete sections. The number of counts corresponding to each type of petrographic features is then multiplied by weighing factors, whose purpose is to balance their relative importance towards the mechanism of distress under analysis (e.g. alkali-silica reaction – ASR) [9, 10]. The latter (i.e. petrographic features × weighing factors) corresponds to the so called "DRI number"; the higher this number, the higher the overall concrete damage. It is worth noting that the factors used in the method have been selected on a logical basis, but relatively arbitrarily; they were recently modified in order to reduce the variability between petrographers performing the test [11, 12]. Ideally, a surface of at least 200 cm^2 should be used for DRI analysis, and it may be greater in the case of mass concrete incorporating larger size aggregate particles. However, for comparative purposes, the final DRI number is normalized to a 100 cm^2 area [6, 7].

The main goal of the DRI is not to "replace" conventional petrographic procedures of concrete, such as ASTM C 856, which may require special techniques or tools (e.g. scanning electron microscopy (SEM) with energy dispersive X-ray analysis (EDS), X-Ray diffraction (XRD), etc.) to appraise the "cause" of damage in concrete. The DRI is rather a complementary petrographic tool aiming to quantify both "past induced expansion" and "damage extent" of affected concrete. Although the DRI has been used by a number of researchers, especially in North America, there is currently no standard test protocol for the method. Otherwise, the method is currently being standardized by Canadian Standards Association (CSA) and it should be a standardized procedure within the next CSA cycle (i.e. 2019–2023).

2.2.2 *DRI Weighing Factors*

Petrographic features (e.g. open or closed cracks, within the aggregate particles and/or cement paste, with or without reaction products) are counted and multiplied by weighing factors (Table 2.1) as per Fig. 2.3. A factor of 2 is given to opened cracks within the aggregate particles and a factor of 3 for cracks in the cement paste, regardless of the presence of reaction products. This has been selected to reduce difficulty in recognizing the presence of reaction products at 15–16x magnification (i.e. DRI magnification). The latter was found to significantly lessen variability amongst operators [12]. Cracks in cement paste indicate a greater importance regarding mechanical properties and/or durability of the material which is why a greater weighing factor is given to these petrographic features. It is worth noting that cracks in the aggregate particles down to 1 mm (i.e. sand particles) are considered in the evaluation as per [6] and [7].

2.2.3 *Using DRI for Concrete Damage Assessment*

Test specimens consist of 100 x 200 mm cylinders fabricated from 23, 35, and 45 MPa concrete mixtures. Each concrete mixture consists of a combination of ASR reactive

Table 2.1 Weighing factors

Petrographic features	Weighing factor
Cracks in coarse aggregate (CCA)	0.25
Opened cracks in coarse aggregates (OCA)	2
Crack with reaction product in coarse aggregate (OCAG)	2
Coarse aggregate debonded (CAD)	3
Disaggregate/corroded aggregate particle (DAP)	2
Cracks in cement paste (CCP)	3
Cracks with reaction product in cement paste (CCPG)	3

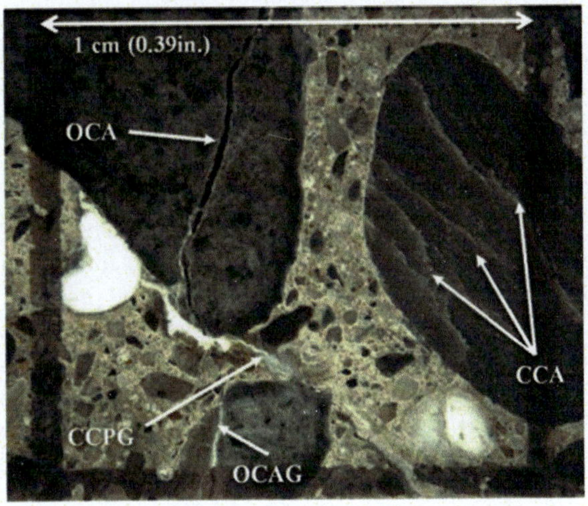

Fig. 2.3 Micrograph shows a 1 cm^2 where most of the petrographic features to be noted in the DRI distance (between the two vertical lines is 1 cm) [10]

coarse aggregate and non-reactive sand or non-reactive coarse aggregate and ASR reactive sand. To accelerate the induced expansion process, reagent grade NaOH was added to bring the alkali content of all mixtures to 1.25% Na_2O_{eq} by cement mass. Cylinders were manufactured, outfitted with stainless steel gauge studs for expansion measurements, placed in buckets lined with damp cloth, and stored at 38 °C and 100% R.H. Cylinder lengths were monitored regularly until they had reached expansion levels selected for further analyses (i.e. 0.05, 0.12, 0.20, and 0.30%). Once expansion levels were reached, specimens were wrapped in plastic film and stored at 12 °C to avoid further ASR development as per [8]. The samples were then first cut in two axially and then one of the flat surfaces obtained was polished. A portable hand-polishing device, which uses diamond-impregnated rubber disks (no. 50 (coarse), 100, 400, 800, 1500 to 3000 (very fine)), was found most suitable for obtaining surfaces of appropriate quality for DRI assessment.

2.2.4 *Test Outcomes*

2.2.4.1 Damage Features as a Function of Expansion

The initial analysis of ASR damage features as a function of induced expansion was performed through absolute counts (i.e. without using weighting factors). Features of deterioration were classified into three groups. Group I consists of closed cracks within the coarse aggregate or the coarse fraction of the sand particles. Group II are opened cracks (with or without reaction products) within the coarse aggregate or coarse fraction of the sand particles. Group III is comprised of cracks in the cement paste, with or without reaction products.

From Fig. 2.4a and 2.4b the following general trends may be observed: (1) the most common feature of "deterioration" in all samples corresponds to Group I cracks (CCA); (2) a progressive increase in the total number of Group I to III cracks is found as a function of increasing expansion for most of the concrete specimens assessed; (3) at similar expansion levels, the total number of counts for Group I to III cracks is always higher for reactive sand mixes (because of higher counts of Group I cracks in the non-reactive coarse aggregate particles), and; (4) the counts of Group II and Group III cracks definitely increased with increasing expansion; these are considered as indicative of ASR-induced expansion and progress.

At early expansion (0.05%), Group I cracks represent between 60–90% of the petrographic features. In general, Group II cracks account for 7–28% but rises to 17–33% in the specimens with reactive fines. The percentage of the cracks belonging to Group III are quite low (<7%). As the expansion increases, the proportion of Group I cracks generally lessens (down to 40–70%), as the relative importance of Group II and Group III cracks increases up to 25–45% and 5–20% respectively, due to ASR progress.

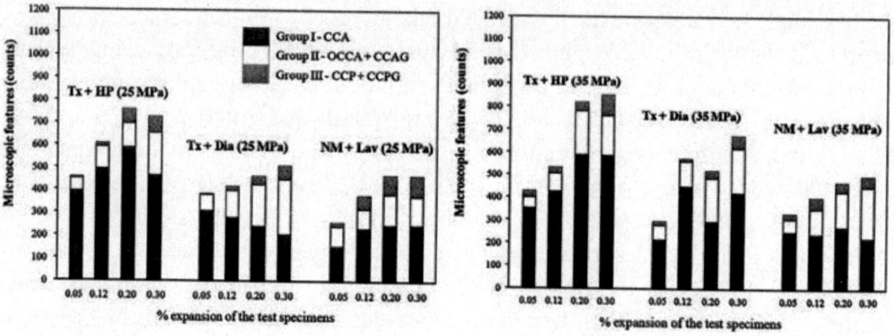

((a)) 25 MPa; Tx+HP and Tx+Dia are reactive sands mixes. NM+Lav is the reactive coarse aggregate mix.

((b)) 35 MPa; Tx+HP and Tx+Dia are reactive sands mixes. NM+Lav is the reactive coarse aggregate mix.

Fig. 2.4 Counts of microscopic damage features normalized for 100 cm^2 specimens [13]

2.2.4.2 Cracks Development as a Function of Expansion

Another interesting analysis that might be conducted through the DRI is the appraisal of crack development within individual aggregate particles as a function of induced expansion. For 3 specimens (25 and 35 MPa for specimens with reactive coarse aggregates and 25 MPa for specimen with reactive fine aggregates), the number particles per section were counted. Then, the number of cracks in the section was divided by the number of particles (ranging from 1 mm to 20 mm) as a function of ASR induced expansion.

The results can bs seen in Fig. 2.5. Generally, this study shows that the higher the expansion, the higher the number of opened cracks (OCA+OCAG) within the aggregate particles from either reactive sand or coarse aggregate. However, the development of closed cracks (CCA) increases as expansion increases for specimens with reactive sand. For specimens with reactive coarse aggregates, the ratio of closed cracks either doesn't change or decreases as a function of expansion.

The results illustrated in Fig. 2.6a, and 2.6b confirm that this is mainly related to a significant increase in the proportion of closed cracks in the non-reactive coarse aggregate particles at higher expansion levels. On the other hand, when the reaction is generated in the coarse aggregate, the proportion of closed cracks in the reactive coarse aggregate particles does not change significantly or is decreasing slightly as a function of expansion. This was expected as the proportion of "active" ASR cracks (i.e. opened cracks, with or without reaction products) is increasing with the progress of expansion in the reactive coarse aggregate particles.

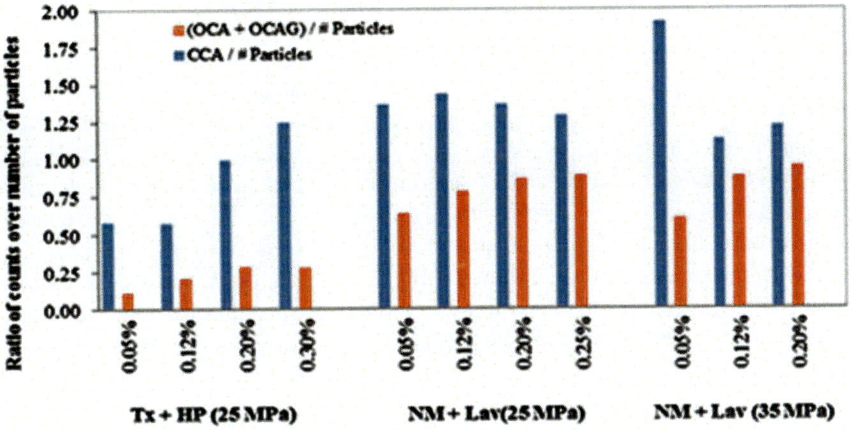

Fig. 2.5 Cracks development in the aggregate particles as a function of expansion for the following mixtures: Tx + HP (25 MPa), NM + Lav (25 MPa) and NM + Lav (35 MPa); where CCA means closed cracks in the aggregate particles and, OCA and OCAG are open cracks within the aggregate particles (with and without reaction products, respectively)

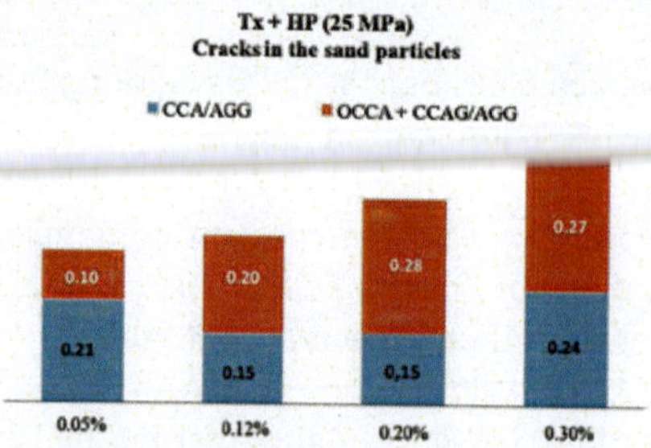

((a)) Microscopic features expressed as counts divided by total number of aggregate particles for opened and closed cracks in reactive sand particles

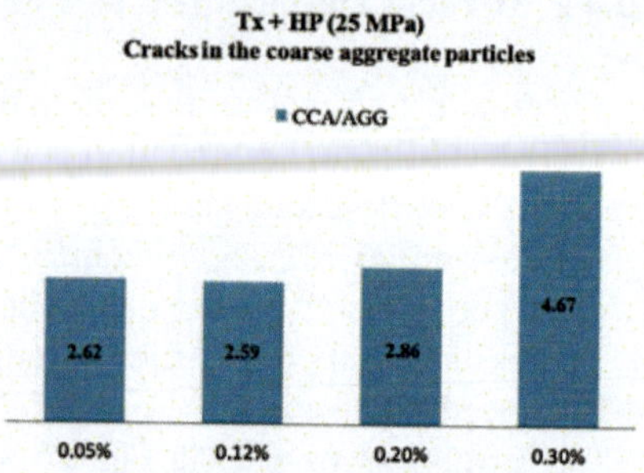

((b)) Microscopic features expressed as counts of closed cracks divided by total number of aggregate particles for non-reactive limestone particles

Fig. 2.6 Crack development in aggregate particles as a function of ASR-induced expansion: **a** number of counts divided by total number of aggregate particles for opened and closed cracks in the reactive sand particles and, **b** number of counts of closed cracks divided by the total number of aggregate particles for non-reactive limestone particles (HP) [13]

2.2.4.3 Crack Density as a Function of Expansion

The crack density of affected concrete may be calculated in the DRI analyses as the sum of the open cracks within the aggregate particles and cement paste (with or without reaction products) divided by the overall area examined. Such cracks are thought to be linked to ASR-induced development. Figure 2.7a and 2.7b, display the results obtained for 25 and 35 MPa mixtures incorporating reactive coarse and fine aggregates (NM and Tx respectively). Analyzing the results, one notices that the higher ASR induced expansion the higher the crack density. Moreover, there is no significant differences in crack density among the various concrete mix designs and different reactive aggregates assessed.

The crack density is calculated as the sum of the open cracks in aggregates and cracks in the cement paste (with or without reaction products) divided by the overall area examined. Such cracks are though to be linked to the development and progress of ASR in the specimens. As is shown if Fig. 2.7a and b, crack density increase is correlated to the increasing expansion. There is no significant differences in crack density between the various concrete mix designs and different reactive aggregates.

2.2.4.4 Crack Width and Length as a Function of Expansion

The maximum crack length (through the aggregate particles, cement paste or both) and width (within the aggregate particles) may also be gathered through the use of the DRI. Figure 2.8 illustrate the results obtained for 25, 35 and 45 MPa mixtures incorporating reactive coarse and fine aggregates (NM and Tx, respectively). Analyzing the results below, one verifies that the higher the induced expansion, the longer/wider the cracks formed.

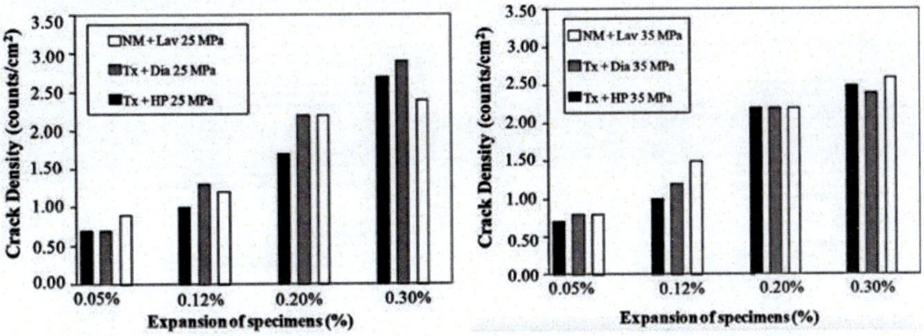

((a)) 25 MPa Concrete ((b)) 35 MPa Concrete

Fig. 2.7 Expansion versus crack density for various concrete strengths [13]

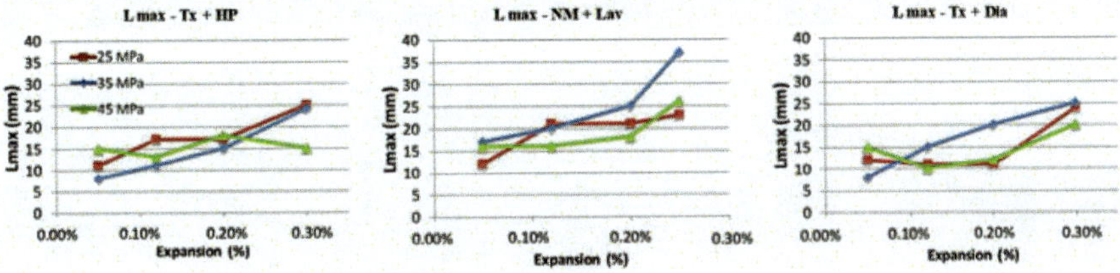

((a))

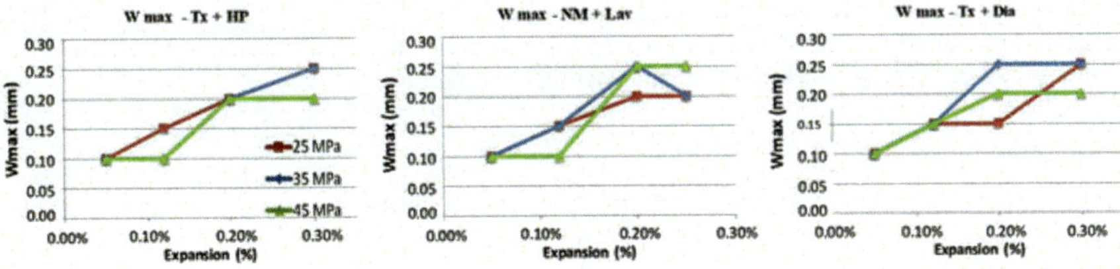

((b))

Fig. 2.8 Maximum width of cracking for all mixtures, as function of expansion degree [7]

2.2.4.5 DRI Number as a Function of Expansion

The detailed results of the semi-quantitative DRI numbers for 25 and 35 MPa mixtures are illustrated in Fig. 2.9, using the latest weighing factors proposed by Villeneuve and Fournier [11] and Villeneuve [12]. Figure 2.10 gives a plot of the DRI values as a function of the expansion for 25, 35 and 45 MPa concrete specimens. The data in Fig. 2.9 confirm that the open cracks within the aggregate particles and cement paste (with and without gel) increase with increasing expansion. Moreover, the DRI numbers correlates quite well (i.e. linearly) with ASR-induced expansion (Fig. 2.10). DRI values ranged from 200 (i.e. low expansion levels—0.05%) to roughly 700 (i.e. high expansion levels—0.30%). Finally, it seems that the DRI values do not

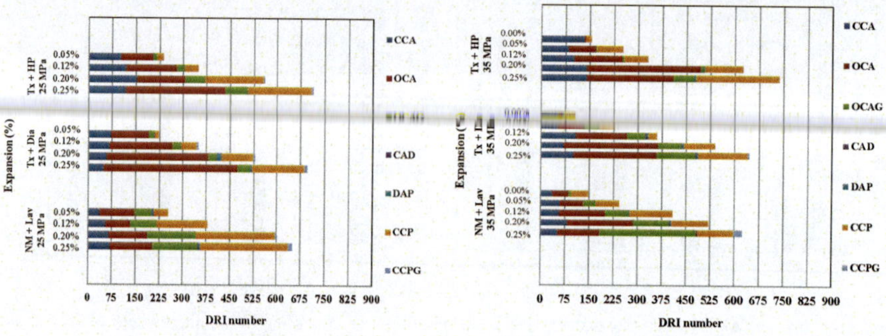

((a)) 25 MPa ((b)) 35 MPa

Fig. 2.9 DRI charts for all the mixtures analyzed over the study: A, 25 MPa mixtures and, + B, 35 MPa mixtures. Both plots contain Tx and NM reactive aggregates

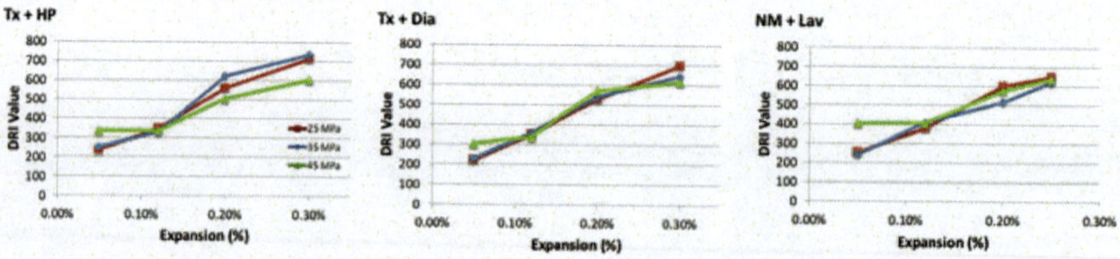

Fig. 2.10 DRI charts for all the mixtures analyzed over the study: A, 25 MPa mixtures and, + B, 35 MPa mixtures. Both plots contain Tx and NM reactive aggregates

significantly change as a function of the concrete mixture features (i.e. mechanical properties and reactive aggregate types and natures).

2.2.4.6 Describing ASR-induced Damage as a Function of Expansion

Based on the analysis of the semi-quantitative DRI number in combination with complementary petrographic observations, such as the measurements of the cracks' length, width, density and the complete analysis of the counts of distress features, Sanchez, Fournier, Jolin, and Duchesne [6] described the development of ASR distress in concrete incorporating "quartz-bearing" reactive aggregates (e.g. greywacke, siliceous limestone, gneiss, schist, argillite, etc.) as a two-step process: (a), the formation of cracking within the reactive aggregate particles in the early stages of the chemical reaction (including ASR "activation" of (pre-existing) closed cracks formed through aggregate processing operations), and (b), the extension of the above cracks into the cement paste to form a cracking network (i.e. connecting reactive aggregate particles with one another) with increasing expansion. Following the minimum energy law, it is easier for the expanding system to propagate the cracks produced

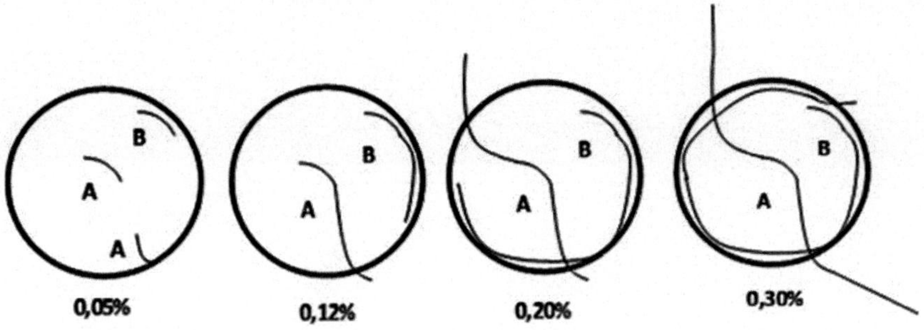

Fig. 2.11 Qualitative microscopic ASR damage model versus expansion level

through step (a) described above, instead of creating a significant number of new cracks. New cracks will always be generated as the alkali reaction keeps developing, but the amount of "new" cracks will be overcome by the increase in length and width of the cracks already formed, thus making the counts of distress features to keep increasing, but at a lower rate, with increasing expansion in the system. The above mechanism is described by the qualitative damage model proposed by Sanchez et al. [6] and presented in Fig. 2.11.

The extensive testing carried out in this study allowed to better understand the correlation between the development of microscopic features of deterioration (visible on polished concrete sections at 15–16x magnification under the stereomicroscope) and changes in the mechanical properties of concrete specimens as a function of ASR expansion; this is illustrated in Fig. 2.12 and summarized in Table 2.2 as a classification of damage degree in concrete due to ASR.

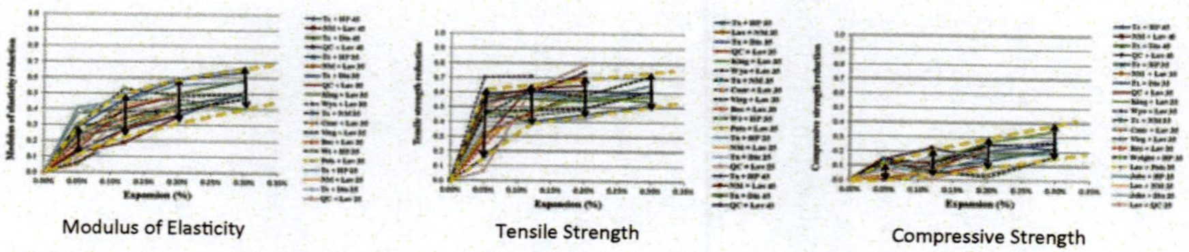

Fig. 2.12 Mechanical properties losses (i.e. compared to a sound concrete presenting the same maturity) as a function of concrete expansions

Table 2.2 Classification of the damage degree in concrete due to ASR

Classification of ASR damage degree (%)	Reference expansion level(%)[1]	Stiffness loss (%)	Compressive loss (%)	Tensile loss (%)	SDI	DRI
Negligible	0.00–0.03	–	–	–	0.06	−100–155
					0.16	
Marginal	0.04 ± 0.01	5–37	(−)10–15	15–60	0.11	−210 to 400
					0.25	
Moderate	0.11 ± 0.01	20–50	0–20	40–65	0.15	−330 to 500
					0.31	
High	0.20 ± 0.01	35–60	13–25	45–80	0.19	−500 to 765
					0.32	
Very high	0.30 ± 0.01	40–67	20–35		0.22	−600 to925
					0.36	

These levels of expansion should not be considered as strict limits between the various classes of damage degree but more as indicators/reference levels for which comparative analysis of petrographic and mechanical data was carried out allowing to highlight significant damage levels in concrete due to ASR progress

2.3 Damage Level Based Assessment

Esperanza Menendez

2.3.1 Introduction

The following is a summary of [14, 15].These papers propose a method to grade the damage and assign a damage level (DL) based on an evaluation of the microstructure. DL is ranked from 0 to 5 with 5 indicating the highest degree of damage. This procedure uses scanning electron microscopy (SEM) combined with energy-dispersive X-ray spectroscopy (EDX) to evaluate the presence and degree of damage. This procedure is also able to determine, if the source of damage is from ASR, internal sulfate attack (ISA) or other mechanisms. Moreover, the DL method is compared to the Damage Rating Index (DRI) to analyze the discrimination of deterioration between both methods when applied to dams affected by ASR and ISA.

2.3.2 Microstructural Evolution of ASR

The external symptoms of ASR are usually cracks on the concrete surface with deposits lining them. But the external manifestation of damage occurs when the internal reactions have advanced, because the formation and evolution of reaction products start at an early stage. At a microstructural level the type and distribution of expansive reaction products can be analyzed by SEM in the backscattering mode combined with EDX using polished concrete.

However, before the microstructural analysis is performed, the cores taken from the structure are analyzed at a macroscopically level to detect sings of alteration; cores cut along the length axis are analyzed by stereomicroscopy to detect the presence of reaction products or cracking. Afterwards, representative samples of the different areas of the concrete cores are prepared to be studied by SEM (backscattering mode) combined with EDX. Also, the preparation of these samples (cutting, drying, epoxy impregnation and polishing) shall be doing carefully to avoid artifacts. In order to be representative, an adequate number of samples have to be analyzed (such as 2–4 samples of 2,5 x 2,5 cm from a half core of 10 x 20 cm). Also, it is very important that the cores taken from the structure are representative of the concrete of the different areas of the structure.

Based on the microstructural analysis of polished samples it is possible to classify them according to one of the five grades of DL. To define each DL it is necessary take into account the characteristics of the products formed, the location of their formation and their abundance. ASR will create gels and micro-crystals that are typically found in the interior of cracked aggregates, at the interface between aggregates and cement paste, in air voids, in capillary pores of the cement paste and in the cracks formed in the cement paste.

Table 2.3 defines the DL, the reaction products formed and their location. The degree of damage attributable to the five DL is low (1), moderate (2), average (3), elevated (4) and extended to the total (5). For each grade of DL the products observed are the ones of the lower grade plus the ones of the defined level.

It is possible to establish a relationship between DL and the expansion in concrete dams (Fig. 2.13). Figure 2.14 gives an example of such a relation. The average expansion of the different blocks in this dam shows a heterogeneous distribution of damage due to ASR. Also, different levels of average expansion of blocks, monitored instrumentally are observed in the dam. There is a clear relation between the DL and dam expansion.

While this correlation is verified in this particular dam, this would have to be confirmed in other structures.

Table 2.3 Definition of damage code as a function of the damage level in concretes affected by alkali-silica reaction and sulfate attack [16]

Damage code	DL-1	DL-2	DL-3	DL-4	DL-5
Damage degree	Pores and open spaces close to paste interface	DL-1 plus products in aggregates	DL-2 plus products in aggregate-paste interfaces	DL-3 plus in the paste	DL-4 plus in the paste and fissures
Reaction products	Smooth gels	Smooth gels and rough gels	Smooth gels, rough gels and micro-crystalline products	Smooth gels, rough gels and micro-crystalline products	Smooth gels, rough gels and micro-crystalline products
Location of products	Pores and open spaces close to paste interface	Pores and open spaces close to paste interface + aggregates	Pores and open spaces close to paste interface + aggregates + aggregate-paste interfaces	Pores and open spaces close to paste interface + Pores + aggregates + aggregate-paste interfaces + paste Pores and open spaces close to paste interface + Pores + aggregates + aggregate-paste interfaces + paste (↑fissures)	

Fig. 2.13 Evolution of the microstructural features and DL in concrete with increasing time and expansion [16]

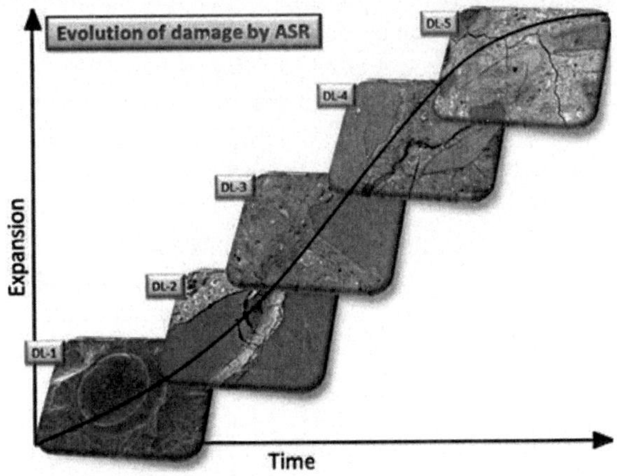

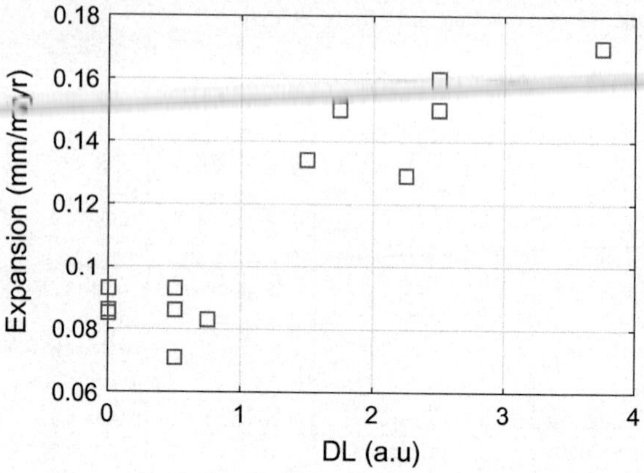

Fig. 2.14 Damage level with respect to average expansion by length by year [16]

2.3.3 DL Vs. DRI Applied to ASR and ISA

The damage level can be applied to different types of damages like ASR, internal or external sulfate attack, because the analysis includes the characterization of the type of reaction products. Also, in some structures, mainly in dams, a simultaneous occurrence of ASR and ISA is observed leading to cracking. ISA is usually associated with delayed ettringite formation or oxidation of pyrites or pyrothite present in aggregates as granitic rocks. On the macroscopic level, it can be difficult to distinguish the two types of reaction. A microstructural analysis with SEM is needed to determine the origin of damage.

The DL can be compared to the DRI. In general, the DRI can be significantly affected by the experience of the petrographer completing the analysis and can vary from one petrographer to another. The same applies to the DL. Consequently, the DL should be analyzed by specialists with deep experience in characterization of expansive process in concrete to guaranty an appropriate identification and assessment of the reaction products. The DRI is effective at providing relative information, when comparing sets of cores from various locations of the same structure, if they are carried out by the same petrographer. Additionally, it is difficult to determine the minimum DRI that indicates significant damage due to ASR. The DRI only evaluates the presence of cracks without considering their origin, while the DL allows the diagnosis of the origin of damage. A comparison between DL and DRI was conducted in [15].Core samples were rated by using both DRI and DL. A range for both DRI and DL values is determined and the results of the two methods are compared. The comparisons between the DL rating and DRI rating are shown in Figs. 2.15a and b. In the former, when data points are located in quadrants 3, 5 or 7, there is good correlation between DRI and DL. When data points are in quadrants 1, 2 or 4, DRI underestimates the damage relative to the DL, while it overestimates the damage in quadrants 6, 8 or 9. Figure 2.15b shows that DRI results correlate well to the DL results for ASR analysis. This is due to the fact that both methods are able to detect

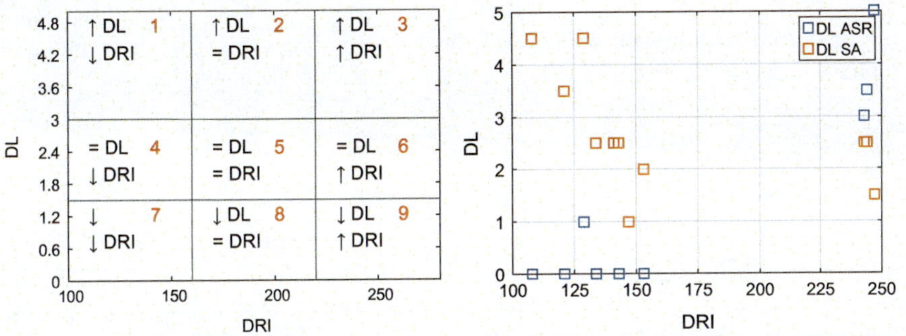

((a)) Control Quadrants

((b)) DL vs. DRI values for ASR and Sulfate Attack applied to different dams.

Fig. 2.15 Comparison between DRI and DL [15]

the evolution of damage. However, for sulfate attack, DRI generally underestimates the amount of damage, since it is difficult to detect the cracks formed by sulfate attack with stereomicroscopy, as such cracks are usually filled by ettringite or other reaction products. DL can be used to define the distribution of damage in different parts of a concrete structure affected by alkali-silica reaction and/or by sulfate attack. The potential heterogeneous distribution of damage may be due to different concrete components, environmental conditions, loads, or other characteristics of the concrete.

2.4 Stiffness Damage Test (SDT)

Leandro Sanchez

2.4.1 Introduction

Stiffness Damage Test (SDT) is an experimental test procedure in which the extent of inner cracking (i.e. distress degree) of a damaged concrete is measured through the use of five loading/unloading compressive cycles. The test is performed on concrete samples (cylinders or cores) with length-diameter ratios of 2.0. The amount of inner cracking is detected through the measure of both the "dissipated energy" and "plastic deformation" parameters. SDT is performed with 40% of the concrete strength, selected from a companion sound concrete with the same composition and maturity of the sample tested [17, 18]. The Stiffness Damage Index (SDI) and Plastic Deformation Index (PDI) were found to appraise the amount of "distress" present in a given concrete material [8, 18]. Moreover, these prior mentioned parameters were seen to correlate quite well with the advancement/progress (e.g. expansion attained to date) of durability related mechanisms such as alkali-silica reaction [8].

2.4.2 Background

In the early 1990's, Crisp and co-workers proposed to use the SDT to quantify the degree of distress in concrete due to ASR. The test method was actually developed by Walsh [19] who observed a good correlation between the crack density and the cycles of loading/unloading (stress/strain relationship) of rock specimens. Crouch [20], following those results, proposed a new test procedure (Stiffness Damage Test—SDT) based on cyclic compression loading of concrete specimens (cylinders or cores). A full review on the SDT development as a diagnostic tool for assessing ASR affected concrete is presented in [17].

Sanchez et al. [17] Sanchez et al. [21] Sanchez et al. [18] Sanchez et al. [8] actually pursued the work of Smaoui, Bérubé, Fournier, Bissonnette, and Durand [22]., by applying the SDT procedure to either specimens cast in the laboratory, presenting different mix-designs (25, 35 and 45 MPa) and incorporating a wide range of reactive aggregates (i.e. coarse vs. fine reactive aggregates, ≠ lithotypes, etc.), or concrete cores extracted from an extremely damaged concrete overpass in Quebec City (Canada). Based on their work, the use of 40% of the design concrete strength seems to be the best approach for characterizing expansion levels. The output parameters, such as the hysteresis area (HA) and the plastic deformation (PD) over the five cycles, as well as the modulus of elasticity (ME) (as an average value of the second and third cycles), seem to be the most diagnostic output results of the test. However, the use of indices (Stiffness Damage Index—SDI and Plastic Deformation Index—PDI) instead of absolute HA or PD values, which take into account the ratio"dissipated energy/total energy" implemented in the system, better represents the real "damage" of the affected materials Fig. 2.16. Actually, this approach decreases the impact of a poor selection of maximum loading level for stiffness damage testing and provides a better understanding of ASR evolution as a function of its expansion [8, 18]. It should be noted that several input parameters, such as the concrete's cure history, the sample's geometry and size, and the sample's location within the structural member, seem to strongly influence the output analyses of the SDT [21].

2.4.3 Samples Preparation and Apparatus

The following section and the following ones are based on the work ofSanchez et al. [17] *Sanchez et al.* [21] *Sanchez et al.* [18] *and Sanchez et al.* [8].

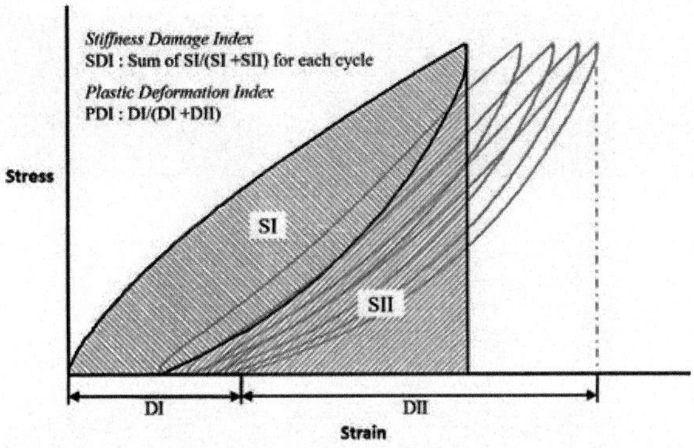

Fig. 2.16 Determination of the stiffness damage index (SDI) and plastic deformation Index (PDI based on the output from the stiffness damage test (SDT) [23]

2.4.3.1 General Guidelines

Concrete samples (laboratory cylinders or cores extracted from aging infrastructure members) with length-diameter ratios of 2.0 shall be used [24, 25]. The location (i.e. concrete member, surface or core, exposed or not exposed to the environment, etc.) shall be clearly identified for further analysis and data comparisons [21]. Prior to testing, the samples ends shall be rectified with either sulfur capping compound or surface grinding in order to avoid any interference over the test [21].

2.4.3.2 Storage & Handling

It is recommended, immediately after extracting the cores, to wrap them with plastic film in order to avoid excessive drying as the humidity level can have an influence on the results [21]. Then, the samples should be stored in the laboratory for at least one week prior to testing (enabling humidity homogenization) [21]. Afterwards, if immediate test is not possible, the specimens should be preserved under either room temperature or slightly less (e.g. results show that 12 °C are able to stop AAR further development). Before SDT test, the samples shall be stored in the moist curing room for a 48 h rewetting [21].

2.4.3.3 Apparatus

Testing machine: use a testing machine capable of imposing a load at the rate and of the magnitude prescribed below.

Compressometer/extensometer: to determine the stress vs strain curve and thus be able to calculate the hysteresis area, plastic deformation and the modulus of elasticity of concrete, one might use a sensing device that measures (to the nearest 5 millionths) the average deformation of two diametrically opposite gauge lines, parallel to the axis and centered about mid-height of the specimen.

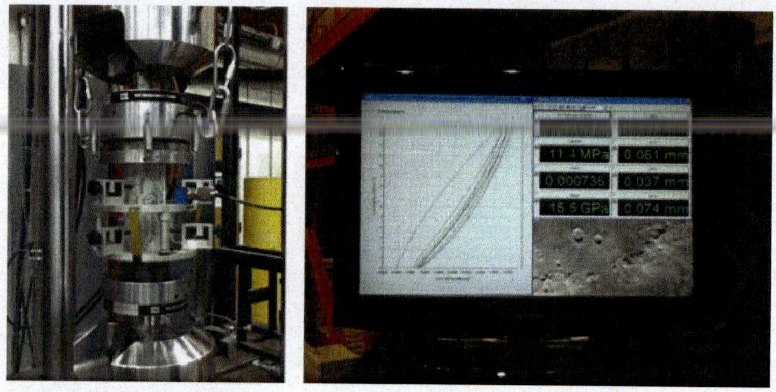

((a)) Test Setup ((b)) Data Acquisition

Fig. 2.17 SDT experimental procedure

2.4.4 Experimental Procedure

In the SDT, concrete test specimens (cores or laboratory-made cylinders) are subjected to five cycles of loading/unloading at a controlled loading rate of 0.10 MPa/s. The samples are tested at a strength level corresponding to 40% of the value obtained through the use of a companion sound/less damaged concrete (presenting the same mix-design and maturity) than the investigated sample. The test takes roughly 20–40 min/sample, depending on the concrete strength. Data collected includes the loading applied and displacements measured on the specimen over time, Fig. 2.17.

2.4.5 Example

2.4.5.1 Impact of Loading Level on the SDT Results

The results of the most critical output parameters (hysteresis area (HA), modulus of elasticity (ME), plastic deformation (PD)) are presented in Fig. 2.18, for stiffness damage tests carried out at the four different loading levels (15 to 40% of the 28-day/design strength) and four expansion levels (up to 0.30%) in the 25 MPa concretes incorporating the reactive Texas sand.

For concrete mixtures incorporating Tx reactive aggregates, clear differences in the HA and PD responses with increasing expansion in the test specimens can be observed only at loading levels corresponding to 30% and (even better!) at 40% of the 28-day strength design, thus suggesting that the above parameters of the test were unable to distinguish the internal concrete damage corresponding to the selected expansion levels up to those loading levels (Figs. 2.18a and c). Moreover, the modulus

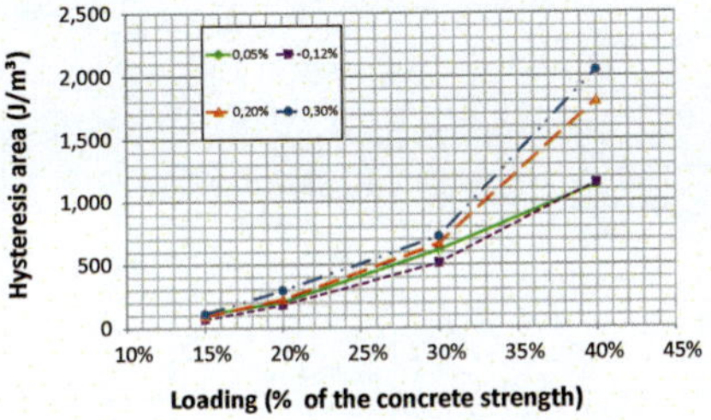

((a)) Hysteresis area (HA)

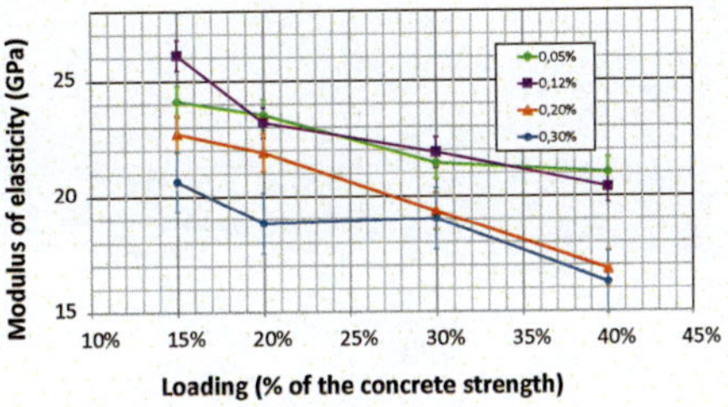

((b)) Modulus of elasticity (ME)

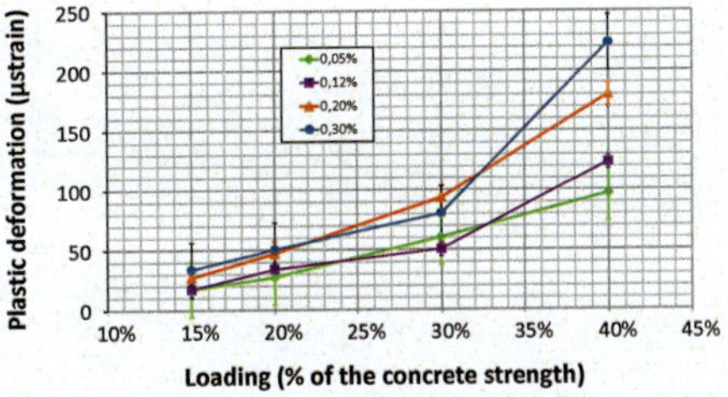

((c)) Plastic deformation (PD)

Fig. 2.18 Responses of selected SDT output parameters as a function of the loading applied (% of the concrete strength) for 25 MPa concrete mixtures incorporating the reactive Tx sand and affected to various degrees by ASR

of elasticity (ME) decreased as a function of the expansion level, relatively similar slopes being observed for each of the expansion levels as a function of the loading level used in the test.

2.4.5.2 Quantification of Damage/AAR Progression Through the SDT

SDT can be quantitatively used to determine the amount of AAR advancement or overall inner damage of the investigated concrete. Damage is measured according to the summation of either "dissipated energy" or the "plastic deformation" during cracks closure over five compression cycles. Typical SDI and PDI curves of AAR-distressed samples are shown in Fig. 2.18. Both SDI and PDI present either a linear or concave trend against AAR expansion levels, depending on the concrete strength and aggregate type/nature. Usually, SDI values range from about 0.08 for sound concretes up to 0.30 for 0.30% of expansion. Likewise, PDI values vary from about 0.05 up to 0.28 for 0.30% of expansion (Fig. 2.19).

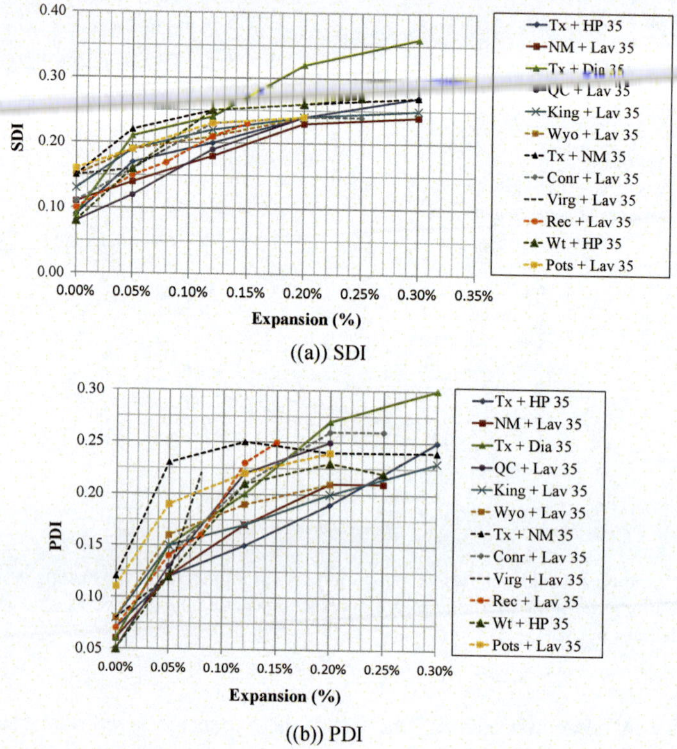

Fig. 2.19 SDI and PDI curves of AAR distressed concrete at 35 MPa mixtures [8]

2.4.6 *Additional Remarks & Limitations*

The SDT was found suitable for a number of applications as presented hereafter:

- SDT appears to be an interesting technique to quantitatively determine the extent of cracking of aging concrete damaged due to different distress mechanisms [26]. It was mainly used to study the damage development of alkali-aggregate reaction (AAR), yet it was successfully applied to quantify/understand damage caused by delayed ettringite formation (DEF) and freezing and thawing cycles (single or coupled with AAR).
- It has been found that the use of a percentage (%) of the compressive strength instead of a fixed loading makes the SDT a quantitative test procedure for assessing concrete presenting different strengths and aggregate types. Moreover, 40% of a companion "sound/less damaged" concrete sample (presenting the same concrete mix-design and maturity) should be used as the likely "actual" strength value of the investigated concrete in practical engineering assessments. In addition, it has been found that the SDT, when tested up to 40% of the compressive strength of

a companion sound concrete, does not present a "destructive character", which enables the reuse of the sample tested for further assessments after testing.

- SDT has been successfully used to quantify damage in concrete cores extracted from aging concrete structures. For decreasing the Variability among companion samples, mainly while the assessment of concrete cores, a 48 h rewetting at a moist curing room shall be performed.
- Although quite promising, the SDT is still quite novel and thus needs to be validated through the assessment of a number of different concrete structures, presenting different damage degrees, exposure classes, reinforcement & confinement conditions.

2.5 Non Destructive Evaluation

Andreas Leemann

NDE methods, e.g., based on elastic waves, can be used to assess damage in concrete. In regard to AAR, linear ultrasound measurements on concrete cores obtained from a dam have shown that the method is not very sensitive to AAR damage progression [27]. This is confirmed by several studies performed on laboratory concrete exposed to an acceleration test protocol [28, 29]. After an initial decrease of ultrasonic pulse velocity due to cracking, a recovery has even been observed due to crack filling by ASR products [2, 30]. On the other hand, nonlinear ultrasounds provide a higher sensitivity to determine changes in material properties when applied on concrete specimens. As such, they allow to determine the degree of AAR-induced damage [29, 31–33]. However, as there is no straight correlation between ASR damage and ASR expansion, especially not on concrete cores retrieved from a confined environment, nonlinear ultrasounds are not suitable to assess past concrete expansion.

2.6 X-Ray Tomographic Microscopy (XTM)

Michele Griffa and Andreas Leemann

X-ray tomographic microscopy (XTM) provides 3D-information of cement paste, mortar and concrete samples. At the laboratory scale, XTM consists in the use of a cone-beam, micro-focus X-ray setup for tomographic imaging. The divergent, cone-beam geometry of the X-ray radiation allows selecting different magnification factors, so that different spatial resolutions can be achieved. Typically, the maximum achievable spatial resolution is of the order of a few microns (excluding synchrotron setups). The maximum sample size could be of the order of ten cm, with a corresponding spatial resolution of the order of 50 microns. As such, XTM can for example characterize the crack network in a sample proving absolute values of crack length, area and volume within the chosen resolution window. It has to be pointed out, that such an analysis requires substantial computing power and the use of advanced 3D image analysis algorithms and respective software implementations allowing the segmentation of the pore space [34–36].

XTM has started to be used as a technique complementary to the optical microscopy techniques [37]. But it cannot substitute them, as it does not achieve the same spatial resolution. Additionally, it does not allow achieving the same high contrast between phases (e.g. porosity, cement paste and aggregates) as obtained in optical microscopy with epoxy impregnated samples containing fluorescent dyes [38].

In regard to past expansion of AAR-damaged concrete, XTM can potentially provide a 3D-crack index. However, the required sample size to obtain representative data, given by the maximum diameter of the concrete aggregates, is in the order of 100–200 cm^3. At this sample size resolution is not sufficient to detect cracks with a width below 50 μm. As AAR damage creates numerous cracks with smaller width, the technique at its current state is not suitable to assess past expansion.

References

1. Fasseu, P., Michel, M.: Détermination de l'indice de fissuration d'un parement de béton; Méthode d'essai LCPC N0. 47. Tech. rep. 47. Paris, France: Laboratoire Central des Ponts et Chaussées, (1997). http://www.ifsttar.fr/fileadmin/user_upload/editions/lcpc/MethodeDEssai/MethodeDEssai-LCPC-ME47.pdf
2. Leemann , A., Griffa, M.: Diagnosis of alkali-aggregate reaction in dams. Tech. rep. Publication 290934. SFOE-Project SI/500863-01. EMPA, Laboratory for Concrete/Construction Chemestry, (2013).http://www.bfe.admin.ch/dokumentation/energieforschung/index.html?lang=de&publication=11174
3. Lindgard, J., Haugen, M., Castro, N., Thomas, M.: Advantages of using plane polished section analysis as part of the microstructural analysis to describe internal cracking due to alkali-silica reaction. In: Proceedings of 14th International Conference on Alkali-Aggregate Reaction (ICAAR). Austin, Texas, USA, (2012)
4. Freere, R.H., Weibel, E.R.: Stereologic techniques in microscopy. J. R. Microscopical Soc. **87**(1), 25–34 (1967)
5. Leemann, A., Merz, C.: An attempt to validate the ultra-accelerated microbar and the concrete performance test with the degree of AAR-induced damage observed in concrete structures. Cem. Concr. Res. **49**, (2013)
6. Sanchez, L., Fournier, B., Jolin, M., Duchesne, J.: Reliable quantification of AAR damage through assessment of the damage rating index (DRI). Cem. Concr. Res. **67**, 74–92 (2015)
7. Sanchez, L., Fournier, B., Jolin, M., Bedoya, M., Bastien, J., Duchesne, J.: Use of damage rating index to quantify Alkali-Silica reaction damage in concrete: fine versus coarse aggregate. ACI Mater. J. **113**(4), 395–407 (2016)
8. Sanchez, L., Fournier, B., Jolin, M., Mitchell, D., Bastien, J.: Overall assessment of Alkali-Aggregate Reaction (AAR) in concretes presenting different strengths and incorporating a wide range of reactive aggregate types and natures. Cem. Concr. Res. **93**, 17–31 (2017)
9. Grattan-Bellew, P., Mitchell, L.: Quantitative petrographic analysis of concrete; the damage rating index (DRI) method, a review. In: Proceedings of Marc-André Bérubé Symposium on AAR in Concrete, CANMET/ACI Advances in Concrete Technology Seminar. Montréal, Canada, (2006)
10. Shrimer, F., et al.: Application and use of damage rating index in assessment of AAR-affected concrete-selected case studies. In: Proceedings of the 11th International Conference on AAR in Concrete, pp. 899–908. Quebec City, Canada. (2000)

11. Villeneuve, V., Fournier, B.: Determination of the damage in concrete affected by ASR; the damage rating index (DRI). In: 15th International Conference on Alkali-Aggregate Reaction (ICAAR). Austin (Texas), Electronic, (2012)
12. Villeneuve, V.: Détermination de l'endommagement du béton par méthode pétrographique quantitative. MA thesis (2011). Québec: Department of Geology and Geological Engineering, Université Laval
13. Sanchez, L., Fournier, B., Jolin, M., Bedoya, M., Bastien, J., Duchesne, J.: Use of damage rating index to quantify Alkali-Silica reaction damage in concrete: fine versus coarse aggregate. ACI Mater. J. **113**(4) (2016)
14. Menéndez, E.: Evaluation and gradation of simultaneous damage in concrete affected by Alkali-Silica reaction and sulfate attack. Inte. Rev. Civil Eng **2**(2) (2011)
15. Menéndez, E., García, R., Prendes, N.: Damage assessment Of dams affected by expansive reactions using stereomicroscopy and electron microscopy. relationship between the damage levels DRI and DL. In: Second International Dam World Conference, 2015
16. Menéndez, E.: Evaluation and gradation of simultaneous damage in concrete affected by Alkali-Silica reaction and sulfate attack. Inte. Rev. Civil Eng. (I.RE.C.E.) **2**(2) (2011)
17. Sanchez, L., Fournier, B., Jolin, M., Bastien, J.: Evaluation of the stiffness damage test (SDT) as a tool for assessing damage in concrete due to ASR: test loading and output responses for concretes incorporating fine or coarse reactive aggregates. Cem. Concr. Res. **56**, 213–229 (2014)
18. Sanchez, L., Fournier, B., Jolin, M., Bastien, J.: Practical use of the stiffness damage test (SDT) for assessing damage in concrete infrastructure affected by alkali-silica reaction. Constr. Build. Mater. **125**, 1178–1188 (2016)
19. Walsh, J.: The effect of cracks on the uniaxial elastic compression of rocks. J. Geophys. Res. **70**(2), 399–411 (1965)
20. Crouch, R.: Specification for the determination of stiffness damage parameters from the low cyclic uniaxial compression of plain concrete cores. Revision A, Mott, Hay & Anderson, Spec. Serv. Div. Intern. Tech. Note (1987)
21. Sanchez, L., Fournier, B., Jolin, M., Bastien, J.: Evaluation of the stiffness damage test (SDT) as a tool for assessing damage in concrete due to ASR: input parameters and variability of the test responses. Constr. Build. Mater. **77**, 20–32 (2015)
22. Smaoui, N., Bérubé, M., Fournier, B., Bissonnette, B., Durand, B.: Evaluation of the expansion attained to date by concrete affected by alkali-silica reaction. Part I: Experimental study. Can. J. Civil Eng. **31**(5), 826–845 (2004)
23. Giannini, E., Sanchez, L., Tuinukuafe, A., Folliard, K.: Characterization of concrete affected by delayed ettringite formation using the stiffness damage test. Constr. Build. Mater. **162**, 253–264 (2018)
24. Chrisp, T., Waldron, P., Wood, J.: Development of a non-destructive test to quantify damage in deteriorated concrete. Mag. Concr. Res. **45**(165), 247–256 (1993)
25. Chrisp, T., Wood, J., Norris, P.: Towards quantification of microstructural damage in AAr deteriorated concrete. (1989)
26. Martin, R., Sanchez, L., Fournier, B., Toutlemonde, F.: "Diagnosis of AAR and DEF: Comparison of Residual Expansion, Stiffness Damage Test and Damage Rating Index. Sao Paulo, Brazil, (2016)
27. Rivard, P., Ballivy, G., Gravel, C., Saint-Pierre, F.: Monitoring of an hydraulic structure affected by ASR: a case study. Cem. Concr. Res. **40**(4), 676–680 (2010)
28. Saint-Pierre, F., Rivard, P., Ballivy, G.: Measurement of alkali-silica reaction progression by ultrasonic waves attenuation. Cem. Concr. Res. **37**(6), 948–956 (2007)
29. Sargolzahi, M., Kodjo, S., Rivard, P., Rhazi, J.: Effectiveness of nondestructive testing for the evaluation of alkali-silica reaction in concrete. Constr. Build. Mater. **24**(8), 1398–1403 (2010)
30. Swamy, R., Wan, W.: Use of dynamic nondestructive test methods to monitor concrete deterioration due to alkali-silica reaction. Cement, Concrete and Aggregates **15**(1), 39–49 (1993)
31. Chen, J., Jayapalan, A., Kim, J., Kurtis, K., Jacobs, L.: Rapid evaluation of alkali-silica reactivity of aggregates using a nonlinear resonance spectroscopy technique. Cem. Concr. Res. **40**(6), 914–923 (2010)

32. Lesnicki, K., Kim, J., Kurtis, K., Jacobs, L.: Assessment of alkali-silica reaction damage through quantification of concrete nonlinearity. Mater. Struct. **46**(3), 497–509 (2013)
33. Ju, T., Achenbach, J., Jacobs, L., Guimaraes, M., Qu, J.: Ultrasonic nondestructive evaluation of alkali-silica reaction damage in concrete prism samples. Mater. Struct. **50**(1), 60 (2017)
34. Schladitz, K.: Quantitative micro-CT. J. Microsc. **243**(2), 111–117 (2011)
35. Maire, E., Withers, P.: Quantitative X-ray tomography. Int. Mater. Rev. **59**(1), 1–43 (2014)
36. Griffa, M., Münch, B., Igarashi, G., Leemann, A., Mokso, R., Jerjen, I., Schuetz, P., Lura, P.: Damage in cement-based materials by the Alkali-Aggregate reaction: detection and characterization by X-ray tomographic microscopy, extended abstract. In: Proceedings of the 1st International Conference on Tomography of Materials and Structures (ICTMS2013), pp. 317–320. Ghent (Belgium), (2013). http://www.ictms.ugent.be/downloads/talks_def_def.pdf
37. Weise, F., Voland, K., Pirskawetz, S., Meinel, D.: Innovative measurement techniques for characterising internal damage processes in concrete due to ASR. In: Proceedings of 14th International Conference on Alkali-Aggregate Reaction (ICAAR). Austin, Texas, USA, (2012)
38. Poole, A., Sims, I.: Concrete petrography. A Handbook of Investigative Techniques. CRC Press (2015)

Chapter 3
Field Assessment of ASR-Affected Structures

Alexis Courtois, Eric R. Giannini, Alexandre Boule, Jean-Marie Hénault, Laurence Jacobs, Benoit Masson, Patrice Rivard, Jerŏme Sausse, and Denis Vautrin

A. Courtois · B. Masson
EDF Direction Ingénierie et Projets Nouveau Nucléaire - Direction Technique, 19 rue Pierre Bourdeix, 69007 Lyon, France
e-mail: alexis.courtois@edf.fr

B. Masson
e-mail: benoit.masson@edf.fr

E. R. Giannini (✉)
RJ Lee Group, La Crosse, WI, USA
e-mail: egiannini@rjleegroup.com

A. Boule
EDF - Direction Industrielle - Département Techniques de réalisation et Expertise en Géosciences et Génie civil, 905 avenue du Camp de Menthe, 13097 Aix-en-Provence Cedex 02, France
e-mail: alexandre.boule@edf.fr

J.-M. Hénault · D. Vautrin
EDF - R&D, 6 quai Watier, 78401 Chatou Cedex, France
e-mail: jean-marie.henault@edf.fr

D. Vautrin
e-mail: denis.vautrin@edf.fr

L. Jacobs
Georgia Institute of Technology, Atlanta, USA
e-mail: laurence.jacobs@coe.gatech.edu

P. Rivard
Department of Civil Engineering and Building Engineering, Sherbrooke, Canada
e-mail: Patrice.Rivard@USherbrooke.ca

J. Sausse
EDF Hydro - Direction Technique Générale - Département Surveillance, 18 avenue Poincaré, 19100 Brive la Gaillarde, France
e-mail: jerome.sausse@edf.fr

© RILEM 2021

V. E. Saouma (ed.), *Diagnosis & Prognosis of AAR Affected Structures*,
RILEM State-of-the-Art Reports 31,
https://doi.org/10.1007/978-3-030-44014-5_3

3.1 Introduction

This chapter deals with the structural monitoring of concrete structures which are affected by alkali-silica reactions (ASR). Its purpose is to present an overview of the main proven monitoring and testing techniques that are available to determine the presence of ASR within a structure and assess the remaining capacity. However, the scope is not limited to the possible damage induced by ASR, but it also addresses other outward and inward manifestations of the pathology. The methods presented are illustrated by some case studies, from operational experiences gained in different countries.

This chapter is organized as follows:

- a short presentation of different, yet complementary, approaches for monitoring civil structures;
- a brief summary of the mechanisms and structural effects of ASR;
- a review of the current techniques and systems that could be used on real structures to monitor ASR effects. This section includes permanent monitoring systems and non-destructive evaluation (NDE) techniques. A distinction is made between "proven" techniques (with their own limitations) and "potentially applicable techniques". Laboratory test methods, even when promising, are not considered proven when no in-situ proof of concept on real-world structures has been reported in the literature;
- a set of recommendations, for current analysis and for further developments, to fill the gap between information needed by engineers to make a prognosis and technical capacities of monitoring devices and systems.

It is obvious that in-situ tests and measurements are performed in order to propose a useful prognosis for the structure/asset owner's maintenance plans. Because service life prediction is not only a matter of monitoring skills, and because knowledge on materials and structures behavior is also involved, this chapter does not address the complete process from the measurement to the aging and serviceability analysis.

3.2 Permanent Monitoring (PM) and Non-destructive Evaluation (NDE)

The objective of this chapter is to provide a snapshot of the current state-of-the-art in monitoring of concrete structures affected by ASR. Different techniques can be used to monitor a structure: Permanent Monitoring (PM) or Non-Destructive Tests/Evaluation (NDT/E).

3.2.1 Permanent or Online Monitoring (PM)

In a permanent monitoring approach, devices and sensors are set up from the start of the construction or during building operations, to monitor some phenomena. Measurement data are periodically collected, analyzed, and stored. The parameters to survey can be related to safety functions, to demonstrate design requirements are met at any time, or to ageing management processes, to anticipate maintenance activities. A specific system, including various multiplexers, data loggers, database, and specific software is installed to collect and process data. Predefined criteria and threshold values are often provided, to be compared to the processed data and raise possible alerts if measurements exceed the expected trends. Such systems need to be maintained and kept functional over time as well, and that is why rugged and reliable components are preferred. Permanent monitoring systems have been installed in many structures all over the world, for example on large dams [1–3] or on other large structures (bridges, LNG storage tanks, tunnels, etc.) [4]. In some countries (like France, South Africa, China, Belgium, Canada, UK, etc.), containment structures of nuclear power plants (NPP) are fit with permanent monitoring [5]. Permanent or online monitoring systems can also be installed well after construction, when distress (e.g. from ASR) is discovered and determined to warrant long-term monitoring. Post-installed systems generally utilize similar techniques and principles as those installed at the time of construction, but are likely to be more focused on monitoring the progress of a particular form of distress.

3.2.2 Monitoring with NDT/E

NDT/E approaches use devices which are not installed permanently but rather probe the structure or part of the structure to assess its properties at a particular point in time [6, 7]. Portable systems are the norm, and typically offer the ability to process and supervise the data sets obtained during measurement. However, data processing and analysis are rarely completed on site, and normally must be undertaken off site. Very often, the basics of these sensing devices rely on mechanical, electrical, or physical properties of the concrete which are indirectly linked with the material properties of interest. For example, permittivity, which can be measured with a capacitive probe, can be correlated with concrete moisture. Calibration of NDT/E methods can be a challenge for monitoring an existing structure, due to the limited availability of samples to carry out calibration tests. However, these methods can measure many concrete properties, do so for locations where no permanent sensor is installed, and can provide relevant information. NDE has to be considered as a meaningful complement to PM when it exists, when a complex behavior is observed. The most successful approaches will combine multiple physics, such as visual inspection, acoustic measurements, and electrical measurements. Other necessary advances in the NDE of cement-based materials include the development of new measurement technologies to make quantitative assessment of multi-scale structures and properties as well as the corresponding analysis techniques to fully understand and interpret the results.

3.3 Overview of ASR

ASR is the most common recognized type of alkali-aggregate reactions (AAR) in hydraulic cement concrete [8]. Although the first described case of ASR was found in a highway structure [9], until the 1990s, ASR damage was mainly associated with hydraulic structures, such as large dams and spillways, and was thought to progress rather slowly in most structures, requiring years for surface manifestations to appear. Because ASR was rarely associated with highway transportation infrastructure, ASR has not been traditionally a consideration in service life models or most regular pavement and bridge inspection programs [4, 10]. In recent decades, ASR has been increasingly identified as a cause of damage in a variety of structures, and in some cases has developed significantly earlier and more rapidly than expected.

3.3.1 Mechanisms

ASR in concrete is a deleterious chemical reaction between alkali hydroxides in the pore solution and reactive silica found in some aggregates. The reaction results in the formation of a hydrophilic gel (ASR gel) that swells in the presence of moisture. This causes expansion and cracking of concrete structures; the surface cracking can leave the concrete exposed to other deterioration mechanisms such as corrosion and frost action. The rate of ASR gel formation and expansion is primarily governed by the dissolution of silica; dissolution rates and solubility limits for silica increase with temperature and the alkalinity of the concrete pore solution [11–15].

The expansion can result in the misalignment of structural elements, closing of expansion joints and/or surface "popouts". ASR was first identified by Stanton 80 years ago as a cause of concrete deterioration [9]. Since that time, ASR has been identified as a cause of deterioration of numerous concrete structures. Although research has yielded considerable success in understanding the mechanism of the reaction and how to minimize the risk of expansive ASR in new construction, knowledge of the structural effects of ASR and how to best assess the extent of damage to existing structures continues to lag, and remains a major topic of ongoing research.

3.3.2 Effects of ASR

The expansion and cracking caused by ASR affect both the concrete and, in reinforced structures, the reinforcing steel. Cracking is the most obvious sign of distress, however similar crack patterns can also be produced by other distress mechanisms, including drying shrinkage, delayed ettringite formation (DEF), and other forms of sulfate attack. Pop-outs, or conical spalls above reacting aggregate particles, staining and discoloration from exuded ASR gel, and closing of expansion joints can also be observed in affected structures.

Although ASR can result in severe degradation of the mechanical properties of concrete, catastrophic failures of affected structures are rare. The expansion of the concrete also stresses the reinforcing bars in reinforced structures; in some cases, this can be sufficient to yield the steel [16]. Steel in tension will correspondingly compress the concrete, setting up a situation similar to post-tensioned concrete, with important implications for structural behavior.

The remainder of this section will discuss the typical crack patterns that result from ASR, the effects on the mechanical properties of concrete, and the implications for structural behavior for reinforced transportation structures.

3.3.3 Cracking

Internal microcracking from ASR is manifested as macrocracking at exposed surfaces, where drier conditions result in less expansion of the concrete; the surface layer is, in effect, torn open by the underlying expanding concrete. The surface crack patterns are very different for plain and reinforced concrete structures, as shown in Fig. 3.1. This figure shows that plain concrete structures typically exhibit random, or "map" cracking patterns due to the lack of restraint, while reinforced structures exhibit cracking that is parallel to the orientation of the primary reinforcement due to the confinement provided in that direction. In pavements where expansion joints have closed, anisotropic restraint also develops, forcing the development of aligned, parallel cracking.

3.3.4 Loss of Mechanical Properties

ASR can cause significant degradation in the mechanical properties of concrete. Not all properties are affected to the same extent, and effects depend on a number of variables in ASR-affected concrete, including the reactive aggregate, structural geometry and reinforcement, and thermal and moisture exposure conditions. The most significant effects tend to be on the stiffness (elastic modulus) and tensile strength properties. The information that follows deals with the unconfined mechanical prop-

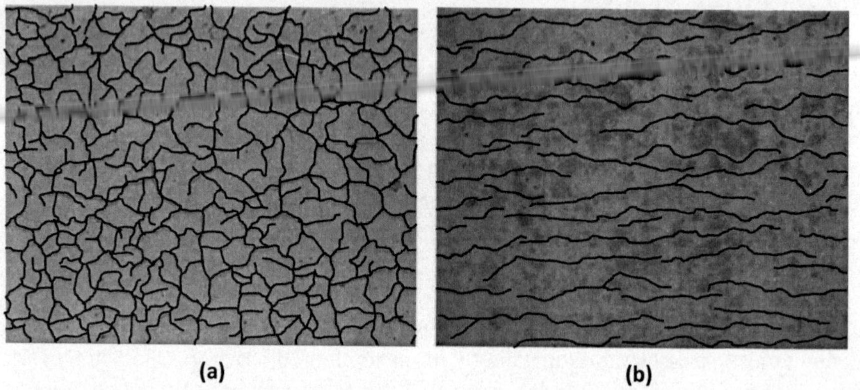

Fig. 3.1 Crack patterns of **a** plain concrete and **b** reinforced concrete [17]

erties, as could be measured using core samples extracted from an ASR-affected structure, and not the effects on load-bearing capacity and serviceability. The latter is discussed briefly in Sect. 3.3.5.

A vast majority of papers agree on the fact the compressive strength is not adversely affected by ASR, unless expansion has reached very high levels [18–23]. With slow/late-reacting aggregate, an increase of compressive stress may even be observed (at least relative to design strengths or strengths measured as part of acceptance testing). For field cases, the major loss of strength is generally observed close to the structure surface where other deterioration processes are also active. In a service hydraulic structure affected by ASR, Rivard et al. [24] found minimal impact on the compressive strength from ASR. It should be cautioned that comparisons to "design strength" or the required 28-day compressive strength may be misleading, because concrete mixtures are designed to exceed this rather than meet it exactly, and because continued hydration in the paste should lead to long-term strength gains.

However, as stated by Giaccio et al. [22]:

> The failure mechanism of concrete in compression is clearly affected by ASR. The shape of the stress-strain curves reflects the presence of internal fissures. The growth and propagation of matrix cracks tend to start earlier due to defects produced by ASR. The period of stable crack propagation is less affected than the period of unstable crack growth which is widely extended, showing that the capability of controlling crack propagation decreases leading to premature failure.

Ultimately, the raw value of compressive strength is not an appropriate indicator of ASR, and should be taken into consideration along with compressive strength values for undamaged concrete of a similar composition and design.

The tensile strength has been reported to be strongly affected by ASR, particularly when determined in direct tension (reduction up to 80%). Fournier and Bérubé [19] suggested considering the tensile-to-compressive strength for a better indicator of internal concrete damage due to AAR.

The literature is in general agreement on the fact that ASR reduces the static elastic modulus of concrete, even at low levels of expansion or when compressive strength is still increasing. According to some papers, the dynamic Young's modulus, calculated from the fundamental resonant frequency, is a very sensitive indicator of the amount of damage associated with ASR [23, 25, 26].

3.3.5 *Structural Capacity and Serviceability*

Most studies of ASR-affected reinforced concrete structures have indicated that the load-bearing capacity is not compromised, despite large expansions, extensive cracking, and the degradation of the strength and stiffness of the concrete [27–29]. Notable exceptions would be studies by Swamy and Al-Asali [18] who reported losses in the flexural strength of single reinforced beams of up to 25% in four-point loading, and den Uijl and Kaptijn [30] who reported a similar reduction of shear strength in beams sawn from in-service bridge decks. It should be noted that the specimens in the Swamy and Al-Asali study contained no shear reinforcement in the central constant moment region, and those in the den Uijl and Kaptijn study contained no shear reinforcement at all. Only a few published studies involve full-scale load tests of in-service structures [31–33]; however, these all indicated that the strength and stiffness of the bridge structures tested were either unaffected or adequate for service loads. Laboratory studies involving shear and flexural load tests of full-scale specimens damaged by ASR have typically come to the same conclusions, with the load capacity either less than expected [34], or lower than less-damaged and undamaged specimens, but still in excess of predicted capacity [35, 36]. A more extensive review of previous load tests can be found in [17, 35].

Reinforcing steel fracture in ASR-affected structures in Japan has attracted significant attention. At least 30 cases of fractured bars have been discovered in structures also damaged by ASR [37]. As shown in Fig. 3.2, brittle fractures have been found at the corners of stirrups, often with many adjacent stirrups all fractured. Because fracture of the bars can lead to a loss of confinement, which is thought to be the saving grace in the performance of ASR-affected structures, this is a major concern. Japan is an active seismic zone, so the concern there is elevated to an even greater degree. Mikata et al. [37] found that when stirrup fracture is combined with corrosion of the reinforcement, the risk of decreased structural performance is increased. Webb [38] provides a more extensive review of the rebar fracture problem in Japan and conducted an investigation into the possibility of fracture with steel grades and reinforcement detailing used in the United States. To date, no cases have been reported outside Japan. However, it is critical to develop an understanding of the cause of the fractures and methods to detect existing or incipient fractures.

The serviceability of concrete structures includes the resistance to excessive deflections, as well as a host of other durability concerns that can shorten the service life of a structure, including frost action and corrosion. Large surface crack widths, and deep penetration of open surface cracks promote the ingress moisture and any

(a) 39

(b) 40

Fig. 3.2 Fractured stirrups in ASR-affected bridge piers in Japan [39, 40].

dissolved aggressive agents, such as chlorides. Additionally, the loss of concrete stiffness and potential for reinforcement yield is a concern for deflections. While the load tests discussed in the previous section gave no indication of excessive deflections, most were live load tests, and did not measure any creep deflections that may develop over a long period of service. Swamy and Al-Asali [18] noted excessive hogging (camber) of singly reinforced beams that developed as a result of expansion gradients in the specimens, which were able to expand more freely on the unreinforced compression face. Deschenes [35] and Larson [36] also noted that during the load tests, new cracks did not form until failure was imminent. Therefore, less warning may be given by ASR-affected structures before failure; the typical shear and flexural cracks that would indicate that the structure was overloaded may not be present in structures affected by ASR.

3.3.6 *Implications for Structural Monitoring*

Structural monitoring often contributes to making a diagnosis and prognosis of observed distress in a structure, and to influence decisions on the possible repairs and/or strengthening to perform, if needed.

While it is understood that ASR involves the dissolution of siliceous minerals from reactive aggregate and the formation of a potentially expansive alkali-silica gel, understanding is incomplete concerning the mechanism of gel expansion, the role of gel composition in its swelling potential, and the apparent lack of correlation between the amount of gel formed and the measured expansion.

These limitations impede the development and improvement of monitoring techniques to detect the field occurrence of ASR and track its progress in affected structures.

The relevance of a measurement system relies to a large extent on a clear and straightforward definition of the quantity to be estimated. However, fundamental examination of concrete structures remains challenging because concrete materials are heterogeneous on multiple scales, hydrated, and opaque. The fact that civil structures can undergo various environmental conditions during their operating life might affect the performance of some techniques. Temperature or humidity may have a direct influence on the accuracy of a measurement system. Further, the gel products themselves are altered when exposed to air and the pattern of damage is known to vary with confinement [41, 42].

However, despite these challenges, engineers are still called upon to make use of documented best practices to monitor structures when distress from ASR is found. The following sections focus on an overview of monitoring and testing techniques, as well as selected case studies where monitoring programs have been implemented. The case study experiences demonstrate that it is possible to monitor civil structures affected by ASR and collect useful data to evaluate the conditions and the serviceability of the structures.

3.4 State-of-the-Art of Measurement Techniques for Monitoring ASR-Affected Structures

This section describes the main proven monitoring techniques that have already been used for structural assessment of concrete buildings affected by ASR. Other promising techniques, successfully tested at laboratory sample scale but not yet in the field, are also mentioned. Then, a classification is proposed at the end of each technique description:

- "POC" for "Proof of Concept", when some in-situ achievements undertaken with the method are available for ASR diagnosis and/or monitoring. A method classified as "POC" has been used to inspect ASR-affected structures or has proven its capability to assess some ASR-related observable (cracks, strain, humidity, temperature, etc.) on real-world structures;
- "PAT" for "Potentially Applicable Techniques", for techniques that either (a) have shown promise in the laboratory but have not yet been tested on real structures, or (b) have been used in the field, but not for the purpose of monitoring ASR-affected structures, as far as the authors of this chapter are aware.

This section is organized by parameters that either (a) drive the development of ASR (moisture, alkali content, and temperature), or (b) are characteristics of the effects of ASR on concrete (expansion, internal stress, cracking, and loss of mechanical properties). The relevance of each parameter to structural assessment will be discussed along with appropriate in-situ measurement techniques. One should keep in mind that monitoring and NDE programs are likely to be most successful when multiple techniques and sources of information are combined (i.e. with data fusion algorithms) [43].

The techniques described in this section are not capable of in-situ petrographic analysis or chemical analysis (e.g. alkali content assessment). This is why core sample tests must be undertaken for a comprehensive diagnosis of a suspected ASR-affected structure; however, core sampling and analysis are not in the scope of this chapter and are addressed in Appendix F.

A summary table of all the methods addressed in this chapter is available in Sect. 3.6, with the classification and an assessment for the accuracy of each NDE technique.

3.4.1 Moisture and Water Content

Moisture is one of the key factors that promotes and sustains ASR in concrete. A distinction between internal relative humidity (RH) and external/ambient humidity has to be made; measurement of the internal RH is the primary challenge for structural assessment. Concrete with an internal relative humidity higher than 70–80% has sufficient moisture to allow for swelling of the ASR gel. Depending on the aggregate reactivity, a threshold can be defined between 80 and 100% of RH for structures or samples that may develop expansion from ASR [44–46]. If the environment of the structure has a relative humidity lower than 80%, the possibility of expansion from ASR decreases dramatically [47]. Then, if concrete moisture monitoring is considered, one should prefer a method capable of maintaining a high degree of accuracy between 70 and 100% RH.

Several methods exist to measure concrete water content or relative humidity with portable probes or in-situ probes placed in the structure. One should keep in mind that if the use of first-class industrial hygrometers allows access to a quantity expressed in % RH, all the other techniques require a calibration between the measured quantity (e.g. permittivity) and the RH or the water content [48].

3.4.1.1 Wood Stick Method

Wood sticks have been used to measure the internal RH of concrete structures [49]. They are inserted inside the structure either near the surface or in the middle of the structure. Figure 3.3 illustrates the insertion of two wood sticks at different depths into the concrete. The objective is to measure RH near the surface (i.e. in the outer 75 mm) of the structure because this is where humidity changes the most. The reason is that the environment (rain for example) primarily affects the saturation of near-surface pores. By measuring the electrical conductivity of the stick, one can evaluate the water content of the stick, and therefore, the relative humidity of the concrete. These parameters are connected and change together.

According to Jensen [49], the wood stick method using sticks made of Ramin wood showed high efficiency for the measurement of high RH values. The accuracy is still good even if values higher than 98% R.H. are reached. This is not the case for

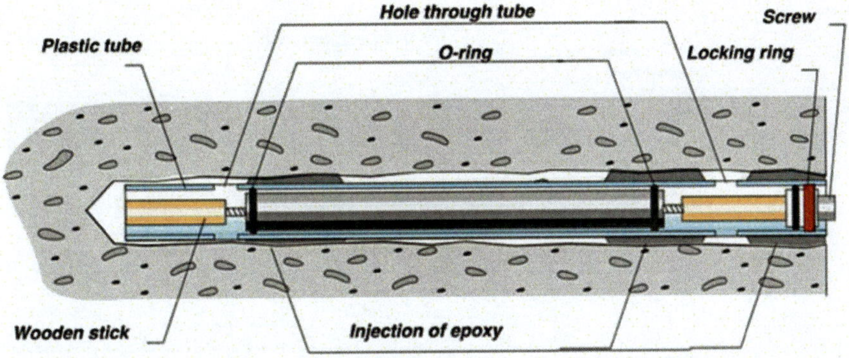

Fig. 3.3 Plastic tube (diameter 2 cm and length 25 cm) mounted with 2 wooden sticks and where the contact between the concrete and the plastic tube has been injected with epoxy [49]

commercially-available electronic measurement devices (such as capacitance probes discussed in a next section) [50]. It is important to note that the stick must be isolated from atmospheric air to reach equilibrium with the surrounding concrete.

However, calibration of the conductivity-RH relationship for each individual stick is needed, and it is recommended to get support from those with experience with the technique when attempting it for the first time.

The wood stick method is classified as POC.

3.4.1.2 RH or Capacitance Probes

A hygrometer is an instrument used for measuring the amount of humidity, which is based on the measurements of some other quantity such as temperature, pressure, or electrical change in the considered medium. By calibration and calculation, these measured quantities can be correlated to humidity. Modern electronic devices use the temperature of condensation (the dew point), or changes in electrical capacitance or resistance, to measure humidity.

The equipment consists of capacitive hygrometers based on electrical capacitance measurement associated with Resistance Temperature Detectors (RTD) (Fig. 3.4). RH measurement relies on a capacitor for which the dielectric of the material (typically a polymer) depends on its moisture content. The temperature and RH measuring ranges of the probes were within the intervals −40 to +80 °C, and 0 to 100% RH, respectively.

A temperature sensor is typically integrated with the moisture sensor to enable determination of RH. The probe is typically installed in a small, sealable hole bored or drilled into the structure (Fig. 3.4), similar to the wood stick installation shown in Fig. 3.3. A recent documented field experience is presented in [51], which concerns the moisture condition within nuclear reactor containment inner walls in addition to other concrete structures within the containments. Field experience on bridge

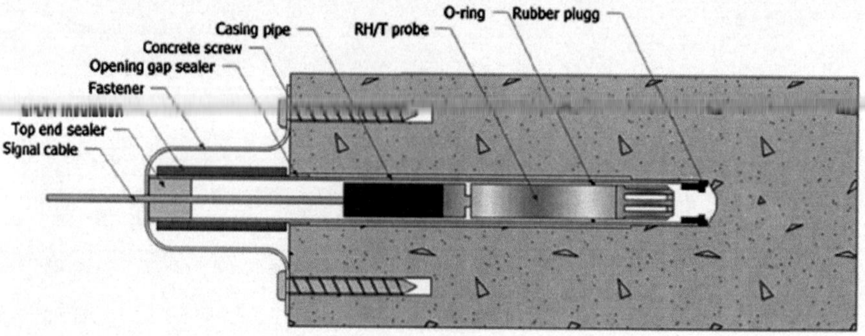

Fig. 3.4 Hygrometer instrumentation setup in a concrete wall [51]

elements can also be found in [52], where relative humidity was monitored to assess the effectiveness of various treatments against ASR.

These types of sensors rely on the water vapor and temperature in the borehole reaching equilibrium with the surrounding concrete in order to obtain a measurement of the RH of the concrete.

Capacitance probes have several limitations. The first one is the variability of the method. Manufacturer specifications generally state that the probe accuracy is reduced above 90%. The stabilization of sensors with high RH values takes more time than with lower RH values [53]. The difference of temperature between the air inside the measurement cavity and within the concrete can affect the values [54]. This limitation is quite similar to this one mentioned previously for the wooden probes. The precision of the measurements changes according to the concrete components. In the case of high-performance concrete (HPC), the moisture cannot flow through the microstructure easily and the borehole and sensors are thus slow to reach equilibrium with the concrete [55]. Additionally, the probes are very sensitive to moisture condensation (e.g. from diurnal temperature changes) or liquid water accumulating in the borehole after rain events [49, 57]. This may limit the ability of these probes to be utilized for continuous measurements and creates challenges for maintaining the boreholes for future use.

RH/capacitance probes are classified as POC.

3.4.1.3 Microwave Techniques

Several techniques are reported:

- Ground Penetrating Radar (GPR);
- Time Domain Reflectometry (TDR) method;
- Open-ended coaxial probe [58].

These techniques are based on an indirect measurement of the permittivity of the tested materials, which depends on water content of the concrete.

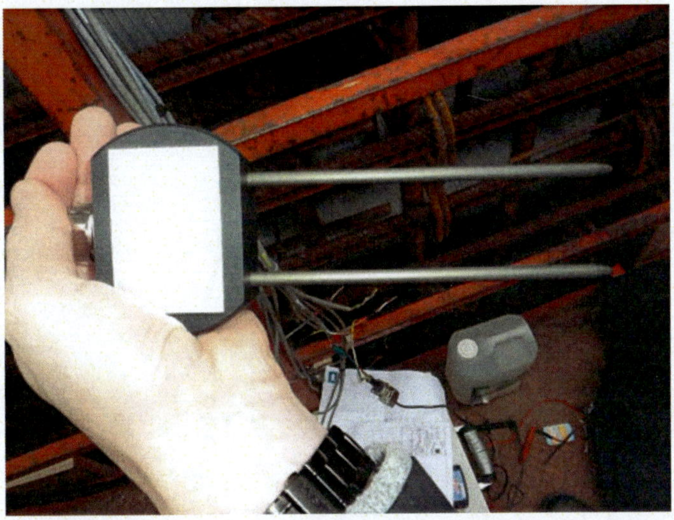

Fig. 3.5 TDR probe before mounting on site (©EDF)

GPR is able to use high-frequency (above 900 MHz) electromagnetic waves to measure water content within one-meter depth from the surface of the structure [59, 60]. For water content assessment, it is recommended to couple the antennas with the structure, rather than using an air-coupled system. GPR is a rather fast technique to survey large structures.

However, the meaningful preparation of GPR data for water content measurements requires good knowledge of the monitoring system and adequate processing tools. Some applications for structural diagnosis are reported in [43]. GPR measurements are also highly affected by the presence of reinforcing steel, so the utility of this technique will be limited when investigating reinforced concrete elements; much of the concrete beyond the clear cover layer will be shadowed by reflections from the steel. Greater depths of concrete (limited primarily by the frequency of the instrument) can be investigated by GPR where no reinforcement is present.

GPR is classified as PAT.

The TDR technique, originally designed to detect discontinuities in electrical cables, can be used for determining the moisture content of concrete [61, 62]. A short-rise time pulse is launched along a transmitter and the reflected wave is analyzed to get a flight time within the probe (Fig. 3.5) which is linked to permittivity and then to water content. The method relies on embedded sensors, and Électricité de France (EDF) tested the method on the containment structures of the EPR Flamanville-3 nuclear power plant [63] and on the "VeRCoRs" mock up [64]. The system is operating, but the preparation can be misleading if specific care is not taken to analyze the signals and to calibrate the system.

TDR is classified as POC.

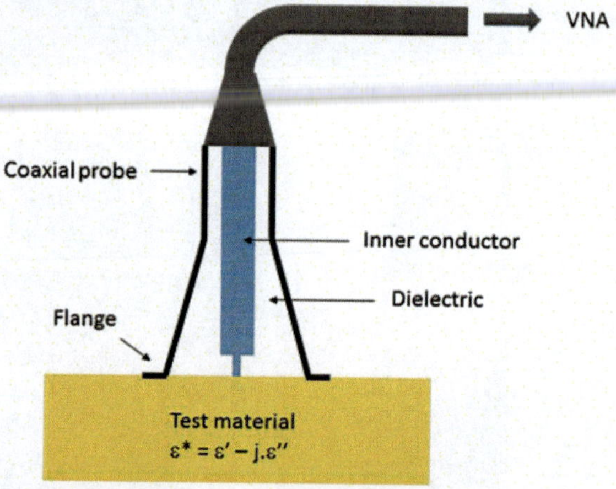

Fig. 3.6 Cross-section of a standard open-ended coaxial probe

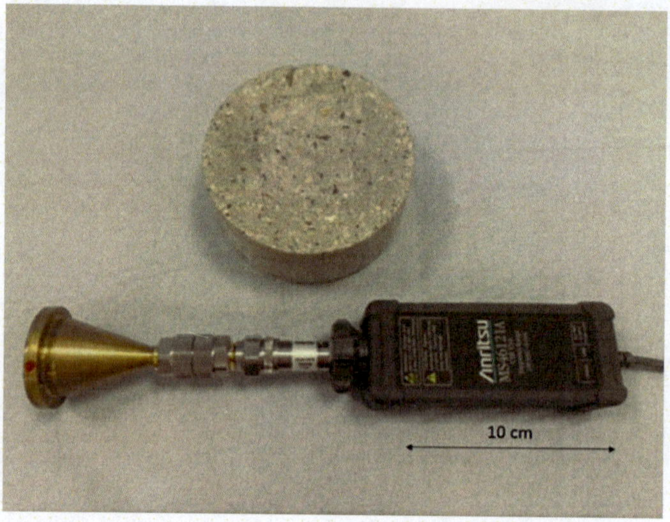

Fig. 3.7 Open-ended coaxial probe and VNA used to monitor concrete water content

Open-ended coaxial probes [58] connected to a vector network analyzer (VNA) seem promising to monitor concrete water content [65, 66]. An open-ended coaxial probe consists of a truncated section of a transmission line. The electromagnetic field propagates along the coaxial line and reflection occurs when the electromagnetic field encounters an impedance mismatch between the probe and the concrete structure. A probe cross-section is schematized in Fig. 3.6 and a device is shown in Fig. 3.7. The reflected signals at different frequencies are measured and then converted into complex permittivity values. Concrete moisture is inferred from pre-established calibration relationships between water content and permittivity, Measurements can be acquired within a specific frequency range and a capacitive model is used to deduce complex permittivity from reflection coefficients acquired by VNA. Hashemi et al. [67] used a VNA to evaluate the sensitivity of different microwave frequencies to the presence of ASR gel with good success. This technique has been successfully used by different teams, but it still lacks industrial feedback from the field.

Open-ended coaxial probes are classified as PAT.

3.4.2 Displacement

The progress and severity of ASR in laboratory tests are typically described in terms of the amount of expansion of the concrete. Determining the rate of expansion and changes in the rate of expansion are vital to making a prognosis for the structure. In assessing field structures, it is common to attempt to determine the amount of expansion to date, which may also be expressed in terms of displacement (particularly if adjacent elements become misaligned), and to measure the rate of continuing expansion. Crack width openings and displacements are related; therefore, it is necessary to perform these measurements on the same structural members, and preferably repeated over time at the same location [68].

This section will focus on some techniques for monitoring the progress of ASR in field structures in terms of expansion or displacement rates, which are:

- Pendulum;
- Permanent or demountable surface extensometer;
- Laser scanning.

Other sensors can be considered to obtain continuous measurements of expansions, such as electrical resistance strain gauges, tapes for diameter measurements of cylindrical structures, micrometers, inclinometers, and surveying against fixed benchmarks near the structure can also be used. They are generally more practical if installed at construction; however, in practice, most structures are not highly instrumented with the expectation to monitor expansion from ASR.

3.4.2.1 Pendulums

Direct and inverted pendulums (see Fig. 3.8) are used to monitor internal lateral displacement of concrete structures such as large dams, tall buildings, or bridge piers [69].

The direct pendulum (plumbline) consists of a stainless steel wire attached to a fix point at a top of structure to be monitored, a weight, and a tank containing damping fluid to damp movements of the weight due to wind and air circulation. Displacements relative to the wire can be measured by a reading table or a reading station for continuous monitoring.

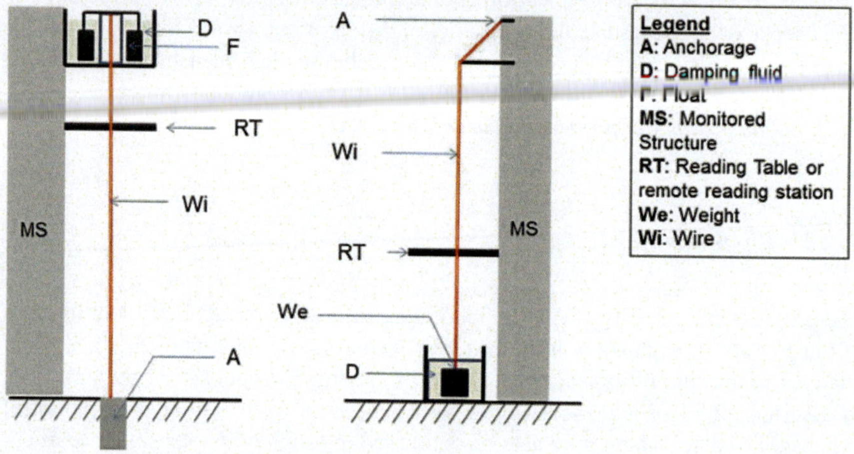

Fig. 3.8 Schematic view of inverse and direct pendulums

The inverted pendulum includes a stainless steel wire anchored in the structure foundation with a float fixed at its upper end. The float, which is free to move in a tank, tensions the wire, and keeps it vertical. Once anchored in a stable point in foundation it measures absolute deformation of the structure. The reading systems are the same as for the direct pendulum. The length of the plumbline is limited to 60 m, for practical reasons. To monitor deflection of a complete profile of a taller building, several pendulums can be installed in a vertical alignment, one above the other, starting with an inverted pendulum anchored in a borehole in the dam foundation.

The measurement uncertainty is about 0.5 mm with a standard reading table and can be improved to 0.2 mm with automatic systems.

Pendulum systems are classified as POC.

3.4.2.2 Permanent and Demountable Surface Extensometers

Various kinds of extensometers can be used to monitor structural displacements. Most of the sensing technologies are associated with the LVDT (Linear Variable Differential Transformer) technology which provides accurate measurements through the movement between a sliding rod and the exterior body of the transducer. To increase extensometer gauge length, the LVDT sensor is associated with Invar wires or rods, in order to limit the effect of thermal expansion on the readings.

Abrishami et al. [70] provides examples of use of this kind of extensometer to monitor the vertical deflections or horizontal displacement of containment structures during the pressure test or due to shrinkage and creep of the concrete. The measurement uncertainty may vary between 0.02 and 0.5 mm, depending on the

Fig. 3.9 Surface strain meter installed on EDF site for measuring horizontal displacement. The gauge length is 1 m (at EDF)

environmental conditions and on gauge length. An example of field implementation is shown in Fig. 3.9.

The Demountable Mechanical Strain Gauge (so-called DEMEC) system involves using a specific extensometer to measure the change in length between embedded gauge points, or targets, installed on a structure. Mayes Instruments, Ltd produces a variety targets, including discs that can be affixed to the surface of the structure, and metal studs that can be installed in pre-drilled holes in the structure (http://www.mayes.co.uk/6.html). Stud-type targets are preferred for instrumenting field structures, and can be fixed in place using rapid-set epoxies.

Targets can also be produced using a drill press, stainless steel bolts or threaded rod, and a countersink drill bit; the drill bit machines a small divot into the end of the bolt or rod. It is critical that the divot is centered and machined cleanly to ensure that the targets can be located within the range of measurement of the gauge when installed on the structure. The extensometer consists of a dial or digital displacement gauge and two conical tips separated by a length of INVAR metal (Fig. 3.10). The tips are shaped to fit precisely into small divots in the targets; one tip is fixed and the other mounted on a knife-edge pivot to allow the gauge to measure displacement from the original gauge length. Digital gauges for the system are capable of measuring the displacement to the nearest 0.001 mm; for a 0.50 m gauge length, this translates to a strain sensitivity of 2 microstrains.

Figure 3.10 shows a DEMEC gauge with a 500-mm gauge length, including detail of the pivoting tip. The advantage of the DEMEC system is that only one measuring device is needed, and can be used to monitor expansion at many locations on multiple structures. A 0.50 m-square grid of DEMEC targets can also be used as reference

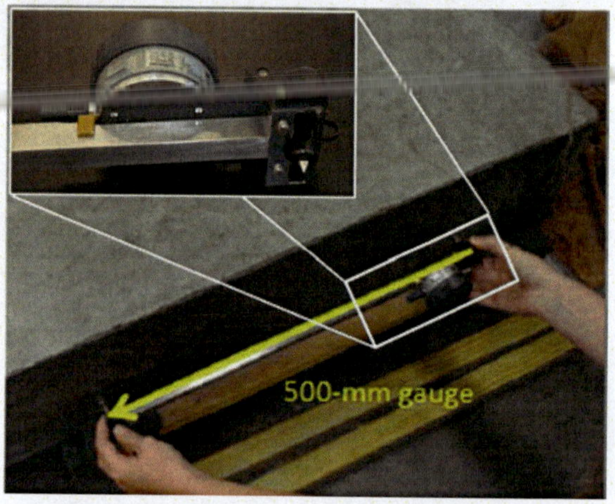

Fig. 3.10 0:500-mm gauge length DEMEC gauge

points for constructing a grid square for measuring the Cracking Index (CI) of the same portion of the structure.

Permanent and demountable surface extensometers are classified as POC.

3.4.3 Concrete Strains (Internal)

3.4.3.1 Vibrating Wire Strain Gages

Vibrating wire strain gauges (VWSG) are very popular for concrete structures monitoring which were successfully tested for dams monitoring in the 1930s. Detailed descriptions of and experiences with these sensors can be found in [63, 71–73].

A standard VWSG consists of two end flanges separated by a stainless steel tube with a tensioned high-strength steel wire clamped into the end flanges and running axially through the center of the tube (Fig. 3.11). An electromagnet mounted at the center of the tube is used to vibrate and measure the vibration frequency of the wire. When the gauge is cast into concrete, the concrete strains are conveyed to the gauge through the end flanges and measured as changes in the wire vibration frequency. An increase in frequency corresponds to increased tensile strain in the wire (expansion of the structure), while a decrease in frequency corresponds to a lessening of tensile strain in the wire (contraction of the structure); the principle is the same as that which is used to tune stringed musical instruments. Special attention shall be paid to the installation of these sensors, which are sensitive to shocks and vibration that could be caused by concrete placements. To secure the VWSG installation, EDF has placed them into small reinforcement cages before setting them into the reinforcement of the structure, which also provide a better precision in positioning (Fig. 3.12). In the UK, the sensors are encapsulated into concrete briquettes. Measurement uncertainty

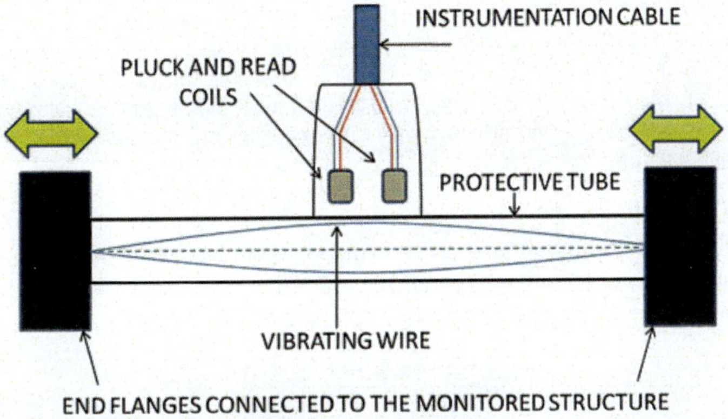

Fig. 3.11 Simplified sketch of a VWSG

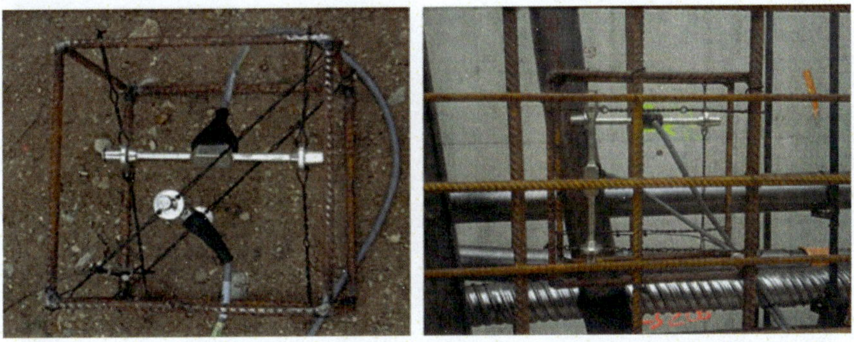

Fig. 3.12 VWSG in the protective cage (left) and put in the reinforcement before placing concrete (right)

is about 1–3% of the full-scale range, when all the influences are taken into account (wiring, data logger specifications, uncertainty on gauge coefficient. . .).

VWSG can also be used to fit multiple point rod-type borehole extensometers to monitor concrete structures and pile foundations [74, 75].

VWSG are classified as POC.

3.4.3.2 Fiber-Optic Sensors (FOS)

A relevant definition of fiber-optic sensors is provided in [76]:

> The term fiber-optic sensor covers a broad range of monitoring devices. Although there are many different types of fiber-optic sensors, they all consist of glass or silica optical fiber cores that are surrounded with a cladding and protective coatings. All fiber-optic sensors work on the same principle: light is transmitted through the optical fiber and internally reflected along the length of the wire and, as the fiber is distorted, the sensors detect a shift in time or frequency of the passing light. Due to their materials, fiber-optic sensors are not affected by

electromagnetic fields and are chemically resistant, making them highly robust monitoring devices. Fiber-optic sensors can be categorized in a number of different ways—intrinsic or extrinsic; short-gauge, long-gauge, or distributed; fiber Bragg grating, Fabry-Perot, or Rayleigh scattering. They can also measure various parameters such as strain, displacement, and temperature.

The main technologies are presented in Fig. 3.13 and a photograph of a distributed FOS installation for containment monitoring is presented in Fig. 3.14.

Measurement uncertainty may vary depending on the technology and the environmental conditions. For Brillouin distributed systems, EDF uses a value of ±20

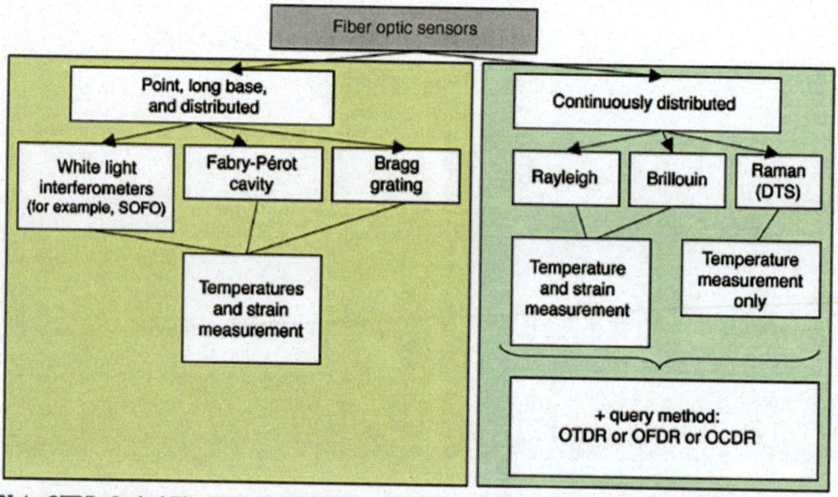

Note: **OTDR: Optical Time Domain Reflectometry; OFDR: Optical frequency domain refectometry; OCDR: Optical Correlation domain reflectometry; SOFO: Surveillance d'Ouvrages par Fibres Optiques; DTS: distributed temperature sensor.**

Fig. 3.13 Various types of fiber-optic sensors that are available and the parameters that they can measure [76]

Fig. 3.14 Fiber-optic cable attached to a containment reinforcement rebar for distributed measurements (Raman, Brillouin, and Rayleigh) ©EDF

μm/m for strain measurement uncertainty [64]. Crack detection, localization and quantification can be achieved using Rayleigh distributed systems. Feasibility in site conditions was recently proven on the VeRCoRs mock-up [77].

EDF and other owners/operators have tested several technologies [64, 76, 78]. Overall, they can be considered as promising, but some time is still necessary to collect more experience on the performance for long-term structural monitoring.

FOS are classified as POC.

3.4.3.3 Snap-Ring Borehole Extensometer (SRBE)

Probes to measure in-situ through-thickness strains are commercially available (manufactured by Geokon). Their Model A-4 Multiple Point Rod Extensometer has Snap-Ring Anchors [79] that can be quickly and easily installed in concrete boreholes. The anchors are pushed to the required depth on the end of setting rods and then a cord is pulled to remove the locking pin which allows two retaining rings on each anchor to snap outward and grip the borehole, Fig. 3.15. Up to eight anchors can be installed, at various depths, in a 76 mm diameter borehole. They are particularly useful in upward-directed boreholes as described in the manual [80].

These probes have been approved by the Nuclear Regulatory Committee (NRC) for installment in nuclear containment structures [81, 82].

SRBE are classified as PAT.

3.4.4 Temperature

The temperature of concrete in ASR-affected structures is of interest for two primary reasons: the effect of temperature on the rate of the reaction and the thermal expan-

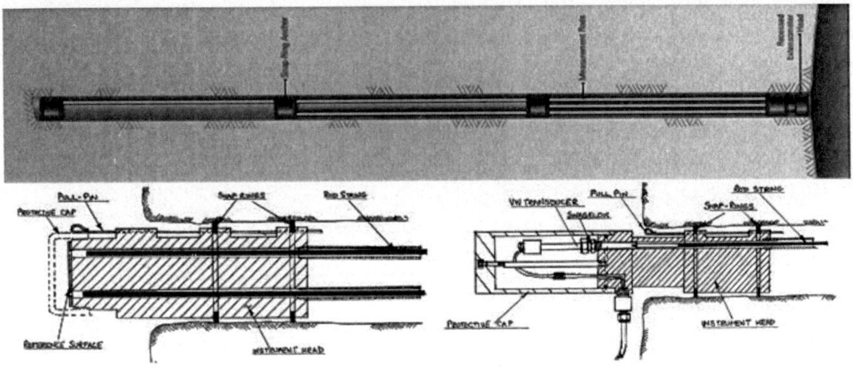

Fig. 3.15 Borehole snap-ring extensometer by Geokon [79, 80]

sion/contraction of the structure and expansion monitoring instruments caused by temperature fluctuations.

The effect of temperature on the rate of silica dissolution, a critical process in the development of ASR, is well-documented [11, 15]. Swiss dams with downstream faces oriented towards the south and thus higher solar radiation are more affected by swelling irreversible displacements [83]. Warmer average temperatures will, in general, accelerate silica dissolution and subsequent formation of ASR gel. Thus, structures in warmer climates and elements facing south and west can expect more rapid development and progress of damage from ASR [84]. In determining the prognosis for the structure, it will be helpful to integrate both information on the local climatology (available from nearby weather stations and government meteorology services) and the internal conditions of the structure.

For practical reasons, temperature measurements should accompany expansion (bulk displacements or internal strain) measurements. If temperature effects are not considered, the thermal expansion/contraction could obscure trends expansion trends caused by ASR, particularly during the early stages of a monitoring program.

Information on the local climatology can be obtained from government meteorology services, nearby weather stations, as well as commercial meteorology companies. For some structures, particularly remote structures, it may be worth investing in a dedicated weather station. Many off-the-shelf models are available, while customized stations can be built with a generic datalogger, power supply, and the desired instrumentation. Solar panels with battery storage systems offer the flexibility to operate "off the grid" and in remote locations. One example of a custom solar-powered system is in place at The University of Texas at Austin's outdoor exposure site [85].

One advantage of a customized weather station is the flexibility to dedicate extra data channels to monitoring internal temperatures in one or more locations inside the structure, as well as other parameters. Internal temperature data can be obtained by embedding resistance temperature detectors (RTD) or thermocouples in the structure [64, 73]. With the use of more generic data acquisition systems, more programming effort is typically required, but the climate and internal temperature data are more easily aligned with other data collected by the system (e.g. embedded or vibrating wire strain gauges, accelerometers for structural health monitoring, or acoustic transducers).

RTDs and thermocouples are classified as POC.

Fiber-optic Raman technologies can monitor temperature along kilometers of fiber length with a good uncertainty (less than 0.5 °C), which can be of interest for the surveillance of large or long structures [78, 86].

Fiber-optic Raman sensors are classified as POC.

3.4.5 *Local Stiffness, Young's Modulus and Concrete Distress*

This section includes on overview of NDE based on acoustic methods. A significant number of published works report a large variety of tests, at laboratory scale or in the field but relatively few are related to ASR-affected structures.

Most of the methods mentioned below have been developed to assess local stiffness through the measurement of in-situ mechanical properties. Because ASR can lead to a drop of stiffness or a decrease of Young's modulus, the principle of acoustic methods seems relevant, and therefore is considered as promising techniques even if available proofs of concept are not specifically oriented towards the issue of ASR.

3.4.5.1 Considerations on Acoustic Wave Methods

These types of NDE are based on wave propagation theory. Three types of waves are produced after an initial impact: compression, shear, and surface waves. Measuring the change in acoustic parameters provides a means to detect various objects and defects such as reinforcement bars, voids, honeycombing, delaminations, and cracks (with current research exploring the potential for carbonation depth and water content assessment). These methods include ultrasonic testing, acoustic emission, and impact-echo.

The preparation of acoustic NDE tests is generally based on the assumption of a linear-elastic behavior of the tested materials, but non-linearity can be considered when more sensitive measurements are sought [7, 43]. However, nonlinear acoustic methods are yet considered as complex and not yet ready for field implementation, even if very promising research works are underway in several countries to enhance the reliability of these methods.

It should be stressed that the potential for acoustic-based NDE to monitor concrete components is exceedingly low when compared to the multiple examples of the successful application of these NDE techniques to metal and composite structures. Part of the difficulty is the large size and scale of most concrete components, but there are also inherent complications due to the nature of cement-based materials. Experimental determination of ultrasonic attenuation and dispersion are difficult tasks in cement-based materials, in part because the manifestations of early-stage ASR damage are often not apparent on the surface of the concrete, which necessitates the use of penetrating characterization methods such as acoustics. Complicating these issues is the fact that traditional low-frequency ultrasonic waves (i.e. ~50 kHz range) have difficulty in differentiating between microstructural features (e.g. porosity and solid phases) and damage (micro-cracks), because these features scatter, slow, and attenuate ultrasonic waves in similar ways. The scattering and dispersion are evidence of the complicated nature of ultrasonic wave propagation in cement-based materials, but on the positive side, this shows that these ultrasonic waves contain quantitative information about the structure through the volume of mortar through which they propagated. Acoustic NDE measurements are also affected by the moisture and temperature conditions of the concrete; these properties should be considered when interpreting results.

3.4.5.2 Ultrasonic Pulse Velocity (UPV)

This test is likely to be the most widely used ultrasonic NDE test. It measures the travel time of an ultrasonic wave pulse (typically a compression, or P-, wave) introduced into a concrete member between a transmitter and a receiver (Fig. 3.16). Simple to use, suitable to thick walls, and supported by standards (ASTM C597 for example), it requires access to both sides of the concrete element to test. It is a point measurement technique, which can be time consuming to gather data over a large area.

Metalssi et al. [88] reports positive results from laboratory tests carried out on ASR-affected samples with different expansion rates, because P-wave velocity is influenced by the magnitude of expansion. Results of monitoring simulated field structures affected by ASR in Giannini et al. [89] demonstrated that (1) the most significant declines in P-wave velocity measured by UPV occurred at the earlier stages of expansion (up to 0.1–0.2% expansion), and (2) UPV values of trimmed core samples were typically higher than in-situ measurements at the same location. Guidance on making moisture and temperature corrections for UPV measurements can be found in [90].

UPV is classified as POC.

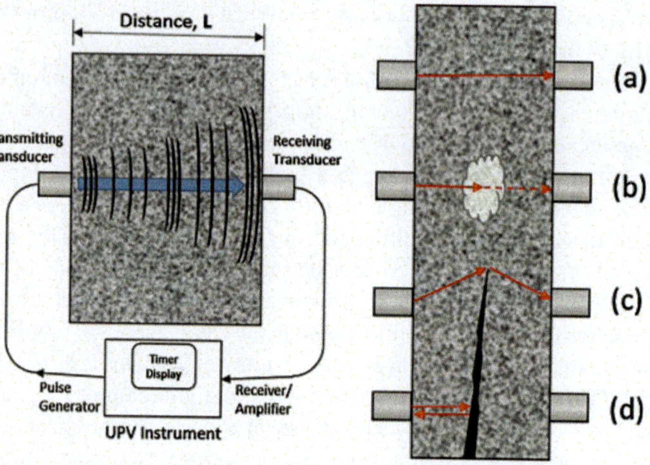

Fig. 3.16 Schematic of through-transmission UPV technique (left); examples of effects of flaws on UPV measurements (right): **a** No flaws, no reduction in velocity; **b** high-porosity or honeycombed region, velocity reduced and attenuated pulse; **c** pulse diverted around crack tip, lower apparent velocity; **d** large open crack, complete reflection of pulse. Flaw detection schematic adapted from [87]

3.4.5.3 Ultrasonic Pulse Echo (UPE)

UPE is based on the generation and reception of ultrasonic waves and measures the amount of energy reflected on an interface or inclusion within the tested structures.

With the UPE technique, the use of shear waves is preferred because they provide better resolution for the same operating frequency; shear waves have shorter wavelengths compared with the compression waves used in conventional pulse echo techniques.

UPE is quite simple to use, offers high-sensitivity measurements to determine the depth of internal flaws and the thickness of section with parallel surfaces, using ultrasonic shear waves and the SAFT algorithm (Synthetic Aperture Focusing Technique). It requires only one accessible surface and may be operated automatically (acquisition and signal processing) [7, 43]. Recent developments are focused on increasing the investigation depth with multi-source and multi-receiver systems (up to 48 pulser/receiver transducers). Wiggenhauser et al. [91] presents a linear array ultrasonic tomography system able to investigate 4-m thick structures (in laboratory conditions).

However, like for other methods, highly skilled and experienced operators are required for interpreting results, especially when the reflected signal is disturbed by reinforcing steel and post-tensioning ducts.

This method has been standardized for other materials (ASTM A114 and ASTM E317) and recent research examples of applying shear waves with SAFT using multiple dry-contact transducers are available for existing reinforced concrete structures. Clayton et al. [92] and Khazanovich et al. [93] have applied SAFT to ASR-damaged plain concrete slab specimens and quantified the damage level using a Hilbert Transform Indicator (HTI), an analytical technique originally developed to characterize freeze-thaw damage [68]. Using HTI values, they were able to distinguish between sound and damaged specimens and map areas of sound versus damaged concrete within each specimen. Payan et al. [43] provide two in-situ tests results on a bridge and on a long external reinforced concrete wall (nearly 50 m long, more than 2 m high, and constructed in 1979). Although the standard effective depth of this technique has been limited to $\sim$30 cm [7], new systems are more efficient and may permit investigation depths of up to several meters [43, 91].

UPE is classified as PAT.

3.4.5.4 Impact-Echo (IE)

In impact-echo testing, an impact is generated with a hammer or spherical impactor and the waves rebound upon encountering the opposite surface (back wall) of the structure, as well as internal defects such as voids, cracks, and delaminations. The vibration of the impacted surface is detected with a transducer and recorded in the time domain; on-board signal processing can filter and process the time-domain signal using FFT to produce a frequency spectrum that it typically more useful for analysis and flaw detection (Fig. 3.17). The location and extent of discrete defects are

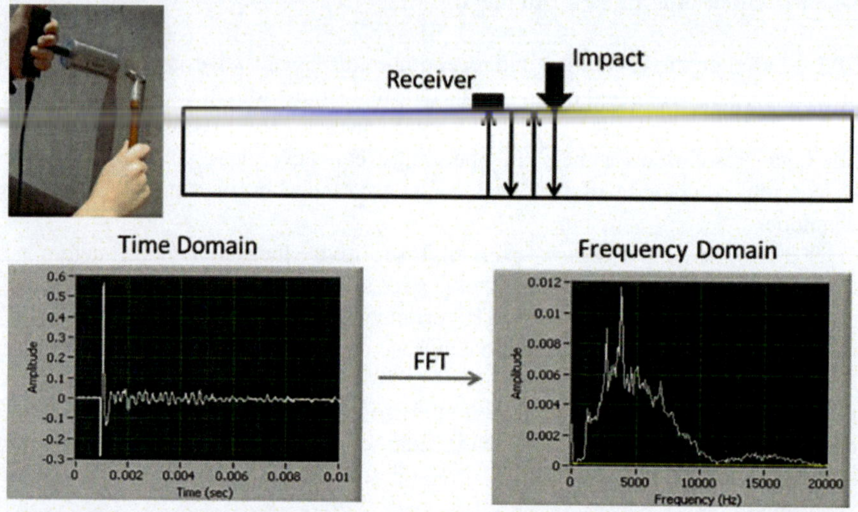

Fig. 3.17 Schematic of the impact-echo method, showing the test setup, acquired time domain signal, and FFT processing to produce frequency domain data for analysis

determined by analyzing the dominant frequencies; the minimum detectable flaw size increases with depth from the impacted surface. Internal microcracking characteristic of ASR may be evident as reduced P-wave velocity (similar to UPV, [89]) or increased apparent thickness of the element; and the increased wave scattering produces a "noisy" frequency spectrum, as shown by Henriksen [94].

IE is relatively simple to perform, but requires an experienced operator to optimize the test setup and to interpret the results. This is particularly important for non-plate-like structural elements such as columns, beams, and piers, or impacts generated near element boundaries.

Payanet al. [43] reports some results collected on real but non-ASR-affected structures, to monitor concrete conditions and actual structural stiffness. Metalssi et al. [88] presents an application dealing with monitoring ASR-affected beams, with the detection of a slight loss of stiffness due to ASR. Metalssi et al. [88] also points out the influence of concrete moisture on IE measurements that should be considered in data preparation.

IE is classified as POC.

3.4.5.5 Acoustic Emission (AE)

AE is a passive method, meaning that acoustic emission data is generated by a rapid release of energy within the structure itself. Therefore, if damage is not actively progressing, acoustic emission data is not generated, and AE is not able to detect existing cracks or distress. Investigations have shown that acoustic emission monitoring may be a useful method for localization and assessment of cracks, leak paths [77], corrosion damage [95], and ASR [96].

For many industries, most notably fiber-reinforced polymeric tanks and pressure vessels, acoustic emission instrumentation and assessment methods have been standardized for decades. Terminology related to acoustic emission is provided in ASTM E1316.

However, considerable work is still necessary to adapt the method to reinforced concrete structures. Although it shows good in-situ performance for crack detection, AE monitoring requires a good knowledge of the features of the structure to monitor damage progression and localize the signal source [7, 43]. Factors to consider in a monitoring system setup include the type of sensor to be used (resonant or broadband) and the layout of the sensor array. These selections are currently made on a case-by-case basis and are often driven by the stage of damage being monitored and assessed (early, meaning non-visible, or later damage states). The influence of ambient noise should also be considered before carrying out an AE survey for structural monitoring.

Initial attempts to apply AE for monitoring ASR-affected structures proved to be inconclusive [97]. More recently, Abdelrahman et al. [96] have made progress demonstrating the potential for AE to monitor ASR crack generation, evolution, and distribution within concrete structures under laboratory conditions, it is clear that more extensive laboratory (and eventually field studies) are needed to demonstrate the viability of AE for monitoring the progress of ASR in concrete structures.

AE monitoring generates very large amounts of raw data, and the data sets may be corrupted by emission records that are not directly associated with damage growth, including internal boundary reflections and external sources. Relevant and effective data processing tools are thus key components of successful preparation of AE signals. Pourahmadian et al. [99] Pourahmadian and Haddar [100] and Soltangharaei et al. [98] provide some examples of relevant or innovative techniques to help interpret AE records, and there is ongoing research in the United States on advancing the state of AE specifically for the purpose of monitoring ASR.

In summary, AE seems to have limited applications to date for monitoring of ASR-affected structures. Although it has the potential to detect microcracking from ASR before the effects of the reaction are manifested at the surface, that requires the installation of a monitoring system before any distress is evident.

AE is classified as PAT.

3.4.5.6 Non-linear Acoustics

While the elastic behavior of most engineering materials is well-described by Hooke's law that asserts a linear stress-strain relationship, all materials are, in fact, not exactly linear to some degree, and are instead nonlinear. This is especially true in multiphase materials with complex microstructures, such as cement-based materials, where non-linear elastic behavior is much more pronounced than in monolithic materials such as metals. In cement-based materials experiencing damage—such as due to ASR—

the non-linearities become more pronounced. The elastic properties at an imperfect interface are strongly nonlinear [101, 102], and the overall nonlinear stress-strain relationship of a material with flexible interfaces can be described in terms of E, the Young's modulus, the nonlinear elastic coefficients, and the characteristic hysteresis observed in concrete and rock [103, 104]. Chemomechanical events such as interfacial weakening and debonding, microcracking, and ASR gel formation produce changes in nonlinear elastic properties, which are generally orders of magnitudes higher than the changes in linear elastic properties [105]. Therefore, experimental quantification of changes in nonlinear elasticity can offer new opportunities for ASR damage assessment.

The nonlinear elastic behavior of damaged materials can be measured efficiently using acoustic (e.g., ultrasonic) waves [106–109]. The nonlinear elastic effects appear in acoustic waves in a variety of ways, depending upon how they are measured and the materials' characteristics and sample geometry.

First, possible interactions between different propagating waves can occur and frequency components that did not exist in the input wave are created by a material's nonlinear response, termed the "nonlinear modulation effect" [110]. This approach is used with success in [111] for ASR-affected concrete specimens.

As a second approach, when a prism of damaged material is excited into its normal-mode vibration, the resonance frequencies of the bar shift downward as the input strain excitation increases [112]. Measured shifts in the resonance frequency are expected to be directly proportional to the amount of ASR damage accumulated in the material, as demonstrated in preliminary studies by Chen et al. [113] and Leśnicki et al. [114].

Although the sensitivity of the observed parameters is improved as compared to linear-based methods [43], the nonlinear acoustic methods remain in a research stage and on-site implementation are still underway. They comprise a promising range of techniques, and further research is needed to better understand the relationships between these measured nonlinearity parameters and the structural, compositional, and eventually mechanical changes resulting from ASR.

Non-linear acoustic techniques are classified as PAT.

3.4.5.7 Diffuse Ultrasonic Techniques

In addition to the coherent ultrasound and nonlinear ultrasound approaches, diffuse ultrasound can be used to quantify material properties and damage. Operating at higher frequency (100 kHz–1MHz), wave-aggregate interactions generate multiple scattering. The multiple-scattering effect in such a medium creates a diffuse acoustic field. Because these diffuse waves travel much longer distances than the direct path, they are more sensitive to small changes occurring in the medium. These incoherent, yet repeatable (under constant conditions), signals are called coda waves and carry rich information about the medium.

The application of quantitative diffuse ultrasonic techniques to characterize the multi-scale structure of cement-based materials is possible. For example, the progression of self-healing of cracked concrete was monitored in situ by diffuse ultrasound [115]. Diffuse ultrasonic measurements were conducted on: uncracked, tensile through-thickness cracked, and flexure partial-thickness cracked concrete specimens, and the effective diffusivity was successfully used to monitor progressive self-healing of concrete. Quiviger et al. [116, 117] demonstrated the use of diffuse ultrasound to detect both real and simulated cracks in concrete beams, although the performance of the method is reduced for actual cracks, probably due to contact points between the edges of the cracks that may cause energy drop of the signal.

Another innovative diffuse ultrasonic technique is Coda Wave Interferometry (CWI). In an ultrasonic CWI measurement, by launching ultrasonic waves and comparing two coda waves measured at different conditions (e.g. at different stress levels), the relative changes in wave velocity could be extracted. The velocity changes may be caused by damage and/or microstructure modifications in the medium at the wavelength scale. Researchers have applied the CWI technique to concrete samples. Larose and Hall [118] and Larose et al. [119] proposed an ultrasonic CWI experimental method based on the use of the stretch technique and tested experimental method for measuring stress-dependent velocity changes and for localizing one crack on a concrete beam. Payan et al. [120] performed CWI measurements on a concrete sample using ultrasonic waves polarized in three perpendicular directions and obtained a complete set of nonlinear elastic parameters.

However, the methods based on diffuse ultrasound and CWI are still in the research stage are not yet ready for an in-situ monitoring campaign for ASR-affected structures.

Diffuse ultrasound techniques are classified as PAT.

3.4.5.8 Surface Wave Techniques

Surface wave methods utilize the dispersive properties of Rayleigh waves (R-waves) to determine the variation of shear wave velocity and to assess the stiffness of the monitored system. Air-coupled sensors can be used so that no specific coupling conditions are required between the structure and the measurement system. However, the signal processing is rather complex. In France, The IEMN laboratory (École Centrale de Lille) developed and tested a portable device tested in the field but not on ASR-affected structure [121]. Ongoing tests are performed on the VeRCoRs mock-up to detect cracks in a 40 cm-thick prestressed cylindrical wall.

More recently, Kim et al. [102, 122] applied the second harmonic generation (SGH) of nonlinear Rayleigh waves to characterize ASR damage under laboratory conditions. They applied this technique to a set of 1220 × 915 × 200 mm (48 × 36 × 8 in.) plain concrete slabs and 3500 × 3000 × 1000 mm (138 × 118 × 39 in.) reinforced concrete slabs. For each set of slabs, some were designed to expand from ASR and one functioned as a non-expansive control specimen. The technique demonstrated a high sensitivity of the nonlinear parameter β (a function of the amplitudes of the fundamental and second harmonics, and wave propagation distance) even at low levels of ASR damage, particularly in a range where visual evidence of damage on the concrete surface is limited.

Rayleigh-wave-based techniques seem promising, in spite of a current lack of demonstration test in the field for ASR characterization.

Surface wave techniques are classified as PAT.

3.4.6 Crack Mapping

Recording crack patterns on the surface of concrete structures is a common method associated with ASR monitoring. For instance, the Federal Highway Administration (FHWA) and the Roads and Transportation Administration in France recommend using crack mapping and indexing in the monitoring process and for rating ASR-affected structures. This method has been used for assessing ASR on numerous transportation structures [123].

However, a US-NRC note [124] reports that, *"the presence and the extent of surface cracking is not a conclusive indication that ASR is present"* and *"It is stated throughout published ASR research that crack mapping is somewhat limited in its application to understanding ASR degradation in concrete"*. A scientific consensus agrees that crack mapping alone provides information limited only at the concrete surface and that other examination method, such as NDE, are needed to get an acceptable and useful understanding of the current state of degradation. Crack mapping/indexing methods are not diagnostic in terms of the cause of the observed distress, and even where ASR has been confirmed through examination of core samples, other mechanisms (e.g. freezing and thawing) may contribute to the observable cracking, or even be the dominant mechanism driving crack growth during the monitoring period.

3.4.6.1 Cracking Index Assessment

The French Institute of Science and Technology for Transport, Development and Network (French acronym "IFSTTAR", formerly "Laboratoire Central des Ponts et Chaussées'—LCPC) developed a method to quantify the amount of surface cracking on ASR-affected structures, which can be used to provide a rough estimate of expansion [125]. However, recent research shows that a relevant correlation between crack index and expansion would need further information on concrete conditions, reinforcement in the structure and environmental conditions [124].

The purpose is to measure crack over a defined grid and to quantify the total amount of cracking periodically. Typically, the widths of all cracks interesting the branches of a specific grid drawn on the structure are summed up and divided by the length over which they were measured, Fig. 3.18. The crack index is determined as a value in millimeters per meter which can be compared to establish criteria linked with action levels for management of the structure. The advantage of this method is to avoid extracting core samples; it does not damage the structure surface. It is recommended

Fig. 3.18 Inspection with crack index assessment on a dam ©EdF

that crack mapping should be repeated every 2–3 years on selected surfaces that are affected by deteriorations and exposed to moisture. The crack mapping is used to calculate the Cracking Index. As mentioned in Sect. 3.4.2.2 DEMEC targets can be placed in the same members as crack mapping and aligned with the grid axes.

Disadvantages of the method include the inability to distinguish between the mechanisms creating the crack growth, the influence of seasonal temperature and moisture variations on the crack growth and opening widths, the potential for error and variability on the part of the operator in measuring crack widths, and the inability to quantify anisotropic deformations (i.e. the cracking data is reduced to a single number with no directionality). CI measurements typically err on the unconservative side and should not be assumed to exactly estimate expansion; they are, however a good relative measure of the severity of ASR in a structure.

Cracking Index assessment is classified as POC.

3.4.6.2 Crack-Meters

Manual or automated crack-meters are suitable to ensure a precise monitoring program for a cracked structure. With good accuracy (about 0.1 mm), crack-meters are a very interesting complement to displacement measurements. Both inductive-type or LVDT-type crack-meters exhibit good reliability with little drift over time. They are well-adapted to the conditions of use, including difficult exposure conditions, and some models can be immersed in water.

Crack-meters are classified as POC.

3.4.6.3 Infrared Thermography (IRT)

Infrared thermography transforms a thermal radiation pattern that is invisible to the human eye to a visual image. It is a non-contact, non-destructive method that measures the temperature distribution on concrete surface. It can be passive (based on natural temperature contrast) or active (the structure to monitor is heated or cooled by the operator for monitoring propose). This technique makes it possible to identify subsurface anomalies such as delaminated concrete or delaminated coatings on the concrete surface. Recommendations for this method are described in ASTM D4788.

This technique measures the surface temperature, so the depth of investigated concrete below the surface is limited. Moreover, it is dependent on ambient conditions (sunlight, wind, rain); this is particularly important if passive measurements are being made, rather than actively heating/cooling the structure to generate defect-induced differences in infrared radiation.

IRT could be considered for ASR-affected structures to detect possible surface delamination due to concrete expansion. Kabir et al. [126] used IRT images, processed with Haar's wavelet transform in order to quantify surface damage of concrete structure. It may also be a convenient tool to detect concrete cracks hidden by coatings (e.g. paints, overlays, FRP repairs). However, there are a lack of reported studies of IRT applied to ASR-affected structures.

IRT is classified as PAT.

3.5 Case Studies

3.5.1 *General Discussion of ASR in Nuclear Power Plant Structures*

Clayton and Cyrus [127] provide an overview of concrete issues that a nuclear power plant (NPP) can face, and how to check the viability of NDE techniques to evaluate and test concrete samples. They reported that: "*There are no NDT methods which have been proven effective in identifying ASR before it is visibly detectable. ASR is usually identified either visibly or by petrographic analysis*". That is why at Seabrook station (New Hampshire, USA), concrete cores were taken from the structure to be tested to diagnose ASR.

A more recent review by Kurtis et al. [128] discusses, in part, the challenges faced by concrete NPP structures with respect to ASR, including the exposure to moisture, elevated temperatures, and radiation. The situation is further complicated by the fact that access is often limited to many NPP structures; yet, given the high stakes, it is vital that suitable assessment and monitoring techniques are developed.

Therefore, in the following examples, the diagnoses are mostly based on permanent monitoring, laboratory tests and visual inspection.

3.5.1.1 Gentilly-2 Nuclear Power Plant (Canada)

Gocevski [129] provides a case study on the concrete degradation in the Gentilly-2 Nuclear Power Plant (G-2 NPP). The life cycle (or licensed operating period) of a nuclear power plant is often initially set for 30–40 years. In the case of G-2 NPP, the first cycle was finishing and a new project had aimed to extend its operating life an additional 25–30 years. This project included an evaluation program to consider the feasibility of the lifespan extension that included visual inspection, in-situ measurements, and laboratory materials tests. The aim of this program was to test the concrete structure affected by AAR and exposed to thermal conditions of Canadian weather. Comprehensive numerical models, fed by monitoring data, were developed and used for this set of studies.

According to Gocevski, AAR was the primary degradation mechanism at G-2 NPP, leading to cracking, structural deformation, and degradation of concrete properties. Most of the analysis inputs for the numerical models were based on assessment of mechanical properties (i.e. structural strain, displacement and stiffness, concrete strength, and residual expansion). There was significant use of data from the permanent monitoring system (i.e. pendulum, vibrating wire strain gages, etc.) for the diagnosis and the numerical model implementation.

Gocevski underlines the fact that "the tests conducted on small samples cannot provide certain information related to real structures such as: heterogeneity of the volumetric expansion, development of stresses in concrete due to the swelling of the silica gel, effect of confinement, etc". It is recommended to carry out mechanical tests on relatively large specimens to get more meaningful conclusions.

Although there was significant discussion of the permanent monitoring system and laboratory tests, almost no information was given on NDE techniques tested on G-2 NPP.

3.5.1.2 Concrete Vessels of the Gas-Cooled Reactor at Saint-Laurent NPP (France)

Two gas-cooled reactors (GCR) have been erected at Saint-Laurent nuclear power station (France) between 1963 and 1971. These plants were operated up to 1990 and 1992, respectively.

GCR prestressed concrete pressure vessels (PCPV) have a safety function similar to the PWR or CANDU containment, with internal diameters ranging from 10 to 24 m, internal height between 15 and 40 m, and wall and cap thickness from 3 to 6 m. The service pressure inside the PCPV ranged between 1.5 and 4.6 MPa in operation.

A dedicated permanent monitoring system was implemented within each PCPV during construction, including embedded extensometers, inverse pendulums, vertical long gage extensometer, and some dynamometers set on tendons that were left ungrouted.

The expected trend for the delayed strain is an overall shrinkage of the structure, mainly due to creep of concrete. However, after only a few years of operation, both PCPV exhibited an increasing "swelling" strain trend, due to AAR [5]. The concrete expansion from AAR led to an increase of the prestressing force, as shown on Figs. 3.19 and 3.20.

Even after the reactors ceased operating, there remained concerns about the integrity of the PCPV through the decommissioning process. An analysis of the stress and strain measurements obtained from the permanent monitoring system with simplified modeling of AAR was undertaken in order to ensure that the integrity of the PCPV will not be jeopardized by an excessive stress in the concrete before the end of PCPV decommissioning [5].

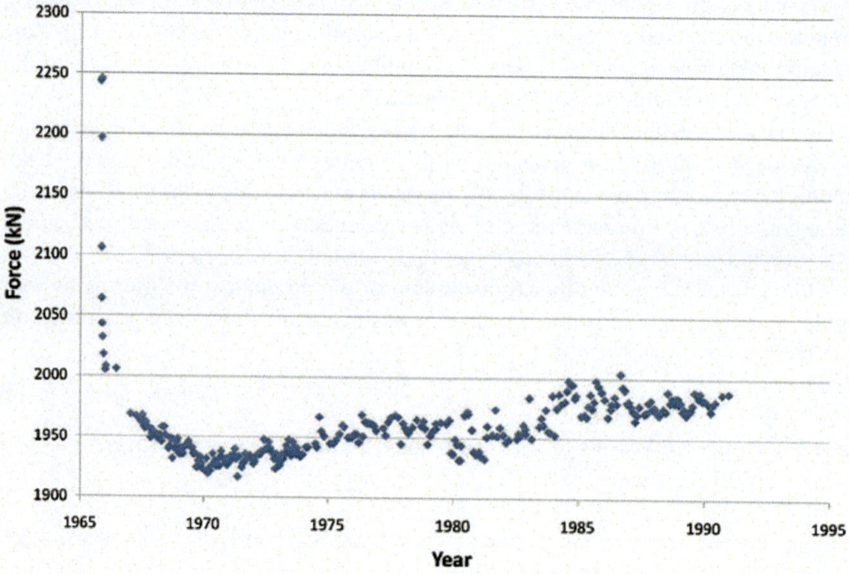

Fig. 3.19 Time history of force readings observed on Saint-Laurent A1 GCR PCPV affected by AAR

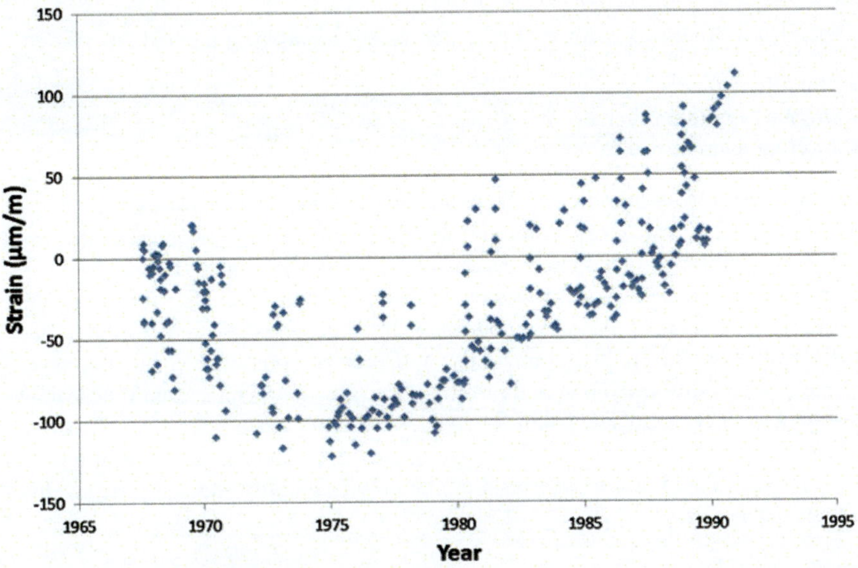

Fig. 3.20 Time history of strain readings observed on Saint-Laurent A1 GCR PCPV affected by AAR (raw data, scatter is due to temperature variation, mainly induced by operation cycles)

3.5.2 Hydraulic Structures and Dams

3.5.2.1 Saunders Generating Station

Danay et al. [130] present a case study of the diagnosis of ASR in the Saunders Generating Station, a hydraulic structure in Canada built between 1954 and 1958. Abnormal behaviour started to be reported in 1972.

Investigations detected unusual behaviour of a turbine. Displacements were measured and results were considered as unacceptable for this kind of machinery: 3.2 mm in one axis, 11.4 mm in the other. Other problems appeared afterwards, including leakage through expansion and construction joints and cracking of stairwells.

Structural investigations were conducted on various components of the dam (tunnels, walls, joints) and included some conventional lab tests (e.g. aggregate expansivity test, compressive stress test). In the 1990s, in-situ investigation and tests to determine the physical state of the structure affected by long-term loads (such as dead loads, thermal cyclic loads, water exposition changes, dewatering, etc.) included:

- crack mapping;
- in-situ rebar stress tests;
- overcoring tests and crack mapping of boreholes.

Concrete swelling was confirmed with permanent monitoring methods:

- pendulum data;
- invar wires data;
- levelling data.

These investigations indicated the presence of ASR, concrete growth, and rising of downstream galleries. Other symptoms observed indicating the presence of abnormal concrete swelling included joints expansion and tiles buckling on floors.

Studies on rebars are also reported in this case study. Results were in conflict with expected data for the study load cases. In-situ rebar strain test results showed that tensile strains were higher than normal. Crack mapping, cores, and TV camera examination of the boreholes yield results coherent with previous tests. Tensile stress in rebars would be associated with concrete cracks but it was not the case. There were two possible reasons: concrete shrinkage or concrete swelling. Because the rebars were in tension and not in compression, the only possibility was concrete swelling (from ASR).

3.5.2.2 Eastern Canada Hydraulic Structure

A study was carried out on a large hydraulic structure located in Eastern Canada by Rivard et al. [24]. It was built in the late 1950s and is affected by ASR to various degrees, detected approximately 20 years after its construction. The structure has therefore been exhibiting signs of ASR for more than 30 years and shows various levels of damage. The ASR in this structure is due to the use of a reactive crushed siliceous limestone as coarse aggregate in the concrete. Horizontal displacements related to the expansion of concrete vary between 0.1 mm/year and 1.6 mm/year.

The program encompassed different components, consisting of:

- stress measurement;
- evaluation of concrete condition by non-destructive methods without drilling (seismic tomography);
- the evaluation of the mechanical, physical, and petrographic properties of the concrete determined from cores recovered from full-length boreholes.

The conclusions were:

- the stresses that were measured are in accordance with the observed damage and typical of stress induced by ASR in a large structure;
- the transverse stress is virtually nil, which is due to the more or less free (unrestrained) expansion in this axis and the relief associated with the crack (found with the tomography);
- higher compression stresses were measured in the confined longitudinal axis, where values were about 2–3 MPa;
- seismic tomography allowed adequate coverage of the structure and enabled a major tensile crack in the gallery to be located and its depth estimated.

Regarding the methods that were used to assess damage on drilled cores, the following conclusions may be drawn:

- mechanical properties were slightly affected in the zone two meters below the surface, and high scatter was observed in the data obtained from this study. The static Young's modulus appears to be the most relevant parameter to quantify ASR damage;
- P-waves and S-waves yielded about the same trend, so S-waves did not appear to be more sensitive to ASR than P-waves;
- petrographic examination should always be conducted on cores since it yields reliable results.

3.5.2.3 Experience Gained on About 150 Dams in France

EDF operates 156 monitored concrete dams, with a mean age of 55 years. Most of these dams are mainly gravity and arch dams [1–3]. Almost all of these concrete dams are equipped by permanent monitoring devices, as described in Sect. 3.2.1.

Among the 156 concrete dams, 93 had sufficient data collected for the characterization of a potential internal swelling reaction (ISR), and 30 were identified as potentially affected by a swelling reaction (Fig. 3.21). The criteria used by EDF to check if a potential ISR is active is a trend for elongation in one direction at a rate greater than 2 μm/m/year.

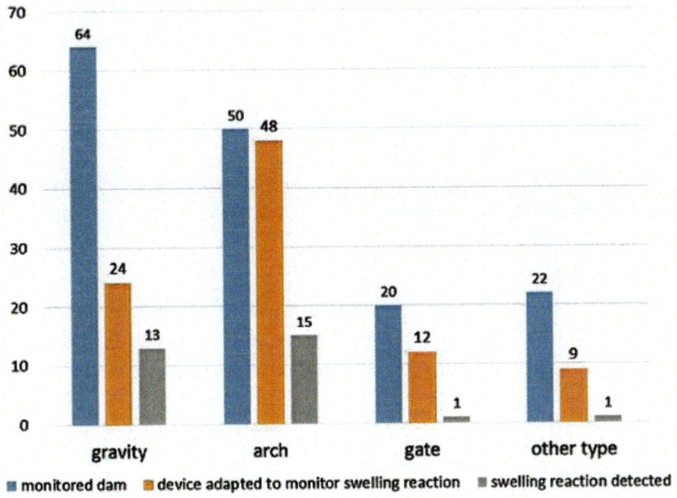

Fig. 3.21 Dams on which sufficient data have been collected to study potential ISR [3]

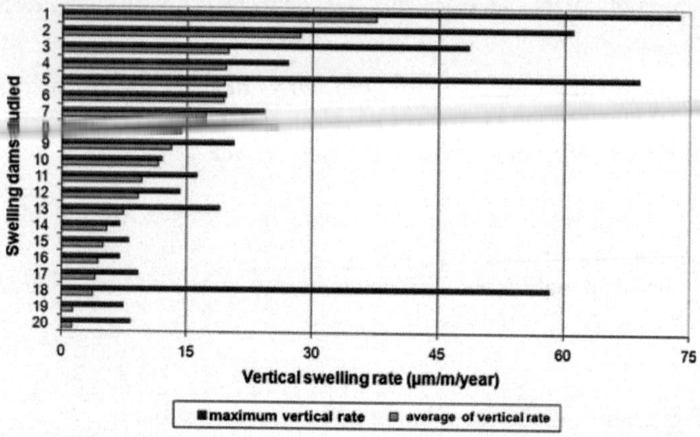

Fig. 3.22 Measured vertical displacing rate (upwards) on the set of EDF analyzed dams [3]

Vertical displacements and arch elongation (for arch dams) have been compared among the different monitored dams (see Figs. 3.22 and 3.23). The distinction between the maximal rate (localized on a part of the dam) and the average rate (average of each rate calculated on each device) underlines possible heterogeneous behavior of ASR within the dams (e.g. only a specific part or parts of a dam is affected by the swelling reaction).

For 8 dams, vertical expansion rates have been plotted versus horizontal crest displacement as shown in Fig. 3.24. The ratio of these values varies between 2 and 4, and may depend on the site conditions (stiffness of the abutments or compressive strength within the structure, for instance).

Guilloteau et al. [3] mentions that, according to this EDF study, gravity dams are more likely to undergo ISR. But other influencing factors are at stake:

- the location of the dams and the geology of the site which imposed the choice of the construction materials. About one-half of the swelling dams are sited in igneous and magmatic rock environments, and many of these rock types are known to be able to release alkalis;
- the construction period: most of the affected dams have been built between 1920 and 1960, before the civil engineering community considered the potential for ISRs such as ASR in technical recommendations and guidelines.

An estimate of the total expansion (strain) undergone by the 30 affected dams since their impoundment is provided in [3], based on a non-linear regression taking into account the observed behavior since the beginning of reaction given by embedded extensometers measurements. Tangential and vertical strains from these dams are given in Fig. 3.25.

- Accumulated vertical strain prediction varied between 200 and 3,500 µm/m according to the model used with the available data set. Among the 30 dams studied, only 7 showed deformations greater than 1000 µm/m. All these dams have

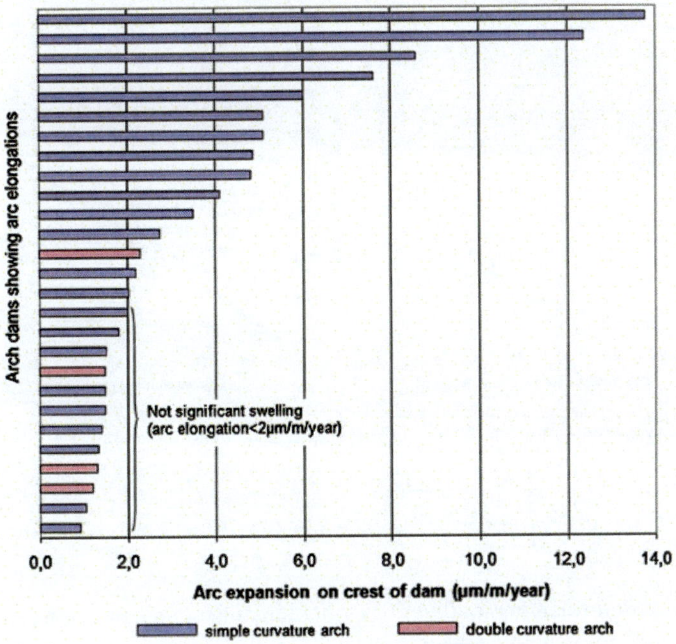

Fig. 3.23 Measured elongation rate on the set of EDF analyzed dams [3]

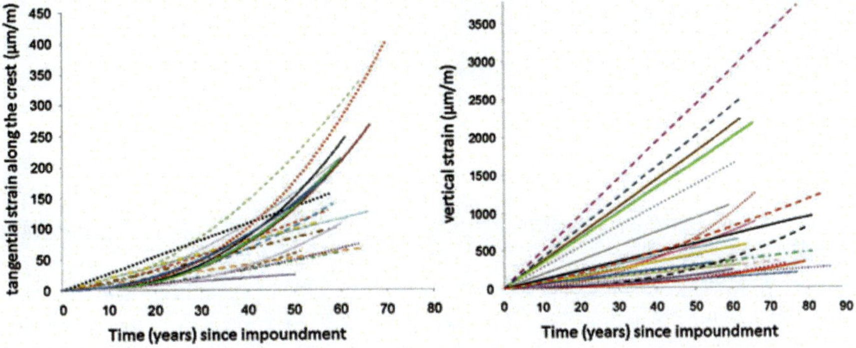

Fig. 3.24 Vertical rate versus horizontal rate of expansion [3]

been equipped with more monitoring devices that can be used to feed numerical models. In some cases, strengthening works have been undertaken on the dams.

- Accumulated predicted tangential strains did not exceed 250 µm/m. These strain values are most likely lower than accumulated vertical strain values because the concrete is in compression resulting from the confinement of the swelling provided by the geometry and the abutment of the structure.

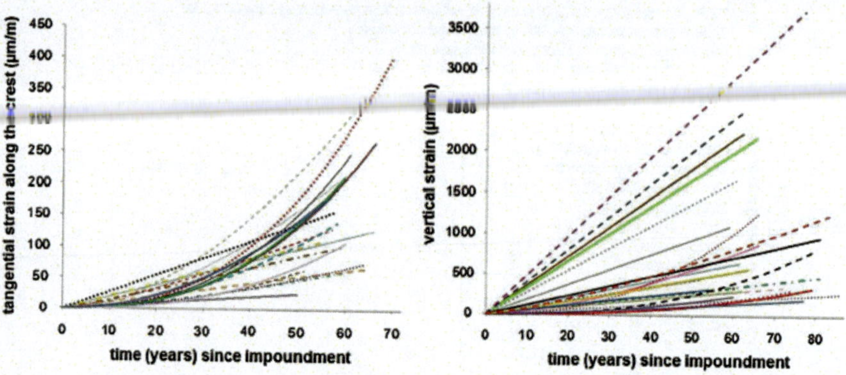

Fig. 3.25 Assessment of tangential and vertical strains since first impoundment on 30 dams [3]

In summary, for an appropriate and valid evaluation of the kinetics of ISR on dams, it is recommended:

- to monitor deformations in directions where the phenomenon can freely expand and generate maximum displacements and strain;
- to design monitoring systems adapted to the structure and to the damage mechanism, keeping in mind the possible heterogeneity of concrete and the magnitude/rate of swelling.

However, the determination of current kinetics is not sufficient to assess the short and long-term consequences of ISR. Visual inspection provides key information to complete the analysis based on monitoring data.

3.5.2.4 Lake Lynn and Yadkin Projects

Gross et al. [131] presented case studies of two hydropower structures. The first one is the Lake Lynn Project on the Cheat River in West Virginia, USA; the second is the Falls Development Dam on the Yadkin River in North Carolina, USA.

The Lake Lynn project is a concrete gravity dam 305 m (1000 ft.) in length and 38 m (125 ft.) in height. AAR was detected in 2004. Damage was located on the spillway deck and worsened with the lack of functional expansions joints. Key elements of the structure were affected by AAR expansion; of note, the trash gate was inoperable during summer for local displacement reasons (binding against the spillway deck). A visual inspection in 2007 of the deck showed that the expansion joints were highly deteriorated due cracking and spalling. Expansion joints were also affected because they were filled with debris, eliminating space for AAR and thermal expansions.

The solution to allow movements between the piers and the decks was to rehabilitate the existing joints in 2007. To do so, joints were first cut from the upstream to downstream face and then in the deck slab itself. Different techniques were used to cut the slab because complicated joints had to be cut. Further details about the cutting process can be found in [131].

The second step was to monitor the evolution of the rehabilitated joints. Structural movements were also monitored during the construction process. Punch marks were installed at joints in steel rails embedded in the deck where they crossed the expansion joints undergoing rehabilitation. Pairs of survey monitoring pins were installed on the upstream parapet wall at the joints and monitored both during and after construction. The collected data helped to understand how the structure moves on both the upstream and downstream sides (amount and direction of the expansion) and to conclude that works had a positive impact on the structure by relieving built-up internal stresses and restoring movement at the joints. Continued monitoring of a crackmeter previously installed at the junction of the spillway deck and powerhouse near the trash gate allowed comparison of long-term trends beginning more than ten years prior to rehabilitation to three years after the rehabilitation. The crackmeter data showed that rehabilitation halted a long-term trend of movement of the deck towards the powerhouse and also reduced the seasonal thermal movements that were exacerbated by a loss of functioning expansion joints.

The Falls Development Dam (part of the Yadkin Project) is a concrete gravity dam 229 m (750 ft.) in length and 34 m (112 ft.) in height. Construction finished in 1919. Beginning in 1994, issues were reported that included binding of the spillway gates, prompting additional inspections and deflection measurements over the next 15 years. Other symptoms discovered included tilting of some piers and cracking of the concrete spillway deck. Further studies showed that the deformation of the spillway concrete piers contributed to the binding of the gates. In 2010, the extraction of two concrete samples from the spillway deck showed some evidence of AAR.

The solution that has been used to address this problem is the construction of two expansion joints in 2010. A monitoring program including the measurement of deflections, uplift pressures, and local movements using piezometers, inclinometers, and crackmeters was implemented. Some instruments (crackmeters and inclinometers) were used to measure the impact of the slot cutting.

Crackmeters (stainless steel pins) were installed on the deck for the first time in March 2010 at different locations of the joints (upstream, middle, and downstream). Data were collected periodically during maintenance operations and showed that after the joint construction and partial closure of the joints from stress relief, the structure was affected by only thermal expansion. Three inclinometers were installed in three spillway piers in March/April 2010 to monitor the long-term effects of the slot cutting. Data were collected weekly before and after the maintenance operations and will continue to be monitored to confirm that the new joints are functioning properly and/or if additional measures are needed to manage AAR in the dam. Only a few months of monitoring after the joint cutting were reported, but the inclinometer data showed lateral movements of the piers consistent with the crackmeter measurements, and no unusual upstream/downstream movements. The long-term monitoring program included establishing action levels for crackmeter measurements that could prompt re-cutting of the joints if they were at risk for fully closing.

3.5.3 *Monitoring of ASR-Affected Highway Structures and Pavements: Results of Field Trials*

As part of the FHWA ASR Development and Deployment Program, nine field trials were conducted across the United States that evaluated and monitored the effectiveness of mitigation measures applied to concrete structures and pavements already exhibiting ASR-induced distress. Thomas et al. [123] documents the field trials and provides recommendations for engineers and researchers engaging in future studies of ASR-damaged structures.

Table 3.1 provides the summary of recommendations for performance monitoring of ASR-affected transportation structures.

In order to avoid high variability in results, measurements are recommended to be carried out twice a year by the same operator or trained operators and under similar weather conditions. Because of the effect of climatic conditions, several years (minimum 3 years, ideally 5 years) of monitoring are required to establish significant trends.

3.6 Summary of Techniques

Table 3.2 provides appraisal ratings of the main techniques mentioned in Sect. 3.4. The appraisal deals with the maturity level of the technique ("Proof of Concept" (POC) or "Potentially Applicable Technique" (PAT), and its accuracy (for localization, detection, features sizing, trueness, precision, repeatability…). For accuracy ranking, "A" is the highest grade, "B" is the medium grade and "C" the lowest. Notice that the ranking could obviously change with time and ongoing research effort to improve NDE techniques for civil structures.

Table 3.1 Summary of recommendations for monitoring of ASR in transportation structures. Adapted from [123]

Visual inspection
Visual examination of treated sections of the structure including regular photographic survey of selected portions; allows monitoring of the progress in damage.
Cracking index (CI) method
The method consists in quantifying surface cracking by recording and summing the crack widths measured along a set of lines drawn (crack map) on the surface of the selected sections. Minimum dimension of crack map: 500 × 500 mm (20 × 20 in.).   • Insert stainless steel studs with DEMEC points at the corners of the crack map grid to allow direct comparison with length changes.
Expansion measurements
Length change measurements are carried out on the same grid (crack map) developed for CI measurements. Thus, the recommended dimension for an expansion grid (vertical and horizontal readings) is 500 × 500 mm (20 × 20 in.). When such a dimension is not possible (e.g., barrier wall in the vertical direction), a smaller size grid (500 × 150 mm) can be used. Other types of measurements, e.g., circular measurements using a Pi-tape, can be used to monitor expansion of structural elements with circular cross-sections (columns).
Temperature and humidity measurements
The method allows the measurement of temperature and relative humidity (RH). Plastic sleeves inserted to different depths, e.g., 25–75 mm (1–3 in.) within concrete elements. The method is useful to monitor the possible beneficial effect of surface treatments to control ASR expansion by reduction of RH in concrete.   • Select monitoring sites where the plastic sleeves are not going to be exposed/subjected to the impacts from cars, ice and snow removal, or be accessible to individuals who could damage the setup.   • In moderately and severely cracked concrete elements, as well as when moisture is available from backfill material (e.g., abutment, wing, and retaining walls), water will likely accumulate in the holes, thus making reliable readings impossible. Moisture can also accumulate in the holes because of water condensation. In such circumstances, the holes should be left open until the holes dry out. An internal RH re-equilibrium period of several hours) is then required before any reliable data can be obtained.
Non-destructive techniques (NDT)
NDT provides an indirect measurement of the concrete condition. The techniques used in the FHWA research were based on the propagation of stress waves, which primarily depend on concrete's Young modulus and density.   • Ultrasonic pulse velocity (UPV) in indirect configuration can be used to assess the surface condition where only one face is accessible. When opposite faces are accessible, UPV will provide an evaluation of the global condition of the concrete.   • Impact-echo can be used to assess the global condition when the opposite face is parallel to the surface (cannot be used on circular columns, for instance).   • Nonlinear acoustics: this method is not recommended at this stage because of (1) the complexity of the method and of signal processing, and (2) the lack of long-term data.   • All NDT methods must be performed by skilled and qualified operators. Data should always be analyzed by experienced engineers.   • In cases of severely cracked massive elements, NDT may not work because of the attenuation of the signals.

Table 3.2 Summary table of techniques for monitoring ASR-affected structures

Method	POC or PAT[a]	Accuracy for ASR diagnosis
Cracking/crack pattern		
Visual inspection	POC	B
Cracking index (CI) method	POC	A
Crack-meters	POC	A
Infrared thermography (IRT)	PAT	C
Structural displacement		
Pendulum	POC	B
Surface extensometer	POC	A
Expansion/deformation		
Vibrating wire strain gauges	POC	A
Fiber-optic systems	POC	A/B/C (depending on the selected system or manufacturer)
Snap-ring borehole extensometer	PAT	A
Temperature		
Temperature probe (RTDs and thermocouples)	POC	A
Distributed fiber-optics system	POC	A
Young's modulus, local stiffness		
Ultrasonic pulse echo (UPE)	PAT	B
Ultrasonic pulse velocity (UPV)	POC	C
Impact-echo (IE)	POC	C
Acoustic emission (AE)	PAT	C
Promising techniques with high resolution and high sensitivity	PAT	A (promising, high sensitivity)
• Nonlinear acoustic		
• Diffusion		
• Surface waves		
Concrete moisture/humidity/water content		
RH/Capacitance probe	POC	C (specific care to avoid possible leaks)
Wood stick	POC	B
Microwave techniques: GPR	PAT	B (needs calibration on the tested concrete)
Microwave techniques: TDR	POC	B (needs calibration on the tested concrete)
Microwave techniques: open-ended coaxial probe	PAT	B (needs calibration on the tested concrete)

[a]POC: Proof-of-Concept for structural monitoring of ASR-relevant parameters on real structures
PAT: Potentially Applicable Technique for monitoring ASR-relevant parameters, but not performed with success yet at the structural level in the field.

3.7 Conclusions

The surface cracking often associated with ASR is a well-known symptom of the reaction, but its onset can occur decades after construction, though internal distress can cause significant expansion and strength loss of concrete structures. The main challenge of monitoring ASR progression in real structures is to detect those changes before any visible evidence and to adapt an adequate maintenance program.

Every project requires a carefully designed monitoring/instrumentation program to provide engineers and owners with the information needed to make decisions regarding the management of the affected structure. The extent of the program primarily depends on the complexity of the structure, the access to the structure, the severity of the distress, and the consequence of a failure of the structure.

For new construction projects, it is highly recommended to consider the installation of embedded sensors to measure temperature, relative humidity, and strains over time.

For existing structures, surface extensometers, and NDE seem to be convenient tools to support a reliable diagnosis. Many techniques and methods exist for monitoring ASR in concrete, and their costs are highly variable. They also vary in terms of accuracy, resolution, and areal/volumetric coverage. Some tools are able to provide information on only one parameter related to ASR, while others are capable of providing formation on more than one parameter. Some only provide surface measurements; others can scan huge volumes of concrete. Some techniques are better suited to diagnosing and assessing the current state of damage, while others will be better suited to monitoring the progress or rate of damage development.

Nondestructive methods measure physical parameters, such as stress wave velocity or nonlinear acoustical parameters, which are *indirectly* related to ASR, but which can also be influenced by features not associated with ASR (e.g. presence of reinforcement, cracks caused by other distress mechanisms, boundary conditions, etc.). Environmental parameters, and primarily temperature, must be monitored along with ASR parameters.

In the cases presented in this chapter, in addition to conventional visual inspection and coring programs, the most relevant parameters that have been measured were:

Surface expansion: typically measured over a period of at least three years using DEMEC-type extensometers and targets located at multiple locations on the structure;

Internal moisture/RH: portable humidity meters and borehole measurements ports are recommended;

Cracking: a systematic quantitative approach, such as the Cracking Index, is highly recommended. However, it must be considered that the observed cracks may not be caused only by ASR, even when the presence of ASR is confirmed petrographically. Other damage mechanisms, in particular freezing and thawing in cold climates, may contribute substantially to the observed cracking, as can seasonal thermal and moisture variations;

Structural movements: in the case of large structures, instruments such as extensometers or inclinometers have been used to track the movements, which can be partially or fully associated with ASR. In some cases, the misalignment of structural elements and/or effects on non structural elements can be more significant than the expansion alone of the structure;

Combination of different NDE techniques: as NDE measurements are not directly related to ASR, it is beneficial to combine information from different techniques and to use complementary sensitivities. However, in this approach, data processing is more complex as correlation between observed quantities are needed. Payan et al. [43] provides some example from the French government-funded research programs.

Although the effectiveness of nondestructive testing and evaluation methods based on acoustic or electromagnetic properties have been demonstrated in the lab, their robustness for in-situ application has to be confirmed. Only a few field cases have been reported in the literature. Some techniques based on linear acoustic properties (e.g. UPV) are easy to perform, but generally provide little useful information. Methods that measure nonlinear behavior typically are much more sensitive to damage initiation, but tend to require more sophisticated signal processing and data preparation, and their use on large-scale structures is very limited to date.

New approaches for in-situ and temporal examination of ASR in bulk samples are needed to provide the understanding of the relationships between material composition/structure, ASR gel formation, and damage propagation. This is necessary to improve testing so that the long-term durability of concrete infrastructure can be ensured. New digital sensing and imaging processing techniques may find use in a visual monitoring campaign. Indeed, to overcome the weaknesses inherent in conventional visual inspection, several studies have proposed methodologies implementing new technologies, such as terrestrial laser scanning (TLS) [132–134], unmanned aerial vehicles (UAVs) [135–141], and 3D Bridge Information Models (BrIM) [142–145].

References

1. Sausse, J., Fabre, J.: Diagnosis of dams affected by swelling reactions: lessons learned from 150 monitored concrete dams in France. In: 6th International Conference on Dam Engineering, Lisbon, Portugal, pp. 156–157 (2011)
2. Mauris, F., Martinot, F., Fabre, J., Bourgey, P., Sausse, J.: Synthesis of hydraulic structures behavior: lessons learned from monitored dams of EDF in France. In: Long-Term Behaviour and Environmentally Friendly Rehabilitation Technologies of Dams, p. 22 (2015)
3. Guilloteau, T., Martinot, F., Sausse, J.: Long-term behaviour of EDF dams regarding concrete swelling structures. In: Swelling Concrete in Dams and Hydraulic Structures, DSC 2017, p. 338 (2017)
4. FHWA: New FHWA program to combat ASR in concrete. Technical report. FHWA-HRT-07-012. Federal Highway Administration (2007)
5. Magne, I., Granger, L., Guinet, P.: Prévision du fluage et du gonflement du caisson UNGG de Saint-Laurent. Revue française de génie civil **2**(1), 123–142 (1998)

6. Naus, D.: Inspection of nuclear power plant structures-overview of methods and related applications. Technical report. Oak Ridge National Laboratory (United States). (ORNL/TM-2007/191) (2009)
7. Snyder, K., Sung, L., Cheok, G.: Nondestructive testing (NDT) and sensor technology for service life modeling of new and existing concrete structures. Technical report. NIST Interagency/Internal Report (NISTIR)-7974 (2013)
8. Thomas, M., Fournier, B., Folliard, K.: Alkali-aggregate reactivity (AAR) facts book. Technical report. FHWA-HIF-13-019. United States. Federal Highway Administration. Office of Pavement Technology (2013)
9. Stanton, T.: Expansion of concrete through reaction between cement and aggregate. Proc. ASCE **66**, 1781–1811 (1940)
10. Folliard, K., Thomas, M., Kurtis, K.: Guidelines for the use of lithium to mitigate or prevent ASR. Technical report. FHWA-RD-03-047. Federal Highway Administration (2003)
11. Walther, J., Helgeson, H.: Calculation of the thermodynamic properties of aqueous silica and the solubility of quartz and its polymorphs at high pressures and temperatures. Am. J. Sci **277**(10), 1315–1351 (1977)
12. Sjöberg, S.: Silica in aqueous environments. J. Non-Cryst. Solids **196**, 51–57 (1996)
13. Brantley, S., Kubicki, J., White, A.: Kinetics of water-rock interaction (2008)
14. Maraghechi, H.: Development and assessment of alkali activated recycled glass-based concretes for civil infrastructure. Ph.D. thesis. The Pennsylvania State University (2014)
15. Rajabipour, F., Giannini, E., Dunant, C., Ideker, J.H., Thomas, M.: Alkali-silica reaction: current understanding of the reaction mechanisms and the knowledge gaps. Cem. Concr. Res. **76**, 130–146 (2015)
16. Nakamura, E., Watanabe, H., Koga, H.: Shear resisting mechanism in RC beams with stirrups fracture due to ASR expansion. In: Proceedings of the 13th International Conference on Alkali-Aggregate Reaction in Concrete (2008)
17. Kreitman, K.: Nondestructive evaluation of reinforced concrete structures affected by ASR. MS thesis. University of Texas, Austin (2011)
18. Swamy, R., Al-Asali, M.: Effect of alkali-silica reaction on the structural behavior of reinforced concrete beams. Struct. J. **86**(4), 451–459 (1989)
19. Fournier, B., Bérubé, M.: Alkali-aggregate reaction in concrete: a review of basic concepts and engineering implications. Can. J. Civ. Eng. **27**(2), 167–191 (2000)
20. Ahmed, T., Burley, E., Rigden, S., Abu-Tair, A.: The effect of alkali reactivity on the mechanical properties of concrete. Constr Build. Mater. **17**(2), 123–144 (2003)
21. Smaoui, N., Bissonnette, B., Bérubé, M., Fournier, B., Durand, B.: Mechanical properties of ASR-affected concrete containing fine or coarse reactive aggregates. J. ASTM Int. **3**(3), 1–16 (2005)
22. Giaccio, G., Zerbino, R., Ponce, J., Batic, O.: Mechanical behavior of concretes damaged by alkali-silica reaction. Cem. Concr. Res. **38**(7), 993–1004 (2008)
23. Rivard, P., Saint-Pierre, F.: Assessing alkali-silica reaction damage to concrete with non-destructive methods: from the lab to the field. Constr. Build. Mater. **23**(2), 902–909 (2009)
24. Rivard, P., Ballivy, G., Gravel, C., Saint-Pierre, F.: Monitoring of an hydraulic structure affected by ASR: a case study. Cem. Concr. Res. **40**(4), 676–680 (2010)
25. Gallias, J.: Comparison of Damaging criteria for testing aggregates by autoclaving treatement. In: Proceedings of the 11th ICAAR, Quebec, Canada, pp. 949–958 (2000)
26. Sargolzahi, M., Kodjo, S., Rivard, P., Rhazi, J.: Effectiveness of nondestructive testing for the evaluation of alkali-silica reaction in concrete. Constr. Build. Mater. **24**(8), 1398–1403 (2010)
27. Chana, P., Korobokis, G.: The structural performance of reinforced concrete affected by alkali silica reaction: Phase II. Technical report. Contractor Report 311, Transport and Road Research Laboratory, Crowthorne, Berkshire, United Kingdom (1992)
28. Bach, F., Thorsen, T., Nielsen, M.: Load-carrying capacity of structural members subjected to alkali-silica reactions. Constr. Build. Mater. **7**(2), 109–115 (1993)

29. Monette, L.: Effects of the alkali-silica reaction on unloaded, statically loaded and dynamically loaded reinforced concrete beams. MS thesis. University of Ottawa (Canada) (1998)
30. den Uijl, J., Kaptijn, N.: Shear tests on beams cut from ASR affected bridge decks. Spec. Publ. **211**, 115–134 (2003)
31. Blight, G., Alexander, M., Schutte, W., Ralph, T.: The effect of alkali aggregate reaction on the strength and deformation of a reinforced concrete structure. In: Proceedings of the Sixth International Conference on Alkali-Aggregate Reaction, pp. 401–410 (1983)
32. Imai, H., Yamasaki, T., Maehara, H., Miyagawa, T.: The deterioration by alkali-silica reaction of Hanshin expressway concrete structures-investigation and repair. In: Proceedings of The 7th International Conference on Concrete Alkali-Aggregate Reactions, Ottawa, Canada (1986). Noyes Publications Publications (1987)
33. Blight, G., Alexander, M., Ralph, T., Lewis, B.: Effect of alkali-aggregate reaction on the performance of a reinforced concrete structure over a six-year period. Mag. Concr. Res. **41**(147), 69–77 (1989)
34. Boenig, A.: Bridges with premature concrete deterioration: field observations and large-scale testing. MS thesis. University of Texas, Austin (2001)
35. Deschenes, D.: ASR/DEF-damaged bent caps: shear tests and field implications. MS thesis. University of Texas (2009)
36. Larson, N.: Structural performance of ASR/DEF damaged prestressed concrete trapezoidal box beams with dapped ends. MS thesis (2010)
37. Mikata, Y., Shimazu, Y., Hatano, Y., Inoue, S.: Flexural and shear capacity of PRC beams damaged by combined deterioration due to ASR and corrosion. In: Proceedings of the 14th International Conference on Alkali-Aggregate Reaction in Concrete, Austin, TX (2012)
38. Webb, Z.: Experimental investigation of ASR/DEF-induced reinforcing bar fracture. MS thesis. University of Texas at Austin (2011)
39. Miyagawa, T., Seto, K., Sasaki, K., Mikata, Y., Kuzume, K., Minami, T.: Fracture of reinforcing steels in concrete structures damaged by alkali-silica reaction - field survey, mechanism and maintenance. J. Adv. Concr. Technol. **4**(3), 339–355 (2006)
40. Torii, K., Wasada, S., Sasatani, T., Minato, T.: A survey on ASR-affected bridge piers with fracture of steel bars on Noto expressway. In: Proceedings of 13th International Conference on Alkali-aggregate Reaction in Concrete, pp. 1304–1311 (2008)
41. Ichikawa, T., Miura, M.: Modified model of alkali-silica reaction. Cem. Concr. Res. **37**(9), 1291–1297 (2007)
42. Ostertag, C., Yi, C., Monteiro, P.: Effect of confinement on properties and characteristics of alkali-silica reaction gel. ACI Mater. J. **104**(3), 276 (2007)
43. Payan, C., Abraham, O., Garnier, V.: 2—Ultrasonic methods. In: Balayssac, J.-P., Garnier, V. (eds.) Non-Destructive Testing and Evaluation of Civil Engineering Structures, pp. 21–85. Elsevier (2018)
44. Nilsson, L.: Pop-outs due to alkali-silica reaction - a moisture problem? In: Proceeding 5th International Conference on AAR in Concrete, Cape Town-South Africa, p. S252 (1981)
45. Nilsson, L.: Moisture effects on the alkali-silica reaction. In: Proceedings of the 6th International Conference on AAR in Concrete, Copenhagen, Denmark, pp. 201–208 (1983)
46. Stark, D.: The moisture condition of field concrete exhibiting alkali-silica reactivity. Spec. Publ. **126**, 973–988 (1991)
47. Pedneault, A.: Development of testing and analytical procedures for the evaluation of the residual potential of reaction, expansion, and deterioration of concrete affected by ASR. MA thesis. Laval University (1996)
48. Courtois, A., Taillade, F., Moreau, G., Clauzon, T., Skoczylas, F., Masson, B.: Water content monitoring for nuclear concrete buildings: needs, feedback and perspectives. In: 5th BIOT Conference on Poromechanics, pp. 10–12 (2013)
49. Jensen, V.: Relative humidity measured by wooden stick method in concrete structures: longterm measurements and reduction of humidity by surface treatment. SP 212-39, Proceedings, Sixth CANMET/ACI Conference on Durability of Concrete. In: ACI International Conferences, Thessaloniki, Greece, pp. 621–636 (2003)

50. Quincot, G., Azenha, M., Barros, J., Faria, R.: State of the art-methods to measure moisture in concrete. In: Projetos De Investigação Científica E Desenvolvimento Tecnológico, Portugal (2011)
51. Oxfall, M.: Climatic conditions inside nuclear reactor containments: evaluation of moisture condition in the concrete within reactor containments and interaction with the ambient compartments (2016)
52. Thomas, M., Folliard, K., Fournier, B., Rivard, P., Drimalas, T., Garber, S.: Methods for evaluating and treating ASR-affected structures: results of field application and demonstration projects-Volume II: Details of field applications and analysis. Technical report. Report No. FHWA-HIF-14-0003. Federal Highway Administration (2013)
53. Andrade, C., Sarria, J., Alonso, C.: Relative humidity in the interior of concrete exposed to natural and artificial weathering. Cem. Concr. Res. **29**(8), 1249–1259 (1999)
54. Grasley, Z.: Internal relative humidity, drying stress gradients, and hygrothermal dilation of concrete. PhD thesis. University of Illinois at Urbana-Champaign (2003)
55. Hedenblad, G.: Measurement of moisture in high performance concrete. In: 5th International Symposium on Utilization of High Strength/High performance Concrete, vol. 2, pp. 1124–1133 (1999)
56. Giannini, E.R.: Field studies of mitigation strategies for alkali-silica reaction in hardened concrete. MS thesis. University of Texas at Austin (2009)
57. Deschenes, Jr., R., Giannini, E.R., Drimalas, T., Fournier, B., Hale, W.M.: Mitigating alkali-silica reaction and freezing and thawing in concrete pavement by silane treatment. ACI Mater. J. **115**(5) (2018)
58. Adous, M., Quéffélec, P., Laguerre, L.: Coaxial/cylindrical transition line for broadband permittivity measurement of civil engineering materials. Measur. Sci. Technol. **17**(8), 2241 (2006)
59. Laurens, S., Balayssac, J.-P., Rhazi, J., Arliguie, G.: Influence of concrete relative humidity on the amplitude of ground-penetrating radar (GPR) signal. Mater. Struct. **35**(4), 198–203 (2002)
60. Sbartaï, Z., Laurens, S., Rhazi, J., Balayssac, J., Arliguie, G.: Using radar direct wave for concrete condition assessment: correlation with electrical resistivity. J. Appl. Geophys. **62**(4), 361–374 (2007)
61. Al-Qadi, I., Riad, S., Mostaf, R., Su, W.: Design and evaluation of a coaxial transmission line fixture to characterize Portland cement concrete. Constr. Build. Mater. **11**(3), 163–173 (1997)
62. Pavlík, Z., Jiřičková, M., Černỳ, R., Sobczuk, H., Suchorab, Z.: Determination of moisture diffusivity using the time domain reflectometry (TDR) method. J. Build. Phys. **30**(1), 59–70 (2006)
63. Coyne, A.: Quelques résultats d'auscultation sonore sur les ouvrages en béton, béton armé ou métal. In: Annales ITBTP, Juillet-Août (1938)
64. Oukhemanou, E., Desforges, S., Buchoud, E., Michel-Ponnelle, S., Courtois, A.: VeRCoRs mock-up: comprehensive monitoring system for reduced scale containment model. In: Proceedings of the Technical Innovation in Nuclear Civil Engineering (TINCE), Paris, France (2016)
65. Filali, B., Boone, F., Rhazi, J., Ballivy, G.: Design and calibration of a large open-ended coaxial probe for the measurement of the dielectric properties of concrete. IEEE Trans. Microw. Theory Tech. **56**(10), 2322–2328 (2008)
66. Guihard, V., Haelewyn, J., Courtois, A., Taillade, F., Balayssac, J., Placko, D., Sanahuja, J.: Relevance and means for measuring water content in concrete structures. An illustration with PWR concrete containment. In: TINCE 2018 Conference Proceedings, Paris (2018)
67. Hashemi, A., Horst, M., Kurtis, K., Donnell, K.M., Zoughi, R.: Comparison of alkali-silica reaction gel behavior in mortar at microwave frequencies. IEEE Trans. Instrum. Measur. **64**(7), 1907–1915 (2015)
68. Freeseman, K., Hoegh, K., Khazanovich, L.: Characterization of concrete at various freeze-thaw damage conditions using SH-waves. AIP Conf. Proc. (AIP Publishing) **1706**(1), 020017 (2016)

69. Roctest: Instruction manual. Direct and inverted pendulums. Models DP and RP. Technical report. E10065060901. Roctest Ltd. (2006)
70. Abrishami, H., Tcherner, J., Barre, F., Borgerhoff, M., Bumann, U., Calonius, K., [illegible], A., Debattista, J.-M., Gallitre, E., Isard, C., [illegible], O., Graves, H., Sircar, M., Huerta, A., White, A., [illegible], P., Kjellin, D., Lillhoek, S., Louhivirta, J., Myllymaeki, J., Vaelikangas, P., Martin, J., Nakano, M., Puttonen, J., Rambach, J.-M., Tarallo, F., Smith, L., Stepan, J., Touret, J.-P., Varpasuo, P.: Bonded or unbonded technologies for nuclear reactor prestressed concrete containments. Technical report. NEA-CSNI-R–2015-5 Nuclear Energy Agency of the Organisation for Economic Co-Operation and Development (2015)
71. Smith, L., Brodt, G., Stafford, B.: Performance assessment and reinstatement of vibrating wire strain gauges in nuclear power plant structures. In: Structural Mechanics in Reactor Technology, vol. 16, Washington DC, USA (2001)
72. Neild, S., Williams, M., McFadden, P.: Development of a vibrating wire strain gauge for measuring small strains in concrete beams. Strain **41**(1), 3–9 (2005)
73. Simon, A., Courtois, A.: Structural monitoring of prestressed concrete containments of nuclear power plants for ageing management. Struct. Mech. React. Technol. **21**, 6–11 (2011)
74. McRae, J.B.: Borehole strainmeter. U.S. Patent No 5,585,555 (1996)
75. Faisal, H., Lee, S.: New pile instrumentation technique for driven and jacked-in prestressed spun concrete piles (2008)
76. EPRI: Program on technology innovation: retrofitted sensors for nuclear containment structures. Technical report. 3002007819. Electric Power Research Institute, Palo Alto, CA (2016)
77. Henault, J., Laviron, P., Desforges, S., Vautrin, D., Courtois, A., Benjamin, M., Legrix, A.: How to characterize the airtightness of containment structures. Overview of monitoring techniques tested on VeRCoRs Mock In: TINCE 2018 Conference Proceedings, Paris (2018)
78. Martinot, F., Martin, G., Henault, J., Moreau, G., Buchoud, E., Beck, Y., Courtois, A.: Distributed fiber optic sensors in concrete of a nuclear containment. In: Proceedings of the International Conference on Structural Health Monitoring of Intelligent Infrastructure, p. 3 (2015)
79. Geokon: Model 1200 (Model A4) borehole extensometer; specification. Product data sheet. Geokon (2002) https://www.geokon.com/content/datasheets/A3_A4_A5_Borehole_Extensometers.pdf
80. Geokon: Model 1200 (Model A4) borehole extensometer; instruction manual. Technical report. Geokon (2002) https://www.geokon.com/content/manuals/1200_Borehole_Exto_Model_A4.pdf
81. MPR: MPR-4273, Revision 0, Seabrook Station—implications of large-scale test program results on reinforced concrete affected by alkali-silica reaction, July 2016 (ML16216A242) (2016). Online; Accessed 15 Oct 2018
82. NextEra Energy: Seabrook, License Amendment Request 16-03—revise current licensing basis to adopt a methodology for the analysis of seismic category I structures with concrete affected by alkali-silica reaction (2016). Online. http://www.nrc.gov/docs/ML1621/ML16216A240.pdf. Accessed 11 July 2018
83. Amberg, F., Droz, P., Leroy, R., Bremen, R.: Rapport du Comité suisse des barrages CSB sur l'état de la problématique de l'expansion du béton dans les barrages Suisses. Technical report. Swiss Committee On Large Dams (2017)
84. Fournier, B., Ideker, J.H., Folliard, K., Thomas, M., Nkinamubanzi, P., Chevrier, R.: Effect of environmental conditions on expansion in concrete due to alkali-silica reaction (ASR). Mater. Charact. **60**(7), 669–679 (2009)
85. Folliard, K.J., Barborak, R., Drimalas, T., Du, L., Garber, S., Ideker, J., Ley, T., Williams, S., Juenger, M., Fournier, B., Thomas, M.D.A.: Preventing ASR/DEF in new concrete Final Report. Technical report. FHWA/TX-06/0-4085-5). Center for Transportation Research
86. Mathieu, J.-P., Charpin, L.,Sémété, P., Toulemonde, C., Boulant, G., Hénault, J.-M., Taillade, F.: Temperature and humidity-driven ageing of the VeRCoRs mock-up. In: Computational Modelling of Concrete Structures: Proceedings of the Conference on Computational Modelling of Concrete and Concrete Structures (EURO-C 2018), February 26–March 1, 2018, Bad Hofgastein, Austria, p. 215. CRC Press (2018)

87. ACI Committee 228: ACI 228.2R-13, Report on nondestructive test methods for evaluation of concrete in structures. American Concrete Institute, Farmington Hills, Michigan, USA (2013)
88. Metalssi, O., Godart, B., Toutlemonde, F.: Effectiveness of nondestructive methods for the evaluation of structures affected by internal swelling reactions: a review of electric, seismic and acoustic methods based on laboratory and site experiences. Exp. Tech. **39**(2), 65–76 (2015)
89. Giannini, E., Folliard, K., Zhu, J., Bayrak, O., Kreitman, K., Webb, Z., Hanson, B.: Non-destructive evaluation of in-service concrete structures affected by alkali-silica reaction (ASR) or delayed ettringite formation (DEF)-final report, Part I. Technical report. FHWA/TX-13/0-6491-1. Federal Highway Administration (2013)
90. Naik, T., Malhotra, V., Popovics, J.: The ultrasonic pulse velocity method. In: Handbook on Nondestructive Testing of Concrete, pp. 182–200. CRC Press (2003)
91. Wiggenhauser, H., Samokrutov, A., Mayer, K., Alekhin, S., Elkin, V.: LAUS—large aperture ultrasonic system-for testing thick concrete structures. In: Proceedings of the International Symposium in Non-destructive Testing in Civil Engineering (NDT-CE), Berlin, Germany (2015)
92. Clayton, D., Khazanovich, L., Salles, L.: Linear array ultrasonic test results from alkali-silica reaction (ASR) specimens. In: ORNL/TM-2016/159, Oak Ridge National Laboratory (2016)
93. Khazanovich, L., Freeseman, K., Salles, L., Clayton, D.: Nondestructive analysis of alkali-silica reaction damage in concrete slabs using shear waves. IAIP Conf. Proc. (AIP Publishing) **1949**(1), 040003 (2018)
94. Henriksen, C.: Impact-echo testing. Concr. Int. **17**(5), 55–58 (1995)
95. Mangual, J., ElBatanouny, M., Ziehl, P., Matta, F.: Corrosion damage quantification of prestressing strands using acoustic emission. J. Mater. Civ. Eng. **25**(9), 1326–1334 (2012)
96. Abdelrahman, M., ElBatanouny, M., Ziehl, P., Fasl, J., Larosche, C., Fraczek, J.: Classification of alkali-silica reaction damage using acoustic emission: a proof-of-concept study. Constr. Build. Mater. **95**, 406–413 (2015)
97. Pour-Ghaz, M., Spragg, R., Castro, J., Weiss, J.: Can acoustic emission be used to detect alkali silica reaction earlier than length change tests? In: 14th International Conference On Alkali Aggregate Reaction in Concrete, Austin, TX (2012)
98. Soltangharaei, V., Anay, R., Hayes, N., Assi, L., Le Pape, Y., Ma, Z., Ziehl, P.: Damage mechanism evaluation of large-scale concrete structures affected by alkali-silica reaction using acoustic emission. Appl. Sci. **8**(11), 2148 (2018)
99. Pourahmadian, F., Guzina, B., Haddar, H.: Generalized linear sampling method for elastic-wave sensing of heterogeneous fractures. Inverse Probl. **33**(5), 055007 (2017)
100. Pourahmadian, F., Haddar, H.: Differential imaging of evolution in elastic domains with unknown micro/macrostructure. SIAM J. Imaging Sci. (2018)
101. Buck, O., Morris, W., Richardson, J.: Acoustic harmonic generation at unbonded interfaces and fatigue cracks. Appl. Phys. Lett. **33**(5), 371–373 (1978)
102. Kim, G., Park, S., Kim, J., Kurtis, K., Hayes, N., Jacobs, L.: Nonlinear Rayleigh surface waves to characterize microscale damage due to alkali-silica reaction (ASR) in full-scale, nuclear concrete specimens. Construction and Building Materials **186**, 1114–1118 (2018)
103. Guyer, R., McCall, K., Van Den, K.: Slow elastic dynamics in a resonant bar of rock. Geophys. Res. Lett. **25**(10), 1585–1588 (1998)
104. Bentahar, M., El Aqra, H., El Guerjouma, R., Griffa, M., Scalerandi, M.: Hysteretic elasticity in damaged concrete: quantitative analysis of slow and fast dynamics. Phys. Rev. B **73**(1), 014116 (2006)
105. Nagy, P.: Fatigue damage assessment by nonlinear ultrasonic materials characterization. Ultrasonics **36**(1–5), 375–381 (1998)
106. Guyer, R., Johnson, P.: Nonlinear mesoscopic elasticity: evidence for a new class of materials. Phys. Today **52**, 30–36 (1999)
107. Van Den Abeele, K., De Visscher, J.: Damage assessment in reinforced concrete using spectral and temporal nonlinear vibration techniques. Cem. Concr. Res. **30**(9), 1453–1464 (2000)

108. Kim, J., Jacobs, L., Qu, J., Littles, J.: Experimental characterization of fatigue damage in a nickel-base superalloy using nonlinear ultrasonicwaves. J. Acoust. Soc. Am. **120**(3), 1266–1273 (2006)
109. Chen, X., Kim, J., Kurtis, K., Qu, J., Shen, C., Jacobs, L.: Characterization of progressive microcracking in Portland cement mortar using nonlinear ultrasonics. NDT & E Int. **41**(2), 112–118 (2008)
110. Van Den Abeele, K., Johnson, P., Sutin, A.: Nonlinear elastic wave spectroscopy (NEWS) techniques to discern material damage, Part I: Nonlinear wave modulation spectroscopy (NWMS). J. Res. Nondestr. Eval. **12**(1), 17–30 (2000)
111. Kodjo, A., Rivard, P., Cohen-Tenoudji, F., Gallias, J.: Evaluation of damages due to alkali-silica reaction with nonlinear acoustics techniques. In: Proceedings of Meetings on Acoustics, vol. 7, no. 1, p. 045003. ASA (2009)
112. Van Den Abeele, K., Carmeliet, J., Ten Cate, J., Johnson, P.: Nonlinear elastic wave spectroscopy (NEWS) techniques to discern material damage, Part II: Single-mode nonlinear resonance acoustic spectroscopy. J. Res. Nondestr. Eval. **12**(1), 31–42 (2000)
113. Chen, J., Kim, J., Kurtis, K., Jacobs, L.: Theoretical and experimental study of the nonlinear resonance vibration of cementitious materials with an application to damage characterization. J. Acoust. Soc. Am. **130**(5), 2728–2737 (2011)
114. Leśnicki, K.J., Kim, J., Kurtis, K., Jacobs, L.: Characterization of ASR damage in concrete using nonlinear impact resonance acoustic spectroscopy technique. NDT & E Int. **44**(8), 721–727 (2011)
115. Sanchez, L., Fournier, B., Jolin, M., Bastien, J.: Evaluation of the stiffness damage test (SDT) as a tool for assessing damage in concrete due to ASR: input parameters and variability of the test responses. Constr. Build. Mater. **77**, 20–32 (2015)
116. Quiviger, A., Payan, C., Chaix, J., Garnier, V., Salin, J.: Effect of the presence and size of a real macro-crack on diffuse ultrasound in concrete. NDT & E Int. **45**(1), 128–132 (2012)
117. Quiviger, A., Girard, A., Payan, C., Chaix, J., Garnier, V., Salin, J.: Influence of the depth and morphology of real cracks on diffuse ultrasound in concrete: a simulation study. NDT & E Int. **60**, 11–16 (2013)
118. Larose, E., Hall, S.: Monitoring stress related velocity variation in concrete with a 2_ 10–5 relative resolution using diffuse ultrasound. J. Acoust. Soc. Am. **125**(4), 1853–1856 (2009)
119. Larose, E., Obermann, A., Digulescu, A., Planes, T., Chaix, J., Mazerolle, F., Moreau, G.: Locating and characterizing a crack in concrete with diffuse ultrasound: a four-point bending test. J. Acoust. Soc. Am. **138**(1), 232–241 (2015)
120. Payan, C., Garnier, V., Moysan, J., Johnson, P.: Determination of third order elastic constants in a complex solid applying coda wave interferometry. Appl. Phys. Lett. **94**(1), 011904 (2009)
121. Kaczmarek, M., Piwakowski, B., Drelich, R.: Noncontact ultrasonic nondestructive techniques: state of the art and their use in civil engineering. J. Infrastruct. Syst. **23**(1), B4016003 (2016)
122. Kim, G., Giannini, E., Klenke, N., Kim, J., Kurtis, K., Jacobs, L.: Measuring alkali-silica reaction (ASR) microscale damage in large-scale concrete slabs using nonlinear Rayleigh surface waves. J. Nondestr. Eval. **36**(2), 29 (2017)
123. Thomas, M., Folliard, K., Fournier, B., Rivard, P., Drimalas, T.: Methods for evaluating and treating ASR-affected structures: results of field application and demonstration projects-Volume I: Summary of findings and recommendations. Technical report. Report No. FHWA-HIF-14-0002. Federal Highway Administration (2013)
124. NRC: In situ Monitoring of Alkali-Silica Reaction (ASR) affected concrete: a study on crack indexing and damage rating index to assess the severity of ASR and to monitor ASR progression. Technical report. ML13108A047. US-Nuclear Regulatory Commission, Position paper (2013)
125. Godart, B., Fasseu, P., Michel, M.: Diagnosis and monitoring of concrete bridges damaged by AAR in Northern France. In: The Ninth International Conference On Alkali-Aggregate Reaction In Concrete, London, pp. 368–375 (1992)

126. Kabir, S., Rivard, P., Ballivy, G.: Neural-network-based damage classification of bridge infrastructure using texture analysis. Can. J. Civ. Eng. **35**(3), 258–267 (2008)
127. Clayton, D., Cyrus, S.: Research in nondestructive evaluation techniques for nuclear concrete structures. In: 40th Annual Review of Progress in Quantitative Nondestructive Evaluation, AIP Publishing, AIP Conference Proceedings, pp. 962–969 (2014)
128. Kurtis, K., Xi, Y., Glinicki, M., Provis, J., Giannini, E., Fu, T.: Can we design concrete to survive nuclear environments? Concr. Int. **39**, 29–35 (2017)
129. Gocevski, V.: Pathologies/degradation mechanisms experienced by Hydro-Quebec during the evaluation of Gentilly-2 NPP (2015)
130. Danay, A., Adeghe, L., Hindy, A.: Diagnosis of the cause of the progressive concrete deformations at Saunders dam. Concr. Int. **15**(9), 25–33 (1993)
131. Gross, M., Vozel, J., Mochrie, B., Schadinger, S., Shiers, P.: Remediation measures implemented to resolve gate operation difficulties related to spillway deck concrete expansion. In: 31st Annual USSD Conference, San Diego, CA, USA, pp. 705–720 (2011)
132. Conde-Carnero, B., Riveiro, B., Arias, P., Caamaño, J.: Exploitation of geometric data provided by laser scanning to create FEM structural models of bridges. J. Perform. Constr. Facil. **30**(3), 04015053 (2015)
133. Truong-Hong, L., Falter, H., Lennon, D., Laefer, D.: Framework for bridge inspection with laser scanning. In: EASEC-14 Structural Engineering and Construction, Ho Chi Minh City, Vietnam, 6–8 January 2016 (2016)
134. Turkan, Y., Laflamme, S., Tan, L.: Terrestrial laser scanning-based bridge structural condition assessment, pp. 1–38 (2016)
135. Ellenberg, A., Branco, L., Krick, A., Bartoli, I., Kontsos, A.: Use of unmanned aerial vehicle for quantitative infrastructure evaluation. J. Infrastruct. Syst. **21**(3), 04014054 (2014)
136. Chan, B., Guan, H., Jo, J., Blumenstein, M.: Towards UAV-based bridge inspection systems: a review and an application perspective. Struct. Monitor. Maint. **2**(3), 283–300 (2015)
137. Irizarry, J., Costa, D.: Exploratory study of potential applications of unmanned aerial systems for construction management tasks. J. Manag. Eng. **32**(3), 05016001 (2016)
138. Mader, D., Blaskow, R., Westfeld, P., Weller, C.: Potential of UAV-based laser scanner and multispectral camera data in building inspection. Inte. Arch. Photogramm. Remote Sens. Spat. Inf. Sci. **41**, 1135–1142 (2016)
139. Gillins, M.: Unmanned aircraft systems for bridge inspection: testing and developing end-to-end operational workflow (2016)
140. Gheisari, M., Irizarry, J.: A user-centered approach to investigate unmanned aerial system (UAS) requirements for a department of transportation applications. In: Conference on Autonomous and Robotic Construction of Infrastructure, p. 85 (2015)
141. Seo, J., Duque, L., Wacker, J.: Drone-enabled bridge inspection methodology and application. Autom. Constr. **94**, 112–126 (2018)
142. Al-Shalabi, F., Turkan, Y., Laflamme, S.: BrIM implementation for documentation of bridge condition for inspection. In: Proceedings of the Canadian Society for Civil Engineering 5th International/11th Construction Specialty Conference,University of British Columbia, Vancouver, Canada, June 2015, pp. 7–10 (2015)
143. McGuire, B., Atadero, R., Clevenger, C., Ozbek, M.: Bridge information modeling for inspection and evaluation. J. Bridge Eng. **21**(4), 04015076 (2016)
144. Costin, A.: A new methodology for interoperability of heterogeneous bridge information models. Ph.D thesis. Georgia Institute of Technology (2016)
145. McKenna, T., Minehane, M., O'Keeffe, B., O'Sullivan, G., Ruane, K.: Bridge information modelling (BrIM) for a listed viaduct. In: Proceedings of the Institution of Civil Engineers-Bridge Engineering, vol. 170, no. 3, pp. 192–203. Thomas Telford Ltd. (2016)

Chapter 4
Monitoring of Dams Suffering from ASR at the Bureau of Reclamation

Jerzy Salamon, William Dressel, and Daniel Liechty

4.1 Introduction

Field monitoring of concrete dams affected by alkali-silica reaction is an essential component of the Safety of Dams evaluation process. Visual observations, instrumentation data recorded in the field, and laboratory concrete testing results combined with structural analyses results are key factors in understanding the past behavior of a dam, as well as assessing the structural integrity of a dam given the current conditions. These activities are critical for predicting the future performance of the structure. Monitoring of dam behavior is not an easy task due to the large size of the structure, limited access to the dam face on the reservoir side, difficult access to the exposed parts of the dam, variation in the material properties, and complexity of the loads acting on the structure. Consequently, the monitoring methods that are successfully applied at other civil structures, bridges, or buildings may not be as easily implemented for concrete dams.

Various techniques have been implemented by the Bureau of Reclamation (Reclamation) for acquiring information and collecting field data on concrete dams affected by ASR. Some of the techniques are typical and commonly used in everyday monitoring of dams; however, other methods are new and have been developed specifically for dams affected by ASR. Reclamation's Dam Safety office has implemented an intensive monitoring program for selected concrete dams affected by ASR. The pri-

J. Salamon (✉) · W. Dressel · D. Liechty
Waterways & Concrete Dams, US Bureau of Reclamation, Denver, CO, USA
e-mail: JSalamon@usbr.gov

W. Dressel
e-mail: WDressel@usbr.gov

D. Liechty
e-mail: DLiechty@usbr.gov

© RILEM 2021
V. E. Saouma (ed.), *Diagnosis & Prognosis of AAR Affected Structures*,
RILEM State-of-the-Art Reports 31,
https://doi.org/10.1007/978-3-030-44014-5_4

mary goal of this pilot program is to develop and validate monitoring techniques specific for concrete dams affected by ASR and then to use the data in the structural condition assessment for the structures.

In general, the methods used to monitor concrete dams affected by ASR can be categorized as follows: visual inspection, unmanned aerial systems (UAS) photogrammetry, monitoring equipment operations, collecting project data, recording dam deformations, crack mapping, nondestructive geophysical surveying, comparison of nondestructive geophysical surveying, and concrete coring. This comprehensive approach to dam monitoring, as currently employed by Reclamation, is described in the following sections of this report.

4.2 Visual Inspection

Observations made during regular visual inspections of dams often provide the first indication of concrete expansion caused by ASR. Typical observed anomalies include relative displacements between monoliths at the dam crest Fig. 4.1a); cracking at the dam crest, spillway piers, and dam galleries (Fig. 4.1b); spalling along lift lines and contraction joints (Fig. 4.1c); and concrete deterioration at a dam face (Fig. 4.1d).

(a) Relative displacement between dam monoliths

(b) Various types of cracks at a dam crest, spillway pier, and dam galley

(c) Spalling along a lift line (left) and a vertical contraction joint (right)

(d) Concrete deterioration on a dam face

Fig. 4.1 Examples of ASR induced damages

4.3 Unmanned Aerial Systems Photogrammetry (UAS)

High-resolution photogrammetric data collection at dam sites using UAS is a technology that produces highly precise information about the physical condition of the structure. When this data is used in time-based comparisons, deterioration of the concrete can be quantitatively and qualitatively ascertained.

Reclamation has started using high-resolution photogrammetry technology in the inspections of dams with UAS. The UAS provided significant improvements in the quality of the visual inspection methods and an easier access to inaccessible parts of the dams. One of the challenges to producing high-resolution photogrammetry is positioning the camera at the most advantageous points required for high accuracy. The recent adoption of UAS provide a solution to this problem by allowing placement of a camera at any location and orientation relative to a dam. Another benefit of UAS is automated positioning, which increases repeatability and reliability.

High resolution photos (approximately 3 mm [mm] per pixel) taken from 9- to 12-m (m) distance are used to create a three-dimensional (3D) dam model with high resolution texture that shows the actual condition of the concrete face, including any surface cracks and/or defects, Fig. 4.2 (the map consists of about 10^9 points, above,

Fig. 4.2 3D map of an arch dam developed using UAS photogrammetry

and the local enlargement of the image of a single 1.5-m-deep concrete lift, below). Performing photogrammetric surveys using UAS on a regular, recurring time interval allows for detailed tracking of concrete deterioration over large portions of the dam.

4.4 Monitoring Equipment Operations

Concrete expansion can cause damage to the embedded parts of mechanical equipment at a dam and, if significant, can limit or even prevent operation of the equipment. Figures 4.3 and 4.4 show examples of various mechanical features that have been damaged or had their operation affected by concrete expansion associated with ASR.

Fig. 4.3 Damage to embedded mechanical equipment: fracture of a gantry crane rail (left), misalignment of the penstock bulkhead guide (center), and crack around the intake trashrack (right)

Fig. 4.4 Changes in spillway gate side seal gap: clearance between seal guard and the pier face per original design (left); gap closure (gate binding) (center); and the resulting ASR-related pier expansion (right)

4.5 Collecting Project Data

Continuous collection of project data is a common practice used in monitoring concrete dams. Among the several types of data acquired, temperature measurements, reservoir water surface level fluctuations, and seepage through the dam provide the primary information used in assessing the structural conditions of dams affected by ASR.

4.5.1 Temperature Measurements

Seasonal ambient temperature changes, internal concrete temperature of the dam, and reservoir water temperature at various elevations provide critical data for modeling concrete expansion caused by temperature change and chemical reactions.

Depending on the orientation of the dam, the ambient temperature on the upstream and downstream side of the dam can vary, due to differences in sunlight exposure. In such cases, it may be valuable to obtain temperature data near the dam surface using either thermistors embedded in concrete (at approximately 0.3-m depth) or temperature gauges installed in a core hole (at approximately 0.3- to 0.6-m depth). A column of several thermistors installed in a drill hole or in dam galleries can be used to measure the internal concrete temperature along a larger depth of concrete at specific depth intervals.

4.5.2 Reservoir Water Surface Level Fluctuations

Continuous monitoring of the reservoir water surface elevation is a common practice at dams and hydro-electric projects. There are several different types of reservoir level gauges capable of providing real-time information. Maintaining a historical record of the reservoir water surface elevation at a dam or hydro-electric project is standard practice. The loading and response of a dam to reservoir elevation is a critical component when analyzing the structural integrity of dams affected by ASR.

4.5.3 Moisture Monitoring

Observed seepage at the downstream face of the dam, drain flows in galleries, and gauges measuring uplift pressure in relief wells indicate the potential for high moisture content in dam concrete, as well as signal the existence of potential cracks in the dam that could be causing seepage. The amount of seepage and moisture in a concrete dam often correlates to the reservoir level fluctuation and rainfall.

Moisture content in concrete can also be determined from concrete cores in a laboratory environment; however, moisture content results obtained using this method may not be representative of the in-situ moisture conditions. The time associated with core preparation and packing, transportation to the laboratory, storage location, and storage duration before testing all can affect moisture content results and should be considered.

Drones equipped with infrared thermography sensors can also be used to detect traces of moisture in a dam. As all objects with a temperature above absolute zero emit infrared radiation, thermal images can be developed using infrared thermography that differentiate the heat signatures from the concrete and reservoir seepage.

4.6 Recording Dam Deformations

Precise monitoring of a dam deformation and its surrounding rock foundation is a key activity that enables a good understanding of the actual behavior of dams affected by ASR. Reclamation has implemented various techniques, primarily to better capture the complex deformations of a dam structure and to cross-check the correctness and accuracy of each implemented measurement method. Dams affected by ASR can demonstrate nonuniform deformations; therefore, it is often necessary to implement multiple measurement methods to fully understand a dam's movement. Additionally, it is important to consider that many dams are located in harsh climates. As a result, the measurement instruments/gauges are often exposed to extreme atmospheric conditions for extended periods of time and may stop functioning or deliver erroneous data.

4.6.1 *Structural Measurement Points*

Vertical displacement and horizontal deflections are measured at multiple locations along the length of the crest and on the downstream face of a dam using various survey techniques including photogrammetry, Light Detection and Ranging (LIDAR), or land survey equipment. Survey prisms with high accuracy and precision have been installed on the crest and the downstream face of the dam such that very minor movements can be detected (Fig. 4.5). Surveys are usually performed once or twice every year for dams affected by ASR.

Collimation points are survey monuments measured in the following manner. A line of sight is established between an instrument and a sighting target control point. Changes in the lateral distances of the collimation survey points, from the lines of sight from survey to survey, indicate a measure of the movement of the points from the initial location relative to the line of sight. These measurements indicate the upstream/downstream deflections and long-term deformations of the dam crest relative to the abutments.

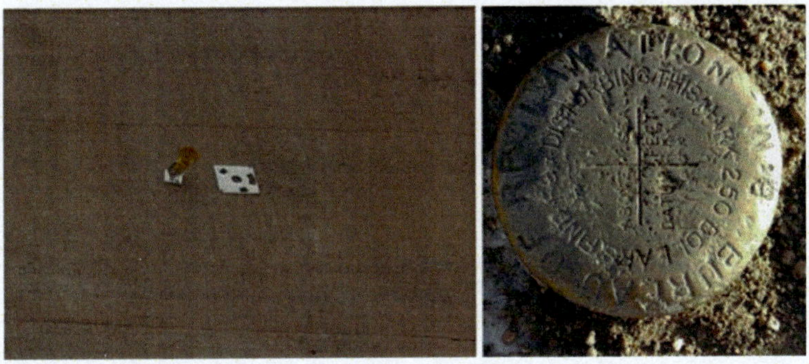

Fig. 4.5 Survey prism adjacent to photogrammetry target (left) and survey monument (right)

4.6.2 *Rope Supported Verticality Surveys*

Rope supported verticality surveys can be performed on either the upstream or downstream side of a dam. The survey itself consists of hanging a plumbline on the face of the dam and measuring the distance from the line to the face of the dam at distinct locations.

4.6.3 *Plumblines*

Typical plumbline installation consists of anchoring a strand of wire at the crest of a dam, suspending it through the dam in a plumbline well approximately 250 mm, and tensioning it with a cylindrical weight at its bottom (just below the base reading station), which is immersed in an oil-filled container to dampen vibrations and movements. For each plumbline installation, relative movements of the wire over time can be monitored at the reading station. In addition to deformations associated with ASR, both changes in reservoir elevation and changes in the temperature of a dam's concrete can cause dam deflections that would be detected by plumbline instruments.

4.6.4 *Tape Extensometers*

Tape extensometers can be installed on a dam crest to monitor deterioration and expansion of the concrete or open cracks. Pairs of 0.3-m-long, stainless steel anchors are installed on both sides of the dam crest at a fixed distance. Distance measurements between each pair of anchors are measured with a steel tape (Fig. 4.6).

Fig. 4.6 Protective casing and closeup of stainless steel tape extensometer anchor

4.6.5 Joint and Crack Meters

Vibrating-wire jointmeters installed across cracks in the rock abutments provide insight into changes to the loading, stress distributions, and abutment reactions (Fig. 4.7, left).

Whittemore gauge points and Avongard crack meters can be used at select cracks or joints to measure changes in openness of the crack or joint due to movement of the dam (Fig. 4.7, right).

4.6.6 Multiple Point Borehole Extensometers

Multiple point (MPBX) installed through a dam crest anchored at various elevations can provide insight into the overall deformations of a dam by measuring movement at distinct points along the borehole (Fig. 4.8).

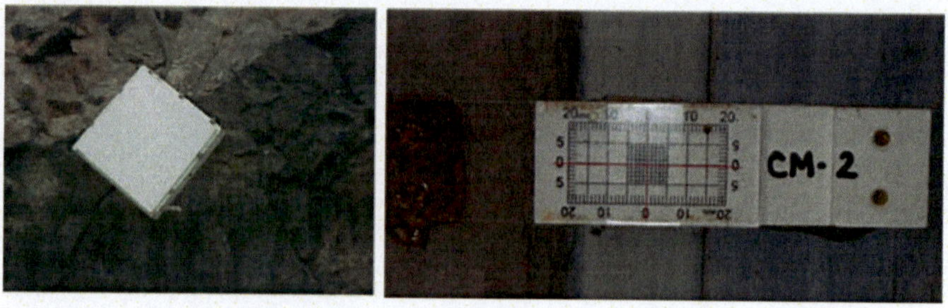

Fig. 4.7 Joint meters within a protective case installed on an abutment (left) and Avongard crack meter at contraction joint (right)

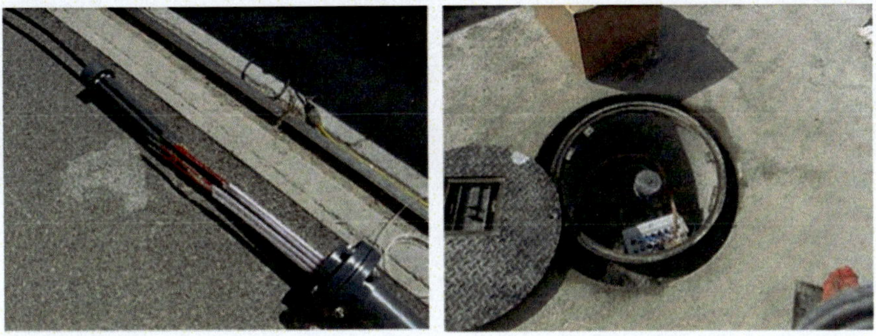

Fig. 4.8 MPBX installation through the crest of a dam

4.6.7 *Inclinometers*

Inclinometers can be installed in existing core sample holes such that complete profiles through the dam can be obtained. The guide wheels for the measurement probe are centered within the inclinometer casing grooves, and the probe is lowered through the entire length of the hole.

4.6.8 *In Situ Stress Measurements*

Reclamation has performed in situ stress measurements in dams affected by ASR by installing Bureau of Mines borehole deformation gages (BDG) and overcoring. This overcoring method typically involves drilling a large-diameter hole (generally between 60 and 120 mm) in the concrete where the stresses are to be determined. The hole is drilled to a sufficient distance from working faces so that the effect of the free surfaces on the stress measurement is minimized. A small pilot hole (usually 38 mm or larger in size) is drilled at the end of the previously drilled hole. The pilot and large-diameter holes must be as concentric as possible. Pilot hole lengths vary between 200 and 500 mm. The BDG is then inserted into the pilot hole, and the large-diameter hole is resumed, which partially or totally relieves stresses and strains within the cylinder of formed material. Changes in strains or displacements are then recorded with the BDG as the Overcoring front proceeds beyond the plane of measurements.

Following overcoring, the recovered overcore can be tested in a biaxial chamber to determine the Young's modulus and Poisson's ratio of the host concrete. Both elastic properties are needed to determine the in-situ stresses from the displacements and strains measured during overcoring.

Flat or solid cylindrical pressure cells can also be installed in boreholes to measure changes in in situ stresses over time. Flat pressure cells are relatively inexpensive to install but are limited to measuring uniaxial stress changes in a direction perpendicular to the cell; however, multiple cells can be used to measure stresses in different directions within the borehole. Solid cylindrical pressure cells allow for the measurement of the 3D stress field in one borehole at distances ranging from 30 to 50 m from the concrete surface.

4.6.9 UAS Photogrammetry and LIDAR Comparison

Reclamation has used both UAS photogrammetry and LIDAR systems to monitor deformation of dams. In addition, Reclamation has conducted research on the accuracy of these techniques at concrete dams and has made several comparisons between the two technologies. The first comparison included the means for data collection and subsequent required processing time. Next, the point clouds generated from each method were analyzed for noise levels and geometrical trueness. Finally, the point clouds were compared (Fig. 4.9).

Both methods required relatively similar amounts of time for data collection and processing. The error in the photogrammetry processing was approximately 1 mm. The georeferencing error for fitting the photogrammetry to the ground control was approximately 9.5 mm. The published accuracy of the Leica ScanStation P40 LIDAR scanner utilized was 9 mm + 10 parts per million, and the registration and georeferencing error was reported as approximately 8.5 mm.

Currently, the accuracy of the photogrammetric and LIDAR may not be satisfactory for deformation measurements of dams affected by ASR; however, with the continued evolution of the technology, both methods are beneficial for understanding

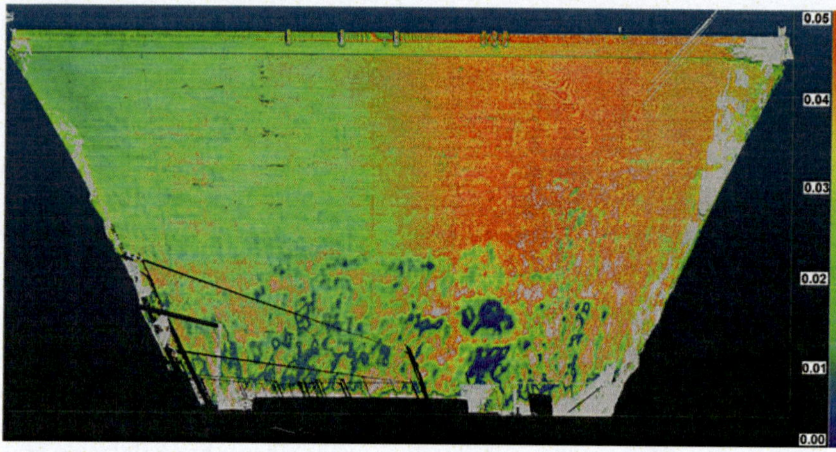

Fig. 4.9 Comparison of the photogrammetric and LIDAR point clouds (scale range: blue-red; 0–15 mm)

dam behavior over an extended period of time and can be used as an evaluation tool to compare the measurement data with other surveying techniques. Reclamation has implemented an extensive research program for evaluating the monitoring and accuracy of dam deformations using various traditional and state of-the-art technologies.

4.7 Crack Mapping

Crack mapping is a dam monitoring method that has evolved over time. Crack mapping has historically been performed through the use of high-resolution photographs, sometimes in conjunction with detailed line surveys. In recent years, Reclamation has begun photogrammetric data collection at dam sites using UAS photogrammetry (Fig. 4.10). While the UAS photogrammetry is not yet accurate enough to detect short-term ASR deformations, the accuracy of the method is sufficient to capture the majority of cracks and determine the overall cracking patterns and changes over time.

4.7.1 Nondestructive Geophysical Surveying

The nondestructive geophysical surveying methods Reclamation has used to investigate the behavior of the dams affected by ASR include: P-Wave Seismic Refraction, Borehole Geophysics, Spectral Analysis of Surface Waves (SASW), Through Dam P-Wave Tomography, and Electrical Resistivity Tomography (ERT). These geophysical surveying methods are discussed below.

4.7.2 P-Wave Seismic Refraction

The seismic refraction surveys consisted of dozens of geophones on the downstream face of the dam (Fig. 4.11). The P-waves are generated with a hammer on the downstream face of the dam, where they traveled into the dam structure to a contact between the overlying slower velocity concrete (damaged concrete that has been gradationally ASR and freeze-thaw affected) and the underlying higher velocity concrete material (measurably less affected or unaffected by ASR and freeze-thaw with respect to seismic velocity). These P-waves are then refracted at this interface, travel along the interface at the faster underlying material's velocity (head waves), and simultaneously travel back to the surface, where they are recorded by geophones. Figure 4.11 illustrates seismic refraction in terms of ray paths and the corresponding first arrival times measured at the surface.

Figure 4.12 illustrates the seismic refraction method in terms of wave fronts at different points in time. As time progresses, the wave front propagates spherically

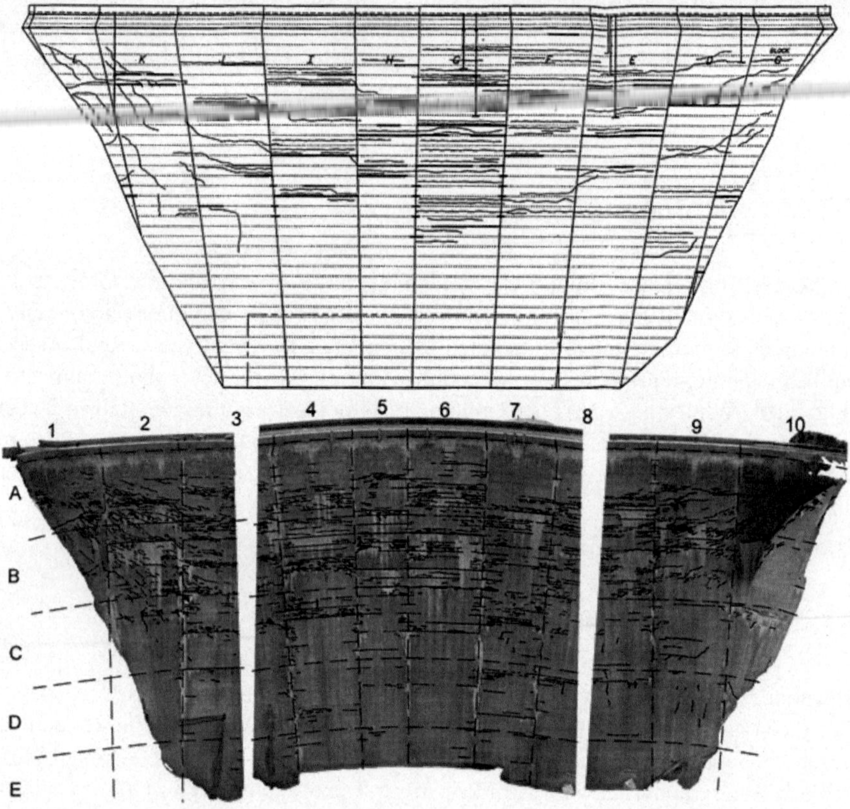

Fig. 4.10 Crack mapping of an arch dam prepared by a traditional method (left) and 20 years later by photogrammetry (right)

in all directions away from the source. When the wave front contacts the boundary between different materials, the wave front (head wave) accelerates in the underlying higher velocity material passing ahead of the surface wave in the slower velocity material. As the wave continues through the high velocity material, it creates a "wake" at the boundary between the upper and lower material layers. This "wake" (most commonly referred to as a "head wave" and seen in dark crimson in Fig. 4.12) is the refracted wave front. It propagates upward through the low velocity material to the surface, where geophones sense its arrival ahead of the surface waves and reflected waves.

The left image in Fig. 4.13 displays an example of a shot gather. Each vertical "wiggle-trace" is a geophone sensor recording of amplitude (horizontal axis) versus time (vertical axis). A wiggle trace record is plotted for each channel. The stacked signals used in final data analysis are plotted in black. The source location is between channel 17 and channel 18. The first arrival picks of P-wave energy at each geo-

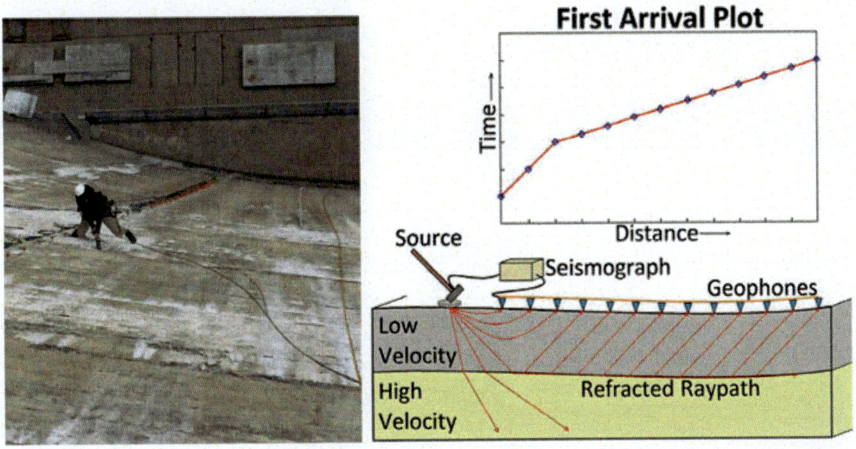

Fig. 4.11 P-wave seismic refraction test setup at Seminoe Dam and raypath seismic refraction concept and travel time plot [1]

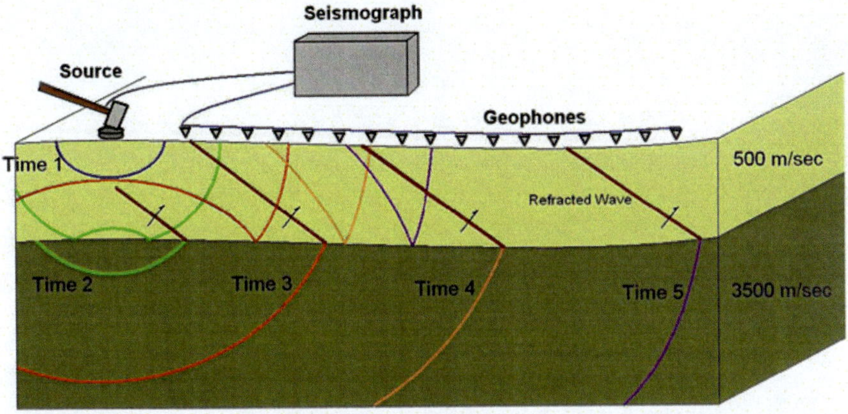

Fig. 4.12 Wave front seismic refraction concept [1]

phone/trace are indicated with red x-marks. Blue x-marks are auto picks generated by the software. After picking the first arrivals of the refracted wave, this P-wave shot gather, along with all the other shot gathers on this survey line, are arranged and used as the data input for inversion.

The right image in Fig. 4.13 displays the resulting velocity tomogram produced from the multiple iteration inversion process. The vertical axis represents elevation, and the horizontal axis represents distance from the upstream face. The tomogram visually displays the distribution of velocity values within the dam. Refraction, as opposed to through dam tomography, has a more limited depth of investigation because it requires the measured seismic waves to penetrate from the surface and

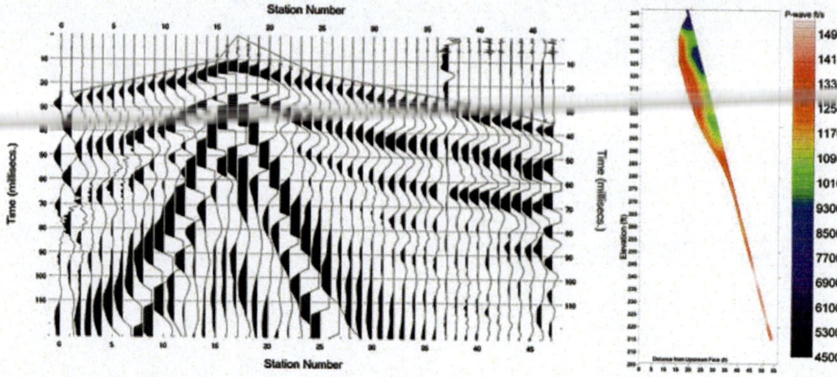

Fig. 4.13 An example P-wave refraction shot gather (X-axis units in 10s of feet, and Y-axis units in time [m/s) (left); resulting P-wave velocity (ft./s) tomogram (right)

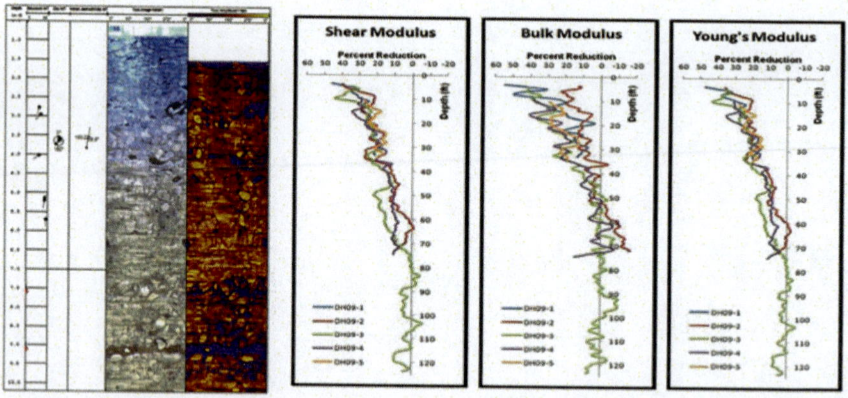

Fig. 4.14 Optical televiewer log (left) and seismic and density borehole logging (right) to calculate shear modulus, bulk modulus, and Young's modulus

return to the surface, rather than travel all the way through the structure to be measured on the other side. For this reason, the data sensitivity is concentrated and limited to the downstream face of the structure (survey side).

4.7.3 *Borehole Geophysics*

Following completion of the drilling of holes from the top of dam, borehole geophysics investigations can be performed including sonic and acoustic logging. Using calculated P-wave and S-wave velocities, along with density values, Bulk, Shear, and Young's Moduli can be determined for each borehole (Fig. 4.14).

4.7.4 Spectral Analysis of Surface Waves

Spectral Analysis of Surface Waves (SASW) method is used for in situ evaluation of concrete quality; therefore, it can be used to identify the ASR-related and freeze and thaw defects in mass concrete. The SASW helps to determine the properties of concrete material without the need for coring.

The SASW method is based on measurements of surface waves propagating in elastic media. The ratio of surface wave velocity to shear wave velocity varies with Poisson's ratio. Reasonable estimates of Poisson's ratio and the mass density of concrete and other materials can be made with only a small effect on the accuracy of the shear wave velocity profile. By using the shear wave velocity, calculations can be made of shear moduli for low-strain amplitudes. Surface wave velocity varies (disperses) with frequency in a deteriorated mass concrete. A plot of surface wave velocity versus wavelength can be expressed by dispersion curves. For sound concrete, the dispersion curve plot is flat (constant velocity) versus wavelength throughout the concrete thickness. If degradation or cracking is present, the dispersion curve will have zones of lower velocity. The depth of the crack or degradation can be estimated from the wavelength range of the dispersion curve.

SASW field tests are conducted using a pair of vibration transducers. The transducers are typically displacement transducers or accelerometers mounted to, or held in contact with, the concrete surface. Impacts are made with a small hammer in line with the receivers. A computerized signal conditioning and data acquisition system collects the waveforms from the receivers and records the signals on a computer hard drive for frequency analyses. The phase information of the transfer function between the two receivers for each frequency is the key spectral measurement. Repeating the above procedure for any given frequency allows for evaluation of the surface wave velocity corresponding to a given wavelength and construction of the dispersion curve.

When a hammer (impact energy source) strikes the surface in line with the geophone unit alignment, a surface wave is created and migrates radially in all directions along the surface of the material away from the source. A range of S-surface wave frequencies is created when the source strikes the surface. This range of frequency waves travels at increasing depths simultaneously and parallel to the surface. High frequency (short wavelength) surface waves travel parallel to the surface at relatively shallow depths. Low frequency (long wavelength) surface waves travel parallel to the surface from the surface down to relatively deeper depths. The depths at which the surface waves travel is a function of their wavelength and the shear strength of the material the survey is conducted in. This wave propagation is diagramed in Fig. 4.15.

The S-wave velocity plot (Fig. 4.15, center) shows how different kinds of undamaged and damaged concrete at varying depths cause the S-wave frequencies at those respective depths to attenuate or slow in velocity (referencing the wave propagation image). Attenuation of a wave may be displayed as lower velocity within the plot, due to lack of data at that depth. Zones that actually slow a surface wave will also report a lower velocity at that depth.

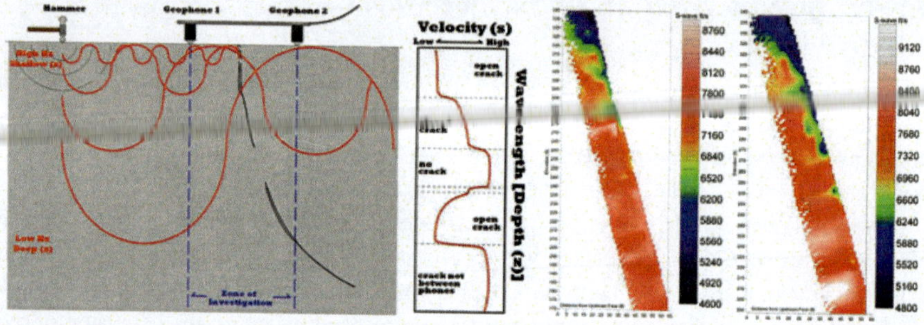

Fig. 4.15 Propagation of the spectrum of surface waves and their interaction with the surveyed material (left) and SASW two-dimensional (2D) S-wave velocity model (ft./s) from interpolated stick logs at Seminoe Sites (right)

Figure 4.15 (right) displays the 2D models produced by combining multiple SASW soundings and their resulting velocity plots into the structure. All of the individual velocity plots or "stick logs" are then interpolated and gridded to produce a cross-sectional, S-wave velocity model of the structure. The models themselves visually display the concrete S-wave velocity value distributions within the structure. Again, data is concentrated along the downstream face of the dam and limited by the depth of investigation of the SASW method, although the depth of investigation is much deeper than that of seismic refraction. This model sensitivity or depth of investigation is twice the distance of the accelerometer separation on the surface.

4.7.5 Through Dam P-Wave Tomography

Through dam P-wave tomography, similar to cross hole tomography, uses a seismic source on one side of the concrete dam to send a seismic wave through the structure, where it is measured by a series of geophones on the opposite side of the structure. Several shot locations (various elevations) are implemented to create a dense raypath coverage through the structure. The raypaths from each source to receiver combination are uniquely affected by the quality of the concrete within the structure, slowing down the seismic wave where concrete is damaged by ASR or freeze-thaw conditions. All of the travel times from each source to receiver raypath combinations are combined into a multiple iteration process to model the entire structure from the upstream face to the downstream face (Fig. 4.16). Water-filled reservoirs introduce inherent challenges for through-dam seismic tomography where dive teams may have to be implemented to aid in geophone placement (hydrophone placement). Through dam tomography can also be implemented by drilling a hole vertically into the dam from the crest downward, then using a borehole probe (either P-wave or S-wave geophones, or P-wave or S-wave source), as well as using the downstream and/or upstream faces as source or receiver array locations. S-wave through dam

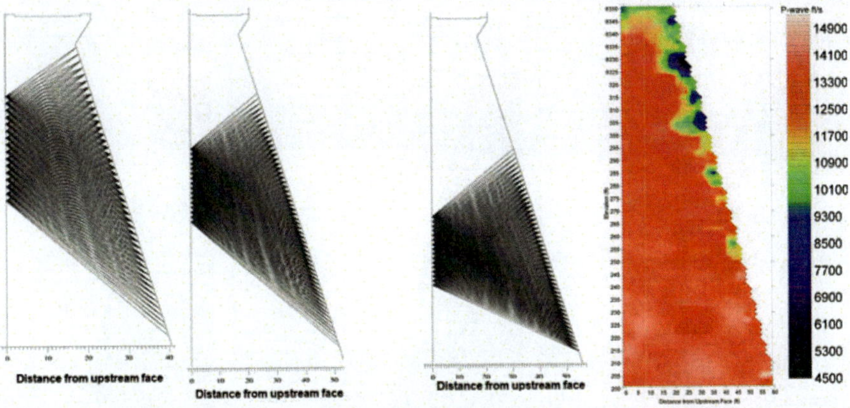

Fig. 4.16 Through dam P-wave tomography source to receiver raypath combinations (left); through dam P-wave tomogram (color scale velocity) (ft/s) (right)

tomography would be impractical to implement if a downhole source is not used because S-wave phones and S-wave sources are unable to create or measure S-waves in water.

Reclamation performed through dam seismic tomography at three cross sections of Seminoe Dam to provide information on the spatial progression of concrete deterioration caused by the alkali aggregate reaction, as displayed in Fig. 4.17.

4.7.6 Electrical Resistivity Tomography

Electrical Resistivity Tomography (ERT) method is an active geophysical method that measures the electric potential differences at specific locations, while a controlled electric current is injected at other locations. The theory of the method holds that in an entirely homogeneous half-space, a resistivity value can be calculated for the subsurface by knowing the current injected and measuring the resulting electric potential at specific locations; however, homogeneity within the subsurface is very rare, and electric current, when introduced, will follow the path of least resistance, concentrating in areas of conductive material and avoiding areas of resistive material. The left image in Fig. 4.18 illustrates the concept of subsurface electric current flow and how current flow is affected by subsurface heterogeneities.

The center image in Fig. 4.18 displays an as-built cross-section of Seminoe Dam and depicts the presence of metal cooling coils. Metal in the subsurface of an ERT survey can negatively affect the outcome because the metal pathways tend to reroute the electrical current, which directly affects the geometric factor. The geometric factor accounts for the geometry of the measurement locations along the survey line, which, in turn, helps convert the calculated resistances to calculated resistivity

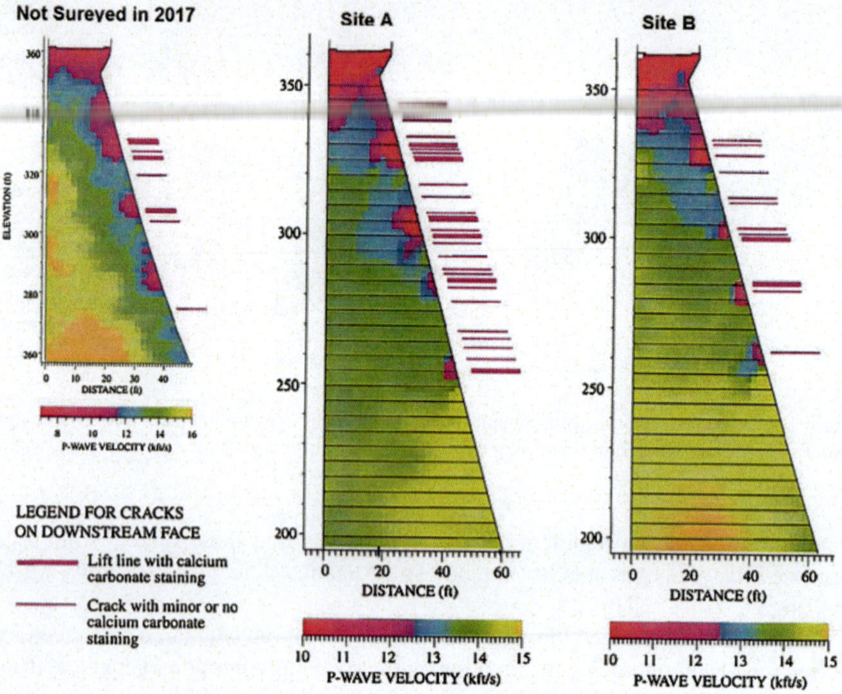

Fig. 4.17 P-wave velocity tomograms for the three cross sections

values in the subsurface prior to inversion. However, the electrical current flow is unpredictably affected, making these particular conditions nonideal for ERT. It is unknown to what extent these cooling coils negatively affected the accuracy of this particular ERT survey because the metal coils are aluminum and, therefore, not as electrically conductive as other types of metal. When comparing the geometric trends of resistivity values to the geometric trends of the P-wave and S-wave values, however, there are good correlations. The right image in Fig. 4.18 displays the ERT survey setup on the face of Seminoe Dam.

The left image in Fig. 4.19 displays the crossplot of measured versus predicted apparent resistivity data, which is an indicator of data processing quality. The right image in Fig. 4.19 is the resulting ERT tomogram, which visually displays the distribution of resistivity values within the structure. This resistivity tomogram correlates fairly well to the through dam seismic tomography, the SASW 2D model, and the P-wave refraction tomography results in terms of obvious anomalies where ASR and freeze-thaw damaged concrete exists.

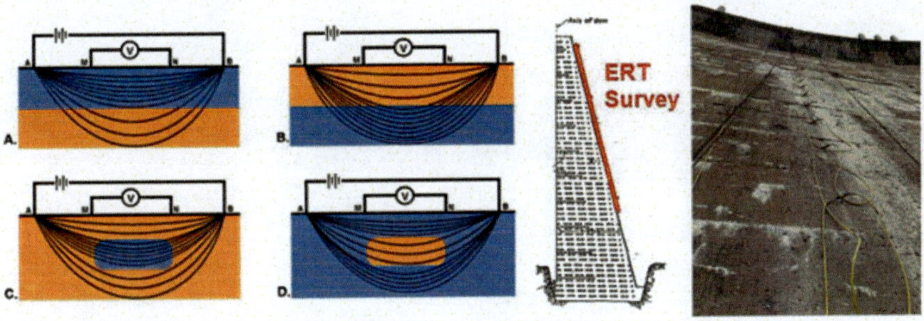

Fig. 4.18 Subsurface electric current flow (left); Seminoe Dam as-built drawing depicting metal cooling coils inside the structure (center); ERT survey setup on the downstream face of Seminoe Dam (right)

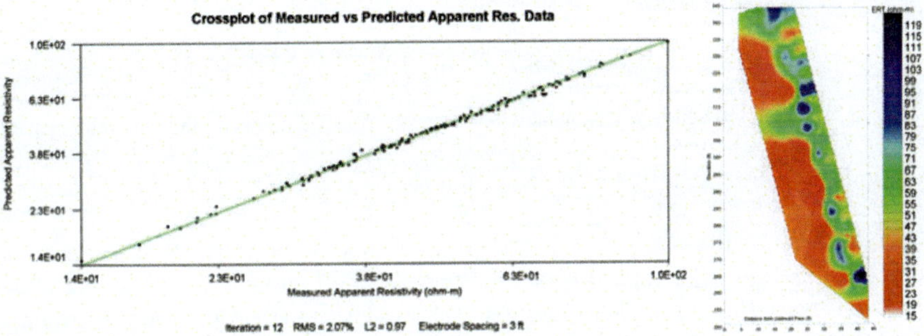

Fig. 4.19 Crossplot of measured versus predicted apparent resistivity data and ERT tomogram displaying distribution of electrical resistivity values (ohm-m)

4.8 Comparison of Nondestructive Geophysical Surveying

of the five nondestructive geophysical survey types Reclamation has implemented, three appear to be the most ideal. While easy to implement, GPR did not result in highly informative models with respect to ASR or freeze thaw material properties. While it has potential for other structures, ERT was not ideal for surveys at Seminoe Dam, due to the unknown effects of electrical current flow from the metal cooling coils within the structure on overall tomogram accuracy. SASW surveys were the easiest to implement and provided S-wave velocity data within the structure; however, they were by far the most processing intensive. Seismic P-wave refraction tomography surveys provided high-quality seismic velocity data but had two drawbacks: (1) a limited depth of investigation (data concentrated near the downstream face only), and (2) a smoothing effect within the tomograms, due to refraction processing assuming fairly horizontal raypath propagation along layer boundaries. Seismic P-wave through dam tomography is the most accurate and provides the greatest depth of investigation (imaging all of the material in a cross-section through the structure).

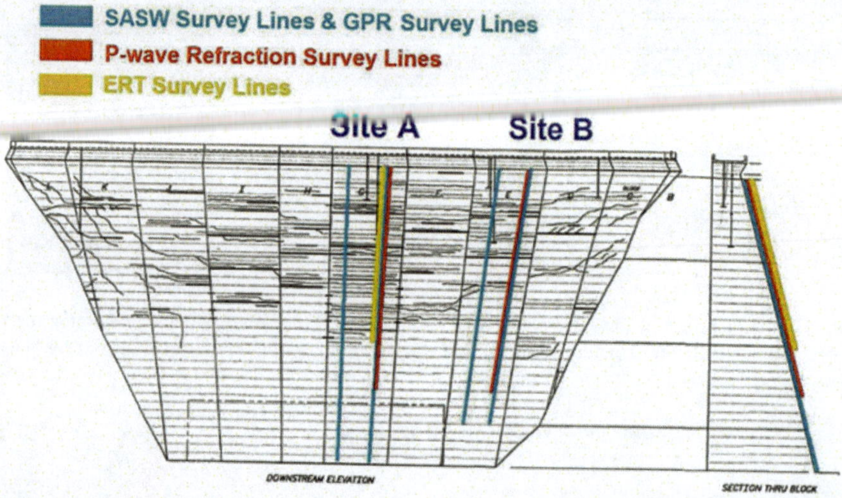

Fig. 4.20 Locations at Seminoe Dam where the geophysical surveying methods were implemented and the results were verified

Figure 4.20 displays three different survey tomograms at the same location on the face of Seminoe Dam per Fig. 4.21. The tomogram on the left is from the 2001 seismic P-wave through dam tomography survey. The middle tomogram is from the 2017 SASW survey. The right tomogram is from the 2017 seismic P-wave refraction survey. All the tomograms are scaled to the same size and transformed so that the Y-axis elevations (ft) are properly correlated spatially.

Cool colors (greens and blues) represent lower P-wave and S-wave velocities (ASR and freeze-thaw damaged concrete), while warm colors (reds and yellows) represent higher velocities (concrete that has not been damaged by ASR or freeze-thaw). Anomalous areas of lower velocity concrete (indicated by large blue anomaly lobes) are labeled by a letter from "A" to "E" in each of the three tomograms and correlated to each other by referencing the anomaly elevation along the Y-axis. Note the strong elevation and spatial correlation in anomalies A, B, and C in all three tomograms, and the correlation of anomaly E in the P-wave through dam and SASW tomograms. In addition, note the increased resolution and ability of the P-wave through dam survey to map velocity anomalies that sharply penetrate into the structure, as opposed to the seismic refraction survey that smooths those anomalies out and lacks the ability to map them deeper into the structure.

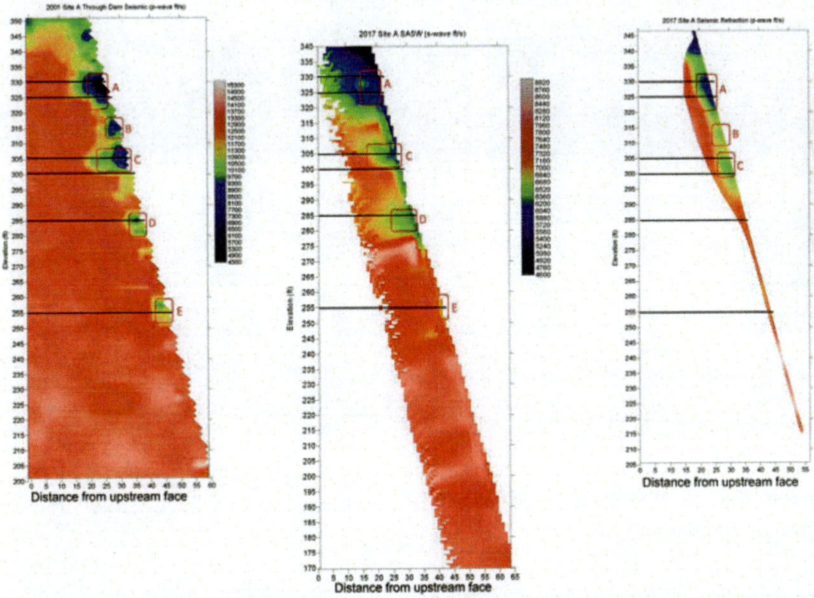

Fig. 4.21 Result comparison between three methods: seismic P-wave through dam tomography (left), SASW (right), and seismic P-wave refraction (right). The results were verified (units: ft and ft/s)

4.9 Concrete Coring

Obtaining good quality cores from a dam affected by ASR can be a difficult task when the concrete is significantly deteriorated because of thermal cracking, freeze-thaw deterioration, and poor bond between the paste and aggregate due to the formation of ASR gel around the reactive aggregates. Sometimes, concrete cores with a depth up to 2–3 m may not be in "testable" condition (see left image in Fig. 4.22). The coring activities may significantly impact the mechanical properties of the samples due to frequent core breaks and aggregate pop-outs. Proper packing and transport of the cores for testing is very important because the drilled cores are no longer confined (begin to expand) and could be further damaged during delivery to a laboratory. Considering the above difficulties, the core condition report prepared by the drillers at the project site is an important record for further evaluation of the concrete properties.

Reclamation has been regularly conducting coring and laboratory testing for dams affected by ASR (at approximately 5-year intervals). On several occasions, 50-m-long and 150-mm-diameter vertical concrete cores, as well as 100-mm horizontal cores, have been extracted). That way, the concrete core test results can be compared with previous concrete coring programs, and the rate of the alkali-silica reaction and

Fig. 4.22 Concrete core sample at depth: 0–1.95 m (left) and 1.95–10.66 m (right) (in U.S. units)

concrete deterioration can be estimated. Taking optical borehole images after core retraction is a very common practice to collect evidence of rubble concrete, cracks, or aggregate pop-outs.

Reference

1. Redpath, B.: Seismic refraction exploration for engineering site investigations. Tech. rep. AD-768 710. Explosive Excavation Research Laboratory, Livermore, California (1973)

Part II
Prognosis; Accelerated Expansion Tests

Chapter 5
Summary of Reported Methods

Leandro Sanchez, Christine Merz, and Victor Saouma

5.1 Introduction

This part of the document discusss a number of techniques used to quantify the residual expansion in concrete affected by alkali-aggregate reaction (AAR). The principal points of each technique (similarities and differences) are presented and an overall discussion is addressed.

Note: Only the basic principles and procedures used in each of the residual expansion techniques are presented. Further details and complete description of the methods can be found in the Supplementary Materials section.

Among the different techniques used worldwide to detect the residual expansion of concrete distressed by AAR, seven methods seem to be the most promising and used:

Toulouse method is a test procedure that estimates the residual expansion of concrete structures affected by AAR. A critical assumption of the method is that the "alkali content" is not the limiting factor of AAR, since alkalis may be replaced by calcium. Hence, it is the "amount of silica" that is assumed to control the final expansion. The test procedure covers the laboratory determination of the swelling potential and the residual free expansion of concrete extracted from structures affected by AAR.

L. Sanchez (✉)
University of Ottawa, Ottawa, Canada
e-mail: Leandro.Sanchez@uottawa.ca

C. Merz
Merz Ingenieurberatung, Möriken, Switzerland
e-mail: merz@merz-ingenieurberatung.ch

V. Saouma
University of Colorado, Boulder, USA
e-mail: saouma@colorado.edu

© RILEM 2021

V. E. Saouma (ed.), *Diagnosis & Prognosis of AAR Affected Structures*,
RILEM State-of-the-Art Reports 31,
https://doi.org/10.1007/978-3-030-44014-5_5

IFSTTAR and Laval University methods cover the laboratory determination of the swelling potential and the residual free expansion of concrete extracted from structures affected by AAR. In the above two procedures, the test is performed without any supplementary alkali addition to the specimens since the test assumption is that the alkali level of real concrete structures decrease over time, which seems to be the factor limiting further expansion.

Japan Method is somewhat similar (at least in concept) to Toulouse Method, since it determines the swelling potential of concrete extracted from structures rather than the residual free expansion of structures affected by AAR. A number of equations to convert laboratory expansion data to overall expansion in field conditions are shown, along with details of constitutive techniques (crack index, alkali-budgets, petrographic examination). Similar concept is also found in EPFL's method, which aims to accelerate AAR expansion in the laboratory and should always be coupled with scanning electron microscopy for quantitative evaluation of the state of the concrete.

EMPA's method determines the expansion rate of concrete cores (designated as "residual expansion potential") extracted from structures affected by AAR. In this test, as in IFSTTAR or Laval University's test procedures, the main assumption is that the level of alkalis of real concrete structures decrease over time, which seems to be the factor limiting further expansion.

English procedure aims to determine the potential for further expansion of concrete in structures with, or suspected of having AAR. Yet, cores should be exposed to similar to those in the field.

5.2 Principles and Methodology

5.2.1 Japan

Long-term version of the accelerated expansion test (80 °C, 1 mol/l NaOH) is used to determine the expansion parameters of Larive's sigmoidal expansion curve (final expansion, latency time and characteristic time). Final expansion thus obtained is corrected for the differences between laboratory and field conditions, i.e. temperature, alkalinity of solution, relative humidity, and water-soluble alkali from aggregate. Overall expansion curve of field concrete is determined by re-fitting three independent data (corrected final expansion, past expansion or crack index at time 1, and crack index at time 2 of core sampling) to the Larive's equation.

5.2.2 Switzerland, France-IFSTTAR, Canada/A

Ifsttar, Laval University (part A) and Empa methods consist of extracting samples from AAR affected structures/structural member(s), store them at environments enabling AAR further expansion (i.e. 100% R.H. at 38 °C) and evaluating the specimens length (longitudinal) and mass changes over time.

5.2.3 Canada/B

Laval University's (part B) procedure's suggest the extraction and immersion of the cores in aggressive alkaline solutions (i.e. either 1N NaOH or the pore solution molarity, see supplementary materials section) at 38 °C in order to evaluate their "maximum potential of further expansion" due to AAR.

5.2.4 France-Toulouse

Toulouse method aims to quantify the swelling kinetic potential for a given structure/structural member. The essence of this method consists of comparing residual swelling measured on mortars made from crushed reactive aggregates with the residual swelling of mortars made from "still unreacted" reactive aggregate to deduce the advancements of AAR in the different classes of aggregates (i.e. different aggregate sizes) of the concrete structure/structural member in question. AAR progress can then be computed into a FEA model to deduce the overall AAR advancement in the given structure/structural member. Finally, residual expansion is obtained from inverse analysis of the behaviour of the damaged structure in which the kinetic is imposed by the results of the mortar tests.

5.2.5 UK

The English method significantly differs from humidity storage methods above in that water is absorbed into the core and the initial shrinkage recovery and any developing AAR expansions are related to the measured water absorption. Alkali leaching is largely eliminated by limiting free water available to 10 g per kg of core. It is also distinct in that six to nine gauge lengths are to measure the characteristic AAR variability of expansions in the core. Test duration is more than one year, but often continued for years in parallel with the monitoring of the structure.

Once cored, specimens are wrapped to avoid moisture losses. Then sets of Demec or similar gauges are fixed longitudinally. The cores are stored individually in plastic containers sealed with a close fitting lid with unabsorbed free water maintained at 10 g per kg of core. Keeping the containers at 30° ensures direct contact of the base of the core with the water. The test conditions of freely available moisture will show trends of long term potential expansion but the ultimate "residual" expansion only becomes apparent after decades with slow reacting aggregates.

5.3 Sampling

5.3.1 Japan

After recording the orientation (Z,X,Y), slender cores are extracted from RC or non RC structures for accelerated expansion test (80 °C, 1 mol/l NaOH). Test specimens should have a diameter of ideally 56 mm with a length of 130 mm, but longer core samples (150–200mm) should be taken including the space to attach two belts with measuring studs of a 100 mm span, along with spare length to discard carbonated surface of concrete.

For a comparative expansion test (40 °C, 100%RH), which is basically similar to IFSTTAR, Laval University and Empa methods but in many aspects modified in Japan, cores (diameter 100 mm, length 230–250mm) are taken. In both cases, cores should be rinsed, tightly wrapped with a cling film keeping moisture on the surface, and stored in the laboratory. Upon receipt, provisional initial reading is made at 20 °C, then expansion tests should start without delay, in order not to miss initial expansion of AAR gel of the early-expansive aggregate, represented by opal/cristobalite/tridymite-bearing andesite and rhyolite of younger geologic ages belonging to the Neogene and Quaternary periods. In the continental regions where the late-expansive aggregates of old geologic ages are widespread, omission of the above procedures may produce little problem.

5.3.2 Switzerland

In Empa's method the pins are glued into small boreholes of the cores and aligned with a jig (axially along the cores length). Stable glue (e.g. organic two-component glue) has to be used that shows no volume changes at the conditions in the reactor. Likewise in the English method a jig is used on the cores to enable gluing 3 longitudinal rows (each of three 50 mm) gauge lengths fitted using the setting bar.

5.3.3 France-IFSTTAR

In Ifsttar, Laval University and Empa methods, cores should be extracted in such a way preventing cracking or steel reinforcement. The diameter of the specimens should be at least 100 ± 10 mm (or three times the nominal maximum size of the aggregates) and its length equal to two/three times the diameter. It has been noted that cores of 150 mm whenever possible are always preferred due to the decrease in alkalis leaching. It is recommended that, at least, the first 50 mm (2 in) of concrete should be avoided due to the likely presence of microcracking at the specimen surface. Just after coring, the cylinders should be washed with clean water, wiped and stored in sealed bags. It is recommended to wait at least one week before the cores are subjected to expansion tests in order to reach a relative mechanical equilibrium (stress release). Before testing, a careful visual inspection of the samples should to be conducted: any singularities such as cracks, deteriorated aggregates, cavities, reaction products, reinforcement bars, etc. must be documented. Careful attention will be paid on avoiding drying of the core during this inspection.

While performing Ifsttar method, whether cylinders of 150 mm in length are selected, 3 measurement basis of 100 mm in length are set at 120° from each other (see Fig. 3 in the supplementary section Ifsttar document). Each measurement basis is placed at the same distance from each cylinder end. In the case of a 250 mm-long cylinder, two measurement basis of 100 mm are set in a row and placed similarly at the same distance from each specimen end. The location discs are glued on the lateral sides of the cylinder. Six (resp. nine) location discs are necessary for one (resp. two) measurement elevation(s) along the vertical axis of the cylinder. Before gluing, the concrete surface is locally dried with compressed air.

5.3.4 Canada

Laval University Method, stainless steel bolts or gauge studs, 13 mm in length by 3 mm in diameter (approximately 0.5 by approximately 0.12 in) are commonly used. They are installed at the end of the samples in small holes drilled dry and cemented with a shrinkage-free cement paste. The drill holes are about 8 mm (approximately 0.3 in) in diameter by 20 mm (0.8 in) deep.

5.3.5 France-Toulouse

In the Toulouse method, cores from both damage and sound locations should be extracted. Then the aggregates shall be recovered through a two-step approach: (1) coarse aggregates (>4 mm) are first extracted through heat treatment by immersing the samples in a Na_2SO_4 solution, followed by at least five cycles of freezing and thawing. This procedure enables the extraction of coarse aggregates from the concrete matrix; (2) fine aggregates are then extracted through an organic chemical attack in order to dissolve them from the cement paste. Salicylic acid solution (1.3 mol/l with methanol as the solvent) is reported to be more efficient than inorganic acid (such as HCl, which may attack reactive silica in the aggregates and thus alter reactivity). Further details regarding chemical extraction and an assessment of chemical advancement can be found in Gao, Cyr, Multon, and Sellier [1].

The aggregates are then crushed and sifted, with only those in the 0.16–3.15 mm range being separated for use. The selected aggregates are subsequently used to prepare two sets of mortar mixes with 1,500 kg/m^3 sand content, 8 kg/m^3 alkali content and a water-to-cement ratio of 0.5. Each mix contains a set of three mortar bar specimens (20, 20, 160 mm).

5.4 Apparatus

Although the test set-up is different depending on the technique performed, the apparatus used display a number of similarities as illustrated hereafter:

1. Extensometer:
   - minimum resolution: 1 μm
   - gauge length: 10 cm
   - including a calibration reference bar
2. Locating discs adapted to the extensometer (generally made of stainless steel)
3. Adhesive resisting to moisture and high temperatures (40 °C; 100% RH)
4. Absorbent paper (thickness: 0.6 mm; fungicidal)
5. PE sheet (thickness: 0.1 mm)
6. Storage container complying with the XP P18–594 (formerly P18–587) standard
7. Storage reactor complying with the XP P18–594 (formerly P18–587) standard
8. Oven set at 105 °C
9. Analytical balance (capacity: in agreement with the weight of the specimens; relative precision: 0.1% of the weight of the specimen)
10. Temperature sensor (readability: 0.1 °C).

5.5 Experimental Procedures

5.5.1 Japan

Slender core samples are stored in the container at 80 °C, immersed in 1 mol/l NaOH solution. For initial reading, core sample is tightly wrapped and stored in the oven at 80 °C for 2 hours to check thermal expansion between 20 and 80 °C by hot-reading, which is performed within 15 s after removal of the core from oven. In this way, new zero point is established for alkali-immersed core samples stored within container in the oven. For a short-term routine test to evaluate reactivity of concrete, length change in NaOH solution is recorded at 1, 3, 7, 14, 21 and 28 days by hot-reading. However, to assess final expansion of concrete, measurement should be continued until at least 6 months, preferably until the expansion stabilizes. In parallel with the expansion test, past expansion of concrete in two different times (core extraction at time 2, and repair/core extraction in earlier time 1) are estimated based on measurement of crack index of core sample and field structure. With a comparative expansion test of fat core (40 °C, 100%RH) measured by hot-reading, intervals for measurement are similar to corresponding methods in other countries, but there is a fundamental difference in the preconditioning. In the modified Japanese method, core samples arrive in wet condition, measurement can be started without delay, there is no need to expose core sample to a 20 °C room for one day before the measurement, and hot-reading adopted to avoid expansion loss.

5.5.2 *Switzerland*

While performing Empa method, the experimental procedure may be divided in 4 steps: (1) initial measurements of dimension and mass before sample conditioning (performed at 20 °C in containers with a thin layer of water at the bottom permitting capillary suction of the cores). The mass of the cores has to be measured daily. This first conditioning step ends when constant mass is reached; (2) the cores are stored (wrapped in plastic film) in the reactor at 38 °C and at 100% RH. And measurements performed (specimens are removed from the reactor and stored at 20 °C for 24 hours before the cores are measured. Afterwards the container are put back into the reactor) until their initial quick expansion levels off (normally 30–60 days), which represents the end of phase 2; (3) same practical procedure from phase 2 (measurements of length and mass), however the expansion vs. time in this phase is found to be quite linear; (4) the cores are removed from the reactor after 160 days. Afterwards, they are dried at 60–70% R.H. until they reach the identical mass as before conditioning started (phase 1). The difference in length change is defined as "irreversible" expansion and enables to distinguish between hygroscopic swelling during conditioning and swelling due the formation of newly formed AAR products. This step requires normally 2–8 weeks.

5.5.3 *France-IFSTTAR*

Regarding Ifsttar and Laval University (part A) procedures, the specimens are weighted and the initial length measurement is made using an extensometer followed by the samples storage in containers containing a water film at the bottom (level of water = 35 ± 5 mm). The container presents a lid with a deflector to inhibit water dropping on the specimens (in Ifsttar method, the specimens are stored at ambient temperature during 24 h prior to initial measurement. In Laval University method, the samples are wrapped and stored at ambient temperature at least 5 days before the initial reading). The container (or bucket) is finally stored at a reactor with temperature of 38 ± 2 °C. The level of water at the bottom of the reactor should constantly be verified; the temperature of the storage room must also be measured and recorded in the test report. Measurements are made after 2, 4, 6, 8, 12, 16, 20, 24, 28, 32, 36, 40, 44, 48 and 52 weeks and then every two months whether necessary. For each measurement, the specimens must be cooled down to a temperature of 20 ± 1 °C, which is done taking the buckets out of the reactor approximately 16 hours before the readings. After measurement, the specimens are placed again in its container (buckets) and in the reactor. Finally after 52 weeks of accelerated aging, the specimens are weighted and a careful visual inspection is performed (note: in Laval University method, the specimens are also weighted in every length change reading).

5.5.4 *Canada*

The cores are stored in the reactor at 38 °C, soaked in solutions of either 1M NaOH or the same concrete pore solution respectively. Measurements are taken immediately before immersion (initial reading), at 24 and 48 h and then every week or month depending on the rate of expansion. The only difference in terms of procedure between these two methods is that for EPFL the measurement apparatus should be located in the same temperature-controlled room as the reactor, to prevent fluctuations from heating and cooling the samples (i.e. samples are taken out of the reactor, wiped, measured and put back one after the other) whereas for Laval University procedure, one may measure using the reactor temperature or even following the procedure illustrated in part A (i.e. take out of the reactor at least 16 hours before reading to avoid temperature changes). Finally in EPFL case, a slice of the core should be sawn and kept for analysis by electron microscopy. This should determine the initial damage state of the material.

5.5.5 *France-Toulouse*

In Toulouse method, mortar bar specimens are cured for 28 days in sealed bags at 20 °C, before conducting accelerated tests at 38 °C and NaOH bath (1 mol/l). Expansion results are plotted as free expansion vs. time for each mortar specimen.

5.5.6 *UK*

In the English method, 10 g/kg sample weight of tap water should be added to the storage container, and the core inserted. It should be sealed and placed in a rack inclined at 30o. For each reading, the full container should be weighed, the core should be removed from the container, shaken so that loose moisture is retained in the container, and placed on its drier (D) end on the scales for weighing. The container and free water should be then be weighed, additional water added if required and reweighed. Readings should be taken immediately and the core reweighed and returned to the sealed container and the whole reweighed. When the weight of free water in the container falls below 7g/kg, additional tap water to bring the free water up to 10g/kg should be added. Length changes should be taken at 1, 2, 3, 7, 14, 21, 28, 42 days, 2, 3, 4, 5, 6, 8, 10 and 12 months. It should continue 3 monthly until both the expansion and weight are stable. The shrinkage recovery and water uptake during the early part of the test need to be clearly differentiated in analysis from long term expansions (see Fig. 5 - supplementary materials section).

5.6 Preparation and End of the Test

5.6.1 *Japan*

In the accelerated concrete core expansion test (80 °C, 1 mol/l NaOH), criteria for judging deleterious reactivity of aggregate in concrete is, 0.1% at 3 weeks for the early-expansive aggregate (volcanic rocks), whereas it could be 0.04% for the late-expansive aggregate. This test is not applicable to chert aggregate. With a comparative expansion test of the fat core (40 °C, 100%RH), expansion >0.1% at 6 months was a deleterious indication of highly reactive, early-expansive aggregate (andesite), but this criterion is not applicable to the late-expansive aggregate. In any case, to assess residual or potential accelerated expansion of concrete, expansion tests should be extended at least to 6 months or longer until expansion curve stabilizes so that Larive's expansion parameters can be determined safely.

5.6.2 *Switzerland*

The method has been used to assess the residual expansion potential of numerous structures in Switzerland, Fig. 5.1. The cores can be classified into three different levels.

The differently colored domains are defined empirically, based on the experience from Swiss concrete structures.

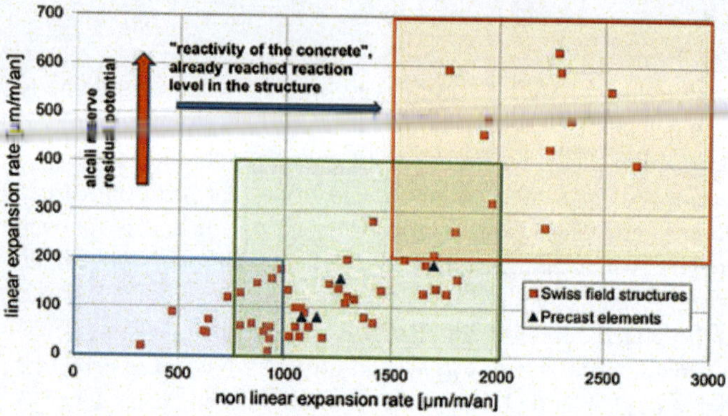

Fig. 5.1 Results of residual expansion measurement of cores extracted from concrete structures in Switzerland

Blue domain: concrete with a residual expansion potential zero or very low.
Green domain: concrete with a moderate residual expansion potential.
Red domain: concrete with a high residual expansion potential.

The domains shown in Fig. 5.1 were defined based on the experience gained from Swiss concrete structures They take into account the degree of damage of the structure at the time of core extraction, age and exposition of the structure and concrete composition (alkali content, cement type, aggregates). As these additional characteristics are considered and influence the assessment of the structure, the domains overlap. The measured values give an indication about range of expansion rates to be expected in a structure. These values are influenced by the expansion that already took place and by the exposition of the studied structure or component. As an example, cores taken from a part of the structure with little AAR-induced damage may show a higher expansion compared to cores taken from a strongly damaged part. This emphasizes that the degree of AAR can vary widely within a structure. It is advantageous if structures are monitored and the measured expansions can be compared with the one determined in the laboratory test. This facilitates the assessment of the structure.

5.6.3 *France-IFSTTAR*

All measurements are presented in a table that gives, for each time step, the expansion of each measurement basis, their mean value and, if applicable, the values corresponding to each level of measurement (for specimens of 250 mm in length). The numerical values are plotted in graphs as well. The mean dimensional evolution during the test is generally divided into three phases:

Table 5.1 Residual expansion qualification criteria, LCPC method

Value	Ranking of the residual expansion
<100 μm/m/year	Negligible
From 100 to 500 μm/m/year	Moderate
>500 μm/m/year	High

Phase a: rapid initial expansion It corresponds to the water uptake of the core. It is mainly linked with the capillary features of the material (porosity, microcracking) and eventually with hydrophilic products (reaction gels). This phase depends as well on the moisture state of the specimen and on the stresses that were applied to the material in the structure. Due to all these parameters, the preparation of the data in this phase is relatively difficult. They are thus not integrated in the evaluation of the residual expansion. By convention, the duration of this phase is 8 weeks.

Phase b: slow and continuous expansion This phase corresponds to the beginning or to the continuation of an internal swelling reaction possibly still active. The mean variation of expansion is calculated by regression on the data between 8 and 52 weeks (if some outliers are detected in the data, they have to be deleted prior to the regression). As a final result, this regression has to be expressed as a mean expansion rate for one year. This value has then to be compared to the threshold given in Table 5.1.

5.6.4 Canada

The short-term length-change results obtained are analyzed in order to determine at what time the core samples reached a relative equilibrium in mass and expansion (after hygrometric equilibrium, which also includes potential additional expansion of preexisting AAR gel). The preconditioning period may range from a few days to a few months, depending upon the storage conditions before testing (e.g., drying). The time and expansion scales are then reset to zero. Parts A and B are normally performed until one year at least after the preconditioning period, and even further when the one-year expansion is relatively low (e.g. <0.025%). The final result is expressed as the average annual rate of expansion over the testing period, excluding the preconditioning period. Linear regression analysis is recommended in order to determine the annual rate of expansion with a better accuracy. preparation of the results can be done with the use of the tables presented below respectively for residual expansion and maximum potential of expansion (Table 5.2).

5.6.5 France-Toulouse

Once the expansion of the set of mortar bars is obtained AAR progress can then be computed into a FEA model to deduce the overall AAR advancement in the given

Table 5.2 Residual expansion results preparation

1-yr exp.1 (percent)	Case	Conclusion about the concrete	Preexisting AAR gel (before test)	Secondary AAR gel (after test)	Result for companion cores tested at 38 °C (100°F) in 1N NaOH	Further expansion expected in the field
<0.004	1	Non-expansive since construction	No/small amounts	No/small amounts	Non-expansive (nonreactive aggregates) or expansive (low alkali concrete content)	Non-expansive (non-reactive aggregates or low-alkali content)
	2	Non-expansive anymore	Yes	No/small amounts	Non-expansive (aggregates nonreactive anymore) or expansive (alkali content not sufficiently high anymore)	Non-expansive (aggregates non-reactive anymore or alkali content not sufficiently high anymore)
	3	Reactive but non-expansive (cracked/porous concrete)	Yes or no depending on age and humidity	Yes	Non-expansive (porous/cracked concrete) or expansive (concrete not as cracked)	Expansive or non-expansive depending on porosity/cracking, humidity, and confinement
	4	Reactive silica completely dissolved in the solution	Yes or no depending on age and humidity	No/small amounts	Non-expansive (low alkalis content) or expansive	Expansive or non-expansive depending on the amount of alkalis, moisture and confinement
	5	Reactive but very low permeable concrete (cracked/porous concrete)	Yes or no depending on age and humidity	No/small amounts	Non-expansive (low alkalis content) or expansive	Expansive or non-expansive depending on the amount of alkalis, moisture and confinement
>0.004	6	Expansive	Yes or no depending on age and humidity	Yes	Expansive	Expansive or non-expansive depending on humidity and confinement

structure/structural member. Finally, the residual expansion is obtained from inverse analysis of the behaviour of the damaged structure in which the kinetics is imposed by the results of the mortar bar tests (see supplementary materials section). After the final length change measurements have been made, each specimen shall be tested for warping and shall be examined.

Warping—The warping, if any, that each specimen has manifested during the test period, shall be determined by placing the specimen on a plane surface and measuring the maximum separation between the specimen and the surface. The specimen shall be placed so that its ends curve down to the reference surface and the measurement made to the nearest 0.3 mm (0.01 in.).

Examination -Items to be noted in the examination include:

- Presence, location, and type of pattern of cracking,
- Appearance of surfaces, surface mottling, and
- Surficial deposits or exudations, their nature, thickness and continuity.

5.6.6 UK

At the conclusion of the expansion test, the "Last Expansion" double set of readings (i.e. core readings separated by at least one hour) should be taken. The lids of containers are then removed to allow slow drying in normal laboratory air conditions (typically 50–60% R.H.). Exact humidity is not critical as aim is slow dry down to initial weight. The surplus water in the bottom of the container will be progressively absorbed and evaporate and any alkali in it will be drawn back into the core as it dries. Weight (including the removed lid) and Demec readings should be taken at 7 day intervals for 28 days to determine the drying shrinkage characteristics with a double set of final readings. If necessary continue drying until core weight is less than initial value. The cores should then be cling film wrapped and stored. FloatBarrier

5.7 Tabulated Summaries

5.7.1 Coarse Grained

Christine Metz

5.7.2 Fine Grained

Victor Saouma

Method	Cores	Measurements	Preconditioning	Conservation	Preparation	Remarks
France IFSTTAR	d= 100 mm; $l = 2 - -3 \times d$	lateral longitudinal on 3 lines on the generated surface bending of cores	wrapped in humid absorbing paper and a plastic sheet no weight control	38 °C, 100% RH	gradient of expansion curve between 8 and 52 weeks the expansion during the first 8 weeks is not considered in the initial version of the test results preparation. Advanced preparation however shows that this is very severe and that the whole expansion curve might be considered.	
Canada	$d \geq 3 \times d_{max}$, recommended d = 150 mm; $l = 2 - -4 \times l$ unsuitable for reinforced concrete structures.	axial longitudinal and 2 diametral directions at right angle mass"hot measurements" correction for thermal dilatation	sealed after coring, avoid drying, test begin at least one week after coring (stress release)no defined saturation of cores	38 °C, 100 %RH	recommended method,first part of expansion curves not considered (first few days up to several month)limit value: 0.03 mm/m/year problem: alkali leaching shows clearly water saturation influence the expansion curves	Carles-Gibergues and Cyr (2002):long term expansion rates at 38 °C /10[illegible] RH: 0.03 - 0.14 mm/m/yea[illegible] for non-reactive aggregat[illegible] 0.04 - 0.3 mm/m/year f[illegible] reactive aggregates…
				1 N NaOH, 80 °C	"unrealistic results"	
				1 N, NaOH, 38 °C	"absolute degree of reactivity"	
				water	alkali dilution; possibly suitable for concretes with w/c <0.35	
Japan		2 belts with 2 studs at 180° on each side of the corelateral longitudinal bending of cores Span of measurements 10 cm "Hot Measurements" Correction for thermal dilations		1 N NaOH, 80 °C	length change at 28 days rather a test for reactivity of aggregate mix in concrete, correlates with content of reactive aggregates	Expansion >0.1 d [illegible] 3 weeks: deleterious for earl[illegible]-expansive aggregates
	$d = 56$ mm $l = 130$ mm $d = 100$ mm, $l = 250$ mm	3 belts with 2 studs at 180° on each side of the core; Total span measurements 20 cm		20° 100% RH when expansion stabilizes 40 °C, 100% RH 6 months		JCI-DD2 expansion[illegible] >0.1% at 6 months: Deleteriou[illegible] for early-expansive aggregate
Switzerland	$d = 100$ mm; $l = 200$ mm	axial longitudinal and 2 diametral directions at right angle; mass smeasurements at 20 °C	Capillary saturation by absorption over a thin water layer at 20 °C during ca. 1 week (mass constancy) all samples at a defined state of saturation	38 °C, 100 %RH	high expansion rates during the first 2–3 months and lin[illegible] expansion rates during the following period up to one year lin[illegible] expansion rateno - low reactivity potential: < 0.01 mm/m/year[illegible]oderate reactivity potential:< 0.03 mm/m/year; high reactivity [illegible]otential: >0.02 up to 0.07 mm/m/year	

	Method name	France toulouse	France IFSTTAR	Canada	Switzerland	UK	Japan	
1	Summary	Measures length change of mortar bars made from recovered gravel and sand.	Measures length and weight change of core cylinders, includes final drying phase.	Measures expansion and weight change of core cylinders with emphasis on controlling for non-ASR effects.	Measures expansion and weight change of core cylinders with emphasis on comparison of relative rates of expansion throughout test.	Measures expansions and weight of core cylinders with emphasis on replicating field conditions.	Test for aggregate reactivity	Estimation of ultimate potential expansion by the accelerated concrete core expansion test with alkali immersion. Measures the length change of concrete core cylinder to estimate future expansion, coupled with knowledge of concrete
2	Sample type	Mortar bars made from recycled aggregates	100 ± 10 mm diameter x 150 to 250 mm cylindrical core	4” or 6” diameter cylinderical core	Cylindrical core, l=200 mm, d=100 mm	70 mm diameter X 200 mm cylindrical core, as for SDT for expansion to date, but adjust to fit structure. Diameters 30 to 100 mm but should be > 3 x Max aggregate	Cylindrical core: $d = 56$ mm. $L = 130$ mm	
3	Time to results	≃200 days	≃ 1 year	> 1 year	≃ 170 days	Initial by 6 months, continue till it stablises for years	21 days (28 days)	Until expansion stabilizes, if possible
4	Results obtained	Residual expansion	Swelling potential and residual free expansion	Swelling potential and residual free expansion	Residual expansion	Potential for further expansion related to water uptake	Potential expansion (at 80 °C, 1N NaOH pore solution)	
5	Provides estimate of expansion rate?	Yes	Yes	No	Yes	test at ≃ 25C is about 10 times rate in wet structure	No	Yes. By fitting with the sigmoidal curve (e.g. Larive’s equation)
6	Provides estimate of historical expansion?	Yes	Yes (indirect determination using reassessment of the structure by modelling)	No	Broad classification only	No but prior SDT to 5MPa on same sample does	No	No. Past expansion = crack index (field inspection + petrography)
7	Provides estimate of time to ultimate expansion?	Yes	No	No	No	No	No	Yes. Ultimate expansion = past expansion + ultimate potential expansion (fitting with sigmoidal curve)

	Method name	France toulouse	France IFSTTAR	Canada	Switzerland	UK	Japan	
8	Provides estimate of magnitude of ultimate expansion?	Yes	Yes (indirect determination using re-assessment of the structure by modelling)	Broad classification only	Broad classification only	An indication, but timescale of AAR is many decades .	No	Yes. Corrections of temperature for TL and Tc, and of Na concentration in the pore water for expansion are necessary. For a mass concrete, correction of aggregate size for expansion is necessary
9	Variables measured	Length	Length	Length, diameter, weight, temperature	Length, diameter, weight	6 or 9 50 mm gauge lengths, weights of sample, free water, steel plate.	Length	
10	Weight measured?	No	No (weighing at the beginning and at the end of the test)	Yes	Yes	Yes	No	
11	How is weight data used?	N/A	N/A	-Determine end of pre-conditioning phase - Obtain corrected axial expansion	Determine end of pre-conditioning phase (mass constance); control of expansion data (avoid dessication and shrinkage)	Weight and expansion plotted v time to guide interpretaion of shrinkage recovery, expansion of pre-existing gel and long term development of new gel and expansion.	Not used	
12	Sample Handling-a	N.A.	After coring, cylinders washed in clean water, wiped, and stored in sealed plastic bags.	Cores should be taken 1 - 3 in. from exposed surface. Skin effects include high microcracking, low alkali due to leaching, or high alkali due to evaporative concentration.	Samples are cut from cores, usually behind the rebars, about 4–5 cm from exposed surface. Samples should not have cracks, holes etc. (visual inspection of samples required)	Samples should be rinsed, surface dried, and allowed to dry in shade for 1 hour before being wrapped plastic.	Keep moisture. After coring (e.g. L =2[illegible]m), rinse the slimes. Wrap the core with a cling film tightly, keeping the moisture of coring inside, not to entrap air under the film to avoid drying and secondary carbonation	
13	Sample Handling-b	N.A.	Drilling water introduces water which does not affect final expansion if test is begun quickly. Mass however, is increased.	After coring, cores should be superficially dried, sealed immediately after by wrapping in heavy duty shrink wrap and storing in sealed polyethylene bags	After coring, cores should be sealed immediately by wrapping in heavy duty shrink wrap and storing in sealed polyethylene bags. Avoid any drying of the cores.	Dry-coring should be avoided as it heats and dries samples.	Record the exuded gel. On unwrapping, exudation of ASR sol/gel that occurred during transportation (looking dark under the cling film) around the reacted aggregate should be recorded with a core scanner	

	Method name	France toulouse	France IFSTTAR	Canada	Switzerland	UK	Japan
14	Sample Handling-c	N.A.	Care taken to avoid drying of sample before testing begins.		Pins are glued into small boreholes, after dardening, the samples are immediatly conditioned. The whole procedure (sampling, transport, cutting, insert the pins) should not exceed more than 3–4 days.	Weight monitored at every stage to detect drying before test	Core cutting. Cut the core at two levels (e.g. 2 cm, 15 cm) from the concrete surface to get a specimen of L13cm. Remaining deeper portion (15–20cm) can be used for petrographic examination if a parallel core sample is not available.
15	Necessary to extract cores unaffected by ASR?	Yes	No (but recommended by Complementary instructions from (Li et al. 2004) in §8.3.1.2)	No	no, but depend on the scope of the measurements, because expansion rate depends on the degree of reaction rate in the samples	Interesting for comparison and gives shrinkage recovery data.	No. For a diagnosis purpose, comparison with unaffected part should be made
16	Special shipping instructions?	No mention	No mention	No mention	keep the cores sealed in vapor-tight polyethylene bags at low temperature (not more than 20 °C)	No mention	The wrapped core should be put in a sealed polyethylene bag to keep moisture in a shipping parcel
17	How important is sample orientation?	Advancement of ASR is independant of orientation	Expansion dependent on coring direction; longitudinal measurement for both vertically and horizontally cored specimens. Improvement has been suggested with axial and transverse measurement.	Expansion dependent on directions of restraint. Note core direction and measure both axes.	No mention	Mentions high variability in expansion from cores in same structure.	Cracking and expansion are dependent of the orientation of restraint in the RC structure. In a concrete slab of affected pavement, parallel cracks develop difficult to obtain a necessary length of core. Important to distinguish the orientation

	Method name	France toulouse	France IFSTTAR	Canada	Switzerland	UK	Japan
18	Sample conditioning	28 Days at 20 °C in sealed bags;38° C in NaOH solutions	38 °C and 100% RH. Samples placed in containers on grills above water. Containers placed in reactor on larger grill above heated water. Drying phase: oven at 105 °C	38 °C and > 95% RH	1. conditioning : capillary saturation at 20° C until mass constance, during ca. 1 week with sample end in tap water 2. storage at 38 °C/100%RH during ca. 160 days, wrapped in plastic film, monthly measurements of weigth and expansion 3. Drying phase: at 20 °C and 60–70 %RH until initial weigth is reached, final measurement of irreversible expansion	Lab ambient temperature (20–25 °C) with sample end in water is standard, but also adjacent to structure and at 38 °C. Stored at 30 deg angle resting in 10g/kg water in sealed container	80 °C 1N NaOH solution (38 or 60 °C is preferable, if possible). Volume ratio of alkali solution to core specimen should be around 4
19	Immersion?	in NaOH solution	No	No	No	Core lower end only	Yes
20	Comments on initial expansion Phase-a	N.A	Rapid initial expansion is due to water uptake in the core. Linked to capillary features (porosity, microcracking), and hydrophillic products (reaction gels). The initial expansion is however hard to differentiate from the ASR expansion and preparation of the test result shall be linked with a re-assessment calculation (see §8.6.2.4.	Short Term variations are caused by: -Thermal expansion or contraction -Expansion due to release of restraining stress -Expansion due to moisture uptake -Free expansion of existing ASR gel due to stress relief -Expansion of existing ASR gel due to water sorption.	Phase 1, Conditioning, is performed at 20 °C with samples absorbing water by capillary action. Conditioning ends when constant mass reached. Usually after 1 week.	Early part of test characterised by shrinkage recovery and water uptake. Shrinkage recovery swelling is reasonably uniform and stabilises after a month. Compare to AAR expansions with the 6 to 9 gauge length expansions highly variable.	Avoid artifact during pre-conditioning. Do not dry the core after sampling, because a drying process retards expansion to start or suppresses expansion irreversibly due to carbonation of ASR gel. Before start of this test, keep the moisture inside by wrapping. Drying the core at 100 °C followed by the immersion into water produces DEF, although high alkalinity and high temperature of this test decomposes ettringite

	Method name	France toulouse	France IFSTTAR	Canada	Switzerland	UK	Japan
21	Comments on initial expansion Phase-b	N.A	Magnitude of initial expansion depends on moisture state, stress history, ASR history. preparation coupled with a re-assessment model (Sect. 8.6.2.4)	Preconditioning lasts days or months. End of preconditioning characterized by points of inflection in mass and elongation curves.	The purpose of the capillary saturation is to ensure all samples have same initial conditions when storage at 38 °C starts. To limit the expansion as much as possible to hygroscopic swelling, the conditioning has to end as soon as mass constance is reached, otherwise expansion of preexisting gel became important	Comparison of elongation to weight change	Do not miss expansive force liberated after coring. Start expansion test as soon as possible, within a few days after coring. Otherwise, pop-out may occur. This first period of expansion should be covered in the measurement, although indistinguishable from hygroscopic swelling of pre-existing ASR gel and expansion of newly formed ASR gel during the test.
22	Comments on initial expansion Phase-c	N.A	Rapid expansion phase assumed to be 8 weeks.	Recommended to wait at least one week before taking zero measurements (to allow fro stress relief).	Storage in reactor at 38 °C and 100%RH: phase 2: strong characteristic expansion phase during 30–60 days, levelling of and followed by phase 3 of constant expansion levelling slowly of	Notes importance of drying after coring. Witout care water absorbed during coring and rinsing may initiate expansion before Demec points fitted.	Measure the thermal expansion first. Attach belts with measuring studs. Wrap with a cling film tightly to avoid evaporation of water from the core, then put it in the oven at 80 °C for 2 hours to measure the thermal expansion before a zero reading is made
23	Comments on initial expansion Phase-d	N.A	Multon (2008) - believes that omitting first 8 weeks of data results in underestimate of expansion potential. He recommends including first 8 weeks data and correcting for temp and water uptake, See §8.6.2.2.		The part of hygroscopic swelling is determined by drying carefully the samples at the end of the test to the initial weigth. Gives the irreversible ASR induced expansion.		Read the length change in a hot state. Read the expansion within 15 seconds in a hot state after removing the test specimen from 80°C. Before reading the length change, do not leave the test specimen in a room of 20°C, e.g. for one day. Waiting one day at every measurement (1, 3, 7, 14, 21, 28 days) misses expansion during this period and will underestimate potential expansion

	Method name	France toulouse	France IFSTTAR	Canada	Switzerland	UK	Japan	
24	Necessary to extract aggregates?	Yes	No	No	No	No	No	Expansion test itself does not need extraction. However, to estimate water-soluble alkali released from the aggregate in assessing the alkali-budgets, to be used for correction of expansion in this test
25	Role of petrography	none	none	Observation of ASR gel (before and after the test) assists in preparation of expansion results.	important to qualify the expansion results	Recommended for verification of UK Chert All cylindrical ends should be retained after sawing for petrographic analysis	Confirmation whether cracking of concrete is due to ASR, and measurements of crack index (past expansion in the microscopiclevel)	
26	Role of microscopy	none	none	none	important to qualify the state of the samples (advancement of ASR and induced degradations)	none	Evaluation of the stage of ASR in concrete, checking of ASR gel composition and reaction site	
27	Finite element calibration required?	Yes	Not required but residual expansion can be used as an input data in macroscopic approaches	No mention	No mention	No mention	No	Provides parameters (τ_L, τ_c and ε^{∞}) necessary for FEM analyses of structural concrete. Checking with monitoring data of expansion and cracking in the future is desirable to refine the expansion curve

	Method name	France toulouse	France IFSTTAR	Canada	Switzerland	UK	Japan	
Questions								
28	Implications	**How does mortar bar expansion relate to concrete expansion?** No direct relation, relation through inverse analysis	**Results of drying phase imply history of expansion. Does a quantitative relationship exist?** The drying phase helps quantifying the reversible drying strain. However, since drying is performed on the fully damaged material, this drying strain could be overestimated.	**Pre-existing ASR gels expand during preconditioning phase due to moisture uptake. Any possibility of quantifying historic ASR by this measure?** Not really ! Performing Damage Rating Index on a companion core (condition before testing), as well as after completion of the expansion test may help for this.	**How reliable can crack indices determined on the concrete surface and in thin section assess history of expansion?** No Reply.	**Expansion variability important for predicting damage. How does laboratory expansion variability relate to field expansion variability?** Structures with significant cracking damage show high variability. Uniform expansions on all 6 or 9 gauge lengths indicate absence of AAR in that core.	**Any exceptions:** Not suitable for chert aggregate, but usable for metamorphic chert. For other rock types, this method applies generally well, even for the late-expansive aggregates.	
29	Applicability					**Can meaningful data be obtained after only 1 year at 20°C?** Yes, with severe AAR abnormal expansions show in 1 to 3 months. With very slow AAR 4% glass in white cement clear results needed three years data.	**Expansion usable?** Yes. Caution: sound field concrete containing reactive aggregate may produce expansion	Yes. Ultimate potential expansion obtained at elevated temperature and boosted alkali should be corrected in the difference from field conditions based on laboratory data.
30	Kinetic data necessary?						**Other data necessary?** No	Yes. A set of two past expansion data, assessed from visual inspection in the field and petrography, should be combined to draw a sigmoidal expansion curve

Reference

1. Gao, X., Cyr, M., Multon, S., Sellier, A.: A comparison of methods for chemical assessment of reactive silica in concrete aggregates by selective dissolution. Cem. Concr. Compos. **37**, 82–94 (2013)

Chapter 6
Accelerated Expansion Test: Japan

Tetsuya Katayama

6.1 Accelerated Expansion Test

6.1.1 Scope

This alkali-immersion accelerated test method (80 °C, 1N NaOH, [1] was originally intended to rapidly identify the swelling potential of concrete core extracted from structure, e.g. [2, Sect. 5.3] . However, its final expansion can be obtained by extending the term of measurement and fitting expansion curve to a sigmoidal curve after Larive [3]. For concrete structure, maximum expansion potential can be assessed by combining the past expansion of the structure based on crack index, and the ultimate expansion corrected for a difference in alkali content. For comparison, it is recommended to perform this test in parallel with a conventional expansion test under a high relative humidity [4] (40 °C RH>95%), formerly [5] (20 °C→40 °C), by further modifying these latter to 100% RH and hot-reading to avoid dewdrops in the storage container and alkali-leaching in one-day pre-cooling at each reading.

6.1.2 Principles and Methodology

The method and its advantage/disadvantage were summarized in Sect. 5.7.2 along with other ones.

T. Katayama (✉)
Taiheiyo Consultant Co. Ltd., Sakura, Japan
e-mail: katayama@catv296.ne.jp

© RILEM 2021

V. E. Saouma (ed.), *Diagnosis & Prognosis of AAR Affected Structures*,
RILEM State-of-the-Art Reports 31,
https://doi.org/10.1007/978-3-030-44014-5_6

6.1.3 Sampling and Sample Preparation

Sampling: Cracking and expansion are dependent on the orientation of restraint in the RC structure. In a concrete slab of affected pavement, parallel cracks develop, making it difficult to obtain a necessary length of core. Take the core (dia. 56 mm, L 200 mm) from the structure, recording the orientation. After coring, rinse the slimes, wrap the core with a cling film tightly, keeping inside the moisture that was absorbed during coring, and expelling air entrapped under the film to avoid drying and secondary carbonation. On transporting, the wrapped core should be put in a sealed polyethylene bag to keep moisture in a shipping parcel.

Cutting: Cut the core at two levels (e.g. 2, 15 cm) from the concrete surface to get a test specimen of L 13 cm. A deeper portion (e.g. level 15–20 cm) can be used for petrographic examination if a parallel core sample is not available. When a large core (e.g. dia. 100 mm, L 200 mm) is supplied, a slender core with nominal size (dia. 56 mm) should be taken by secondary coring in the laboratory. Attach two stainless belts with measuring studs, set at two opposite directions 180° apart with a span of 100 mm on each belt, to read the warping of the specimen that may occur during storage.

Keeping moisture: Do not dry the core after sampling, because drying delays the start of expansion or suppresses expansion irreversibly due to secondary carbonation of ASR gel. Before starting this test, keep the moisture inside by wrapping with a cling film. A worst case is to dry the core at 100 °C followed by immersion into water. This produces DEF even though high alkalinity and high temperature of this test decomposes ettringite.

6.1.4 Apparatus

- Oven is used to store a stainless container at 80 °C ± 2 °C. Temperature of the oven should be homogenized by a fan and thermostatically controlled.
- Stainless container with a tight lid is used to keep core specimens inside. The container with enough volume is filled with 80 °C 1N NaOH solution, Fig. 6.1b.
- The shape of the container should be vertically long rather than horizontally to minimize evaporation loss of NaOH solution and unexpected condensation of alkali during storage.
- Stainless belts, with attached measuring studs (gauge plug) separated at 180°, are used to attach to the core specimen of diameter 56 mm.
- Each measuring stud (diameter 9–10 mm) contains embedded stainless bead (dia. 1.5 mm). Instead, a prefabricated belt with embedded stainless beads (points) can be used (Fig. 6.1a), or studs can be attached with epoxy bond on the belt
- Contact gauge, with a measuring span of 100 mm, is used to read the length change of the core specimen, down to minimum of 0.001 mm, Fig. 6.1c.

(a) Prefabricated belt with embedded points (arrows)

(b) storage of core specimens within 1N NaOH solution [2]

(c) reading of length change in a hot state within 15 seconds after removal of a specimen from 80°C NaOH solution

Fig. 6.1 Apparatus for testing

6.1.5 Experimental Procedures

Recording as received: On unwrapping, exudation of ASR sol/gel that occurred during transportation (looking dark under the cling film) around the reacted aggregate should be recorded with a core scanner.

Preconditioning: Start the testing immediately within a few days after coring. Do not waste time such as waiting for stabilization of the initial expansion due to liberation of accumulated stress by ASR gel in field concrete (pop-out may occur). Expansion in this first period should be covered in the measurement, although indistinguishable from hygroscopic swelling of pre-existing ASR gel and expansion of newly formed ASR gel during the test. The water to test specimen volume ratio should not be less than 4.

Phase 1 (Initial length at 20 °C): Experimental procedures should be started as soon as possible after sampling core because pop-outs occur within 1 week after coring due to liberation of expansive force in the case of early-expansive aggregate, such as opal-bearing shale or limestone, cristobalite-bearing andesite, and glassy rhyolite rich in hydrated glass. This expansion, called liberated expansion, occurs irrespective of the moisture uptake. The present test method measures total expansion of liberated expansion and accelerated expansion without distinguishing them. Attach the belt on the core specimen and read the initial length of the test specimen (both sides of 180° apart) in the longitudinal direction at 20 °C.

Phase 2 (Thermal expansion at 80 °C): Wrap the test specimen with a cling film tightly to keep the moisture inside that was absorbed during coring. Store it in a dry oven for 2 hours at 80 °C to reach thermal equilibrium without losing moisture. After removing the test specimen from the oven, immediately unwrap the cling film then read the thermal expansion (20 °C → 80 °C) of the test specimen (both sides of 180° apart) at a hot state within 15 s. This length is a zero-reading for further testing to be performed at 80 °C in alkali-immersion.

Phase 3 (Accelerated expansion at 80 °C): Store the test specimen in NaOH solution (80 °C, 1N) and read the length change until 4 weeks (1, 3, 7, 14, 21 and

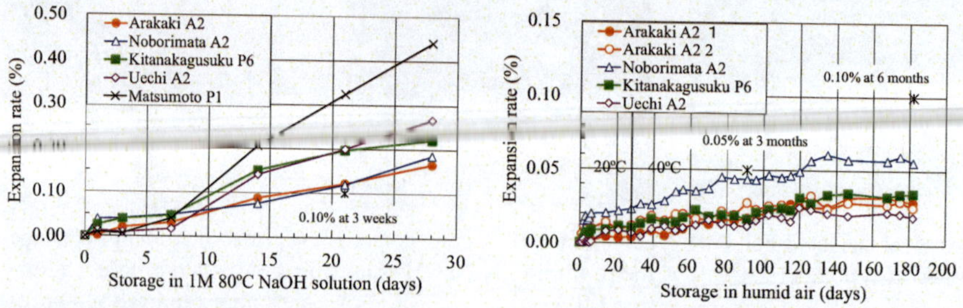

(a) Alkali-immersion expansion test (80°C, 1N NaOH, dia. 5 cm, L 13 cm) identified deleterious nature of the late-expansive aggregates (>0.1% at 3 weeks.)

(b) conventional JCI-DD2 humid chamber expansion test (40°C, dia. 10 cm, L 25 cm, >95% RH) unable to identify deleteriousness (>0.1% at 6 months) of the same aggregates due to late-expansion alkali leaching and drying up of core during storage.

Fig. 6.2 Comparison of concrete expansion tests [6]

28 days), Fig. 6.2a. Read the length change of the test specimen (both sides of 180° apart) in a hot state within 15 s to avoid thermal contraction due to rapid drop of temperature. Before reading the length change, do not leave the test specimen in a room of 20 °C and wait a day at each measurement, because this will miss expansion during this period and will underestimate potential expansion. Record any changes noted during the storage period, e.g. cracking, dissolution of aggregate particles and solution of silica sol/gel into NaOH solution. For a routine test to examine potential reactivity of the aggregate in concrete, measurement can be terminated at 4 weeks. However, for the purpose of assessing ultimate expansion of the late-expansive aggregate, measurement of expansion is necessary at least 6 months, more preferably, until expansion stabilizes to draw an asymptotic curve.

6.1.6 Preparation

6.1.6.1 Evaluation of the Potential Reactivity of Aggregates in Concrete

Potential reactivity of aggregates in concrete is determined from the expansion rate at 3 weeks, while measurement is continued until 4 weeks, Table 6.1. The criterion distinguishing reactive aggregate is based on the correlation with the extent of field deterioration of existing concrete structures in Japan, e.g. [2].

Additionally, reactivity of concrete can be classified as follows:

- High: highly reactive minerals/rocks are contained in a pessimum proportion
- Moderate: reactive minerals are contained
- Low (sound): reactive minerals are absent or present in an acceptable level.

Table 6.1 Potential reactivity of concrete in the alkaline environment (80 °C, 1N NaOH) as correlated with the severity of damage of structures undergoing early-expansive ASR of volcanic aggregates in Japan [2]

Concrete core	Expansion at 3 weeks	Potential reactivity
dia. 56 mm, length 130 mm	>0.1%	Deleterious
	<0.1%	Innocuous*

*Chert aggregate does not produce deleterious expansion due to dissolution into NaOH solution

Table 6.2 Factors that affect expansion of concrete core stored at 80 °C 1N NaOH (Katayama unpublished results)

<table>
<tr><th rowspan="2">field ASR</th><th rowspan="2">Expansion at 3 weeks</th><th rowspan="2">potential reactivity of concrete</th><th colspan="2">Reactive aggregate</th><th colspan="4">Inhibitor of ASR cracking in field</th></tr>
<tr><th>early-expansive</th><th>late-expansive</th><th>low-alkali cement, Na_2Oeq < $3 kg/m^3$</th><th>slag fly ash</th><th>stress</th><th>alkali leaching</th></tr>
<tr><td rowspan="4">Yes</td><td>>0.25%</td><td>high</td><td rowspan="2">andesite, rhyolite, shale, containing opal, glass, cristobalite and tridymite,</td><td></td><td rowspan="2">highly reactive aggregate in pessimum proportion</td><td rowspan="2">not used in effective amount, with highly reactive aggregate</td><td></td><td></td></tr>
<tr><td>0.1-0.25%</td><td>moderate</td><td rowspan="2">welded tuff, sandstone, shale, slate, quartzite, schist, gneiss containing micro-crystalline quartz</td><td></td><td></td></tr>
<tr><td>0.04-0.1%</td><td>low**</td><td></td><td></td><td></td><td></td><td></td></tr>
<tr><td><0.04%</td><td>***</td><td></td><td>chert***</td><td></td><td></td><td></td><td></td></tr>
<tr><td rowspan="4">No*</td><td>>0.25%</td><td>high</td><td rowspan="3">andesite, rhyolite, opaline shale</td><td></td><td rowspan="2">reactive aggregate</td><td></td><td rowspan="3">prestressed concrete (girder) stressed concrete (lower dam)</td><td rowspan="3">leaching of alkali by flowing water (canal)</td></tr>
<tr><td>0.1-0.25%</td><td>moderate</td><td rowspan="2">welded tuff, sandstone, shale, slate, quartzite, schist, gneiss</td><td></td></tr>
<tr><td><0.1%</td><td>low</td><td>less reactive aggregate</td><td>used in effective amount</td></tr>
<tr><td><0.1%</td><td>low</td><td></td><td></td><td></td><td></td><td></td><td></td></tr>
</table>

*Younger concrete (<3 years old) may not develop cracks
**Older concrete, may contain reactive minerals mostly consumed (>40 years old)
***This method is unsuitable for non-metamorphic chert which dissolves into NaOH solution
Important expansion limits, commonly found in some accelerated mortar bar tests, are highlighted

For the late-expansive aggregate, expansion limit could be >0.04% at 21 days, Table 6.2. In the alkali-immersion concrete prism test (7.5 × 7.5 × 30 cm, 1N NaOH, 80 °C), late-expansive aggregates produce expansion >0.04% at 24 days [7].

Several factors affect the expansion rate of the concrete core specimen, Table 6.2. Some field deteriorated concretes produce small expansion in this test: Non-metamorphic chert dissolves into NaOH solution, causing small expansion. By immersion into 1N NaOH solution, core specimens from sound concretes either containing reactive aggregate with low-alkali cement or undergoing leaching of alkali by flowing water (e.g. canal wall), produce large expansion. Core specimens from pre-stressed concrete (e.g. girder) or highly stressed concrete (e.g. lower level of large dam) may produce large expansion as a result of liberation from constraint.

Blastfurnace slag or fly ash concretes (e.g. massive concrete of footing or dam) may not produce deleterious expansion in this test, even if they present weak petrographic evidence of ASR (reaction rim).

6.1.6.2 Evaluation of the Final Maximum Swelling Potential

Ultimate swelling potential of concrete (maximum potential expansion) can be assessed from the expansion curve when it shows an asymptotic nature after a long period, e.g. 6 months or longer, or by the fitting to some suitable sigmoidal curve. In principle, ultimate expansion does not seem to change by the temperature at a certain range, e.g. 20 °C or 80 °C. 1N NaOH solution is equivalent to pore water of concrete (w/c = 0.5) with cement alkali of Na_2O 1.44% (Sect. 6.4.2). Hence, ultimate expansion obtained by the boosted alkali shall be corrected for the difference in the alkali content between the expansion test and field concrete, using both data on the released alkali from the aggregate and calibration from CPT data for the aggregate.

6.1.7 Examination at End of Test

After the final length change measurements, each specimen shall be checked for the following items:

Warping: if the length of the core specimen in two directions shows a marked difference during the test period, expansion rates in two directions should be reported, Fig. 6.3a.

Dissolution of aggregate: if any particles of aggregate show surface dissolution or discoloration, record this either by scanning the core specimen after removing the belt, or by close-up photography, Fig. 6.3b.

Cracking of aggregate and cement paste: it would be informative to check the pattern of cracking and record the cracking or dissolution of aggregate. A core scanner is useful.

Petrographic examination: polarizing microscopy of polished thin section is useful in confirming what rock type and reactive minerals have reacted to produce ASR gel and expansion cracks in the core specimens tested. When reactive phase is difficult to identify under the polarizing microscope, SEM observation, coupled with EDS analysis on the same polished thin section is useful in identifying the reacted phases in this test.

Besides this, petrographic examination is necessary to check the presence, location, and pattern of cracking, and appearance of deposits or exudations, their nature, thickness and continuity.

(a) Warping due to severe cracking and different expansion rates (Δ=0.33%) at 600 days, (image by K Mukai)

(b) Dissolution of the surface of chert aggregate at 28 days, image by K. MukatiWarping due to severe cracking and different expansion rates (Δ=0.33%) at 600 days, (image by K Mukai)

Fig. 6.3 Concrete core specimens after NaOH immersion test, Katayama (unpublished results)

Color of NaOH solution: record any changes in color of the solution, and exudation of siliceous sol/gel if there is any.

Evaporation of NaOH solution: record this if anomalous by weighing the container with filled solution.

6.1.8 Report

Report the following information:

1. Type of structure and appearance (cracking, pop-outs, exudation, etc)
2. Environment of structure (availability of water and moisture, chloride ion, freezing and thawing)
3. Grade of deterioration due to ASR of the structure (e.g. stage I, II, III, IV), position of core sampling in the structural member
4. Time until first measurement of the length was made
5. Source of aggregate
6. Type, source of Portland cement (alkali content), and concrete mix, if possible
7. Table of reading length changes and average length change in percent
8. Any significant features of the core specimens (exudations, warping, cracking, rim-formation, dissolution of aggregate, etc.) during and after test
9. Any pozzolanic materials used (fly ash, slag) and their formulated content in the construction record.

6.1.9 Precision and Bias

Precision shall be considered satisfactory if the differences in the value for percentage expansion of any given specimen in a group from the average value for percentage expansion of the group do not exceed 15%. Thermal contraction of the core (dia. 55 mm) within 1 min after removal from the NaOH solution is around 0.006 and the range of reading lengths after one minute is within 0.015% [8].

Inter-laboratory test [9] revealed that the ratio (Y) of expansion (x) of the core specimens to the standard core size (dia. 56 mm) at 14 days was $Y = -0.0028x + 1.16$. This means that the reduction of core size from 56 to 44 mm corresponds to an increase in the relative expansion of 4%, owing to an increased absorption of NaOH solution by the core specimens. In the same concrete block, expansion rate at 28 days is within 15% of the average of each specimen, but it increases in the aged field deteriorated concrete in which ASR is advanced and cracking of aggregate was pronounced and heterogeneous.

6.1.10 Limitations and Applicability to Analysis

1. Testing condition (80 °C, 1N NaOH) is harsh to chert and flint aggregates due to dissolution, and is not applicable to concrete with appreciable amounts of these rock types. Other expansion tests (humid air) are recommended for them.
2. The method could produce deleterious expansion of a concrete core from sound field structures made with reactive aggregate and low-alkali cement, which needs caution.
3. When the accelerating condition is moderated to 40 °C, 0.5N NaOH, alkali-immersion test better correlates with the severity of field deterioration of structures [10]. However, it produces extremely slow expansion unsuitable for assessing final expansion (ε^{∞}).
4. Fitting short term expansion curves could produce large errors of final expansion (ε^{∞}) of concrete core (1N NaOH, 80 °C) and apparent activation energy (U^{∞}) according to the rock type.
5. To assess final expansion (ε^{∞}), this test should be performed for a long term (>180 days), preferably until expansion stabilizes, and result should be fitted with Larive equation. Brunetaud equation should not be used for a short-term data when expansion is still continuing, because this could highly overestimate expansion.

6.2 Assessment of the Past and Future Swelling Potential Based on Field Expansion and Laboratory Core Expansion Test

Several approaches are possible to assess residual expansion of the ASR-affected field concrete structure. One of the easiest ways is to postulate a sigmoidal expansion curve following Larive's equation, using three parameters (ε^{∞}, τ_L, τ_C) identified by a combination of visual inspection in the field, petrographic examination, and accelerated concrete core expansion test, Fig. 6.4. To estimate the past expansion of concrete, crack indices can be used from field inspection [11], microscopic examination and both [12]. Cautions must be taken in interpreting the result and the cracking is due to expansion by ASR [13]. To confirm this Petrographic Examination should be performed.

After identifying ASR and the type of the reacted aggregate by petrographic examination, i.e. early-expansive or late-expansive, it is necessary to obtain expansion data to fix the expansion curve (Fig. 6.5a). preparations of the sigmoidal expansion curves are given in Appendix D, along with relevant references, e.g. [14].

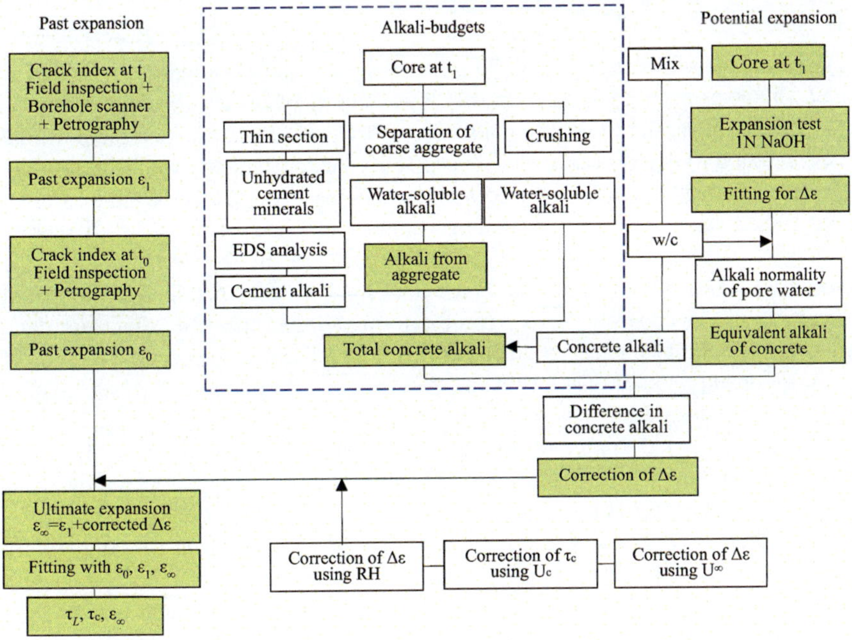

Fig. 6.4 Flow chart to estimate the ultimate expansion of concrete based on accelerated concrete core expansion test, visual inspection of structure, petrography and laboratory test for alkali-budgets, Katayama (unpublished results)

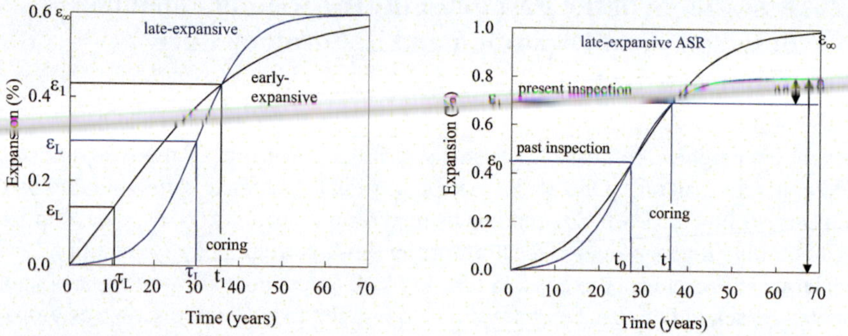

(a) Sigmoidal expansion curves following Larive's equation to show the early- and the late-expansive ASR in concrete

(b) To determine the curve, two past expansion data (τ_L, ε_L) (t_0, ε_0) and asymptotic expansion of core $\Delta\varepsilon$ are necessary (late-expansive ASR). (Katayama unpublished results)

Fig. 6.5 Determination of past expansion

6.2.1 Determination of the Past Expansion (ε_L, ε_0)

In order to determine the shape of the expansion curve in field conditions, Table 6.3, at least three points on the curve preferably spanning over the inflection point (ε_L, τ_L), or a combination of two past expansion data (t_1, ε_1, and t_0, ε_0) and the potential final (maximum) expansion ($\Delta\varepsilon$) deduced from the accelerated expansion test, are necessary, Figs. 6.5b and 6.6b. Using the core samples from the damaged portion of the structure has a great merit; because they contain the same materials as the structure (aggregate, alkali content, mix proportion, etc.), there is no need to obtain the same unused aggregate. The essential part of the crack indices can be obtained during visual inspection of the concrete surface (time t_1). To estimate the past expansion data, (time t_0), reading the additional displacement from the former inspection, records of crack width and photographs for image analysis, and examination of preserved core specimens are useful.

It should be noted that conventional concrete prism test produces smaller final expansion at higher temperature, because concrete prism is much affected by alkali leaching and drying up than concrete cylinder with smaller surface area/diameter ratio in the concrete cylinder test. If a correction of ultimate expansion for temperature is deemed necessary, apparent activation energy based on more reliable concrete cylinder test ($U_\infty = 690$ K, according to conventinal expression) should be used with Arrhenius equation (Table 6.3). Care should be taken, however, that application of some fitting equation to immature S-shaped expansion curve (e.g. Brunetaud equation at early stage or lower temperature) could produce excessively larger final expansion (Fig. 6.7a).

Table 6.3 A practical approach to predicting ASR expansion in concrete structure (Katayama unpublished results)

<table>
<tr><td colspan="2" rowspan="2"></td><td colspan="4">structure (deteriorated part)</td></tr>
<tr><td colspan="3">field +laboratory</td><td>laboratory</td></tr>
<tr><td rowspan="3">expansion</td><td>time (year)</td><td>present t_1</td><td>past t_0</td><td>future t_∞</td><td>future potential</td></tr>
<tr><td>expansion (%)</td><td>past ε_1</td><td>past ε_0</td><td>final ε_∞</td><td>potential expansion $\Delta\varepsilon = [\varepsilon_\infty]$</td></tr>
<tr><td>method</td><td>crack index</td><td>crack index or reading displacement</td><td>$\varepsilon_\infty=\varepsilon_1+\Delta\varepsilon$</td><td>accelerated expansion test
(e.g. 80°C 1N NaOH)
temperature: the lower, the better</td></tr>
<tr><td rowspan="2">expansion curve</td><td>sigmoid</td><td colspan="4">$\varepsilon t=\varepsilon_\infty \times [1-\exp(-t/\tau c)]/\{1+\exp[-(t-\tau_L)/\tau c]\}$ Larive's equation</td></tr>
<tr><td>constants</td><td colspan="4">determination of time constants (τ_L, τc) of field concrete
using two known data (t_1, ε_1),(t_0, ε_0) and estimated final expansion (ε_∞)</td></tr>
<tr><td rowspan="7">corrections for expansion</td><td rowspan="2">temperature</td><td colspan="3">not necessary
(if special correction is necessary to fit the field condition)</td><td>alkali-immersion test:
apparent activation energy of finsl expansion U_∞
positive to negative, depending on rock types</td></tr>
<tr><td colspan="4">if final expansion of core changes by temperature, use Arrhenius equation
for temperatures K_1, K_2 with apparent activation energy U_∞
$\Delta\varepsilon_{\infty1}/\Delta\varepsilon_{\infty2}= \exp U_\infty[(1/K_2)-(1/K_1)]$ (see Table 6.6)</td></tr>
<tr><td rowspan="2">alkali content Na_2Oeq (kg/m^3)</td><td colspan="3" rowspan="2">total alkali of concrete (R_1)
= cement alkali + water-soluble alkali from aggregates and other sources
(alkali-budgets)</td><td>equivalent alkali of concrete core (R_2)
(immersed in NaOH solution)</td></tr>
<tr><td>correction for the difference R_2-R_1
$(\Delta\varepsilon)'=\Delta\varepsilon - (R_2-R_1)\times F$, F=0.06</td></tr>
<tr><td rowspan="2">degree of humidity or water saturation</td><td colspan="3" rowspan="2">dry part of the structure produces small or negligible expansion with large (τ_L, τc) and small final expansion (ε_∞)</td><td>no drying or alkali-leaching occurs in alkali solution. corresponds to relative humidity 100%</td></tr>
<tr><td>from this data, expansion for lower humidity or water saturation can be assessed</td></tr>
<tr><td>stress</td><td colspan="3">highly stressed part of the structure produces small or negligible final expansion (ε_∞), which can be confirmed by crack index</td><td>based on free expansion</td></tr>
<tr><td>limitation</td><td></td><td colspan="3">no limitation</td><td>not applicable to chert aggregate which dissolves and produces small expansion</td></tr>
</table>

* if correction of final expansion for temperature is deemed necessary for concrete prism data, apparent activation energy of final expansion $U_\infty = -690$ K (instead of 0), can be used with Arrhenius equation, although the reality may be smaller.
Important parameters and equations used in this chapter are highlighted

6.2.2 *Estimation of Maximum Potential Expansion ($\Delta\varepsilon$)*

An accelerated expansion test (80 °C, 1N NaOH solution, dia. 5.6 cm by L 13 cm, [15], is performed on the concrete core specimen taken from the damaged structure to estimate final expansion [ε^∞] by fitting the expansion curve with Larive's equation Fig. 6.7. This is used as potential residual expansion ($\Delta\varepsilon$) to estimate final expansion ε^∞ of the field concrete: $\varepsilon^\infty = \varepsilon_1 + \Delta\varepsilon$.

In principle, application of lower temperature (e.g. 38, 60 °C) and lower alkalinity of the solution (e.g. 0.5N–0.75N NaOH) is better for the concrete core test, Fig. 6.9b but late-expansive aggregates in concrete do not produce converging expansion within one year.

Unlike the elaborated method proposed by Sellier, Bourdarot, Multon, Cyr, and Grimal [16], this method does not need to perform a set of accelerated mortar bar test at a high alkaline condition (60 °C, Na_2O_{eq} 8 kg/m^3) for the coarse and fine aggregates that are extracted from the sound area of the structure, assumptions of several kinetic parameters, nor mathematical fitting to find the reaction rate of reactive quartz in the aggregate.

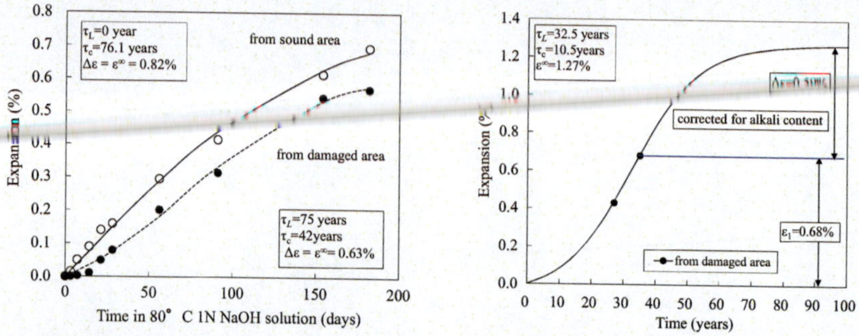

(a) Accelerated expansion test of concrete cores from the sound and damaged part of the structure, adapted from Katayama, Sarai, Higashi, and Honma [15]

(b) Reconstructed expansion curve using data from the damaged part and concrete expansion test. Past expansion was assessed by visual inspection in the field and petrographic examination, Katayama (unpublished result)

Fig. 6.6 Combination of field expansion and accelerated expansion test

6.3 Problems Encountered in Alkali-Immersion Test

6.3.1 Expansion Parameters of Concrete (τ_L, τ_C, ε^∞)

Expansion curves of concrete can be drawn by fitting with Larive Eq. 6.1which defines a sigmoidal curve with three expansion parameters (τ_L, τ_c, and ε^∞) [3]: where ε^∞: ultimate expansion (final asymptotic expansion); τ_L: latency time (duration before the inflection point of S-shaped curve); τ_C: characteristic time (inverse of the expansion rate); t: time; $\varepsilon(t)$t: free expansion at time t (free strain, originally in volume which is three time that of linear expansion).

For a continuously growing expansion over a long period, most typically in DEF, a Brunetaud version (Eq. 6.2) with five parameters (τ_L, τ_c, and ε^∞, Φ, δ) including correction terms (Φ and δ; $\Phi < \delta$)) could be used [17].

$$\varepsilon(t, T) = \frac{1 - e^{-\frac{t}{\tau_C(T)}}}{1 + e^{-\frac{(t-\tau_l(T))}{\tau_C(T)}}} \quad \text{Larive Equation} \quad (6.1)$$

$$= \varepsilon^\infty \frac{1 - e^{-\frac{t}{\tau_C(T)}}}{1 + e^{-\frac{(t-\tau_l(T))}{\tau_C(T)}}} \left(1 - \frac{\Phi}{t + \delta}\right) \quad \text{Brunetaud Equation} \quad (6.2)$$

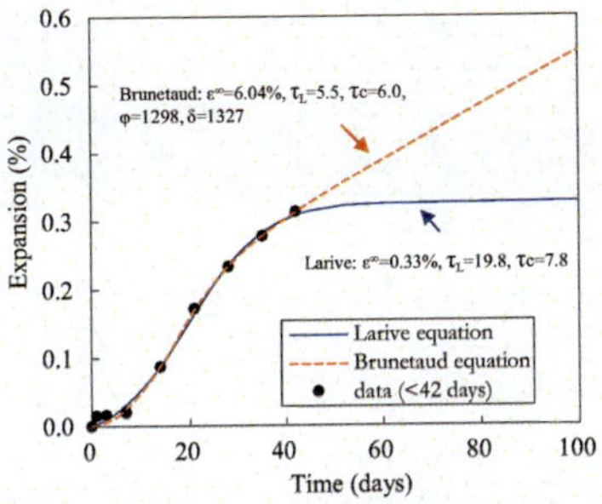

(a) Brunetaud equation fitted to the early period expansion data (expansion stage 0.5) highly overestimates the ultimate expansion (ε^{∞}), approaching 10 times. Drawn using data Mukai [8]

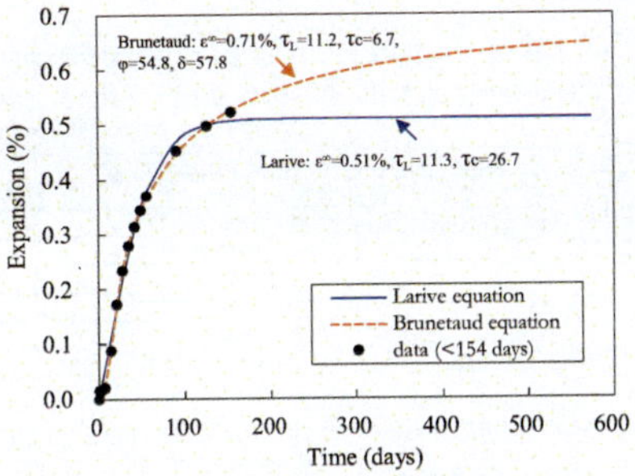

(b) Brunetaud equation fitted with the middle period expansion data (expansion stage 0.9) overestimates ultimate expansion by 20% than the real asymptotic expansion (ε^{∞} 0.59%). Drawn using data from Mukai [8]

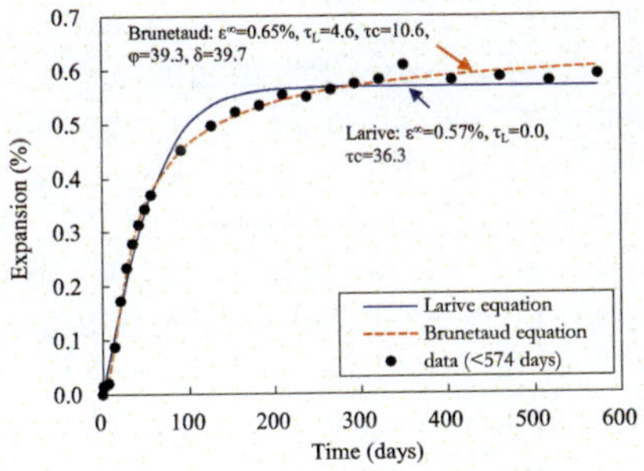

(c) Brunetaud equation fitted with the long-term expansion data (expansion stage 1.0) tells ultimate expansion only 10% over the real asymptotic expansion (ε^{∞} 0.59%), while Larive equation underestimates only by 3%. Drawn using data from Mukai [8]

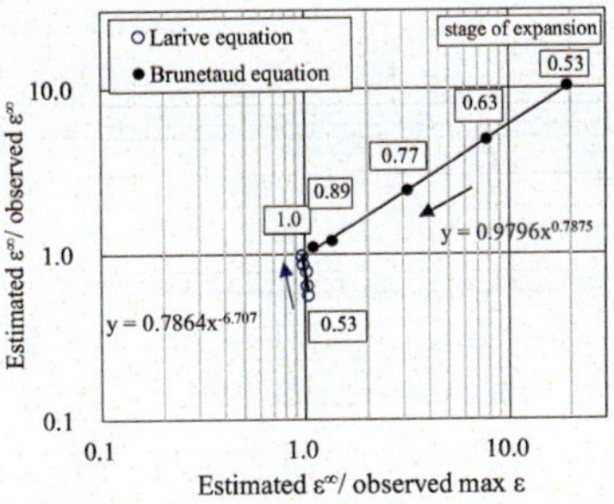

(d) Estimation errors of the ultimate expansion (ε^{∞}) fitted to the Larive equation and the Brunetaud equation, as a function of the stage of expansion. Drawn using data from Mukai [8]

Fig. 6.7 Estimation errors of the fitting equations based on the same sigmoidal expansion curve (observed ε^{∞} = 0.59%) of a core specimen (1N NaOH immersion, 80 °C, quartzite), Katayama (unpublished data)

6.3.2 Estimation Errors of Fitting Equations Applied to Core in 1N NaOH Solution

Estimation errors of Larive equation and Brunetaud equation were compared at a different maximum time, based on the same S-shaped expansion curve of a core specimen with known asymptotic expansion ε^{∞} = 0.59%, Fig. 6.7a–c. Brunetaud equation considerably overestimates the ultimate expansion (ε^{∞}), e.g. as much as

Table 6.4 Estimation errors of ultimate expansion (ε^{∞}) based on the fitting with expansion data at shorter ages against observed ultimate expansion ε^{∞} (0.59%), Katayama (unpublished data)

	time (days)	42	56	91	154	[illegible]
stage of expansion	observed max ε/observed ε∞	[illegible]	[illegible]	0.[illegible]	0.89	1.0
Brunetaud equation	estimated ε∞ (%)	6.04	2.86	1.42	0.71	0.65
	estimated ε∞/observed ε∞	x10.2	x4.85	x2.41	x1.20	x1.10
	estimated ε∞/observed max ε	x19.3	x7.75	x3.14	x1.36	x1.10
Larive equation	estimated ε∞ (%)	0.33	0.38	0.46	0.51	0.57
	estimated ε∞/observed ε∞	x0.56	x0.64	x0.78	x0.86	x0.97
	estimated ε∞/observed max ε	x1.04	x1.03	x1.02	x0.98	x0.97
Weighted average	ε∞ from two equations B:L=1:4	0.59	0.57	0.58	0.54	0.59

Important equations and parameters used in this chapter are highlighted

10 times when applied to early stage expansion curve (expansion ratio $\varepsilon/\varepsilon^{\infty} = 0.5$) (Table 6.4, Fig. 6.7a). By contrast, Larive equation underestimates the ultimate expansion but converges fast just below the maximum expansion observed (Fig. 6.7a). At an advanced stage of expansion, both equations converge to the observed ultimate expansion ($\varepsilon^{\infty} = 0.59\%$, Fig. 6.7c).

Even with the same ultimate expansion, the above estimation error could occur with the S-shaped expansion curves of concrete prism test (CPT) at different temperatures. Therefore, application of the Brunetaud equation to the early expansion curve at lower temperature might lead to conclude that the ultimate expansion at lower temperature is several times larger than that at higher temperature.

Published data of alkali-wrapping CPT (Na_2O_{eq} 3 kg/m^3), which was possibly less affected by alkali leaching and drying up of concrete on cold reading, indicates that the ultimate expansion (ε^{∞}) at 20 °C based on the Brunetaud equation was more than 3 times that obtained at 60 °C, giving a large activation energy for expansion ($U\varepsilon^{\infty}$) in the temperature range 20 °C-60 °C [18, 19]. However, this could be a superficial effect of the fitting, because the curves of data points in Figs. 6.7a–c broadly resembles the shape of expansion curves of CPT obtainable at 20, 40 and 60 °C within one year. Large difference in the ultimate expansion by temperature most likely disappears when a correction is made in a way following Table 6.4, Eq. 6.3. Final expansion ($\varepsilon^{\infty} = 0.59\%$) can be approximated by the weighted average of log expansion at given time (t_1), choosing a suitable ratio 1:4 in this case from Brunetaud and Larive equations.

$$\ln \varepsilon^{\infty} = \left[\ln \varepsilon^{\infty}((\text{Brunetaud}, t_1) + 4 \times \ln \varepsilon^{\infty}((\text{Larive}, t_1)\right] / 5 \qquad (6.3)$$

6.3.3 *Temperature Dependence of Final Expansion in 1N NaOH Solution*

At higher temperature, many reactions occur, including ASR (reaction of silica and formation of ASR gel), pozzolacic reaction related to clay minerals which consumes reactive silica thereby reducing ASR expansion, and mobilization of ASR gel and

Table 6.5 Apparent activation energy of final expansion (U_∞) of concrete cylinders and prisms immersed in 1N NaOH solution, calculated by Katayama (in preparation)

type	rock type	region	U_∞(K) Larive eq.	Brunetaud eq.	average L:B=1:1	concrete conditions	size	Na_2Oeq (kg/m³)	remarks
early-expansive	andesite (6:4 non-reactive)	Japan	1720	1860*	1790	40,60,80°C <26 weeks	cylinder dia.7.5, L15cm	8.0 (NaCl added)	marked pessimum [20]
late-expansive	sandy dolostone	Quebec	1370	-760	310	38°C <57 weeks	CSA prism 7.5x7.5x30cm	3.9 (NaOH added)	[7]
	meta-greywacke	Quebec	-120	-370	-250				
	quartz sandstone	Quebec	-400	-2510	-1460				
	greywacke-siltst.	Ontario	-720	-2720	-1720	80°C <9 weeks			
	greywacke	New Brunswick	-4770	-420	-2600				
	chert (gravel)	Japan	-3250	-5200	-4230	40, 80°C <57 weeks	cylinder dia.10, L20cm	1.53 (cement only)	dissolution, softening [21]
	chert		-3740	-6790	-5270				

*Non-linear due to pessimum phenomenon; Data sources: [20] Kuroda, Inoue, Yoshino, Nishibayashi, and Miyagawa (2008); [7] Berube, Dupont, Pigeon, and Stoian (1992); [21] Iwatsuki and Morino (2006)

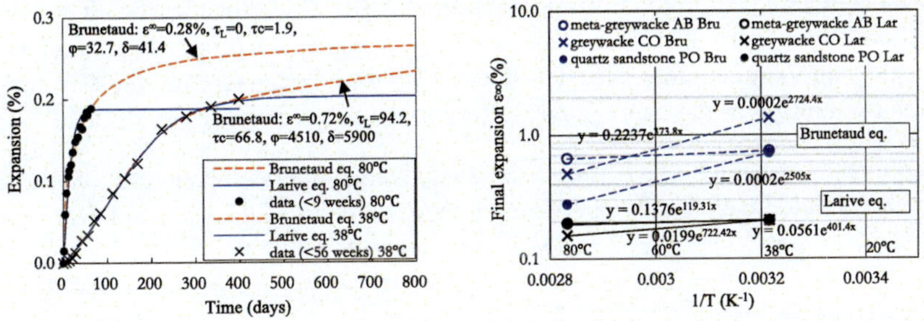

(a) Expansion (ε^∞) of alkali-immersed concrete prisms of a quartz sandstone fitted to two equations, drawn using data from Bérubé, Dupont, Pigeon, and Stoian [7]

(b) Final expansion (ε^∞) of alkali-immersed concrete prisms of sedimentary rocks fitted to two equations, drawn using data from Bérubé, Dupont, Pigeon, and Stoian [7]

Fig. 6.8 Expansion of concrete prisms (Na_2O_{eq} 3.9 kg/m^3, NaOH) of late-expansive sedimentary aggregates immersed in 1N NaOH solution (Katayama, in preparation)

its dissolution into alkaline solution. As a result, some rock types produce decreased expansion at higher temperature, while others produce increased expansion.

Apparent activation energy (U_∞) of laboratory concrete specimens immersed in 1N NaOH solution was calculated here from fitting published expansion curves [7, 20, 21] with Larive equation and Brunetaud equation, Table 6.5. Results show that U_∞ depends on the rock type of the aggregate. However, accuracy of estimation for Canadian aggregates is low owing to short duration of testing at 80 °C, Table 6.5 resulting in a great difference in two equations, Fig. 6.8a, b. Hence, to convert the estimated final expansion (ε^∞) at 80 °C into the one at a lower temperature, some intermediate values of U_∞ from two equations should better be used.

- Early-expansive aggregate: highly reactive andesite has positive values of U_∞, indicating that final expansion (ε^∞) increases by temperature. This means that ASR

gel is formed vigorously overwhelming dissolution into NaOH solution despite the expected decrease in viscosity of ASR gel at higher temperature.
- Late-expansive aggregate: greywacke, sandstone and siltstone have negative values of U_∞, showing that ε^∞ decreases by temperature. This is due possibly to a pozzolanic reaction that takes place concurrently with ASR in more argillaceous rocks, which forms non-expansive CASH gel and hydrogarnet in addition to ASR gel.
- Ideally, certain rock types with $U_\infty = 0$ unlisted in Table 6.5 are expected to have final expansion that is unaffected by temperature. Expansion of the testing data is necessary.
- Chert has the smallest values of U_∞, i.e. smallest expansion at 80 °C. This is a result of migration and dissolution of ASR gel into surrounding NaOH solution, which prevail over production and fixation of ASR gel within concrete. It has been reported that core specimen containing chert aggregate produces smaller final expansion than concrete cylinder, because of direct exposure to NaOH solution and dissolution of the aggregate resulting in the drop of rigidity [21].

Therefore, expansion limits in the accelerated alkali-immersion concrete core expansion test should be changed depending on the rock types (Tables 6.1 and 6.2).

6.3.4 *Alkali Dependence of Final Expansion in NaOH Solution*

No data has been published of alkali-immersion concrete expansion test performed at different temperatures for a long-term (>1 year). Only available data is for a highly reactive andesite in concrete cylinder up to 26 weeks [20], which can be interpreted as follows:

- NaOH solution with higher concentration (2N) than pore water of concrete produces larger expansion at higher temperature ($\varepsilon^{80^\circ C} > \varepsilon^{60^\circ C} > \varepsilon^{40^\circ C}$). On the Arrhenius plot, gradient of expansion over temperature is positive, i.e. negative slope on Fig. 6.9a.
- Alkali solution with lower concentration (0.5N) than pore water produces smaller expansion at higher temperature ($\varepsilon^{40^\circ C} > \varepsilon^{60^\circ C} > \varepsilon^{80^\circ C}$). Gradient of expansion over temperature is negative., i.e. positive slope on Fig. 6.9a
- The andesite aggregate presents pessimum phenomena at an intermediate alkali concentration and temperature.

Using available expansion curves up to 26 weeks [20], final expansion (ε^∞) and its apparent activation energy (U_∞) were calculated here based on Arrhenius plot Fig. 6.9a, b. The apparent activation energy (U) of concrete is proportional to the log (square root) of NaOH concentration, Fig. 6.9b, which suggests that ASR expansion is governed by diffusion. Normality of NaOH solution that does not affect final expansion by temperature ($U_\infty = 0$) appears to be 0.5–0.6N, while that at 26 weeks

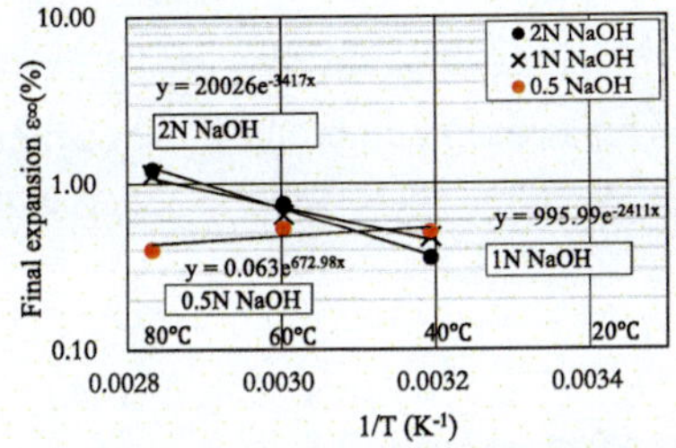

(a) Final expansion (ε^{∞}) of concrete cylinders (andesite) after Larive's equation, drawn using data from Kuroda, Inoue, Yoshino, Nishibayashi, and Miyagawa [20]

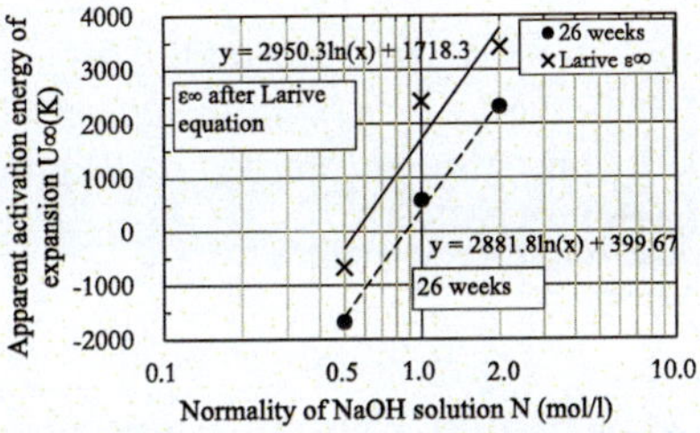

(b) Apparent activation energy of final expansion (U_{∞}) of concrete cylinders (andesite) immersed in NaOH solution, drawn using data from Kuroda, Inoue, Yoshino, Nishibayashi, and Miyagawa [20]

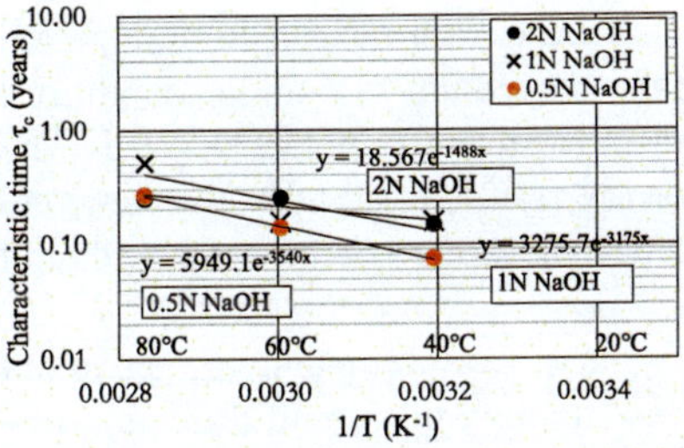

(c) Characteristic time (τ_C^{∞}) of concrete cylinders (andesite) after Larive's equation, drawn using data from Kuroda, Inoue, Yoshino, Nishibayashi, and Miyagawa [20]

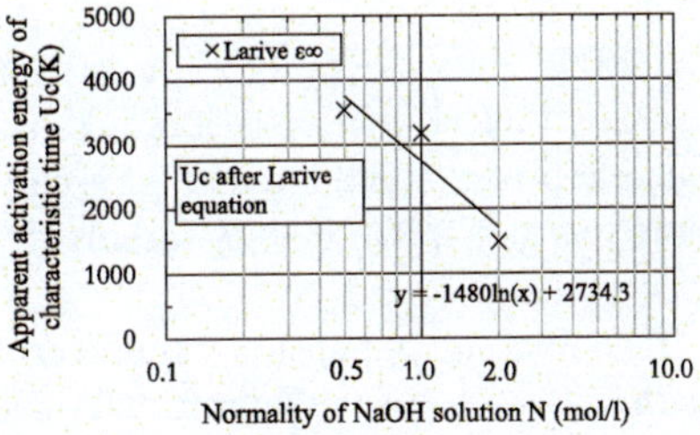

(d) Apparent activation energy of characteristic time (U_C^{∞}) of concrete cylinders (andesite) immersed in NaOH solution, drawn using data from Kuroda, Inoue, Yoshino, Nishibayashi, and Miyagawa (2008) using data from Kuroda, Inoue, Yoshino, Nishibayashi, and Miyagawa [20]

Fig. 6.9 Normality of NaOH solution, final expansion and characteristic time of concrete cylinders (Na_2O_{eq} 8 kg/m^3, NaCl) containing highly reactive andesite aggregate (Katayama in preparation)

is 0.8–0.9N. This means that application of 1N NaOH solution at 80 °C produces excessive final expansion (U_{∞}) in the long term and that correction is necessary in estimating final expansion in field condition (e.g. 16 °C). For convenience, parameters growing at higher temperature are defined as positive. Then, andesite in concrete immersed in >0.56 mol/l NaOH solution has positive U_{∞} (Eq. 6.4) whereas U_C is invariably positive.

For andesite in concrete immersed in N mol/l NaOH solution, U_{∞} is positive:

$$U_{\infty}(K) = 2,950 \ln(N) + 1,720 \tag{6.4}$$

$$U_C(K) = -1,480 \ln(N) + 2,730 \tag{6.5}$$

For other rock types, see Table 6.5.

In the case with a highly reactive andesite, with a pessimum proportion at a high alkali concentration (Na_2O_{eq} = 8 kg/m^3) adjusted with NaCl, result is not usable straightforwardly.

6.4 Problems Encountered in Using Parameters of Concrete Prisms/cylinder Tests

6.4.1 A Simple Estimation of Expansion Parameters

After converting results of accelerated alkali-immersion test (80 °C, 1N NaOH) to concrete prism conditions (e.g.38 °C, NaOH concentration nearest to pore water, RH 100%, etc.) (Sects. 6.3.3 and 6.3.4) further conversion to outdoor conditions (e.g. 20 °C, RH 40%) is necessary based on expansion parameters derived from concrete prism/cylinder data (Table 6.6). Likewise, results of comparative concrete core expansion test (e.g.38 °C, arbitral Na_2O_{eq}, RH 100%, etc.), if there is any, should be corrected .

To obtain the expansion parameters (τ_L, τ_C) at a given temperature, activation energy (U_L, U_c) of the aggregate is necessary. Although this could differ by rock type, well-known values of French limestone aggregate (τ_L 9,400 K, τ_C 5,400 K: [3, 22] have been used extensively. To correct the time constants (τ_L, τ_C) for concrete alkali, total alkali content estimated in the alkali-budgets should be used, instead of the total alkali content from the cement only. For convenience, signs of apparent activation energy should be distinguished. Positive: values of Arrhenius plot have a negative slope and increase at higher temperature. Negative: values of Arrhenius plot have positive slope and decrease at higher temperature. This means that apparent activation energies of final expansion (U_∞) of concrete prism/cylinder tests are negative, although they are positive in the conventional usage in Table 6.6. These parameters have been known for a few rock types, hence some of them have to be used as a substitute.

6.4.2 Estimation of the Amount of Alkali from Immersed NaOH Solution

Theoretically, alkali ion concentration of interstitial pore water of concrete is not affected by the unit mass of cement in concrete, but by the water-cement ratio.

1. Theoretically, if 100% of cement alkali is converted to water-soluble alkali, normality of alkali ions should be as follows, Table 6.7:

Table 6.6 Corrections of expansion parameters in the concrete prism/cylinder test (compiled by Katayama)

temperature (K) $K=T°C+273$		alkali content (R) (Na_2Oeq kg/m^3)		apparent water saturation degree (S) 0-1.0	relative humidity (H) 0.6-1.0
original CPT data					
PARTNER project (2006)	LCPC	JCI committee (1989, 1991)		LCPC	Univ.Marne la Vallee
av. 9 European rock types	French limestone	av. 5 Japanese andesites	av.5 greywackes UK and Canada	French limestone	Belgian limestone
38°C, 60°C concrete prism Na_2Oeq 5.5kg/m^3	23-58°C concrete cylinder, Na_2Oeq 5.1kg/m^3	40°C concrete prism Na_2Oeq 2.8-7.4kg/m^3	38°C concrete prism Na_2Oeq 1.7-7kg/m^3	38°C concrete cylinder Na_2Oeq 5.1kg/m^3	(20°C) mortar cylinder Na_2Oeq 12.9kg/m^3
$\tau_L=1.13\times10^{-9}e^{(7780/K)}$ $\tau_C=2.52\times10^{-9}e^{(7770/K)}$	$\tau_L=4\times10^{-12}e^{(9412/K)}$ $\tau_C=6\times10^{-7}e^{(5373/K)}$	$\tau_L=163e^{-0.25R}$ $\tau_C=360e^{-0.24R}$		$\tau_L=172e^{-1.08S}$ days $\tau_C=40e^{-0.41S}$ days	
		$V_L=-0.25$kg/m^3 $V_C=-0.24$kg/m^3	$\varepsilon_\infty=0.063$ xNa_2Oeq -0.16		$\varepsilon_\infty=0.0108e^{3.28H}$
activation energy $U_L=7780$(K) $U_C=7770$(K) $U_\infty=1900$(K)*	activation energy $U_L=9400$(K) $U_C=5400$(K)	R2: alkali of prism R1: to be used for alkali of structure	R2: alkali of alkali-soaked core R1: to be used for alkali of structure	$\tau_L1/\tau_L2=\exp1.08(S2-S1)$ $\tau_C1/\tau_C2=\exp0.41(S2-S1)$	
$\tau_L1/\tau_L2=\exp U_L[(1/K1)-(1/K2)]$ $\tau_C1/\tau_C2=\exp U_C[(1/K1)-(1/K2)]$ $\varepsilon_\infty1/\varepsilon_\infty2=\exp U_\infty[(1/K1)-(1/K2)]$	$\tau_L1/\tau_L2=\exp U_L[(1/K1)-(1/K2)]$ $\tau_C1/\tau_C2=\exp U_C[(1/K1)-(1/K2)]$	$\tau_L1/\tau_L2=\exp V_L[((R1-R2)]$ $\tau_C1/\tau_C2=\exp V_C[(R1-R2)]$	$\varepsilon_\infty'=\varepsilon_\infty-(R2-R1)\times0.06$(%) formulated using data by Sibbick & Page (1992), Blackwell et al.(1996), Fournier et al. (2000)	apparent water saturation** / water absorption in water 1.0 / >0.5% humid 1.0-0.2 / 0.5-0.1% sealed 0.2-0 / 0.1-0% drying 0 / <0%	$\varepsilon_\infty1/\varepsilon_\infty2=\exp3.28(H1-H2)$
recalculated from Kawabata et al. (2013)	Larive (1998), Li et al.(2000)	recalculated from Kawabata et al. (2013)		** a rough assumption using data of Li et al. (2000)	calculated from Poyet et al.(2004)

*large value, due to drying up and alkali leaching of concrete prism at higher temperature, 1 kJ/mol = 120.3 K

**A change from water saturation (S = 1) to drying condition (S = 0) at 38 °C results in extension of τ_L in 2.9 times and τ_C in 1.5 times, according to data by Li et al. [22]

Important equations and parameters used in this chapter are highlighted

Table 6.7 Alkalinity of pore water in concrete prism

	w/c	$[OH^-]$ (mol/L) = $[Na^+]$ (eq/L) = [NaOH] (mol/L)	$[OH^-]$ (mol/L) = $[K^+]$ eq/L) = [KOH] (mol/L)
theoretical	not specified	= 0.3226 x Na_2O (%)/(w/c)	= 0.2123 x K_2O (%)/(w/c)
Diamond (1989)	0.5	= 0.699 x Na_2O (%) + 0.017	
Helmuth (1993)	not specified	= 0.339 x Na_2O (%)/(w/c) + 0.022	

$$[Na^+](eq/L) = \{(Na_2O(\%)/100)x1000/(w/c)\}/\{[Na_2O]/2\} = 0.3226 \times Na_2O(\%)/(w/c) \tag{6.6}$$

$$[K^+](eq/L) = \{(K_2O(\%)/100)x1000/(w/c)\}/\{[K_2O]/2\} = 0.3226 \times K_2O(\%)/(w/c) \tag{6.7}$$

At w/c = 0.5, 1N of NaOH concentration in the pore water in concrete corresponds to the alkali content of cement of Na_2O_{eq} 1.55% (= 1/(0.3226 × 2)). At w/c = 0.5, 1N of KOH in the pore water corresponds to the alkali content of cement of K_2O 2.36% (=1/(0.2123 × 2)), with its Na equivalent being 0.658 × 2.36 = 1.55 (%). This value gives the maximum concentration of equivalent alkali when a concrete specimen is immersed within alkaline solution.

2. Diamond [23] presented an empirical formula at w/c = 0.5. In his regression line, when OH^- concentration (=NaOH=Na^+_{eq}) is 1N, cement alkali Na_2O_{eq} corresponds to 1.41%. This means that 91% of alkali is retrieved (=1.41 × 100/1.55). $[Na^+]$ (eq/L) = 0.699 × Na_2O (%) + 0.017. This means that 91% of alkali is retrieved (=1.41 × 100/1.55).

$$[Na^+](eq/L) = 0.699 \times Na_2O(\%) + 0.017 \tag{6.8}$$

3. Helmuth et al. [24] presented a general empirical formula, according to Stark [25]:

$$[OH^-](eq/L) = [Na^+](eq/L) = 0.339 \times Na_2O/(\%)/(w/c) + 0.022 \tag{6.9}$$

Result of calculation is shown in Fig. 6.10a, along with conditions of concrete prism tests with and without immersion in NaOH solutions. For convenience, the content of cement alkali is regarded as being increased by the alkali immersion of concrete (w/c = 0.5), although the amount of alkali that is absorbed by the concrete prism should be calculated using the value of water absorption or wet chemical analysis of alkalis in concrete.

6.4.3 Relationship Between Alkali Content and Final Expansion

The slope of the expansion curves, Fig. 6.10b appears to approach flat in the alkali range over 7 kg/m^3, but this does not mean that the influence of alkali content on the expansion can be neglected. Corrections of expansion in the concrete prism tests for alkali content between Na_2O_{eq} 3 and 7 kg with a steep slope in the Fig. 6.10b, should be made. Similar relationships but with different intercepts have been known with several rock types, e.g. Japanese andesite [26] and chert, Canadian limestone and greywacke [27], UK flint and greywacke [28, 29], etc.

6.4.4 Problem of Alkali-Wrapping with Condensed NaOH Solution

Modified versions of CPT and ACPT have been applied locally, such as wrapping or immersion of concrete prism specimens with water or NaOH solutions to avoid artifacts that reduce long-term expansion of concrete prism caused by drying and leaching of alkalis.

For example, when 1N NaOH solution is used to immerse the prism (Na_2O_{eq} = 3.9 kg/m^3, w/c = 0.55) that contains pore water with a similar normality of alkali ions

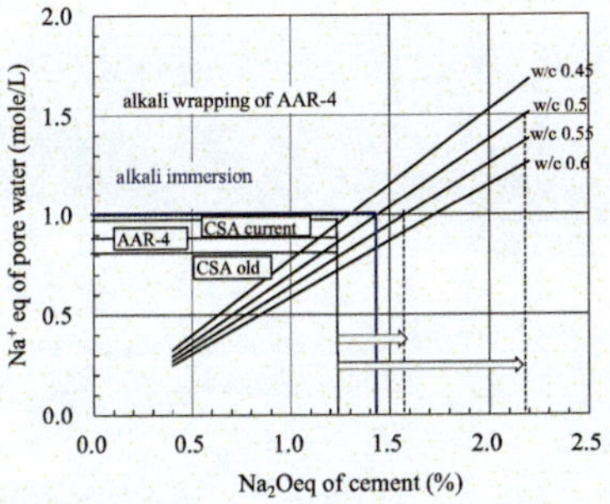

(a) Relationship between cement alkali content and molarity of NaOH (=OH=Na) in the pore solution for different w/c ratios, calculated from the Helmuth equation [24]. Conditions of concrete prism tests are indicated.

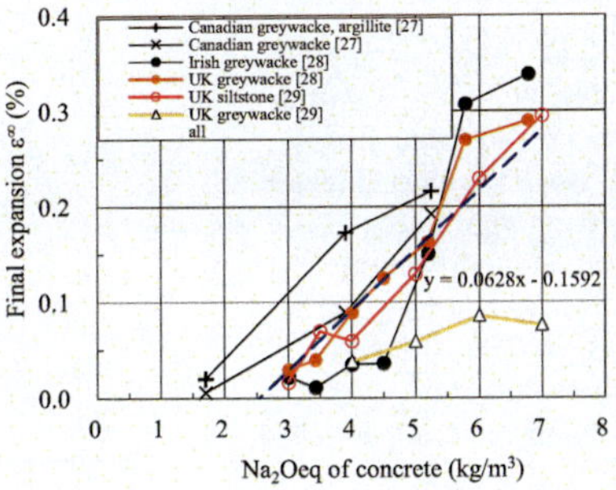

(b) Final expansion vs alkali level of concrete prism containing greywacke (38°C) with average slope, blue line $\varepsilon^{\infty} = 0.063 \times$ Na_2O_{eq} (kg/m^3)-0.16, for correction. Drawn using data from: [27–29]

Fig. 6.10 Alkali correction (Katayama unpublished results)

(0.79N NaOH, Helmuth equation), similar expansion is obtained without producing drying and alkali leaching and excessive expansion [7]. In the concrete prisms containing cement alkali of Na_2O_{eq} 1.25% at w/c = 0.5 (CPT at 38 °C, Na_2O_{eq} 5.25 kg/m^3; ACPT at 60 °C, Na_2O_{eq} 5.5 kg/m^3), concentration of alkalis in the pore water will be 0.87N, Fig. 6.10a, closer to 1N. Immersion of concrete prism into 1N NaOH solution transforms the alkalinity of pore water to 1N NaOH. In this case, according to Helmuth equation, 1N NaOH solution corresponds to cement alkali of Na_2O 1.44% at w/c = 0.5 (1N [Na^+] (mol/L) = 0.339 × Na_2O (%)/(w/c) + 0.022), and concrete alkali of 6.37 kg/m^3 (= 440 × 1.44/100). At higher temperature (e.g. 80 °C), some specimens may release reactive silica within 1N NaOH solution [7].

When highly concentrated alkali solution (e.g. 1.5N NaOH, alkali-wrapping concrete prism test being used locally in Japan)) [1] is used for wrapping concrete prism (e.g. RILEM AAR-4: w/c = 0.5, Na_2O_{eq} = 5.5 kg/m^3), pronounced penetration of Na ions and acceleration of ASR take place near the concrete surface, Fig. 6.11 resulting in excessive expansion. 1.5N NaOH solution corresponds to pore solution in concrete made with cement alkali of Na_2O = 2.18% (1.5N NaOH = 0.339 × Na_2O (%)/0.5 + 0.022) and concrete alkali of Na_2O = 9.59 kg/m^3 (= 440 × 2.18/100). Such a concentrated NaOH solution dissolves highly reactive volcanic rocks (andesite) and liberates alkali silica gel into the solution, thus producing relatively small expansion in ACPT [26]. However, in the course of reaction, some part of alkali ions in the pore water of concrete may be removed by the formation of ASR gel and rosette-like crystals, if water-soluble alkali is not supplied from alkali-bearing minerals in the aggregate. Conversely, concrete prism stored in alkali-free water produces smaller expansion than that in a humid condition due to leaching of alkali from the prism into surrounding water [7]. It is therefore important to select a concentration of NaOH

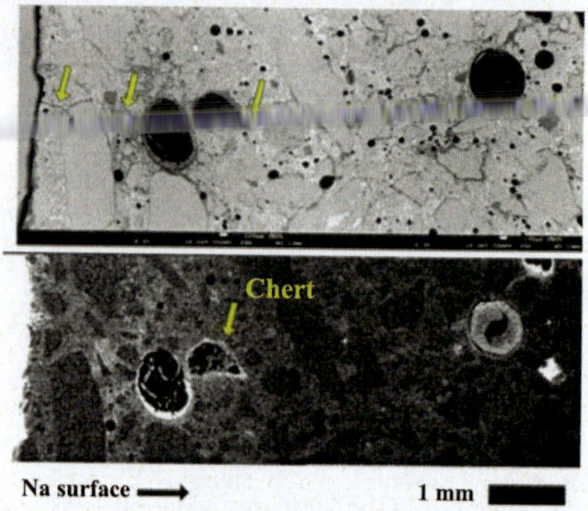

Fig. 6.11 Alkali-wrapping concrete prism test showing ingress of Na^+ ions from 1.5N NaOH solution, denser than the pore solution (Japanese modification of RILEM AAR-4 CPT: 60 °C, Na_2O_{eq} 1.25%, 5.5 kg/m^3, cement 440 kg/m^3, w/c = 0.5), Katayama (unpublished data)

solution as close as possible to that of the pore solution of the concrete prism and core.

6.4.5 Correction of Alkali Content of Concrete for Expansion ($\Delta\varepsilon'$)

There are two things to know before a correction is made: R_1 and R_2.

R_1: In this case, the amount of cement alkali and concrete are assumed to be 1.07% (=1.3 × minimum clinker alkali 0.82% from past data [15] and 3.21 kg/m^3 (= 1.3 × 0.82 × assumed concrete mix 300 kg/m^3), respectively, Table 6.8. When a possible contribution of water-soluble alkali from the aggregate around 0.5 kg/m^3 is taken into account based on the reacted rock type (greywacke plus gneiss, table C.3, total concrete alkali of concrete (R_1) will be Na_2O_{eq} 3.71 kg/m^3.

R_2: Since the concrete core specimen is immersed in 1N NaOH solution in the accelerated expansion test (Table 6.8), OH^- ion concentration of pore water in the core is thought to become 1 mol/L, equivalent to the same concentration of Na^+_{eq} ion. Figure 6.10(b) shows relationships between the final expansion *vs.* alkali level of concrete prism containing greywacke (38 °C) with a regression line (blue)

$$\varepsilon^{\infty} = 0.063 \times Na_2O_{eq}(kg/m^3) - 0.16\% \tag{6.10}$$

A rounded value of the slope (0.06) is used as a coefficient (F) of $(R_2 - R_1) \times F$ to correct a boosted expansion of the accelerated concrete core expansion test for the boosted alkali content of the core. Alkali content of cement corresponding to 1N NaOH pore solution is equivalent to Na_2O_{eq} 1.44% at w/c = 0.5 (=(1 − 0.022) × 0.5/0.339) from Helmush Eq. 6.9. At the same cement mix, total equivalent alkali of concrete (R_2) will be Na_2O_{eq} 4.32 kg/m^3 (= (1.44/100) × 300 kg/m^3). R_2–R_1 is "boosted alkali content" of the concrete immersed in NaOH solution, i.e 0.61 kg/m^3

Table 6.8 Past expansion and potential expansion of concrete based on concrete core expansion test with water-soluble alkali releasable from reactive aggregate in concrete, (Katayama unpublished results)

	field concrete					concrete core											
						alkali-budgets						solution chemistry			expansion test		
	inspection + petrography		expansion curve			petrography	total alkali of concrete including water-soluble alkali from aggregate								storage 80°C, 1N NaOH		
	27 years	35 years	fitting			clinker	cement			aggre-gate	con-crete	w/c	pore water		fitting		
	ε_0 %	ε_1 %	ε^{∞} %	τ_L years	τ_C years	Na_2Oeq %		kg/m^3	Na_2Oeq kg/m^3				normality	NaOH*	ΔE	T_L days	T_C days
original concrete						0.82	1.07	300	3.21			0.5	0.75N				
adding alkali from aggregate						0.82	1.07	300	3.21	0.5*	3.71	0.5	0.86N		0.59		
immersion in NaOH						0.82	1.07	300	3.21	0.5*	4.32	0.5	1.00N	1N	0.63	75	42
field expansion	0.43	0.68	1.27	33	10												

Important equations and parameters used in this chapter are highlighted

(= 4.32 − 3.71). As a result, the boosted expansion is only minor, 0.04% (0.61 × 0.06) hence the corrected expansion ($\Delta\varepsilon'$) is 0.59% (= 0.63 − 0.04).

Correction of $\Delta\varepsilon$:

$$(\Delta\varepsilon') = \Delta\varepsilon - (R_2 - R_1) \times F = 0.63 - (4.32 - 3.71) \times 0.06 = 0.59(\%) \quad (6.11)$$

When expansion of concrete is not affected by temperature, then the final expansion ε^{∞} of this concrete structure was estimated to be 1.27%

$$\varepsilon^{\infty} = \varepsilon_L + \Delta\varepsilon' = 0.68(\%) + 0.59(\%) = 1.27(\%) \quad (6.12)$$

This may be higher than the final expansion of concrete as would generally be expected from the concrete prism test. However, the range of expansion over 1% is not rare in the field deteriorated concrete structures in Japan. With hydraulic structures containing similar late-expansive aggregates in the same region with this case, past expansion estimated from on-site crack index, Crack index ranged 0.5–2.2%, even addition of data from the core expansion test and crack index by microscopy was not made [12]. According to the on-site monitoring of highway structures undergoing early-expansive ASR under the influence of de-icing salts without the evidence of DEF [2], local expansion of up to 2.8% (over-ranged) has been recorded in the decade that follows 20–25 years after construction [12, 30]. Hence, the above expansion rate estimated here is not exceptional to ASR.

6.4.6 *Alkali Dependence on Other Expansion Parameters (τ_L, τ_C)*

This relationship has been known for Japanese aggregates (mostly early-expansive andesite) based on early data of concrete prism test, more or less affected by alkali leaching and drying up of concrete. It is a power function of the difference in alkali content in the range Na_2O_{eq} 3–7kg/m^3, having a gentle slope of $V_L = -0.25$ kg/m^3 for τ_L, and Vc = −0.24 kg/m^3 for τ_C [31]. In principle, it is desirable to correct the shape of the expansion curve of concrete prism and concrete cylinder.

$$\tau_{L1}/\tau_{L2} = \exp[V_L(R_1 - R_2)] \tag{6.13}$$

$$\tau_{C1}/\tau_{C2} = \exp[Vc(R_1 - R_2)] \tag{6.14}$$

However, in the concrete core expansion test of core specimen taken from aged field structures, e.g. 20–60 years old, aggregate in concrete has already reacted passing the latency time of the induction stage (τ_L) on the original expansion curve, and hence the slope of the curve characterizing the reaction rate (τ_C) has been distorted from the original state. Hence, correction of parameters from core specimen (τ_L, τ_C) for alkali content is not performed here. Instead, only final expansion of the core ($\Delta\varepsilon$) is used to deduce final expansion of field concrete (ε^∞), and combining with past field expansion (ε_0, ε_1), S-shaped curve is drawn by fitting with Larive equation Fig. 6.6.

6.4.7 *Apparent Expansion Loss of Concrete Cured at Humid Air at Higher Temperature*

There has been a long-standing question about whether the ultimate expansion of concrete (typically, concrete prism test, CPT) could differ by temperature. Early results of CPT (Sect. 7.5 cm × 7.5 cm, humid curing) presented markedly smaller expansion at higher temperature, Table 6.9 [3, 7, 22, 29, 32–45]. For instance, accelerated CPT at 60 °C (limestone) produced about a 50% decrease of expansion at 38 °C [46]. This is generally attributed to artifacts due to drying out of prism surfaces and alkali-leaching [46, 47] enhanced by condensed trickling in the cooling process of prisms from 60 °C to 20 °C for cold-reading of length change [33, 48, 49].

By contrast, concrete cylinder test (CCT) gives more reliable results than CPT. Larive [3, pp. 156–158] indicated that asymptotic expansion of the cylinders (dia. 13 cm, L 24 cm, humid curing, limestone) was practically the same, e.g. 0.22% at 23 °C (average of 11 samples) and 0.20% at 38 °C (average of 29 samples) with a gap of 10%. A typical example illustrated by Larive, Laplaud, and Joly [48], stored in 100% RH and measured by hot-reading at 38 °C, showed essentially the same asymptotic expansion as that of 23 °C: with a small surplus of 4–7% at 38 °C, when fitted with Larive equation or its modified Brunetaud version (Fig. 6.12a). The same magnitude of long-term expansion of concrete prisms at 23 and 38 °C (100% RH) has also been reported for quartzite and reactive basalt in Australia (Table 6.9) [43].

However, with humid curing at 60 °C in the reactor (limestone), accelerated CCT by [3] presented a drop of about 30% asymptotic expansion and a gain of 25% weight than those at 38 °C. Such a large loss of expansion at 60 °C can be improved

Table 6.9 Expansion parameters of representative rock types in the concrete prism and concrete cylinder tests (Na_2O_{eq} mainly 5.1–5.5 kg/m^3) fitted with Larive equation (compiled by Katayama)

	rock type	reactive mineral	region	38°C sigmoidal curve			38°C 0.04 %	60°C sigmoidal curve			60°C 0.04 %	apparent activation energy (expressed as positive)			alkali	sample	geologic age
				ε^{∞} %	τ_l days	τ_c days	x days	ε^{∞} %	τ_l days	τ_c days	x days	U_L K	U_C K	U_ε K	Na_2O_{eq} kg/m^3		
early-expansive	opal	op						0.24	4.1	0.53	2				5.0	at 36% of coarse [44]	
			Texas	0.38	57.9	19.6	33	0.42	11.9	5.8	3.5			x	5.25	sand [45]*	
				x				x						790		sand [45]*	
	andesite	op, cr	Japan	0.13	"0"	44.1	5	0.14	"0"	10.8	1		8440	x	5.5	Tohoku, at 30% of coarse, pessimum [32]	Neogene
								0.17	18	10	12				5.5	No.5[33]	
		tr	Japan	0.36	90.7	64.9	33								5.57	H2 19 [34]#	Cenozoic
		cr	Japan	0.23	78.3	30.7	41								5.8	K 376 [34]#, at 50%	Cenozoic
	rhyolitic tuff	mq	Quebec	0.30	89.6	42.3	43								5.1	TU-2A[7]	
moderate-expansive	soda glass	gl						0.86	10.8	23.1	5				5.0	as sand, 100%RH [44]	
	andesite	gl	Japan	0.35	119.8	56.0	53								5.13	T2 404 [34]#	Miocene
	limestone		France	x				x+						580	5.13	[3]**	Carboniferous
				0.19	66.7	21.5	45	0.19+	204+	60+	140+			x		[42]**, North France, 100% RH, hot-reading	
				0.23	64.8	22.0		0.12	8.9	6.6		9400 [6]	5400 [6]			[3]**	
								0.35	78	22.2						[3]**, 38°C in water	
								0.22+	225+	47.7+						[3]**, 23°C	
		cq,mq	Belgium	0.28	90	57.9	50	0.15	"0"	20.6	14		4870		5.5	B1 [35]	Paleozoic
		ch,cq	Ontario	0.22	97	49	52	0.16	17	14	14	8280	5790		5.25	Spratt av.(5) [36]	Ordovician
	quartzite		Australia	0.18	126	70	75	0.20+	794+	259+	390+			x	6.9	Queensland [43]	
			France					0.24	15.7	34.4	18				5.0	100% RH [44]	
	cataclasite	cq,mq	Norway	0.48	186.7	142.6	65	0.18	34.4	22.7	19	7980	8670		5.5	N1, av.(2) [35]**	Proterozoic
late-expansive	meta-rhyolite	mq	Norway	0.09	67	77	92								5.13	Moflåt [37]**	Proterozoic
	chert	ch, cq	Japan	0.09	55.8	101.6	100	0.06	"0"	13.3	24		10590		5.5	Chubu, gravel [32]	Paleo-Mesozoic
			Japan	0.14	121.5	54.6	96								5.8	SC 237 [34]#	
	greywacke /siltstone	mq	Nova Scotia	0.28	138	42	75								5.1	Rocky Lake, NE-2A [7]	E.Paleozoic
		mq	Ontario	0.17	192	94	111								5.1	CO-2A [7]	
	greywacke	mq	New Brunswick	0.23	157	44	100								5.1	Springhill, NB-2A[7]	Silurian
				0.54	191	64	74								5.25	Springhill, NB [41]	
				0.54	233	116	75								5.25	Mactaquac [41]	
		mq,cq	UK	0.39	233	53	148	0.18	36	21	22	8950	4370		5.5	Bally Barnes UK1,av.(2) [35]***	Permo-Triassic
	sandstone	cq,mq	Norway	0.27	211	71	106								5.5	N2, av.(2) [35]***	L.Precam.
	quartz sandstone	qo	Quebec	0.19	218	131	130								5.25	Potsdam [38]	Cambrian
	gravel: greywacke, argillite	mq	Ontario	0.22	282	97	160	0.19	46	41	35	8560	4040		5.25	Sudbury gravel av.(5) [36]	Pre-cambrian
	meta-greywacke	mq	Quebec	0.21	217	110	124									Notre-Dame-du-Nord [7]	Pre-cambrian
	pelit. hornfels	mq	Japan					0.09	"0"	87	91				5.5	Chubu, No.14 [33]	Mesozoic
	biotite schist	mq	Switzerland	0.13	111	68	118	0.15	123	68	120			x	1.2%	C 40°C, in water [39]	
	green schist		Quebec	0.16	156	122	130								5.25	Sherbrooke [38]	
	quartzite	oq	Norway	0.05	97	47	140								5.13	Tana [37]	L.Precamb
			Brazil	0.05	"0"	138.9	330	0.06	67	29	90		7390	x	5.25	Ilha Solteira [40]	
	mylonite$	mq	Norway	0.07	168	72	200								5.13	Ardal [37]***	Proterozoic
	granite-gneiss		Brazil	0.06	54	104	202	0.10	72	39	72		4630		5.25	Recife [40]	Proterozoic
	strained granite	mq	UK	0.11	204	61	162								5.5	Anglesey [29]***	Pre-cambrian
	granite		Brazil	0.04	13	74	330	0.05	82	23	112		5510	x	5.25	Embu [40]	L.Precamb.
	gravel: quartzite, chert	mq,cp	UK	0.13	119	211	227	0.09	38	33	50	5380	8750		5.5	gravel [35], Thames Valley UK2,av.(2)***	
	gravel: rhyolite, quartzite	mq	Norway	0.45	920	399	225	0.14	55.5	27.5	43	13250	12620		5.5	gravel [35], N5 60°C av.(2)***	Proterozoic

concrete cylinder: * dia.10cm x L20cm; **dia.13cm x L24cm; + 23°C; # 40°C; *** expansion not converge within 1 year; $ quartz diorite
Reactive mineral: op opal; cr cristobalite; tr tridymite; gl glass; ch chalcedony; cq cryptocrystalline quartz; mq microcrystalline quartz; oq quartz overgrowth

(light green) extremely late-expansive aggregate, (magenta) 40°C, 60°C (1.5N NaOH-wrapping), (orange) water-wrapping or in water
(dark green) nearly the same ultimate expansion in two temperatures (difference <10%)

Data sources: [32] Tanaka, Yamada, Sagawa, and Ogawa (2015); [33]Yamada, Karasuda, Ogawa, Sagawa, Osako, Hamada, and Isneini, (2014); [34]JCI (1991), [7]Berube, Dupont, Pigeon, and Stoian, (1992); [3]Larive (1998); [22] Li, Ulm, Coussy, Larive & Fan (2000); [35]Nixon & Lane (2006); [36]Fournier, Chevrier, de Grosboir, Lisella, Folliard, Ideker, Shehata,Thomas & Baxter (2004); [37]Jensen (1993); [38]Durand (2000); [39] Haha, Gallucci, Guidoum & Scrivener (2007); [40]Sanchez, Kuperman, Helene & Kihara (2008); [29]Sibbick & Page (1992); [41]Hayman, Thomas, Beaman & Gilks (2010); [42]Larive & Coussy (1996); [43]Shayan, Diggins & Ivanusec (1996); [44]Moisson,Cyr, Ringot & Carles-Gibergues (2004); [45]Stacey, Folliard, Drimalas, & Thomas (2016)

by a modified CCT with lateral sealing and water-capping (dia. 10–15 cm, L 15–20 cm), which produced only 13% smaller expansion (1 year) than that at 38 °C [45]. When expansion data are confined to concrete cylinders with the size easy to handle (dia. 10 cm, L 20 cm), a scatter of long-term expansion between 38 and 60 °C is greatly diminished, approaching nearly identical (for a selected pair, Fig. 6.12b. This modified CCT produced about 50% (38 °C) to 30% (60 °C) larger expansion than CPT with the same mix.

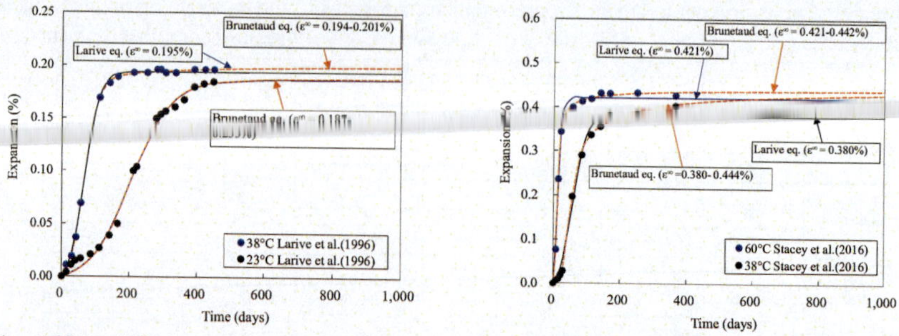

(a) 100% RH curing and hot-reading (dia.13cm, L24cm, limestone), expansion calculated from data by Larive, Laplaud, and Joly [48]

(b) Sealed curing with water cap and cold-reading (dia.10cm, L20cm, reactive sand), drawn using data selected from Stacey, Folliard, Drimalas, and Thomas [45]

Fig. 6.12 Similar ultimate expansion of concrete cylinders at a temperature range of 23–60 °C fitted with Larive equation and Brunetaud equation, Katayama (unpublished data)

From the above CCT data carefully taken, it can be assumed that asymptotic expansion (ε^{∞}) of concrete is independent from the temperature in a certain range, if the aggregate, humidity and mix proportion are the same [48], whereas other expansion parameters (τ_L and τ_C) depend on temperature following the Arrhenius law [50, 51]. Concrete with a large section is less affected by the expansion loss at higher temperature or low humidity. This is because with large concrete (1) expansive ASR gel is unlikely to escape from the concrete surface even if higher temperature lowers the viscosity of the gel (Katayama, in preparation) and (2) drying of interstitial water from the interior of concrete which promotes transformation of expansive ASR gel into non-expansion rosette crystals is generally retarded.

Another method to avoid alkali leaching and drying is to perform a CPT wrapped with alkali solution (e.g. 1.5N NaOH: JCI-TC115FS Sect. 4.3 [1, 32]). This option may produce essentially similar ultimate expansion at 40 °C (0.13%) and 60 °C (0.14%) with early-expansive andesite (Table 6.9), but expansion was smaller than expected due to dissolution of the early-expansive volcanic aggregate into condensed alkali solution [26].

6.4.8 Apparent Activation Energy for Ultimate Expansion (U_{∞})

This parameter is still controversial. Apparent activation energy (U_{∞}) for ultimate expansion in CPT could be calculated using the Arrhenius law, based on expansion of concrete prisms at different temperature [18, 31]. Tables 6.6 and 6.9 show positive values of U_{∞} as reported by the original authors, but hereafter they will be shown as negative values, because parameters (ε^{∞}) increasing at high temperature should have positive sign, as stated in Sects. 6.3.3 and 6.3.4.

Calculated value of U_{∞} from alkali-wrapping CPT for a Japanese andesite is considerably large absolute value (−2,890 K for 20–60 °C, [18], with conversion 1 kJ/mol = 120.3 K), which is due possibly to overestimation of ε^{∞} at 20 °C by the fitting with Brunetaud equation.

Conventional CPT expansion data are affected by alkali leaching and drying up. Hence calculated value of U_{∞} (−1900 K) between 38–60 °C [31] for averaged 9 European aggregates [35], even fitted with Larive equation, also seem to be incorrect.

It should be noted, however, that more reliable values of the apparent activation energy U_{∞}, calculated here from CCT data with small gaps of ε^{∞} between 23–60 °C, almost free from alkali leaching and drying, are considerably small, in average U_{∞}= −690 K (−790 K at 38–60 °C for Texas sand; −580 K at 23–38 °C for French limestone). When selected data of smaller expansion gap (nearly zero) of ε^{∞} at different temperatures is used, U_{∞} approaches zero.

6.4.9 *Correction of Temperature for Final Expansion (ε^{∞})*

There has been argument that ultimate expansion of concrete differs by temperature, and hence expansion parameters obtained at higher temperature should be corrected to those at ambient temperature. If this correction is necessary, then potential expansion of field concrete can be estimated using apparent activation energy (U_{∞}) based on Arrhenius equation [31], like the correction of τ_L, τ_C for temperature [50].

If a conservative value of U_{∞} = −690 K from the concrete cylinder test (average of French limestone and Texas sand) is used, then potential final expansion ($\Delta\varepsilon^{\infty}$) of concrete core at ambient temperature (notation 1: 16 °C field) extrapolated to 80 °C (notation 2: accelerated core test) would be 1.54 times the latter. This means that final expansion (ε^{∞}) of this concrete structure will be 1.61%, i.e. about 30% increase (=1.61 × 100/1.27) when a temperature effect on ultimate expansion of core between 80 °C and 16 °C is postulated. However, actual increase in expansion is possibly smaller, because there is evidence that U_{∞} for selected expansion data is smaller than 690 K, even not zero.

$$\Delta\varepsilon_1^{\infty}/\Delta\varepsilon_2^{\infty} = \exp U_{\infty}[(1/K_2) - (1/K_1)] \tag{6.15}$$

$$\begin{aligned}\Delta\varepsilon_1^{\infty} &= (\Delta\varepsilon_2^{\infty}) \times \exp U_{\infty}[(1/K_2) - (1/K_1)] = \\ &\quad 0.63(\%) \times \exp(-690) \times \{[1/(273+80)] - [1/(273+16)]\} \\ &= 0.97(\%)\end{aligned} \tag{6.16}$$

$$\left(\Delta\varepsilon_1^{\infty}\right)' = \Delta\varepsilon_1^{\infty} - (R_2 - R_1) \times F = 0.97(\%) - (4.32 - 3.71) \times 0.06 = 0.93(\%) \tag{6.17}$$

$$\varepsilon_1^{\infty} = \varepsilon_1 + \left(\Delta\varepsilon_1^{\infty}\right)' = 0.68(\%) + 0.93(\%) = 1.61(\%) \tag{6.18}$$

6.4.10 Correction of Final Expansion at 1N NaOH for Temperature

Andesite aggregate: If the coarse aggregate is highly reactive andesite with a pessimum proportion, final expansion of core at 1N NaOH (=1 mol/l) solution after Larive equation, will become about a half that at 80 °C (0.63% → 0.33%), after conversion from 80 °C to 38 °C using apparent activation energy of andesite (U_∞ = 1720 K). Then the final expansion corrected for the difference in alkali content of concrete will be slightly lower (0.33% → 0.29%).

$$\Delta\varepsilon_1^\infty = (\Delta\varepsilon_2^\infty) \times \exp U_\infty \left[(1/K_2) - (1/K_1)\right] = 0.63(\%) \times \exp(1720) \times \{[1/(273+80)] - [1/(273+38)]\} = 0.33(\%) \tag{6.19}$$

$$(\Delta\varepsilon_1^\infty)' = \Delta\varepsilon_1^\infty - (R_2 - R_1) \times F = 0.33(\%) - (4.32 - 3.71) \times 0.06 = 0.29(\%) \tag{6.20}$$

Subsequent procedures follow the calculation of expansion of the concrete prism/cylinder test. If the expansion has a temperature dependence but a small apparent activation energy (U_∞)(−690 K) is used, then the final expansion of concrete equivalent to that of the concrete prism/cylinder from 38 to 16 °C will result in a negligible increase in expansion (0.29% → 0.34%). As a result, the total expansion of the concrete structure, estimated by adding the past expansion of concrete to the above expansion, will be smaller than that of the former estimation (1.02% vs 1.61%). If no further correction is necessary, expansion will remain even smaller (0.97% vs 1.02%).

$$\Delta\varepsilon_1^\infty = (\Delta\varepsilon_2^\infty) \times \exp U_\infty \left[(1/K_2) - (1/K_1)\right] = 0.29(\%) \times \exp(-690) \times \{[1/(273+38)] - [1/(273+16)]\} = 0.34(\%) \tag{6.21}$$

$$\varepsilon^\infty = \varepsilon_1 + (\Delta\varepsilon)' = 0.68(\%) + 0.34(\%) = 1.02(\%) \tag{6.22}$$

$$\varepsilon^\infty = \varepsilon_1 + (\Delta\varepsilon)' = 0.68(\%) + 0.29(\%) = 0.97(\%) \tag{6.23}$$

Other aggregate: some rock types have small activation energy (U_∞), possibly being around 0 or smaller, Table 6.5. If a small value of U_∞ (e.g. -300K) (Table 8.5) is used, expansion will increase slightly when correction is made for temperature from 80 to 38 °C. If necessary, further corrections as above can be made.

$$\begin{aligned}\Delta\varepsilon_1^\infty &= (\Delta\varepsilon_2^\infty) \times \exp U_\infty \left[(1/K_2) - (1/K_1)\right]\\ &= 0.63(\%) \times \exp(-300)) \times \{[1/(273+80)] - [1/(273+38)]\}\\ &= 0.71(\%)\end{aligned} \tag{6.24}$$

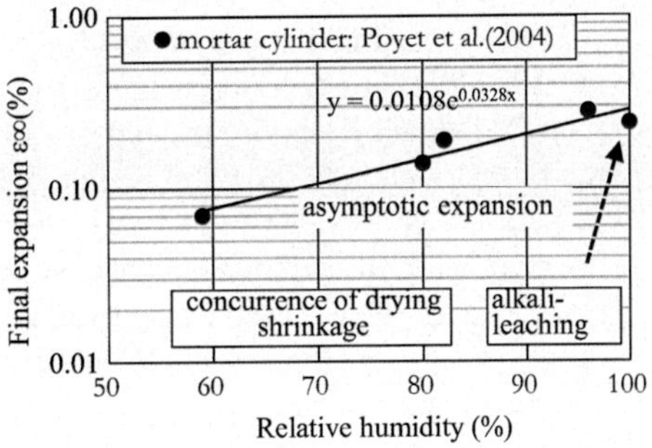

Fig. 6.13 Influence of humidity on the expansion: Expansion of mortar cylinder (limestone), drawn using data from Poyet et al. [52]; Katayama (unpublished data)

6.4.11 Correction of Relative Humidity

Final expansion of concrete is corrected for relative humidity assuming environmental factor of concrete structure. A simplest way is to use the Eq. (6.25), obtained using data by Poyet et al. [52]. This was based on mortar cylinder expansion test for Belgian limestone in a relative humidity range H= 0.6–1.0 at 20 °C. After normalizing expansion to 20 °C (Fig. 6.13):

$$\Delta\varepsilon_2 = \Delta\varepsilon_1^{\infty} \times \exp -3.28(H_1 - H_2) \tag{6.25}$$

For instance, correction for relative humidity from: H_2 = 100%→ H_1 = 85%: H_2 = 1.0, H_1 = 0.85, $\varepsilon_{\infty 2}$ = 0.36% and $\varepsilon_{\infty 1}$ = 0.22%.

6.4.12 Limitations and Applicability to Analysis

Influence of external alkali dosage: In this core expansion test, difference in the alkali concentration between the immersing alkali solution (1N NaOH) and concrete pore water is corrected for expansion. However, the amount of alkalis absorbed by the reactive minerals in the core specimen may increase until an apparent equilibrium is attained well before reactive minerals are consumed. This aspect needs checking by comparing with results of estimation from other expansion test performed under a high relative humidity (e.g. JCI-S-011) [4].

Alkali dependence on expansion: the shape of the regression line between the alkali content and final expansion, originally based on the concrete prism test at 38 °C, varies considerably by rock types and pessimum phenomenon of the aggregate. Because correction of this alkali-dependence for temperature to 80 °C is not made here, this may be a source of error.

Overestimation of expansion: this may occur when ultimate expansion of core is corrected for the temperature difference between field and 80 °C, using an apparent activation energy (U_{∞}) that was derived from early CPT data affected by alkali-leaching and drying up. Overestimation of final expansion of concrete (ε^{∞}) could also occur when a growing expansion curve at early stage is fitted with Brunetaud equation, instead of Larive equation. Estimation error related to the fitting equations shown in Sect. 6.3.2.

Underestimation of expansion: this may occur when contribution of long-term alkali release from the aggregate is underestimated, e.g. by using short-term extraction of water-soluble alkali of aggregate in concrete Sect. C.4.7, instead of long-term extraction of alkalis in saturated $Ca(OH)_2$ and alkali solutions.When substantial part of reactive aggregate is chert, this alkali-immersion test should not be used, because expansion of core is reduced.

References

1. JCI-TC115FS §4.3.: Method of measuring expansion ratio of core specimen immersed in alkali solution (immersion method using alkali solution). In: Committee Report on Diagnosis of ASR and its Ideal Future, pp. 316–319. Tokyo, in Japanese (2014)
2. Katayama, T., Tagami, M., Sarai, Y., Izumi, S., Hira, T.: Alkali-aggregate reaction under the influence of deicing salts in the Hokuriku district. Jpn. Mater. Charact. **53**, 105–122 (2004)
3. Larive, C.: Apports Combinés de l'Experimentation et de la Modélisation à la Comprehension del'Alcali-Réaction et de ses Effets Mécaniques. Doctoral thesis, Laboratoire Central des Ponts et Chaussées, Paris, 1998. https://hal.inria.fr/docs/00/52/06/76/PDF/1997TH_LARIVE_C_NS20683.pdf
4. JCI-S-011-2017.: Method of measuring expansion ratio of core sampled from concrete structure, 10 pp. Tokyo, Japan (in Japanese) (2017)
5. JCI-DD2.: Determination of expansion ratio of AAR-damaged structure using concrete core. Jpn. Concr. Inst. 5–9 (in Japanese) (1989)
6. Katayama, T., Oshiro, T., Sarai, Y., Zaha, K., Yamato, T.: Late-expansive ASR due to imported sand and local aggregates in Okinawa Island, southwestern Japan. In: Proceedings of 13th International Conference on Alkali-Aggregate Reaction (ICAAR). Trondheim, Norway, pp. 862–873 (2008)
7. Bérubé, M., Dupont, N., Pigeon, M., Stoian, A.: Expansion test methods for mass concrete exposed to alkali-aggregate reaction. In: International Conference on Concrete Alkali-Aggregate Reactions in Hydroelectric Plants and Dams. project CEA No.715 G 687. Fredericton, New Brunswick, Canada (1992)
8. Mukai, K.: Influence of reading time on core expansion in the alkali immersion test. Technical Report. Unpublished data, Taiheiyo Consultant (in Japanese) (2014)
9. Endo, H., Tanaka, H., Goto, T.: Inter-laboratory test of the residual expansion tests of concrete core specimens. In: Proceedings, Japan Society of Civil Engineers 517, pp. 1033–1034 (in Japanese) (2013)
10. Nomura, M., Torii, K.: Evaluation of ASR residual expansivity of cones by accelerated curing testing. Cem. Sci. Concr. Technol. **59** 139–145 (in Japanese) (2005)
11. Fournier, B., Bérubé, M., Folliard, K., Thomas, M.: Report on the diagnosis, prognosis, and mitigation of Alkali-Silica Reaction (ASR) in transportation structures. Technical Report, FHWA-HIF-09-004. 154 p. Federal Highway Administration, 2010
12. Katayama, T.: An attempt to estimate past expansion of concrete based on petrographic stage of alkali-silica reaction. In: Proceedings of 39th International Conference on Cement Microscopy, pp. 217–236. Toronto, Canada (2017)

13. Buford, A.: In situ monitoring of alkali-silica reaction (ASR) affected concrete: a study on crack indexing and damage rating index to assess the severity of ASR and to monitor ASR progression. Technical Report Position Paper: Memorandum to Trapp, J.M. Washington D.C.: United States Nuclear Regulatory Commission (2013)
14. Saouma, V., Martin, R., Hariri-Ardebili, M.A., Katayama, T.: A mathematical model for the kinetics of the alkali-silica chemical reaction. Cem. Concr. Res. **68**, 184–195 (2015)
15. Katayama, T., Sarai, Y., Higashi, Y., Honma, A.: Late-expansive alkali silica reaction in the Ohnyu and Furikusa headwork structures, central Japan. In: Proceedings of 12th International Conference on Alkali-Aggregate Reaction (ICAAR), pp. 1086–1094. Beijing, China (2004)
16. Sellier, A., Bourdarot, E., Multon, S., Cyr, M., Grimal, E.: Combination of structural monitoring and laboratory tests for assessment of alkali-aggregate reaction swelling: application to gate structure dam. ACI Mater. J. **106**(3), 281–290 (2009)
17. Brunetaud, X.: Étude de l'influence de différents paramètres et de leurs interactions sur la cinétique de l'amplitude de la réaction sulfatique interne au béton. Ph.D. thesis. Chôtenay-Malabry, Ecole centrale de Paris, 2005
18. Kawabata, Y., Yamada, K., Ogawa, S., Sagawa, Y.: Prediction of ASR expansion of field-exposed concrete based on accelerated concrete prism test with alkali-wrapping. Cem. Sci. Concr. Technol. **69**, pp. 496–503 (in Japanese) (2015)
19. Kawabata, Y., Yamada, K., Ogawa, S.: Modeling of environmental conditions and their impact on the expansion of concrete affected by the alkali-silica reaction. In: Sellier, A., Grimal, E., Multon, S., Bourdarot, E. (eds.) Swelling Concrete in Dams and Hydraulic Structures, pp. 163-175. ISTE Ltd., UK (2017)
20. Kuroda, T., Inoue, S., Yoshino, A., Nishibayashi, S., Miyagawa, T.: Effects of accelerated test conditions on ASR expansion of concrete core. In: Proceedings of 13th International Conference on Alkali-Aggregate Reaction (ICAAR), pp. 195-203. Trondheim, Norway (2008)
21. Iwatsuki, E., Morino, K.: ASR expansion behavior of concrete specimens containing chert aggregate immersed in NaOH solution. In: Proceedings of Japan Concrete Institute, vol. 28, pp. 827–832 (in Japanese) (2006)
22. Li, K., Ulm, F., Coussy, O., Larive, C., Fan, L.: Chemoelastic modelling of alkali-silica reaction in concrete. In: Proceedings of 11th International Conference on Alkali-Aggregate Reaction (ICAAR), pp. 989–998. Quebec, Canada (2000)
23. Diamond, S.: ASR-Another look at mechanisms. In: Proceedings of 8th International Conference on Alkali-Aggregate Reaction (ICAAR), pp. 83-94. Kyoto, Japan (1989)
24. Helmuth, R., Stark, D., Diamond, S., Moranville-Regourd, M.: Alkali-silica reactivity: an overview of research. Technical Report. National Research Council, 1993. http://onlinepubs.trb.org/Onlinepubs/shrp/SHRP-C-342.pdf
25. Stark, D.: Immersion test to identify cement alkali levels and pozzolans to prevent ASR. In: Proceedings of 10th International Conference on Alkali-Aggregate Reaction (ICAAR), pp. 355-361. London, UK (1996)
26. Kawabata, Y., Yamada, K., Ogawa, S., Martin, R., Sagawa, Y., Seignol, J.F., Toutlemonde, F.: Correlation between laboratory expansion and field expansion of concrete: prediction based on modified concrete expansion test. In: Proceedings of15th International Conference on Alkali-Aggregate Reaction (ICAAR). Sao Paolo, Brasil (2016)
27. Fournier, B., Bérubé, M., Frenette, J.: Laboratory investigations for evaluating potential alkali-reactivity of aggregates and selecting preventive measures against alkali-aggregate reaction (AAR)—what do they really mean? In: Proceedings of 11th International Conference on Alkali-Aggregate Reaction (ICAAR), pp. 287–296. Quebec, Canada (2000)
28. Blackwell, B., Thomas, M., Pettifer, K., Nixon, P.: An appraisal of UK greywacke deposits and current methods of avoiding AAR. In: Proceedings of 10th International Conference on Alkali-Aggregate Reaction (ICAAR), pp. 492–499. Melbourne, Australia (1996)
29. Sibbick , G., Page, C.: Susceptibility of various UK aggregates to alkali aggregate reaction. In: Proceedings of 9th International Conference on Alkali-Aggregate Reaction (ICAAR), pp. 980-987. London, UK (1996)

30. Nomura, M., Komatsubara, A., Kuroyanagi, M., Torii, K.: Evaluation of petrographic features of reactive aggregates in the Hokuriku district and residual expansion of core. Conc. J. (Japan Concrete Institute, JCI) **33**(1), 953–958 (in Japanese) (2011)
31. Kawabata, Y., Yamada, K., Ogawa, S., Osako, M.: An approach to prediction of ASR expansion of concrete using accelerated concrete prism test. In: Proceedings of 13 JSMS Symposium on Concrete Structures Scenarios, pp. 435–458. Kyoto, Japan (in Japanese) (2013)
32. Tanaka, A., Yamada, K., Sagawa, Y., Ogawa, S.: Influence of curing temperature and alkali content on the expansion behaviour in the accelerated concrete prism test. In: Proceedings of Japan Concrete Institute, vol. 37, pp. 895–900 (in Japanese) (2015)
33. Yamada, K., Karasuda, S., Ogawa, S., Sagawa, Y., Osako, M., Hamada, H., Isneini, M.: CPT as an evaluation method of concrete mixture for ASR expansion. Constr. Build. Mater. **64**, 184–191 (2014)
34. JCI: Committee report on the testing of alkali-aggregate reaction by means of concrete prism method, Japan Concrete Institute (JCI), 85p. Tokyo (in Japanese) (1991)
35. Nixon, P., Lane, S.: Experience from testing of the alkali reactivity of European aggregates according to several concrete prism test methods. In: Partner Report 3 (2006)
36. Fournier, B., Chevrier, R., de Grosbois, M., Lisella, R., Folliard, K., Ideker, J., Shehatad, M., Thomas, M., Baxter, S.: The accelerated concrete prism test: variability of the test method and proposed expansion limits. In: Proceedings of 12th International Conference on Alkali-Aggregate Reaction, pp. 314–323. Beijing, China (2004)
37. Jensen, V.: Alkali aggregate reaction in southern Norway. The Norwegian Institute Technology, University of Trondheim, Doctoral thesis (Technicae) (1993)
38. Durand, B.: More results about the use of lithium salts and minerals admixtures to inhibit ASR in concrete. In: Proceedings of 11th International Conference on Alkali-Aggregate Reaction, pp. 623–632. Quebec, Canada (2000)
39. Haha, M.B., Gallucci, E., Guidoum, A., Scrivener, K.L.: Relation of expansion due to alkali silica reaction to the degree of reaction measured by SEM image analysis. In: Cement & Concrete Research, vol. 37, pp. 1206–1214 (2007)
40. Sanchez, L., Kuperman, S.C., Helene, P., Kihara, Y.: Trial to correlate the accelerated mortar bar test, the standard and the accelerated concrete prism tests. In: Proceedings of 13th International Conference on Alkali-Aggregate Reaction, pp. 1131–1141. Trondheim, Norway (2008)
41. Hayman, S., Thomas, M., Beaman, N., Gilks, P.: Selection of an effective ASR-prevention strategy for use with a highly reactive aggregate for the reconstruction of concrete structures at Mactaquac generating station. In: Cement & Concrete Research, vol. 40, pp. 605–610 (2010)
42. Larive, C., Coussy, O.: Behavior of AAR-prevention concrete, modelling. In: Proceedings of 10th International Conference on Alkali-Aggregate Reaction, pp. 662–669. Melbourne, Australia (1996)
43. Shayan, A., Diggins, R., Ivanusec, I.: Long-term effectiveness of fly ash in preventing deleterious expansion due to alkali-aggregate reaction in concrete. In: Proceedings 10th International Conference on Alkali-Aggregate Reaction (ICAAR), pp. 538–545. Melbourne, Australi (1996)
44. Moisson, M., Cyr, M., Ringot, R., Carles-Gibergues, A.: Efficiency of reactive aggregate powder in controlling the expansion of concrete affected by alkali-silica reaction (ASR). In: Proceedings of 12th International Conference on Alkali-Aggregate Reaction, pp. 617–624. Beijing, China (2004)
45. Stacey, S., Folliard, K., Drimalas, T., Thomas, M.: An accelerated and more accurate test method to ASTM C1293: The concrete cylinder test. In: Proceedings of the 15th International Conference on Alkali-Aggregate Reaction (ICAAR). paper No. 130. Sao Paolo, Brazil (2016)
46. Folliard, K., Ideker, J., Thomas, M., Fournier, B.: Assessing aggregate reactivity using accelerated concrete prism test. In: Proceedings of 7th CANMET/ACI International Conference on recent Advances in Concrete Technology, Supplementary Papers, pp. 269–283. Las Vegas, US (2004)
47. Rogers, C., Hooton, R.: Reduction in mortar and concrete expansion with reactive aggregates due to alkali leaching. Cem. Concr. Aggreg. **13**(1), 42–49 (1993)

48. Larive, C., Laplaud, A., Joly, M.: Behaviour of AAR-affected concrete, experimental data. In: Proceedings of 10th International Conference on Alkali-Aggregate Reaction (ICAAR), pp. 670–677. Melbourne, Australia (1996)
49. Lindgard, J., Thomas, M., Sellevold, R., Pedersen, B., Andic-Cakir, O., Justness, H., Ronning, T.: Alkali-silica reaction (ASR) performance testing: influence of specimen pre-treatment, exposure conditions and prism size on alkali leaching and prism expansion. Cem. Concr. Res. **53** (2013)
50. Ulm, F., Coussy, O., Li, K., Larive, C.: Thermo-chemo-mechanics of ASR expansion in concrete structures. ASCE J. Eng. Mech. **126**(3), 233–242 (2000)
51. Gomes, J., Batista, A., Oliveira, S.: Analysis of concrete dams under swelling processes. In: Proceedings of 12th International Conference on Alkali-Aggregate Reaction (ICAAR), pp. 1148-1157. Beijing, China (2004)
52. Poyet, S., Sellier, A., Capra, B., Foray, G., Torrenti, J., Cognon, H., Bourdarot, E.: Modelling of alkali-silica reaction in concrete, part 2: influence of water on ASR. In: Proceedings of 12th International Conference on Alkali-Aggregate Reaction, pp. 185–190. Beijing, China (2004)

Chapter 7
Accelerated Expansion Test: Switzerland

Andreas Leemann, Christine Merz, and Stéphane Cuchet

7.1 Scope

This test method covers the laboratory determination of the expansion rate of concrete cores (designated as "residual expansion potential" in the following context) extracted from structures affected by alkali silica reactions.

This test differs considerably from methods where aggregates are extracted from concrete extracted from structures and reused in accelerated mortar tests with added alkali. In the concrete structures the decreasing alkali level in the pore solution seems to be the factor limiting expansion and not the availability of reactive minerals as in the accelerated mortar tests.

This method does not encompass an assessment of the expansion taken place before the extraction of the cores.

A. Leemann (✉)
Empa, Swiss Federal Laboratories for Materials Testing and Research, Laboratory for Concrete and Construction Chemistry, Überlandstrasse 129, 8600 DÜbendorf, Switzerland
e-mail: andreas.leemann@empa.ch

C. Merz
Merz Ingenieurberatung, Mőriken, Switzerland
e-mail: merz@merz-ingenieurberatung.ch

S. Cuchet
Stéphane Cuchet, Holcim Suisse, Laboratoire des matériaux, Eclépens, Switzerland
e-mail: stephane.cuchet@lafargeholcim.com

© RILEM 2021

V. E. Saouma (ed.), *Diagnosis & Prognosis of AAR Affected Structures*, RILEM State-of-the-Art Reports 31,
https://doi.org/10.1007/978-3-030-44014-5_7

7.2 Principles and Methodology

The method described below is based on the method LCPC 44 Essai [1]. It has been used for various structures including the ones investigated in a project in Switzerland financed by the Federal road authorities [2]. Some of the results of this project are published in [3].

The residual expansion potential is determined on cores taken from structures. They are stored at 38 °C and 100% RH for several months up to more than one year. The measured parameters are mass change, diametral and longitudinal expansions Fig. 7.1, measurements every 4 weeks at 20 °C). The longitudinal expansion is used to assess the residual expansion potential.

The principle of the method is described in the following paragraphs. It is not a manual for conducting the test, but describes its major characteristics.
Advantages:

1. no alkalis added.
2. connection between length and mass change.
3. classification of the residual expansion potential in three categories.
4. concrete test.

Disdvantages:

1. some alkali leaching, although minimized by protective measures (wrapping)
2. how can the measured expansion be used in models?

The method is a test for concrete extracted from structures. It is not an aggregate test and differs considerably from methods where aggregates are extracted from concrete and reused in accelerated mortar tests with added alkali.

In the concrete structures the decreasing alkali level in the pore solution seems to be the factor limiting expansion.

Fig. 7.1 Core (diameter = 100 mm, length = 200 mm) used for the expansion test. The longitudinal expansion is measured on the length axis using two stainless steel pins fixed with glue in small boreholes. The diametral expansion is measured at two positions perpendicular to the length axis

7.3 Sampling and Sample Preparation

Immediately after being taken from the structure, the cores are sealed in plastic bags. The cores are cut in the lab to appropriate size (typical diameter and length: 100 and 200 mm). Usually, three cores are tested per coring site. Small holes are drilled into the cores fitting the pins. As soon as the drying state of surface permits, the pins are glued (e.g. Araldite 2014-1) into the the small boreholes of the cores and aligned with a jig. Stable glue (e.g. organic two-component glue) has to be used that shows no volume changes at the conditions in the reactor.

7.4 Apparatus

The apparatus used for the test (based on AFNOR P18–454 and employed in Swiss standard SIA MB 2042) is shown in Figs. 7.2. Three cores are stored in one container. There is water at the bottom of the containers, but as the cores stand on a grid no

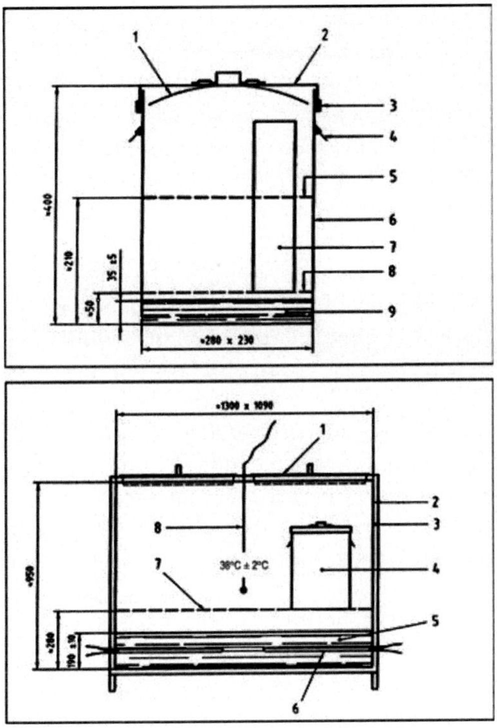

1 Inlet preventing condensed water to drop on samples
2 Cover
3 Sealing
4 Handle with joint
5 Stainless steel grid keeping samples in upright position
6 Stainless steel container
7 3 samples (Ø ~ 100mm, L ~200 mm)
8 Stainless steel grid with opening width of ~ 10 mm
9 Water

1 Cover
2 Stainless steel
3 Thermal isolation
4 Container
5 Water
6 Immersion heater
7 Stainless steel grid
8 Thermocouple for temperature control

Fig. 7.2 Stainless steel container for Empa test

direct contact with water is possible. The containers themselves are placed into the reactor with a climate of 38 °C and 100% relative humidity.

7.5 Experimental Procedures

The expansion of the cores can be divided into three different phases, Fig. 7.3:

Phase 1: The initial measurements of dimension and mass are performed before sample conditioning. Sample conditioning is done at 20 °C in containers with a thin layer of water at its bottom permitting capillary suction of the cores. The mass of the cores has to be measured daily. Conditioning ends when constant mass is reached. This is usually achieved after one week. Immediately after reaching constant mass, the cores have to be moved into the reactor. The water saturation is accompanied with a first swelling of the cores (hygroscopic swelling).
Remark: *Conditioning is mandatory, because otherwise the expansion related to moisture uptake cannot be distinguished from expansion related to ASR, after the samples have been placed into the reactor. This is the only way to ensure identical conditions as a starting point for cores of different structures making possible comparisons and classifications for the residual expansion.*

Phases 2 & 3: The cores are stored in the reactor at 38 °C and at 100 % RH. After the cores are wrapped in a plastic film (minimizes mass loss and alkali leaching), they are moved into the containers. The containers are placed into the reactor. For the length measurements the containers are removed from the reactor and stored at 20 °C for 24 h before the cores are measured. Afterwards the containers are put back into the reactor.
The same apparatus as for the French performance-test AFNOR P18–454 is used. At the beginning of the storage at 38 °C and 100% RH a strong but quickly levelling off expansion during 30–60 days is usually observed (phase 2). In the following phase 3 the expansion rate is constant. However, it can also level off in certain cases.
Remark: Mass control is very important, but not required in all published methods [e.g. 1]. An preparation of the length change is difficult without having the data for the mass change available. The mass change is used as a control parameter in phases 2 and 3. It allows detecting unintentional changes of RH in the reactor or containers and assessing the reasons for irregularities in the expansion behavior of cores.

Phase 4: The cores are removed from the reactor after 160 days. Afterwards, they are dried at 60–70% RH until they reach the identical mass as before conditioning started (phase 1). The difference in length change is defined as "irreversible" expansion and enables to distinguish between hygroscopic swelling due to the formation of newly formed ASR products. This step requires 2–8 weeks.
So far, the longitudinal expansions during phases 2 and 3 are used to assess the residual expansion potential of the cores. However, a first assessment is usually

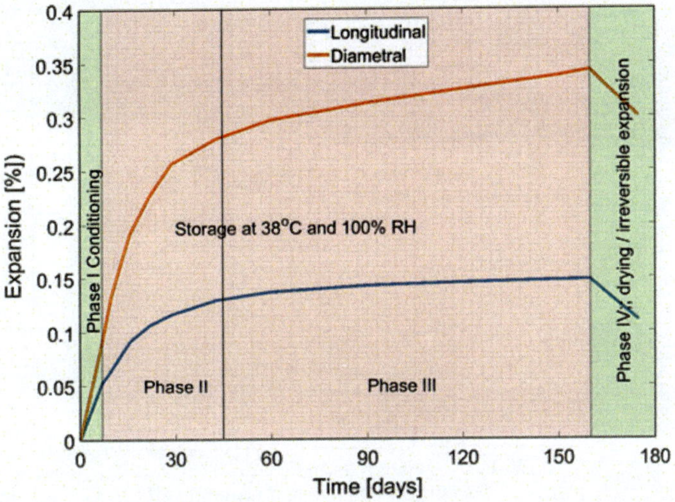

Fig. 7.3 Different phases of expansion during the residual expansion measurements in longitudinal and diametral direction of the cores

already possible at the end of the non-linear expansion (phase 2). The determination of the irreversible expansion in phase 4 is used as a control parameter. It gives a rough indication about the magnitude of hygroscopic swelling and ASR-induced expansion.

7.6 Preparation

Calculate the difference between the initial length of the specimen and the length at the end of each period. The mean of the longitudinal length change at the beginning of phase 2 (non linear expansion rate) and the mean of following linear length change (phase 3, linear expansion rate) are used to assess the residual expansion potential Fig. 7.4. The diametral length changes are used as a control parameter. They can indicate an inhomogeneity of expansion over the length of the cores.

Constant relationship between longitudinal and diametral expansion has been observed during expansions test by several authors as Smaoui, Bérubé, Fournier, and Bissonnette [4], Larive [5] Multon, Leclainche, Bourdarot, and Toutlemonde [6] and ourselves. For instance we use the diametral expansion as control parameter and also to detect an eventual warping of the cores.

The method has been used to assess the residual expansion potential of numerous structures in Switzerland, Fig. 7.4. The cores can be classified into three different levels.

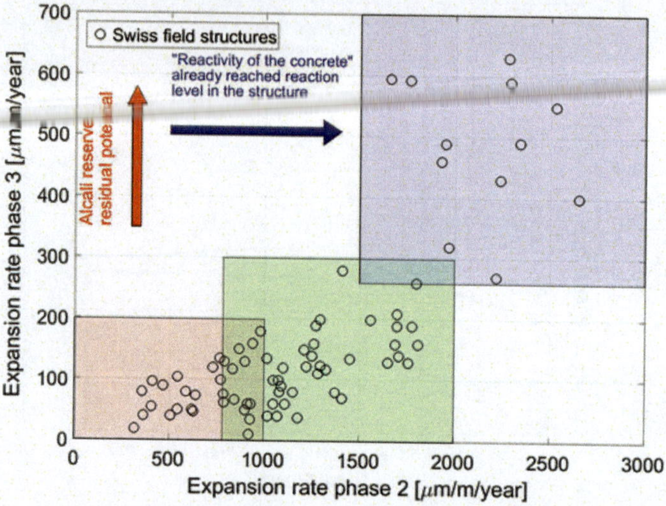

Fig. 7.4 Results of residual expansion measurement of cores extracted from concrete structures in Switzerland

The differently colored domains are defined empirically, based on the experience from Swiss concrete structures.

Blue domain:	concrete with a residual expansion potential zero or very low.
Green domain:	concrete with a moderate residual expansion potential.
Red domain:	concrete with a high residual expansion potential.

The domains shown in Fig. 7.4 were defined based on the experience gained from Swiss concrete structures They take into account the degree of damage of the structure at the time of core extraction, age and exposition of the structure and concrete composition (alkali content, cement type, aggregates). As these additional characteristics are considered and influence the assessment of the structure, the domains overlap. The measured values give an indication about range of expansion rates to be expected in a structure. These values are influenced by the expansion that already took place and by the exposition of the studied structure or component. As an example, cores taken from a part of the structure with little ASR-induced damage may show a higher expansion compared to cores taken from a strongly damaged part. This emphasizes that the degree of ASR can vary widely within a structure. It is advantageous if structures are monitored and the measured expansions can be compared with the one determined in the laboratory test. This facilitates the assessment of the structure.

7.7 Examination at End of Test

After the final length measurements, the condition of the wrapping, the stability of the pins and the characteristics of each core are examined.

This includes:

1. Presence, location, and pattern of cracking,
2. Appearance of surfaces, surface mottling, and
3. Surficial deposits or exudations, their nature, thickness and distribution.

7.8 Report

The report has to contain the following information:

1. Distribution and severity of damages in the structure, preferably including crack-indices.
2. Location of the coring sites.
3. Microscopic investigation including petrography of aggregates, type of cement, mineral additions, microcracks with crack-indices on the microscopic scale, altered aggregates, presence of ASR products and information about other processes like leaching, ettringite formation, carbonation etc.
4. Alkali content of the concrete as mass percent of potassium oxide (K_2O), sodium oxide (Na_2O), and calculated sodium oxide (Na_2O) equivalent. Refer to used method for extracting alkalis (water soluble and/or acid soluble alkalis).
5. If possible: source of aggregate/concrete.
6. If possible: Type and source of Portland cement/mineral additions with their respective alkali contents.
7. Any relevant information concerning the preparation of concrete core.
8. Development of length change of each set of three cores during phases 1–4.
9. Any significant features revealed by examination of the specimens during and after the test.

7.9 Limitations and Applicability to Analysis

The method characterizes the kinetics of expansion at the conditions in the reactor (temperature of 38 °C and RH of 100%). As such, it does not provide a value for residual expansion of the tested concrete.

The expansion measured in this test shows a strong positive correlation to the crack-index determined in strongly damaged parts of structures with an age in the range of 40 years [3]. This correlation seems to justify using a parameter determined in this test for finite element analysis. Based on the experience with the method up to now, the nonlinear expansion in phase 2 is influenced by the degree of damage of the concrete at the time of coring. The linear expansion in phase 3 does not show such an influence and consequently better represents solely the kinetics of expansion. Therefore, it is recommended to use the approximately constant expansion rate in phase 3 for finite element analysis. It is likely that the expansion rate in the structure is lower due to the lower temperatures. However, a correction for the temperature at the location of the structure seems inappropriate, because the decrease of expansion by decreasing the temperature from 38 to 20 °C is concrete- and aggregate-specific, as shown in Merz and Leemann [3]. As such, the use of the expansion rate in phase 3 enables the modelling of a worst case scenario and should allow identifying the areas of the structure for which expansion is the most critical.

7.10 Precision and Bias

The dominating factor for the uncertainty of the expansion is the variation of the concrete cores. The magnitude of error of the measurement itself is negligible (±0.002 mm). Depending on the homogeneity of the concrete at the coring site, significant variations in the expansion of single cores within a set of three cores are possible. Therefore, all cores have to be considered, even if there is an outlier.

References

1. Fasseu, P.: Alcali-Reaction Du Beton-Essai D'Expansion Residuelle Sur Beton Durci-Projet De Methode D'Essai. Technical report No. 44. Laboratoire Central des Ponts et Chaussées (LPC) (1997). http://www.ifsttar.fr/fileadmin/user_upload/editions/lcpc/MethodeDEssai/MethodeDEssai-LCPC-ME44.pdf
2. Merz, C., Leemann, A.: Validierung der AAR-Prüfungen für Neubau und Instandsetzung. Technical report 648. Bundesamt Fuer Strassenbau (ASTRA)/Office Federal Des Routes (OFROU) (2011)
3. Merz, C., Leemann, A.: Assessment of the residual expansion potential of concrete from structures damaged by AAR. Cem. Concr. Res. **52**, 182–189 (2013)
4. Smaoui, N., Bérubé, M., Fournier, B., Bissonnette, B.: Influence of specimen geometry, orientation of casting plane, and mode of concrete consolidation on expansion due to ASR. Cem., Concr. Aggreg. **26**(2), 1–13 (2004)
5. Larive, C.: Apports Combinés de l'Experimentation et de la Modélisation à la Comprehension del'Alcali-Réaction et de ses Effets Mécaniques. Ph.D. thesis. Paris: Laboratoire Central des Ponts et Chaussées (1998). https://hal.inria.fr/docs/00/52/06/76/PDF/1997TH_LARIVE_C_NS20683.pdf
6. Multon, S., Leclainche, G., Bourdarot, E., Toutlemonde, F.: Alkali-Silica reaction in specimens under multi-axial mechanical stresses. In: Proceedings of CONSEC 4 (Concrete Under Severe Conditions), pp. 2004–2011. Seoul (2004)

Chapter 8
Accelerated Expansion Test: France, IFSTTAR

Renaud-Pierre Martin, Bruno Godart, and François Toutlemonde

8.1 Scope

This test method covers the laboratory determination of the swelling potential and the residual free expansion of concrete extracted from structures affected by alkali silica reactions.

The method is not an aggregate test like in the methods where aggregates are extracted from concrete cored from structures and reused in accelerated mortar tests with added alkali.

This test method only indirectly encompasses the assessment of the reaction advancement in the structure. Indeed, the method allows assessing the reaction advancement (at least in an indirect way) by assessing the $\varepsilon_c(t_0^c)$ parameter which corresponds to the expansion developed until coring of the specimen in the structure.

This document is mainly a translation of [6] (based itself on [2–5]. The experimental technique described hereafter is not a standard and has been developed as an experimental diagnosis tool by the French Public Works Research Laboratory.[1]

[1]LCPC became IFSTTAR in 2011.

R.-P. Martin (✉)
Université Paris Est, Materials And Structures Department,
Urban and Civil Engineering Testing And Modeling Laboratory (EMGCU),
IFSTTAR, French Institute of Science and Technology for Transport,
Development and Networks, 77447 Marne la Vallée, France
e-mail: renaud-pierre.martin@ifsttar.fr

B. Godart · F. Toutlemonde
Université Paris Est, Materials and Structures Department, IFSTTAR,
French institute of science and technology for transport, development and networks,
77447 Marne la Vallée, France
e-mail: bruno.godart@ifsttar.fr

F. Toutlemonde
e-mail: francois.toutlemonde@ifsttar.fr

© RILEM 2021

V. E. Saouma (ed.), *Diagnosis & Prognosis of AAR Affected Structures*,
RILEM State-of-the-Art Reports 31,
https://doi.org/10.1007/978-3-030-44014-5_8

The preparation of the results obtained at the end of the test (Sect. 8.6.1) has been given in order to obtain a classification between concretes having different residual expansions. IFSTTAR and/or any of its employees cannot be held liable of any other conclusions based on this preparation of the obtained results.

Comments were added by the authors of this contribution based on their up-to-date experience in using this method and on recent research advances from the literature. They were added for sake of comprehension and are mentioned as such in the following.

The residual expansion test described in this document is a part of a more complete structure management methodology proposed to deal with AAR problems in France [1, 7]. Thus, this test should not be used separately but as one of the elements of a general methodology for expert investigations; in other words, a single residual expansion measurement cannot be considered as representative enough to conclude on the problem of an AAR-affected structure.

8.2 Principles and Methodology

This method consists in getting a sample from a core drilled out from (a part of) a concrete structure. This core is equipped with measurement basis allowing to monitor the length evolution of the specimen during the test. During this test, the sample is stored in a water saturated atmosphere in a reactor stored at 38 °C.

8.3 Sampling and Sample Preparation

8.3.1 Sampling

8.3.1.1 Prescription of LPC Method No. 44

Coring of the cylinders has to be performed under water cooling conditions. As suitable, the coring position will avoid cracks or steel reinforcement. Based on the usual size of the aggregates in the civil engineering structures (typically <25 mm), the diameter of the core will be 100 ± 10 mm. Its length will be equal to two or three times the diameter if one or two levels of measurement bases are set along the axis of the cylinder respectively. Just after coring, the cylinders are washed with clean water, wiped and stored in sealed bags.

8.3.1.2 Comments of the Present Authors

According to [8], since drilling is usually carried out under water, coring will largely change the water content in the periphery of the tested concrete, according to its porosity. If the potential residual expansion test begins quickly, the water supply will not affect the final expansion. However, it will change the measured mass variation. Moreover, these authors indicate that the cores have to be sealed immediately after coring in a plastic bag until being tested in order to prevent any drying.

For dams where the maximum diameter of aggregates can be significantly higher than in other civil engineering structures, the prescribed diameter of the core should have to be increased to get a core representative of concrete of the structure.

Exposure conditions of the structural part should be taken into account before coring as well.

According to [9] coring has to be performed in both apparently sound and affected zones. The residual expansion test is then performed on the cylinders coming from the different zones. This approach allows to have various residual expansion histories and makes it easier to determinate the magnitude and kinetics of expansion as presented hereafter.

Reference [8] provides some advice regarding the coring details in a structure:

- The results presented in this reference prove that expansions may be highly variable. This implies that AAR expansions have to be characterized with a sufficiently large number of specimens in order to account for scatter of the results and determine reliable mean expansion values. To diagnose an affected structure, it is thus necessary to take enough cores to determine a reliable residual expansion value. Moreover, the residual expansions cannot be directly generalized to the whole expertised structure, particularly to parts unexposed to water. Thus, samples must be cored according to a careful analysis of possible water supply in the structure. In particular, cores drilled from deep parts less affected by surface moisture content variation, should be favored.
- Anisotropy of AAR expansion implies that the coring direction may have an influence on the determination of residual expansion. Thus different expansion values can be obtained according to coring direction. The presence of existing cracks in the cores can also have large effects on the final residual expansions. Thus, to account for this phenomenon, measuring expansions both axially and transversally should be done, at least for a limited number of samples.
- Variability of expansion within a single core necessitates to install several measurement bases on the core.

Finally, in [10], a residual expansion procedure inspired from that presented in this document has been applied. They used cores of various diameters (50 and 100 mm) and concluded that the cores have to be as large as possible to limit the leaching of alkalis during ageing of the specimens. They suggest that in structures with little reinforcement, the diameter of the cores may be increased up to 150 mm.

8.3.2 Cylinder Preparation

The ends of the drilled core are cut to get a cylinder with a length of 150 or 250 mm for one or two measurement bases (preferably two) along the vertical axis respectively.

8.3.3 Visual Inspection

The position of the core in the structure has to be referenced. After this, a careful visual inspection of the sample has to be done: any singularity such as cracks, deteriorated aggregates, inclusions, cavities, reaction products, etc. must be documented.

A careful attention will be paid on avoiding drying of the core during this inspection.

8.3.4 Extensometer and Instrumentation

For cylinders of 150 mm in length, 3 measurement bases of 100 mm in length are set at 120° from each other, Fig. 8.1. Each measurement basis is placed at the same distance from each end face of the cylinder. In the case of a 250 mm-long cylinder,

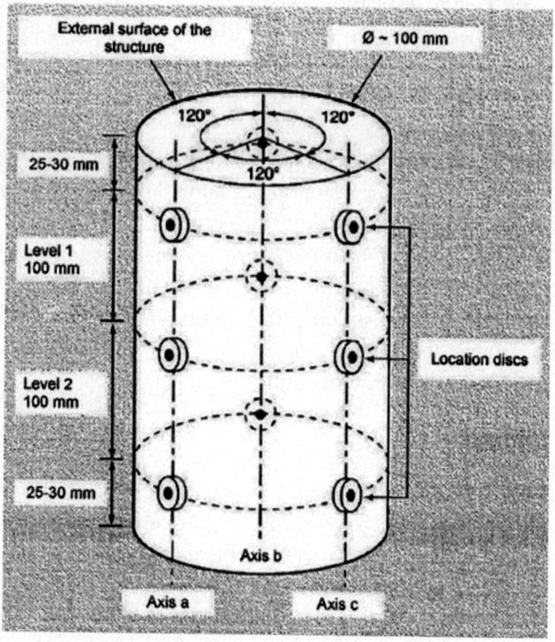

Fig. 8.1 Instrumentation of the specimen (2 levels of expansion measurement)

two measurement bases of 100 mm are set in a row and placed similarly at the same distance from each end face of the specimen.

The location discs are glued on the lateral side of the cylinder. Six (resp. nine) location discs are necessary for one (resp. two) measurement elevation(s) along the vertical axis of the cylinder. Before gluing, the concrete surface is locally dried with compressed air.

The specimen is then stored at ambient temperature during 24 h.

8.4 Apparatus

1. Extensometer:
    - minimum resolution: 1 μm
    - gauge length: 10 cm
    - including a calibration reference bar
2. Location discs adapted to the extensometer (from experience, they should be generally made of stainless steel)
3. Adhesive resisting to moisture and high temperatures (as detailed hereafter ~40 °C; 100% RH)
4. Absorbent paper (thickness: 0.6 mm; fungicidal)
5. Polyethylene sheet (thickness: 0.1 mm)
6. Container complying with [11] (formerly [3]) (see Fig. 8.2a)
7. Reactor (recommended dimensions provided by [11] (formerly [3]) standard. This standard requires a reactor with electrical heaters ensuring a temperature of 38 ± 2 °C and a moisture as near as possible of 100% RH, Fig. 8.2b.
8. Oven set at 105 °C
9. Analytical balance (capacity: in agreement with the weight of the specimens; relative precision: 0.1% of the weight of the specimen)
10. Temperature sensor (resolution 0.1 °C).

8.5 Experimental Procedures

8.5.1 Storage of the Specimens

The specimen shall be weighed. The lateral sides are then wrapped with a sheet of absorbent paper (item 4 of Sect. 8.4) impregnated with distilled water. A PE sheet is then placed on the lateral sides of the cylinder (item 5 of Sect. 8.4) with openings at the position of the location discs.

The absorbent paper and the PE sheet will be maintained on the specimen with tape. The end faces will be directly exposed to the ambient air.

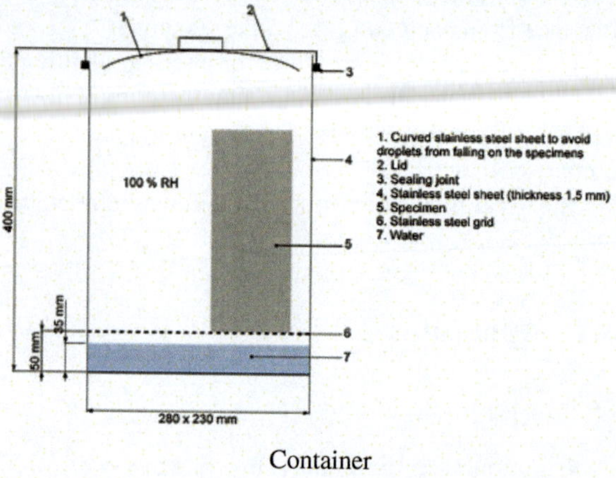

Fig. 8.2a Container (note: the dimensions presented in this figure are those recommended in NF P18-594:2015 which substitutes to former standard XP P18-594). (See NF P 18-454:2004)

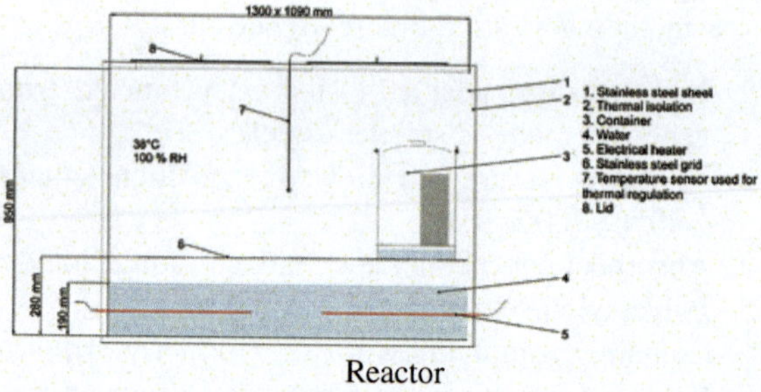

Fig. 8.2b Reactor (note: the dimensions presented in this figure are those recommended in XP P18-594; NF P18-594:2015 which substitutes to former standard XP P18-594 requires a reactor using electrical heaters ensuring a temperature of 38 ± 2 °C and a moisture as near as possible of 100% RH) (See NF P 18-454:2004)

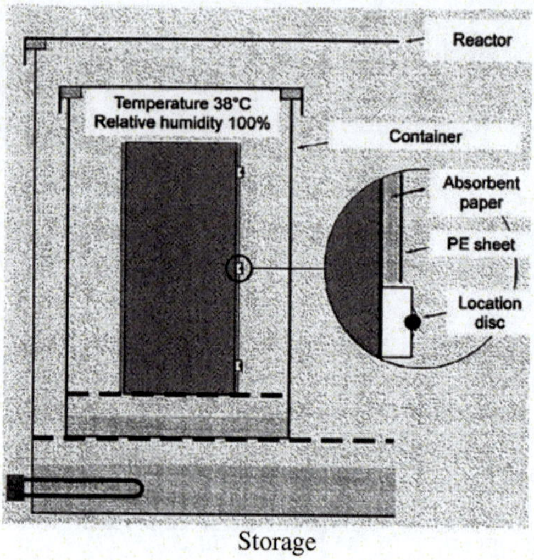

Fig. 8.2c Storage of the specimens (note: the initial publication of the LPC nr. 44 method required a reactor and a container conform to the standard P18-587 presently substituted by NF P18-594:2015)

Each measurement basis shall be marked permanently on the coating of the cylinders.

After this, the initial length measurement has to be made using the extensometer (item 1 of Sect. 8.4). The temperature of the storage room must be measured and recorded in the test report.

The specimen is then stored in the container (item 6 of Sect. 8.4), directly on the grid above the level of water. The bottom of the container has to be filled with water (level of water = 35 ± 5 mm). Each container can contain 2 specimens. After storage of the cylinder, the container is closed with its lid, this latter having a deflector to avoid water dropping on the specimens.

The container is then stored in the reactor (item 7 of Sect. 8.4). Its internal temperature is set at 38 ± 2 °C. The level of water in the bottom of the reactor shall be checked: it has to be above the electrical heater and below the grid where the containers are laid, Fig. 8.2c.

8.5.2 Measurements

The measurements are made after the following exposure durations: 2, 4, 6, 8, 12, 16, 20, 24, 28, 32, 36, 40, 44, 48 and 52 weeks.

For each measurement, the specimens shall be cooled down to a temperature of 20 ± 1 °C. To do this, the day before the measurement, the specimens are taken from their container and are stored at least 16 h in a transfer container at 20 ± 1 °C with 35 ± 5 mm of water at the bottom.

Before the length measurement, the temperature in the transfer container is checked with a temperature sensor (item 10 of Sect. 8.4).

For all the measurements, no mass monitoring is performed: due to the use of an absorbent paper to maintain a high moisture around the specimen, the weight variations exhibit high variability during the ageing procedure.

Prior to any measurement, the extensometer is used to set the "zero" measurement using the calibration reference bar. The length measurements are then performed. The specimens are placed once again in their container at 38 °C right after the measurement. After measurement, the specimen is placed again in its container in the reactor.

After the 52 weeks of accelerated ageing, the absorbent paper and the PE sheet are removed from the surface. The specimens are then weighed. A careful visual inspection is done as well; every singularity is recorded (crack, gel, etc.) and precisely situated on the specimen (the measurement basis can be used as references).

The specimen is then stored in the oven at 105 °C (item 8 of Sect. 8.4). During the whole drying process, the shrinkage is measured with the extensometer until its stabilization. The shrinkage measurement is performed once the cylinders have been cooled down at 20 °C.

8.6 Data Analysis

8.6.1 Presentation of Results

All measurements are presented in a table, Table 8.3 that gives, for each timestep, the expansion of each measurement basis, their mean value and, if applicable, the values corresponding to each level of measurement (for specimens of 250 mm in length). The numerical values are plotted in graphs as well. An example of test report is given in Appendix. Examples of residual expansion curves are shown in Fig. 8.5.

8.6.2 Data Analysis Methodology

8.6.2.1 General Approach

Nota The following data analysis of the test results is based on the feedback available while writing this document. Since few experience has been accumulated according to this testing procedure, the preparation has to be considered as indicative and provisional.

The mean dimensional evolution during the test is generally divided into three phases:

Phase a: rapid initial expansion It corresponds to the water uptake of the core. It is mainly linked with the capillary features of the material (porosity, microcracking) and eventually with hydrophilic products (reaction gels).
This phase depends as well on the moisture state of the specimen and on the stresses that were applied to the material in the structure.
Due to all these parameters, the data analysis of the data in this phase is relatively difficult. They are thus not integrated in the evaluation of the residual expansion. By convention, the duration of this phase is 8 weeks.

Phase b: slow and continuous expansion This phase corresponds to the beginning or to the continuation of an internal swelling reaction possibly still active. The mean variation of expansion is calculated by regression on the data between 8 and 52 weeks (if some outliers are detected in the data, they have to be deleted prior to the regression). As a final result, this regression has to be expressed as a mean expansion rate for one year. This value has then to be compared to the threshold given in Table 8.1.

Phase c (optional): drying shrinkage It corresponds to the opposite of phase a, however, during this phase, the initial and final hydrous states of the specimen are well controlled. The duration of this phase varies between 1 and 2 months.
The test results recorded during this phase bring valuable information regarding the microstructural state of the material while the moisture content is varying.

However, due to the lack of feedback, no classification criteria have been established so far concerning these data.

8.6.2.2 Discussion from the Present Authors Concerning Initial Water Uptake

Based on [8]

Not considering the expansions occurring during the 8 first weeks of the residual expansion test leads to underestimate residual expansion potentials. Indeed, [8] compared residual expansions determined as prescribed in the original version of the residual expansion test procedure to the whole expansion curve of the same concrete: they found out that the calculated residual expansion underestimated the real expansion potential of the material.

To take this into account, the thermal expansion (which can be corrected from the data by starting the measurements on cores with a temperature of 38 °C) and the expansion due to water uptake must be accounted for. It is thus necessary to estimate the expansions due to water absorption and to deduce them from the whole expansion data to get the AAR residual expansion.

8.6.2.3 Discussion Concerning the Use of Residual Expansion in a Global Structural Investigation from the Present Authors

Based on [10]

Due to the influence of the size and geometry of the specimen on expansion, to the effect of location and orientation of the cores, of the stress and moisture state in the structure, residual expansion values should be used with care while using these results to analyze the mechanical behavior of affected structures and/or modeling them.

Reference [10] suggest that the residual expansion test should be used in combination with other diagnosis tools such as:

- The cracking index [12].
- Longtime monitoring of the structure.

Table 8.1 Residual expansion qualification criteria

Value	Ranking of the residual expansion
<100 μm/m/year	Negligible
From 100 to 500 μm/m/year	Moderate
>500 μm/m/year	High

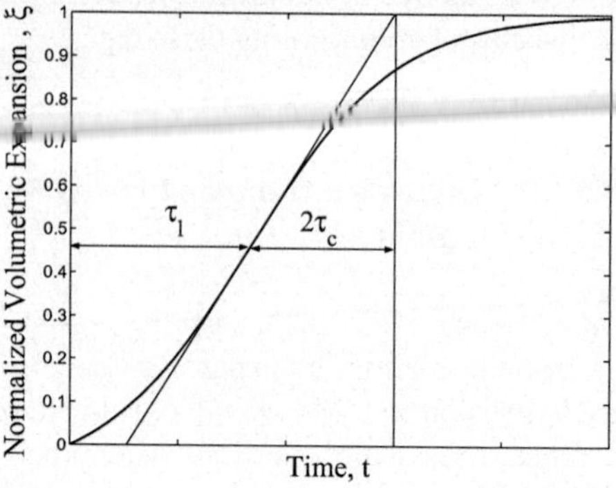

Fig. 8.3 Modelling of the expansion curve, [13]

Moreover, they suggest that more feedback is needed from field structures to better analyse the residual expansion test.

8.6.2.4 Further Comments from the Present Authors for Data Analysis of the Results

Based on [9]

The following methodology has been used in expert investigations conducted at Ifsttar to interpret the results of the residual expansion test.

Reference [13] introduced the S-shaped free expansion curve of a concrete affected by AAR modeled with the following equation:

$$\varepsilon(t) = \frac{1 - e^{-\frac{t}{\tau_c}}}{1 + e^{-\frac{(t-\tau_l)}{\tau_c}}} \varepsilon^{\infty} \tag{8.1}$$

illustrated by Fig. 8.3. Where ε^{∞} is the asymptotic expansion τ_c is the characteristic time and is linked with the expansion rate at the inflection point of the curve, τ_L is the latency time and corresponds to the age of concrete from its exposure to accelerated AAR conditions at the inflection point.

Based on [9] approach, the determination of the residual expansion of a concrete affected by AAR is associated with a long term monitoring of a structure. These authors propose the following procedure to assess the residual expansion.

Expert reassessment of a structure needs a structural monitoring and a residual expansion measurement (for examples of application, see [6, 14]). For both of these tools, at the time of investigation, the previous structural and material behaviors are not known. Reference [9] describes mathematically this process by writing:

$$\text{Find} \quad \boldsymbol{\kappa} \in \mathcal{G}_c : f_s(\boldsymbol{\kappa}) \rightarrow \min \tag{8.2}$$

where:

$\boldsymbol{\kappa}$ Vector of values $(\varepsilon^{\infty}, \tau_L, \tau_c)$ that describes well the residual expansion measured on the core

$\mathcal{G}_c$ Set of the solutions $\boldsymbol{\kappa}$.

f_s Function comparing the calculated structural behavior based on $\boldsymbol{\kappa}$ (by using a reassessment model such as RGIB [15]) and the monitored structural behavior.

In other words, the possible residual expansion parameters are determined on the residual experimental measurements and the optimal set of Larive's parameters is determined by choosing the one that best fits with the structural behavior.

The expansion of a residual expansion core $\varepsilon_c(t)$ can be written as follows:

$$\varepsilon_c(t) = \begin{cases} \text{unknown} & 0 \le t \quad < t_0^c \\ \varepsilon_c(t_0^c) + \varepsilon_{res}(t) & t_0^c \le t \le t_1^c \end{cases} \tag{8.3}$$

where:

$\varepsilon_c(t_0^c)$ is the expansion developed by the core at the coring time.
$\varepsilon_{res}(t)$ is the expansion measured during the residual expansion test
t_0^c is the coring time
t_1^c is the time by the end of the residual expansion test

Step 1: A possible value of $\varepsilon_c(t_0^c)$, called ε_0, is chosen $\varepsilon_c(t_0^c, t_1^c) = \varepsilon_0 + \varepsilon_{res}(t_0^c, t_1^c)$ is then calculated. The set of Larive's parameters $(\varepsilon^{\infty}, \tau_L, \tau_c)$, (leading to an estimation of $\varepsilon(t)$) that best fits with $\varepsilon_c(t)$ during the interval $\left[t_0^c; t_1^c\right]$ is then determined. This set of parameters corresponds to one $\boldsymbol{\kappa}$ vector composing $\mathcal{G}_c$. $\mathcal{G}_c$ is built by renewing this fitting process n times with n different ε_0 values, n being determined by the operator. In other words, $\mathcal{G}_c$ contains n possible solutions $(\varepsilon_0, \varepsilon^{\infty}, \tau_L, \tau_c)$ for the residual expansion.

Step 2: Using each of the n possible solutions of $\mathcal{G}_c$, the mechanical behavior of the affected structure is calculated. Each calculation is then compared to the structural monitoring. The solution leading to a calculated structural behavior that best fits with the monitoring is considered as the optimum set of parameters that represents the residual expansion of the material.

8.6.3 Conclusions

The concrete residual expansion test is one of the tools that can be used to diagnose and assess the structures affected by the expansion of concrete. Its aim is to provide elements regarding the future evolution of the expansion of concrete (prognosis).

The absence of residual expansion can indicate that the concrete is either non-reactive (due to its composition) or that its reactivity has been exhausted. In both cases, such a result is an indicator that the behavior of the investigated structure might be stable, regardless of its exposure conditions (except from an external alkali supply).

The assessment of a significant residual expansion rate indicates that the concrete is prone to expand in the future if the exposure conditions are favorable (especially regarding moisture).

At this stage, it is impossible to predict quantitatively the behavior of an affected structure based on the only residual expansion. Indeed, a lot of parameters are involved in the structural field that are not taken into account in the residual expansion test such as:

- the concrete of a structure is under a given stress state that does not exist in the residual expansion test;
- compared to the field exposure of a structure, the expansions in the residual expansion test are highly triggered by the exposure at 38 °C and 100% of relative humidity;
- the environmental conditions are different for the structures and for the cores;
- the residual expansion test does not take into account the influence of rebars.

Note: From experience gained by the authors in implementing the method, it can be stated that some of these items can be accounted for when the residual expansion test result is used to determine input values in a coupled chemo-mechanical model. Namely, the residual expansion helps quantifying prescribed strains induced by AAR.

8.7 Examination at End of Test

Addendum to LPC nr. 44 method suggested by the RILEM TC 259

After the final length change measurements have been made, each specimen shall be tested for warping and shall be examined.

Warping (Ifsttar has no experience in this kind of measurement)—The warping, if any, that each specimen has manifested during the test period, shall be determined by placing the specimen on a plane surface and measuring the maximum separation between the specimen and the surface. The specimen shall be placed so that its ends curve down to the reference surface and the measurement made to the nearest 0.3 mm (0.01 in.).

Examination-Items to be noted in the examination include:

- Presence, location, and type of pattern of cracking,
- Appearance of surfaces, surface mottling, and
- Surficial deposits or exudations, their nature, thickness and continuity.

Table 8.2 Example of differences between residual expansion test conditions and field conditions

Parameters	Field	Laboratory
Temperature	Variable with time. Possible temperature gradients	Constant with time, 38 °C, except cooling at 20 °C for each measurement. Temperature field almost homogeneous in the core
Humidity	Variable with time. Possible moisture gradients	Constant with time, ~ 100 % RH. Moisture almost homogeneous in the core
Stresses	3D complex stress field. Likely to evolve as a function of time (due to service loads, evolution of AAR, etc.)	Free expansion
Alkali leaching	Different alkali leaching regimes depending on the water supply conditions within the structure	High moisture likely to enhance alkali leaching and thus reduce expansion. Experimental setup designed to limit this aspect though

8.8 Limitations and Applicability to Analysis

In this residual expansion test, the cores extracted from the investigated structures are exposed to conditions prone to accelerate the swelling of concrete. As such, the representativeness of the test with regard to the concrete field performance must be questioned; one must keep in mind that the test conditions are simplified as compared to the field. Table 8.2 illustrates some significant differences between laboratory and field approaches.

Consequently, this residual expansion test is used as part of a more complete structure management procedure as described in details in [6, 7] and must not be used as a single test with direct conclusions based on the test output. In particular, it is associated with a long-term structural monitoring (collecting data for a long period makes easier the identification of conventional structural behavior, such as seasonal variations, and structural evolution due to AAR). Moreover, the use of several cores (different locations in the structure, different orientations, etc.) provides multiple information that may be useful while comparing residual expansion test results and structural monitoring, and because also of the possible heterogeneity of concrete in the structure. The combined use of these tools allows providing a diagnosis-prognosis approach for the structure manager as illustrated in Fig. 8.4.

Residual expansion test results are to be used to assess input parameters of structural reassessment models. A possible methodology is provided in Sect. 8.6; the sooner the sample is cored with regard to the AAR development, the more information is available to assess the expansion parameters with the residual expansion test and the more reliable is the approach. When using residual expansion tests to assess the input parameters of reassessment models, attention must be paid to the differences between laboratory testing conditions and field exposure (see Table 8.2)

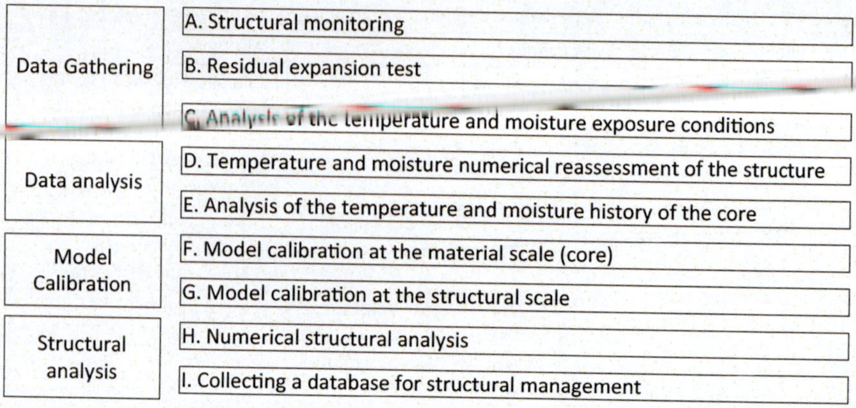

Fig. 8.4 Use of the residual expansion test in the structural reassessment methodology

and how they are taken into account in the reassessment process. In other words, equation implemented in the reassessment tools must transcribe the simple exposure conditions of the residual expansion test to the complex exposure conditions of the field.

8.9 Report

Addendum to LPC nr. 44 method suggested by the RILEM TC 259

Report the following information:

- Type and source of aggregate/concrete if possible.
- Type and source of Portland cement/binder composition.
- If possible: Alkali content of cement/binder as percent potassium oxide (K_2O), sodium oxide (Na_2O), and calculated sodium oxide (Na_2O) equivalent. Refer to relevant method for extracting alkalis from cement/binder (water soluble and acid soluble alkalis).
- Average length change in percent at each reading of the specimens.
- Any relevant information concerning the preparation of concrete core samples.
- Any relevant information concerning the preparation of aggregates if the method uses extraction aggregates from cores taken from structures.
- Any significant features revealed by examination of the specimens during and after test.
- Type, source, proportions, and chemical analyses, including Na_2O and K_2O, of any pozzolans in the concrete.
- Coring outline.
- Coring conditions.
- Coring direction.
- Any difference while performing the test as compared to the procedure described in this document.

Table 8.3 Tabulation of experimental data

Storage conditions	Ageing time (weeks)	Date	Concrete temperature(°C)	Dimension monitoring (µm)							Correction	Expansion (µm/m)						Mean expansion (µm/m)			Residual expansion (µm/m/year)
				Calibration reference bar	Level I			Level II				Level I			Level II			Level I	Level II	Global	
					Basis a	Basis b	Basis c	Basis a	Basis b	Basis c		Basis a	Basis b	Basis c	Basis a	Basis b	Basis c				
Expressions:				L0	Ia	Ib	Ic	IIa	IIb	IIc	C0	dIa	dIb	dIc	dIIa	dIIb	dIIc	MI	MII	Mean	ER
38°C / 100% RH	0																				
	2																				
	4																				
	6																				
	8																				RE = µm/m/year
	12																				
	16																				
	20																				
	24																				
	28																				
	32																				
	36																				
	40																				
	44																				
	48																				
	52																				
105°C 0% RH	56																				
	60																				
	64																				
	68																				

Calcultations : C0 = L0(week i) - L0(week 0) + α*(tb(week i) - tb(week 0))*L(the thermal dilation coefficient α is considered to be equal to 10.10^{-6}°c^{-1}; L is the gauge length)

dIa ... dIIc = [(Ia...IIc(week i) - Ia...IIc(week 0)) - c0]/L

MI = (dIa+dIb+dIc)/3

MII = (dIIa+dIIb+dIIc)/3

Mean = (MI + MII)/2

RE = Slope of the regression line of the expansion data between 8 and 52 weeks of ageing

Correction RPM :

Coefficient de dilatation à prendre en compte dans l'expression de la correction thermique

Ajout de la longueur de base pour le calcul de la déformation

8.10 Precision and Bias

Addendum to LPC nr 44 method suggested by the RILEM TC 259

Although implemented since 1997, the method has not deserved systematic metrological evaluation. From the present practice, it can be considered that precision shall be considered satisfactory if the differences in the value for percentage expansion of any given specimen in a group from the average value for percentage expansion of the group does not exceed 20 %.

Interlaboratory test data for the determination of the repeatability of the method. Repeatability shall be considered satisfactory if the percentage expansion of each specimen from the same concrete is within 15 % of the average.

8.11 Appendix

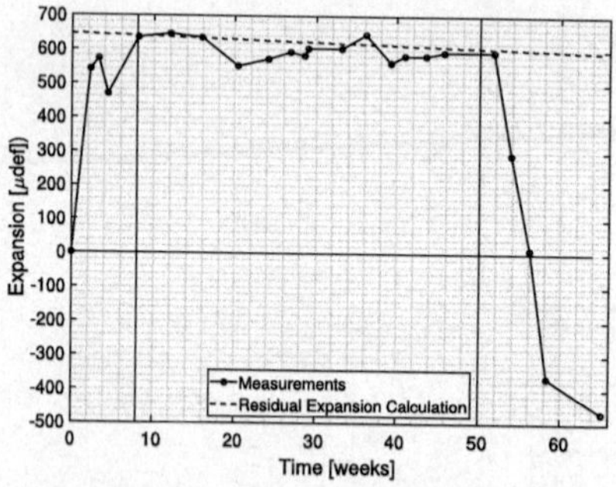

((a)) Concrete with no residual expansion potential that was never affected by AAR; RE=-68 μm/m/year

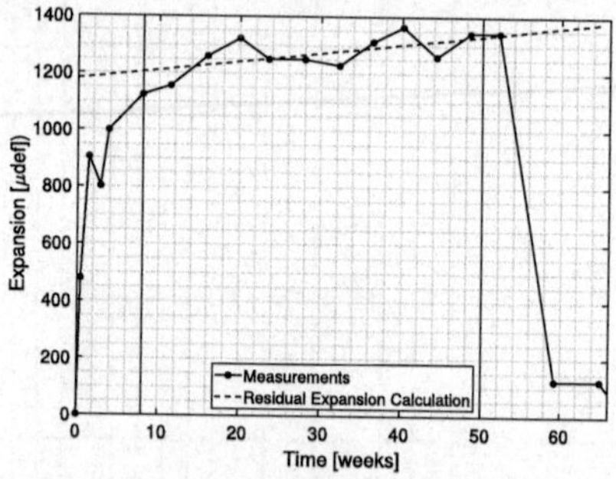

((b)) Concrete with slight residual expansion potential that was previously affected by AAR; RE=196 μm/m/year

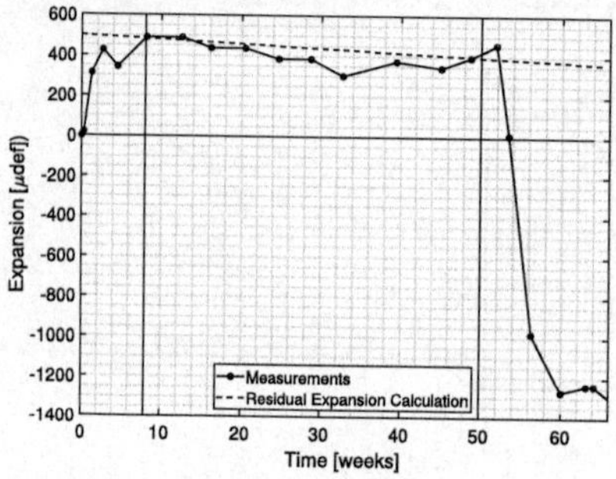

((c)) Concrete with no residual expansion potential that was previously affected by AAR; RE=-112 μm/m/year

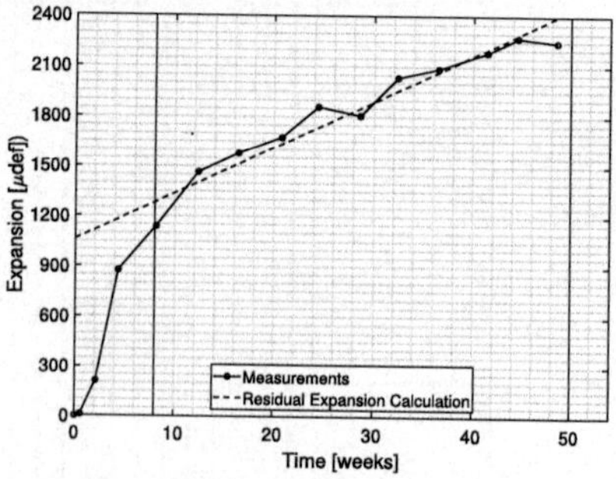

((d)) Concrete with a high residual expansion potential; RE=1,327 μm/m/year

Fig. 8.5 Examples of residual expansion curves

References

1. Fasseu, P.: Alcali-Reaction Du Beton-Essai D'Expansion Residuelle Sur Beton Durci-Projet De Methode D'Essai. Technical report No. 44. Laboratoire Central des Ponts et Chaussées (LPC) (1997). http://www.ifsttar.fr/fileadmin/user_upload/editions/lcpc/MethodeDEssai/MethodeDEssai-LCPC-ME44.pdf
2. British Cement Association.: The diagnosis of alkali-silica reaction. Report of a working party. Technical report, British Cement Association (1992)
3. AFNOR.: Granulats-Stabilité dimensionnelle en milieu alcalin-Essai sur béton. AFNOR, p. 18–587 (1990)
4. Annales de l'ITBTP. les Alcalis-Réactions, points de la question aujourdh́ui, prévention, diagnostic, responsabilités. In: Annales de l'ITBTP 1st part nr. 517, October, 2nd part nr. 518, November (1993)
5. Godart, B., Fasseu, P., Michel, M.: Diagnosis and monitoring of concrete bridges damaged by AAR in Northern France. In: The Ninth International Conference On Alkali-Aggregate Reaction In Concrete, pp. 368–375. London (1992)
6. Fasseu, P., Mahut, B., Aide à la gestion des ouvrages atteints de reactions de gonflement interne. Technical report, 66 Pages, Paris, France: Laboratoire Central des Ponts et Chaussèes (2003). http://www.ifsttar.fr/fileadmin/user_upload/editions/lcpc/GuideTechnique/GuideTechnique-LCPC-GONFLIN.pdf
7. Godart, B., Mahut, B., Fasseu, P., Michel, M.: The guide for aiding to the management of structures damaged by concrete expansion in France. In: Proceedings of the 12th ICAAR, pp. 1219–1228. Bejing, China (2004)
8. Multon, S., Barin, F., Godart, B., Toutlemonde, F.: Estimation of the residual expansion of concrete affected by alkali silica reaction. J. Mater. Civ. Eng. **20**, 54–62 (2008)
9. Li, K., Coussy, O., Larive, C.: Modélisation chimico-mécanique du comportement des bétons affectés par la réaction d'alcali-silice et expertise numérique des ouvrages d'art dégradés. Technical report OA43. 202 pages. Laboratoire Central des Ponts et Chaussées (2004). http://www.ifsttar.fr/fileadmin/user_upload/editions/lcpc/ERLPC/ERLPC-OA-LCPC-OA43.pdf
10. Merz, C., Leemann, A.: sAssessment of the residual expansion potential of concrete from structures damaged by AAR. Cem. Concr. Res. **52**, 182–189 (2013)
11. AFNOR NF.: Méthodes d'essai de réactivité aux alcalis. AFNOR, pp. 18–594 (2015)
12. Fasseu, P., Michel, M.: Détermination de l'indice de fissuration d'un parement de béton; Méthode d'essai LCPC N0. 47. Technical report 47. Paris, France: Laboratoire Central des Ponts et Chaussées (1997). http://www.ifsttar.fr/fileadmin/user_upload/editions/lcpc/MethodeDEssai/MethodeDEssai-LCPC-ME47.pdf
13. Larive, C.: Apports combinés de l'expérimentation et de la modélisation à la compréhension de l'alcali-réaction et de ses effets mécaniques. Technical report OA28, 404 pages, Laboratoire Central des Ponts et Chaussées (1998)
14. Godart, B.: Méthodes de suivi dimensionnel et de suivi de la fissuration des structures: Avec application aux structures atteintes de réaction de gonflement interne du béton: Guide technique (2009)
15. Seignol, J., Baghdadi, N., Toutlemonde, F.: A macroscopic chemomechanical model aimed at re-assessment of delayed ettringite formation affected concrete structures. In: 1st International Conference on Computational Technologies in Concrete Structures CTCS'09, pp. 422–440 (2009)

Chapter 9
Accelerated Expansion Test: Canada, Laval University Method

Benoît Fournier and Leandro Sanchez

9.1 Scope

This test method covers the laboratory determination of the swelling potential and the residual free expansion of concrete extracted from structures affected by alkali-silica reaction.

The method is not an aggregate test like in the methods where aggregates are extracted from concrete cored from structures and reused in accelerated mortar tests with added alkali.

This method does not encompass the assessment of the reaction advancement in the structure.

9.2 Principles and Methodology

The procedures used for testing cores from ASR-affected structures vary greatly from one study to another [1]. Two test procedures are recommended hereafter, where the cores are: (1), tested in air at >95% R.H. and 38 °C (100 °F), and (2), tested in 1N NaOH solution at 38 °C (100 °F). The first test is considered the most realistic for evaluating the *residual concrete expansivity* and the potential for further expansion whereas the second approach accounts for the *maximum expansion potential* of the concrete [2]. This chapter describes the principles of the above two methods.

B. Fournier
Universié Laval, Quebec City, QC, Canada
e-mail: benoit.fournier@ggl.ulaval.ca

L. Sanchez (✉)
University of Ottawa, Ottawa, ON, Canada
e-mail: Leandro.Sanchez@uottawa.ca

© RILEM 2021

V. E. Saouma (ed.), *Diagnosis & Prognosis of AAR Affected Structures*, RILEM State-of-the-Art Reports 31,
https://doi.org/10.1007/978-3-030-44014-5_9

9.3 Sampling and Sample Preparation

9.3.1 Core Diameter

The diameter of the cores should be at least three times the maximum size of aggregates, and their length should be two to three times their diameter. For concrete with aggregates of nominal size smaller than 35 mm (1.4 in), a core diameter of 100 mm (4 in) is the most practical. However, as discussed hereafter, a core diameter of 150 mm (6 in) is always recommended when testing in humid air (>95% R.H. and 38 °C (100 °F)), which reduces alkali leaching during the test and postpones the consequent leveling off of the expansion curve.

9.3.2 Surface Effects and Depth of Cores

Concrete skin (from the exposed surface down to about 25–75 mm (1–3 in) in the concrete) is usually more macro-cracked and may have suffered from significant alkali leaching or, conversely, alkali concentration through evaporation or supply of de-icing chemicals. Therefore, this concrete may not be representative of the ASR-affected member under study regarding the evaluation of potential for further expansion. In particular, wetting/drying and freezing/thawing cycles may have greatly contributed to the development of cracking in the surface concrete. It is recommended that, at least, the first 50 mm (2 in) of concrete be avoided.

9.3.3 Number, Location, and Orientation of Cores

Large variations in mixture proportioning, exposure conditions (humidity, temperature, etc.), stress conditions, and internal deterioration (expansion and microcracking), may occur within a single structure or a single member of a structure, leading to variations in the test results. It is therefore important to take cores from various members of the concrete structure under investigation, or from different locations within a single component but subjected to different exposure conditions, in order to have a good coverage of such variations.[1]

For a given type of concrete member, it is recommended to test a minimum of three specimens in humid air at >95% R.H. and 38 °C. More than one test specimen

---

[1] Reference [3] reported that ASR expansion is greatly reduced in the direction of the main restraints (reinforcements, prestressing, postensioning, loading, …), while restraining concrete samples in one or two directions can actually contribute at increasing expansion in the unrestrained (or less restrained) direction(s). Moreover, a number of studies also indicated that the ASR expansion of non-reinforced concrete is greater in the direction perpendicular to the casting plane, which phenomenon is amplified in the presence of flat and elongated reactive aggregate particles [4].

can be taken from cores of sufficient length, provided that the above considerations about concrete variability are taken into account. At the time of coring, it is important to note the orientation of the cores in the concrete member(s) under study

9.3.4 Storage of Cores Before Testing

9.3.4.1 Initial Adjustments

indexInitial adjustment After coring, the volume of the core samples may change to reach a relative equilibrium with respect to the new stress and environmental (temperature and moisture) conditions to which the concrete is now exposed. These short-term variations can be related to:

- Thermal expansion or contraction, whether the temperature in nature was lower or higher than the testing conditions.
- Expansion due to release of restraining stresses due to reinforcement, prestressing, pretensioning, loading, etc. (with the possibility that contraction takes place in the less restrained directions).
- Expansion due to moisture uptake or contraction due to drying shrinkage (particularly if the cores were not sealed after coring), whether the humidity conditions in concrete were lower or higher than in nature.
- Free expansion of ASR gel already present in the core samples, as a result of lower internal stress on this gel with respect to the conditions prevailing in nature.
- Expansion of ASR gel already present in the core samples due to additional water sorption, as a result of higher humidity with respect to the conditions prevailing in nature.

9.3.4.2 "Pre-Testing" Conditions

In order to minimize the effects of drying (shrinkage, microcracking) and expansion due to rewetting at the beginning of the tests, immediately after coring, the core samples should be superficially dried to remove the excess water and then sealed immediately by wrapping them in heavy duty shrink wrap and storing them in sealed polyethylene bags. This also prevents carbonation as it was observed that cores tested in 1N NaOH at 80 °C (176 °F), which were allowed to dry for one month in the laboratory before being tested, presented higher expansions and mass increases at the beginning of the tests, i.e. during the so-called *preconditioning* period, than companion cores kept sealed since coring [5]. However, after correction for these short-term effects, which are not related to further ASR expansion in the new storage conditions, the expansion results were similar for both series of cores. Nevertheless, it is always recommended that all cores be wrapped immediately after coring in order to prevent drying. This could actually contribute at significantly reducing the

Table 9.1 Suggested length of core samples tested as a function of core diameter

Core diameter	Recommended length*	Minimum length*
50 (2 in)	150 (6 in)	100 (4 in)
75 (3 in)	225 (11 in)	150 (6 in)
100 (4 in)	300 (12 in)	200 (8 in)
150 (6 in)	300 (12 in)	300 (12 in)

*Approximately (max ± 5 mm)

duration of the preconditioning period, which can be particularly long (up to 2–3 months) when cores that were allowed to dry have to be tested in humid air.

It is recommended to wait at least one week before the cores are subjected to expansion testing in order to reach a relative mechanical equilibrium. However, the volumetric changes taking place during this *pre-testing* period, which is likely attributed to stress release, could also be monitored by installing gauge reference studs and taking initial length shortly after coring, following the instructions described hereafter.

9.3.5 Preparation of Cores

9.3.5.1 Sawing of Cores

The core samples are sawn perpendicular to their axis using a diamond saw. As already mentioned, their length should be two to three times their diameter (see Table 9.1). This step is not absolutely required when only lateral and diametrical measurements are made, but the presence of flat ends facilitates the handling and the storage of the cores.

9.3.5.2 Installation of Gage Reference Studs

Stainless steel bolts or gage studs, e.g. 13 mm in length by 3 mm in diameter (approximately 0.5 by 0.12 in), with a machined *demec point* at the end, are commonly used for that purpose. They are installed in small holes drilled dry and cemented with a shrinkage-free cement paste. The drill holes are about 8 mm (approximately 0.3 in) in diameter by 20 mm (0.8 in) deep.

Because of the influence of the directions of principal stresses, restraints and casting plane on expansion due to ASR, gage studs could be installed in order to monitor the longitudinal and diametrical expansion of the cores, including one direction perpendicular to the casting plane.

Axial (longitudinal) measurements They are more recommended than lateral (longitudinal) measurements because they generally result in lower experimental vari-

Table 9.2 Suggested distance between gage reference studs for lateral measurements on cores [mm]

Length of cores	Distance between studs	Distance between studs and end portions of cores
90–140 (0.36–5.6 in)	50 (2 in)	20–45 (0.8–1.8 in)
140–190 (5.6–7.6 in)	100 (4 in)	20–45 (0.8–1.8 in)
190–240 (7.6–9.6 in)	150 (6 in)	20–45 (0.8–1.8 in)
240–290 (9.6–11.6 in)	200 (8 in)	20–45 (0.8–1.8 in)
≥290 (≥ 11.6 in)	250 (10 in)	≥ 20 (≥ 0.8 in)

ability [3, 5]. Gage studs are centrally installed at both ends of the core, and the measurements can be performed using an arch-type device or other appropriate equipment. Axial measurements can also be performed with the measuring device used for the Concrete Prism Test [6], provided that appropriate gage studs are fixed at both ends of the cores.

Lateral (longitudinal) measurements They are also possible, however less recommended than axial measurements because of the larger variability often observed between the individual measurements performed on the same core [5]. For such measurements, at least two diametrically opposite lines of two gage studs are installed on the cores tested, with 50–250 mm (2–10 in) For such measurements, at least two diametrically opposite lines (or three lines at 120o) of two gage studs are installed on the cores tested, with 50–250 mm (2–10 in) between the studs, depending upon the length of the core specimens and the measuring device available.[2] The studs should be placed not less than 20 mm (0.8 in) from the ends of the cores (50 mm (2 in) from the end corresponding to the exposed concrete surface), Table 9.2.

Diametrical measurements They are performed by installing, at mid-length of the cores, two diametrically opposed studs. Measurements can also be made near (≥ 20 mm) (0.8 in) both ends of the cores [3]. They are performed using the same arch-type measuring device (or outside micrometer) as for axial measurements Fig. 9.1.

9.3.6 *Periodic Measurements (Expansion, Mass and Temperature)*

In order to help in the preparation of possible short term volumetric changes unrelated to the residual/further ASR, the mass of the core samples is measured in addition to their volumetric changes (diameter, length). Moreover, the temperature of the

[2]Reference [7] recommends that nine gage studs be fixed to each core along three equally spaced lines with two 50 mm (2 in) gage lengths each (6 readings per core).

(a) Axial reading using an arch-type measuring device (outside micrometer)

(b) Diametrical measurement using the same measuring device

(c) Lateral reading using standard conventional length-change measuring device used for freeze-thaw / creep testing

Fig. 9.1 Expansion measurements

storage room (test in humid air) at the time of measurement is also recorded in order to normalize all length values with respect to the nominal test temperature of 38 °C (100 °F). The correction factor used is 0.001%/°C which is an estimate of the thermal expansion coefficient of conventional concretes. The temperature correction is particularly critical when testing cores in humid air at >95% R.H., because small expansions are generally obtained [5].

9.4 Apparatus

1. Extensometer:
   - minimum resolution: 1 μm
   - gauge length: 10 cm
   - including a calibration reference bar
2. Stainless steel bolts or gage studs (e.g. 13 mm in length by 3 mm in diameter);
3. Shrinkage-free fast set cement;
4. Absorbent paper (thickness: 0.6 mm; fungicidal);
5. PE sheet (thickness: 0.1 mm);
6. Sealed plastic pails with wicks inside (or appropriate alternative containers)
7. Temperature controlled chamber or alternative reactor (to maintain temperature at 38 °C (100 °F);
8. Analytical balance (capacity: in agreement with the weight of the specimens; relative precision: 0.1% of the weight of the specimen).

9.5 Experimental Procedures

9.5.1 Experimental Procedure (in Air at >95% R.H. and 38 °C)

9.5.1.1 Test Method

- The initial mass and dimensions (length, diameter) of the core samples are first measured.
- The core samples are then placed vertically, above water (>95% R.H.), in sealed plastic pails with wicks lining the inside walls (or in alternative appropriate containers), which are stored in a room (or in alternative reactor) maintained at 38 °C. These conditions are those used in the Concrete Prism Test [6].
- Initial expansion and mass measurements are taken one day after storage at >95% R.H. and 38 °C (100 °F), as thermal equilibrium is attained. Periodic measurements of expansion, mass and temperature of the storage room are then taken after 1, 2, 3, 4, 6, 8, 10, 12, 16, 20, 24, 30, 36, 44 and 52 weeks, then once every 2 months, if necessary.
- The specimens are always measured hot. Immediately after the plastic pail (or alternative container) has been removed from the storage room (or alternative reactor) and after a core sample has been taken out from its container, the longitudinal and/or diametrical deformations of the core sample are measured, then its mass. This overall process should take less than one minute. Because of the low expansion rates usually obtained, it is highly recommended that the diameter and length measurements be corrected for the actual temperature in the storage room (or reactor) at the moment of measurement, as described before.
- The short-term length-change results obtained are analyzed in order to determine at what time the core samples reached a relative equilibrium in mass and expansion (after hygrometric equilibrium, which also includes potential additional expansion of preexisting ASR gel and is not related to potential further expansion of the concrete due to ASR). This so-called *preconditioning period* may range from a few days to a few months, depending upon the storage conditions before testing (e.g., drying). The time and expansion scales are then reset to zero when such pre-testing equilibrium has been reached.
- The measurements are made until one year at least after the *preconditioning period*, and even further when the one-year expansion is relatively low (e.g., <0.025%).
- The final result is expressed as the average annual rate of expansion over the testing period, excluding the *preconditioning period*. Linear regression analysis is recommended in order to determine the annual rate of expansion with a better accuracy.

9.5.1.2 Expansion Limit Criterion

Because the alkali content of field concrete is normally lower than that of laboratory concrete made in accordance with [6], relatively low expansions are usually obtained. However, an expansion as low as 0.003% per year, which is the lower limit considered in the methodology proposed in the Appendix I of [8], may be of great importance for the existing structure under study. In fact, expansion rates in the range of 0.002 to 0.005% per year are common in the case of AAR-affected concrete structures[3] [9]. On the other hand, very low expansions, e.g., $\leq 0.005\%$ per year, are not statistically significant considering the experimental variation despite the reading precision of the expansion measurements is about $\pm$ 0.0005. In order to improve the statistical significance of the results, (1), the tests are often extended over the usual one-year period; (2), the measurements are more frequent than in the standard Concrete Prism Test [6], and (3), linear regression analysis is recommended in order to better assess the annual rate of expansion.

9.5.2 *Experimental Procedure (Soaked in 1M NaOH RH and 38 °C)*

9.5.2.1 Test Method

- The initial mass and dimensions (length, diameter) of the core samples are first measured.
- The core samples are then placed vertically in sealed plastic pails, soaked in a 1M NaOH solution and stored in a room (or in alternative reactor) maintained at 38 °C (100 °F).
- Initial expansion and mass measurements are taken one day after storage at 38°, as thermal equilibrium is attained. Periodic measurements of expansion, mass and temperature of the storage room are then taken after 1, 2, 3, 4, 6, 8, 10, 12, 16, 20, 24, 30, 36, 44 and 52 weeks, then once every 2 months, if necessary.
- The specimens are always measured hot. Immediately after the plastic pail (or alternative container) has been removed from the storage room (or alternative reactor) and after a core sample has been taken out from its container, the longitudinal and/or diametrical deformations of the core sample are measured, then its mass. This overall process should take less than one minute. Because of the low expansion rates usually obtained, it is highly recommended that the diameter and length measurements be corrected for the actual temperature in the storage room (or reactor) at the moment of measurement, as described before.

[3] For instance, in the study by [2] on the effectiveness of sealers against the ASR developed in median barriers located in Québec City (Québec, Canada), an expansion rate of 0.005% per year has been measured for the unsealed control sections of barriers.

- The short-term length-change results obtained are analyzed in order to determine at what time the core samples reached a relative equilibrium in mass and expansion (after hygrometric equilibrium, which also includes potential additional expansion of preexisting ASR gel and is not related to potential further expansion of the concrete due to ASR). This so-called *preconditioning period* is normally obtained throughout one week. The time and expansion scales are then reset to zero.
- The measurements are made until one year at least after the *preconditioning period.*
- The final result is expressed as the average annual rate of expansion over the testing period, excluding the *preconditioning period.* Linear regression analysis is recommended in order to determine the annual rate of expansion with a better accuracy.

9.5.2.2 Expansion Limit Criterion

Expansion levels of 0.04% at one year seem to well distinguish reactive and non-reactive behaviour [5].

9.6 Preparation

The expansions measured in the tests may have been underestimated if the concrete specimens tested were abnormally fractured or porous compared to the overall concrete member under investigation. In such a case, some ASR gel produced during the test may have expanded freely in voids without causing additional expansion. The preparation of the test results is thus not always easy.[4]

9.6.1 Assessment of the Final Swelling Potential

9.6.1.1 Early Thermal and Moisture Adjustments ("Preconditioning")

When testing length changes of cores, a preliminary expansion phase can be observed due to various mechanisms unrelated to the residual/further ASR expansion:

- Thermal expansion (concrete), since the test temperature (38 °C) (100 °F) is normally higher than the temperature at which the core samples were stored before being tested.

[4]Petrographic examination of the concrete cores before the tests (cracking, primary reaction products, etc.), the measurement of the water-soluble alkali content of the concrete, the verification of the presence of secondary reaction products after the tests, and the expansion results obtained for companion cores immersed in 1N NaOH at 38 °C (100 °F) may greatly help in the preparation of the results of tests in humid air.

- Stress release (concrete and preexisting ASR gel) when the samples are tested early after coring, as a result of lower stress with respect to the conditions prevailing in nature.
- Moisture uptake (concrete and preexisting ASR gel), since the test humidity conditions are normally higher than that found in nature (cores sealed immediately after coring), or if the core samples have been allowed to dry after coring. In some cases, the expansion due to rewetting may reach 0.1% or more, while ASR expansion afterwards might be negligible.

The only way to properly account for these short term variations consists in conditioning the samples under the same conditions as for the tests, until a relative equilibrium in mass and expansion is reached. Then, the time and expansion scales are reset to zero. The time at which this equilibrium is reached is recognized by the presence of points of inflexion on corresponding mass and expansion curves. The points of inflexion are sometimes difficult to locate precisely, particularly on the expansion curve, but it is better to take the zero reading later rather than too early; the most critical point is to ensure that all initial adjustments are completed; otherwise the expansion due to residual/further ASR would be overestimated. On the other hand, exceeding the equilibrium point just delays the completion of the tests.

9.6.1.2 Determination of the Residual and Maximum Expansion in the Field

According to the [10], the maximum ASR expansion that a concrete member may develop in the field or its "residual expansion" corresponds to the maximal expansion obtained from concrete cores tested in humid air (>95% R.H.) at 38 °C (100 °F). Otherwise, the immersion of concrete cores into a 1N NaOH solution seems to evaluate the "maximum expansion potential" of the aggregates in a given concrete.

9.6.1.3 Determination of the Ultimate Expansion in the Field

According to [10], the maximum ASR expansion that a concrete member may attain in the field or the residual expansion corresponds to the maximal expansion obtained for concrete cores tested in humid air (>95% R.H.) at 38 °C (100 °F). However, the maximum expansion is likely to be underestimated this way since cores may suffer from significant alkali leaching under the above test conditions; consequently, the expansion tends to level off after a certain time, which is not necessarily due to consumption of reactive mineral phases or alkalis [11]. In this respect, the greater the core diameter, the lesser the alkali leaching during the expansion test in humid air and the higher the expansions in the long term [1]. The alkali leaching during expansion tests in humid air is also influenced by the concrete permeability (i.e., water-to-cement ratio), which varies from one concrete to another.

Otherwise, the second version of the method (where concrete cores are soaked into 1M NaOH solution) seems to evaluate the *maximum expansion potential* of the aggregates in a given concrete. However, due to its very aggressive environment, it is difficult to correlate the results from this procedure obtained with the expected behaviour of the same concrete in the field.

9.7 Examination at End of Test

After the final length change measurements have been made, each specimen shall be tested for warping and shall be examined.

Warping The warping, if any, that each specimen has manifested during the test period, shall be determined by placing the specimen on a plane surface and measuring the maximum separation between the specimen and the surface. The specimen shall be placed so that its ends curve down to the reference surface and the measurement made to the nearest 0.3 mm (0.01 in.).

Examination Items to be noted in the examination include:

- Presence, location, and type of pattern of cracking, crack opening,
- Appearance of surfaces, surface mottling, and
- Surficial deposits or exudations, their nature, thickness and continuity.

9.8 Report

Report the following information:

- Type and source of aggregate/concrete if possible.
- Type and source of portland cement/binder composition.
- If possible: alkali content of cement/binder as percent potassium oxide (K_2O), sodium oxide (Na_2O), and calculated sodium oxide (Na_2O) equivalent. Refer to relevant method for extracting alkalis from cement/binder (water soluble and acid soluble alkalis).
- Average length change in percent at each reading of the specimens.
- Any relevant information concerning the preparation of concrete core samples.
- Any significant features revealed by examination of the specimens during and after test.
- Type, source, proportions, and chemical analyses, including Na_2O and K_2O, of any pozzolans in the concrete.

9.9 Limitations of the Tests

The residual expansion obtained on cores tested at 38 °C (100 °F) and RH > 95% has shown over time to underestimated ASR residual expansion since cores may suffer from significant alkali leaching during the test; consequently, the expansion tends to level off after a certain amount of time, which is not necessarily due to consumption of reactive mineral phases or alkalis [11]. In this regard, the greater the core diameter, the lesser the alkali leaching during the expansion test in humid air and the higher the expansion levels in the long-term [12]. The alkali leaching during expansion tests in humid air is also influenced by the concrete permeability (i.e. water-to-cement ratio), which varies from one concrete mixture to another. Conversely, the immersion of cores extracted from structures into 1N NaOH solution aims to prevent significant leaching during the expansion test; yet, the very aggressive environment of the procedure makes the correlation of laboratory findings with field behaviour quite difficult. Furthermore, a number of field conditions such as confinement and/or anisotropy effects are not accounted for in none of the current residual expansion procedures, which certainly influences the test outcomes and its correlation with field behaviour.

9.10 Precision and Bias

The dominating factor for the uncertainty of the expansion is the variation of the concrete cores. The magnitude of error of the measurement itself is negligible (±0.002 mm). Depending on the homogeneity of the concrete at the coring site, significant variations in the expansion of single cores within a set of three cores are possible. Therefore, all cores have to be considered, even if there is an outlier.

References

1. Bérubé, M., Smaoui, N., Coté, T.: Expansion tests on cores from ASR affected structures. In: Proceedings of the 12th International Conference AAR, pp. 821–832. Beijing, PRC (2004)
2. Bérubé, M., Frenette, J., Pedneault, A., Rivest, M.: Laboratory assessment of the potential rate of ASR expansion of field concrete. Cem., Concr. Aggreg. **24**(1), 13–19 (2002)
3. Smaoui, N., Bérubé, M., Fournier, B., Bissonnette, B.: Influence of specimen geometry, orientation of casting plane, and mode of concrete consolidation on expansion due to ASR. Cem., Concr. Aggreg. **26**(2), 1–13 (2004)
4. Smaoui, N., Bérubé, M., Fournier, B., Bissonnette, B., Durand, B.: Evaluation of the expansion attained to date by concrete affected by alkali-silica reaction. Part I: Experimental study. Can. J. Civ. Eng. **31**(5), 826–845 (2004)
5. Bérubé, M., Frenette, J., Landry, M., McPhedran, D., Pedneault, A., Ouellet, S.: Evaluation du potentiel résiduel de réaction et d'éxpansion du béton en service atteint de réactivité alcalis-silice. Technical report Final report No. CO-93-04 to Hydro-Quebece (1994)

6. ASTM C1293.: C1293-18a Standard Test Method for Determination of Length Change of Concrete Due to Alkali-Silica Reaction. Annual Book of ASTM Standards (2018)
7. BCA.: The Diagnosis of Alkali-Silica Reaction-Report of a Working Party. Technical report 44 p. Wexham Springs, Slough (UK), SL3 6PL (1992)
8. Fournier, B., Bérubé, M., Folliard, K., Thomas, M.: Report on the Diagnosis, Prognosis, and Mitigation of Alkali-Silica Reaction (ASR) in Transportation Structures. Technical report FHWA-HIF-09-004. 154 p. Federal Highway Administration (2010)
9. CSA A864-00.: Guide to the Evaluation and Management of Concrete Structures Affected by Alkali-Aggregate Reaction. Canadian Standards Association (CSA) (2000)
10. ISE.: Structural Effects of Alkali-Silica Reaction-Technical Guidance Appraisal of Existing Structures. Technical report 11 Upper Belgrave Street, London SW1X 8BH: Institution of Structural Engineers (ISE) (1992)
11. Rogers, C., Hooton, R.: Reduction in mortar and concrete expansion with reactive aggregates due to alkali leaching. Cem., Concr. Aggreg. **13**(1), 42–49 (1993)
12. Bérubé, M., Tremblay, C., Fournier, B., Thomas, M., Stokes, D.: Influence of lithium-based products proposed for counteracting ASR on the chemistry of pore solution and cement hydrates. In: Taylor, H.F.W. (ed.) Cement and Concrete Research 34.9. Commemorative Issue, pp. 1645–1660 (2004). ISSN: 0008-8846. https://doi.org/10.1016/j.cemconres.2004.03.025. http://www.sciencedirect.com/science/article/pii/S0008884604001528

Chapter 10
Accelerated Expansion Test: France, LMDC-EDF

Alain Sellier, Stéphane Multon, Pierre Nicot, and Etienne Grimal

10.1 Scope

Toulouse method [1] is a methodology for estimating the residual expansion in concrete structures affected by AAR. A critical assumption herein is that alkali content is not the limiting factor in AAR, since alkalis may be substituted by calcium. Hence, it is the finite mass of silica that controls termination of the expansion.

This test method covers the laboratory determination of the swelling potential and the residual free expansion of concrete extracted from structures affected by alkali silica reaction.

The method is an aggregate test where aggregates are extracted from concrete and reused in accelerated mortar tests with added alkali. The mortar tests are used

A. Sellier (✉) · S. Multon
Civil Engineering, Paul Sabatier University, and Construction and Durability Materials Laboratory of Toulouse (France),Toulouse, France
e-mail: alain.sellier@insa-toulouse.fr

S. Multon
e-mail: stephane.multon@insa-toulouse.fr

P. Nicot
Transfer Division, Paul Sabatier University, and Construction and Durability Materials Laboratory of Toulouse (France),Toulouse, France
e-mail: pierre.nicot@insa-toulouse.fr

E. Grimal
Electricité de France, Centre d'Ingénierie Hydraulique, EDF-CIH Technolac, 73373 Le Bourget du Lac Cedex, France
e-mail: etienne.grimal@edf.fr

© RILEM 2021

V. E. Saouma (ed.), *Diagnosis & Prognosis of AAR Affected Structures*, RILEM State-of-the-Art Reports 31,
https://doi.org/10.1007/978-3-030-44014-5_10

to determine the advancement of the different classes of aggregate sizes used in the structure. Residual expansion is evaluated from inverse analysis of data available on the damaged structure (displacements, cracking measurements).

10.2 Principles and Methodology

At this point, the authors presented a method for quantifying the swelling kinetic potential at a given site. The essence of this method consists of comparing residual swelling measured on mortars made from crushed reactive aggregates with the residual swelling of mortars made from sound crushed aggregate to deduce the advancements of ASR in the different class of reactive aggregate size of the structure concrete. The advancements can then be computed to deduce an advancement of ASR in the structure concrete and used in finite element calculations. Residual expansion in structure is finally deduced from inverse analysis of the behavior of the damaged structure in which the kinetic is imposed by the results of the mortar tests.

This method can thus be broken down into three parts:

1. Field work.
2. Laboratory tests of mortar.
3. Inverse finite element simulation.

10.2.1 Analysis of Mortar Tests

The objective of the mortar test is to provide the fraction of relative silica already reacted in the structure. On purpose the aggregates of the affected concrete are extracted with a chemo-mechanical attack of the specimen, the obtained aggregates are sifted by class and each class crushed to a same size. Hence, all the mortars tested have the same size and the same amount of reactive aggregate. So if the aggregate mineralogy is the same, the difference of swelling between a mortar and another one can be due only to a difference in the consumption of reactive silica before the extraction of the aggregate from the concrete. This state can be used to classify the residual reactivity of each aggregate, but also, thanks to a comparison with mortar made with aggregates extracted from dry zone or career, to obtain the advancement of the reactive silica consumption. This advancement, in relation with the age of the structure, allows finally assessing the average kinetic of the reaction in the structure per aggregate size.

To properly understand this procedure, a simplified model is used to analyze the expansions on laboratory mortars. It assumes that a reactive aggregate of volume V_a^i can create a gel volume V_g^i proportional to its own volume:

$$V_g^i = f^i V_a^i \tag{10.1}$$

where f^i is the asymptotic fraction of gel created per volume of aggregate. It depends on both the attacked reactive silica content of the aggregate and ASR-gel molar volume [2]. The parameter f^i can be determined through inverse analysis of mortar expansion tests. In laboratory test, each mortar uses only one class of aggregate size i used in the structure and crushed to a unique size (for example the sand, the mid-aggregate or the largest aggregate are crushed separately to obtain three type of fine aggregate for the mortar, the three fine aggregate are the used alone in three series of mortar bars).

The stress-free asymptotic swelling of 1 mortar test i, ε_i, is approximated by:

$$\varepsilon = n_i \langle V_g^i - V_p^i \rangle \tag{10.2}$$

With n_i, the number of aggregate per m^3 of mortar. It is calculated in assuming a spherical shape for aggregates:

$$n_i = \frac{3\phi_i C_a}{\pi R_a^3} \tag{10.3}$$

With ϕ_i the volume fraction of reactive aggregates with mean radius R_a and C_a the volume fraction of all the aggregate per m^3 of mortar.

As, in mortar test, all the reactive aggregates are crushed and sieved to obtain only one aggregate size independently of the real size of the original aggregate in the structure, the reactive aggregate size is unique.

In Eq. 10.2, V_p^i is the volume of gel which does not cause expansion at the beginning of the ASR (initially called connected porosity in previous simplified theories [1]). It is supposed to depend only on aggregate size and thus is the same for all mortar. With the assumption that aggregate can be supposed to be spherical, this volume can be assessed by:

$$V_p^i = \frac{4}{3}\pi \left[\left(R_a^i + t_{0-exp} \right)^3 - (R_a^i)^3 \right] \tag{10.4}$$

with t_{0_exp} an average thickness to quantify the volume of the gels inefficient for expansion.

10.2.2 Advancement of ASR in Structure Concrete

The second step is to deduce the advancement of ASR in the different class of reactive aggregate size of the structure concrete at the time when the cores were drilled from the structure. It is assumed that the degree of chemical advancement of the reaction (hence gel expansion) is directly related to the degree of silica attacked in each aggregate. For a given aggregate size, we can thus define the advancement at the time of coring $A(t = t_c)$:

$$A(t = t_c) = \frac{\mathrm{Si}^{\mathrm{reacted}}(t = t_c)}{\mathrm{Si}^{\mathrm{reactive}}(t = 0)} \tag{10.5}$$

where $\mathrm{Si}^{\mathrm{reacted}}$ is the mass of silica consumed at time of coring in the aggregate size i and $\mathrm{Si}^{\mathrm{reactive}}(t = 0)$ the total reactive silica content. When the structure is built $\mathrm{Si}^{\mathrm{reacted}}(t = 0)$ is equal to 0.

For $A(t) \approx 1$, the residual swelling of corresponding mortar bars will be close to zero, and conversely for $A(t) \approx 0$ the residual swelling of mortar bars will be maximum. The former case corresponds to"old" concrete and the latter case to "young" concrete. Thus, at a same moment of coring, it can be expected that the reactive silica already consumed by ASR in the smallest aggregate is important and close to the total initial reactive silica. In this case, the advancement will be close to 1. For larger aggregate, the reactive silica should be less attacked, and advancement should be smaller. If the advancement is equal to 1, the ASR is finished and this aggregate size should not contribute to ASR expansion anymore. At the opposite, an aggregate size with an advancement of 0, should contribute to ASR in the structure for a long time if the environmental conditions make this possible. The engineer really needs to know the value of $A(t)$. In a first approach, the advancement can be determined by proportionality:

$$A(t = t_c) = \frac{f_i^{reacted}}{f_i^{reactive}} \tag{10.6}$$

where $f_i^{reacted}$ is the asymptotic fraction of gel created per volume of aggregate in an aggregate size i calibrated on mortar tests performed on damaged aggregate taken from the structure and $f_i^{reactive}$ is the asymptotic fraction of gel created per volume of aggregate in the aggregate size i, calibrated on mortar tests performed on sound aggregate.

As described before, the mortar tests are only used to determine the advancement and not the residual expansion in the structure. As a consequence, the functions f_i are not used in the structural calculations. Only the advancement $A_i(tc)$ will be used to assess the kinetic of the reaction.

10.2.3 *Inverse Analysis of ASR-Damaged Structures*

The advancements can then be computed to deduce a unique advancement of ASR $A(t)$ in the structure concrete and used in finite element calculations as exposed in the preparation part. Residual expansion in structure is finally deduced from inverse analyze of the behavior of the damaged structure.

As has been well established, chemical advancement depends on the degree of saturation Sr and temperature T; consequently, the derivative of chemical advancement with respect to reaction time t is given by:

$$\frac{\partial A}{\partial t} = \alpha_{\text{ref}} \exp\left[-\frac{E_a}{R}\left(\frac{1}{T} - \frac{1}{T_{\text{ref}}}\right)\right] \times \frac{\text{Sr} - \text{Sr}^{\text{threshold}}}{1 - \text{Sr}^{\text{threshold}}}(\text{Sr} - A) \tag{10.7}$$

Sr can in turn be determined by numerically solving the nonlinear mass transfer equation. α_{ref} is a constant controlling the kinetics, E_a the AAR activation energy (estimated by the authors at 47 kJ/M), R the universal gas constant (8.31 J/M/K), T_{ref} a reference temperature chosen arbitrarily, T the current absolute temperature, Sr the degree of saturation of concrete porosity, and $\text{Sr}^{\text{threshold}}$ the minimum degree of saturation necessary to allow for evolution of the chemical (found to be 0.3-0.5 by [2]). The integration of the previous differential form 10.7 can then be exploited to assess the kinetic as follows:

$$\frac{1}{\alpha_{ref}} = \frac{1}{A(t_c)} \int_{t=0}^{t=t_c} \exp\left[-\frac{E_a}{R}\left(\frac{1}{T} - \frac{1}{T_{\text{ref}}}\right)\right] \times \frac{\text{Sr} - \text{Sr}^{\text{threshold}}}{1 - \text{Sr}^{\text{threshold}}}(\text{Sr} - A)dt \tag{10.8}$$

If T_{ref} is chosen equal to the mean temperature at the drilling location, Sr taken equal also to the value at the drilling location, then the kinetic of the reaction is characterized, for the aggregate size i by

$$\tau_r ef^i = \frac{1}{\alpha^i_{ref}} = \frac{-tc}{(\ln(Sr - A(tc))} \left(\frac{Sr - Sr^{\text{th}}}{1 - Sr^{\text{th}}}\right) \tag{10.9}$$

Once τ^i_{ref} assessed, the finite element model can be used to perform an inverse analysis of the structure aiming at fitting the swelling potential.

10.3 Sampling

The first component of field work entails typical structural monitoring, which (to the greatest extent possible) should track displacements, cracking, crack opening displacements, temperature and moisture.

The second component consists of core recovery. In the example presented below, the structure was composed of two types of concrete, C250 and C350 (with 250 and 350 kg/m^3 of cement respectively), and as is often the case the entire structure was not undergoing AAR expansion. Hence, cores were recovered from each of the two zones, as well as from zones with non-reactive concrete (Steps 1 and 2) as proposed in [3]. Observations of crack patterns, of degree of saturation and SEM analysis on concrete samples can be used to determine zones without ASR (sound zones are usually dry zones and should not present any ASR-gels) and zones with a lot of ASR.

10.4 Apparatus

1. Extensometer to measure expansion on mortar specimens.
2. Container to store mortar specimens in alkali solution.

10.5 Experimental Procedures

1. First of all, both intact and degraded aggregates must be very carefully recovered (Steps 3 and 4 in Fig. 10.1). A two-step approach is proposed: Coarse aggregates (>4 mm) are first extracted through heat treatment by immersing the samples in a Na_2SO_4 solution, followed by at least five cycles of freezing and thawing [4]. This procedure will facilitate the extraction of coarse aggregates.
2. Fine aggregates are then extracted through what the authors refer to as an *organic chemical* attack in order to dissolve them from the cement paste. Salicylic acid solution (1.3 mol/l with methanol as the solvent) is reported to be more efficient than inorganic acid (such as HCl, which may attack reactive silica in the aggregates and thus alter reactivity). Further details regarding chemical extraction and an assessment of chemical advancement can be found in [5].

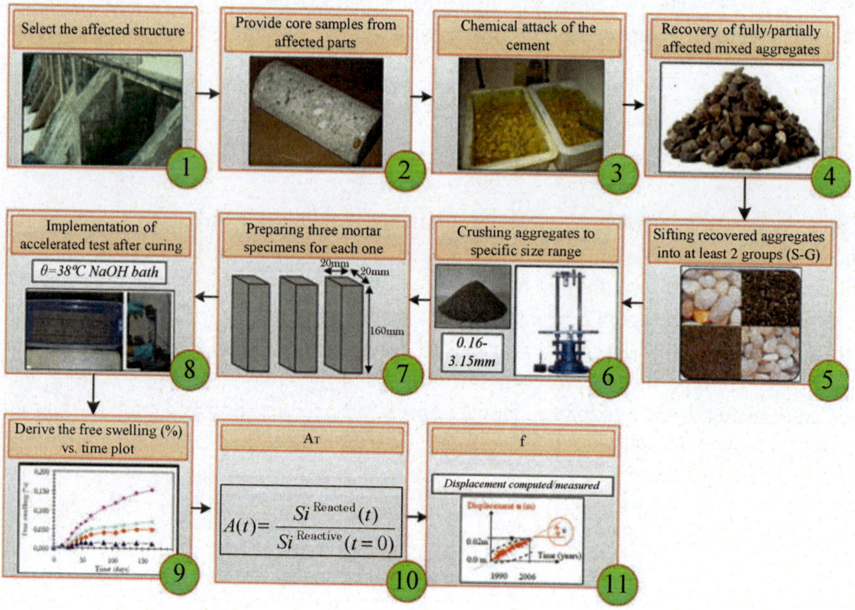

Fig. 10.1 Schematic description of the [1] method

Table 10.1 Mortar bars prepared from crushed aggregate

in-situ cement content Kg/m^3	Size range mm	% by weight	Label
250	0–5	32	S250 ◊
	5–16	18	G250 □
	16–100	51	
350	0–5	49	S350 △
	5–30	51	G350 ×

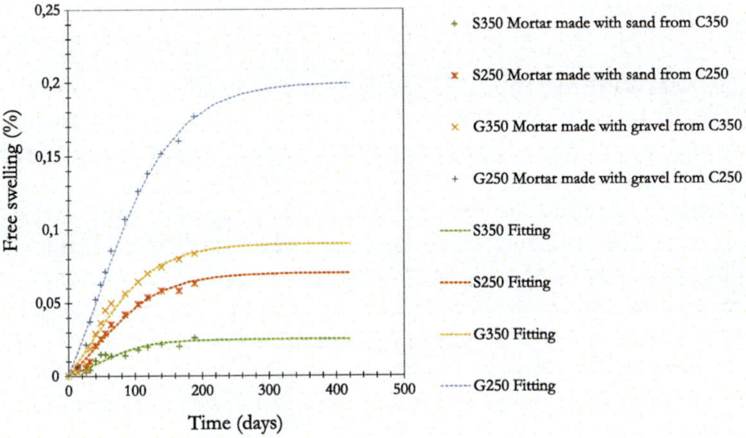

Fig. 10.2 Swelling of mortar specimens [1]

The aggregates are then divided into two of three groups, as shown in Table 10.1 (Step 5). The aggregates are then crushed (Step 6) and sifted, with only those in the 0.16–2.5 mm range being selected. These selected crushed aggregates are subsequently used to prepare two sets of mixes with 1,500 kg/m^3 sand content, 8 kg/m^3 alkali content and a water-to-cement ratio of 0.5. The proportion of reactive aggregate in mortar can be determined from reactive silica content as recommended in [6]. Chemical tests to determine the reactive silica content can be carried out according to procedures discussed in [4]. Next, each mix is cast in at least three mortar specimens (20, 20, 160 mm) (Step 7).

The specimens are cured for 28 days in sealed bags at 20 °C, before conducting accelerated tests at 38 °C in 1 mol/l NaOH solution, to attack all the reactive silica in aggregate (Step 8). Results are plotted as free expansion vs. time for each mortar specimen (Steps 9), Fig. 10.2. These results lead to the following observations:

1. Mortar made from gravel exhibits greater swelling than specimens made from sand. This finding is explained by the fact that in situ smaller aggregates are more likely than larger ones to have depleted their propensity to react.
2. Swelling capability decreases with cement content of the concrete (generally for a certain percentage of free swelling: G250 > G350 and S250 > S350). Since the

amounts of alkali and hydroxyl ions in concrete are proportional to the cement, the action of C350 concrete was more advanced than for C250; henceforth, in the laboratory, the potential aggregate expansion of C250 > C350.

10.6 Preparation

10.6.1 Assessment of the Reaction Advancement in the Structure

As mentioned earlier, $A(t)$ (Eq. 10.5) is the primary unknown being sought, along with f (Eq. 10.1). These determinations will be made in the present section.

Silica first reacts *in situ* and then in the laboratory (via the accelerated test, Fig. 10.3).

Two types of aggregate are extracted. Ideally, the intact aggregates are extracted from a dry zone in the structure where AAR has not evolved to a great extent [3]. This should be verified by SEM analysis of concrete samples drilled from these zones. Concrete used to extract intact aggregates should not present any ASR-gels. This practice is adopted in order to determine the total reactive silica in the aggregate ($Si^{reactive}$). On the other hand, the affected aggregates can be extracted from zones that have clearly been exposed to AAR, allowing the residual reactive silica to be expressed as:

$$Si^{residual} = Si^{reactive} - Si^{reacted} \tag{10.10}$$

In assuming that the time of coring and laboratory testing is t_c relative to construction completion (usually the time between coring and laboratory testing is small compared to structure time), then the chemical advancement can be expressed as (Step 10):

$$A(t_{\text{lab}}) = \frac{\text{Si}^{\text{reacted}}}{\text{Si}^{\text{reactive}}} = 1 - \frac{\text{Si}^{\text{residual}}}{\text{Si}^{\text{reactive}}} \tag{10.11}$$

However, since the rate of chemical reaction evolution is independent of the absolute expansion of the structure, it is possible to determine $A(t_c)$ from laboratory experiments and this value would correspond to the amount of silica consumed in situ (Figure 10.2). These advancements are determined by calibration of the mortar expansion tests using Eq. 10.6 (for example calibration of expansion curves presented in Fig. 10.3—step 9 of Fig. 10.1.

To avoid using the different advancement computed by Eq. 10.9, it can be convenient to compute an averaged advancement taking into account the different advancement assessed by the mortar tests for each aggregate size, [2].

For this calculation, a first approach consists in making the same simplification than for the mortar analysis. Thus, the concrete expansion is supposed to be the sum of expansions ε_i caused by each aggregate according to its proportion in the concrete

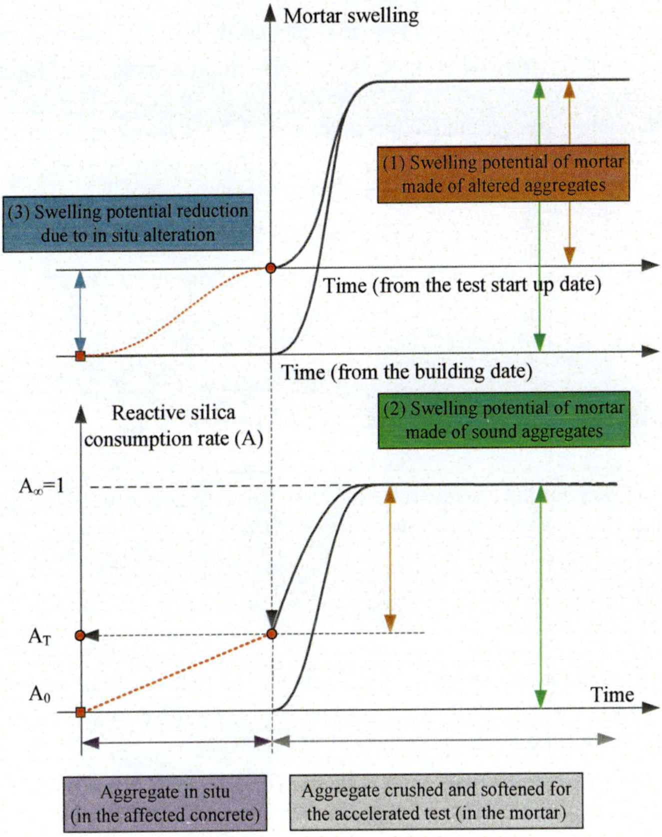

Fig. 10.3 Swelling of mortar specimens [1]

ϕ_i:

$$\varepsilon_{concrete} = \sum_{i=1}^{n} \phi_i \varepsilon_i \tag{10.12}$$

With n, the number of reactive aggregate classes for which mortar test were performed, ϕ_i, the fraction of each class of aggregate (given in the example in Table 10.1), and ε_i the expansion of each class of aggregate size which can be determined from Eq. 10.2.

It is important to note that in this case, reactive aggregates have different sizes (the true size of aggregate in the structure) and Eq. 10.2 can be rewritten

$$\varepsilon_i = n_i (V_g^i - V_p^i) \tag{10.13}$$

With $V_g^i = f^i V_a^i$. f^i the value obtained with mortar test, V_a^i the real size of the aggregate, V_p^i the "connected porosity" computed with the real size of the aggregate in the concrete.

This calculation gives a unique advancement of ASR A(tc) in the structure concrete:

$$A(t_c) = \frac{\varepsilon_{concrete}(t)}{\varepsilon_{concrete}(\infty)} \tag{10.14}$$

Which can be interpreted with Eq. 10.9 to obtain the "averaged" kinetic α_{ref} for the concrete.

10.6.2 *Assessment of the Final Swelling Potential*

Lastly, since free swelling has been assumed to be directly proportional to the reaction progress, Eq. 10.7, swelling can now be derived from

$$\frac{\partial \varepsilon^{AAR}}{\partial t} \neq \varepsilon^{\infty} \frac{\partial A}{\partial t} \tag{10.15}$$

with the only remaining unknown now being ε^{∞} for the structure (which has the same role than f in the mortar tests), which has to be determined through curve fitting using *in situ* measurements (Step 11 in Fig. 10.1).

10.7 Examination at End of Test

After the final length change measurements have been made, each specimen shall be tested for warping and shall be examined.

Warping The warping, if any, that each specimen has manifested during the test period, shall be determined by placing the specimen on a plane surface and measuring the maximum separation between the specimen and the surface. The specimen shall be placed so that its ends curve down to the reference surface and the measurement made to the nearest 0.3 mm (0.01 in.). This could lead to discrepancy in the measurement of mortar expansion and thus lead to discrepancy on kinetics determination.

Examination Items to be noted in the examination of mortar specimens include:

- Presence, location, and type of pattern of cracking,
- Appearance of surfaces, surface mottling, and
- Surficial deposits or exudations, their nature, thickness and continuity.

If large differences are noted between specimens of different types (different aggregate sizes, different concrete), it could lead to wrong determination of rela-

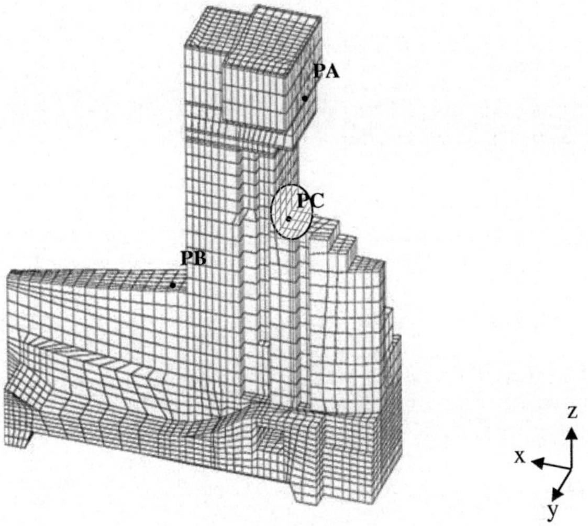

Fig. 10.4 PC the point where the displacement in the field is fitted to assess free swelling of concrete knowing the characteristic time determinate in the lab [1]

tive advancements between aggregate sizes or concrete, particularly due to different ASR-gel leaching between specimens, and lead to discrepancy on the final kinetics determination. Using chemical tests to have a first assessment of the reactive silica content in aggregate could lead to a better optimization of mortar composition for expansion tests [6]. Chemical tests can be carried out according to procedures discussed in [4].

10.8 Limitations and Applicability to Analysis

Once the lab test performed, the kinetic of the reaction in the field is known and characterized by the characteristic time of the reaction ($\tau_{ref,i}$ in Eq. 10.9); the last parameter to be fitted is the final amplitude of free swelling (ε^{∞} in Eq. 10.15). This parameter is a constant characterizing the swelling potential in the field, it cannot be determined in the lab for the reason that the silica gel formed in the lab is not the same than in the field. But as ε^{∞} is used in the finite element model to compute the displacement of the structure, it is fitted on a structural displacement.

So, the method to pass from the lab to the field in the Toulouse method consists to use the characteristic time obtained with Eq. 10.9, and to find the final amplitude of free swelling ε^{∞} able to fit the kinetic of deformation at a given point of the structure.

For instance for the dam pile illustrated in Fig. 10.4, the displacement of point PC is used to fit ε^{∞}. The value of ε^{∞} is kept only if the displacement of other points of the structure (PA and PB in this case) can be recovered without changing ε^{∞} (Fig. 10.5).

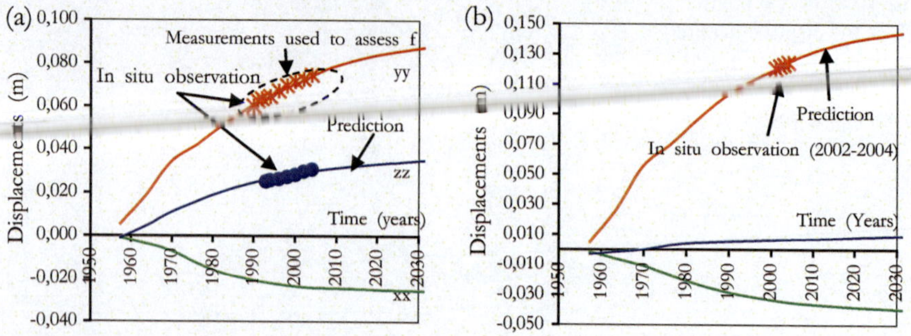

Fig. 10.5 Fitting of free swelling amplitude on the trend of the displacement curve (a red curve), prediction validation (a blue curve and b) [1]

10.9 Report

Report the following information:

- Type, source, proportions of each size, and chemical analyses of aggregates used in the concrete of the structure.
- Type and source of Portland cement/binder composition.
- If possible: Alkali content of cement/binder and pozzolans (if necessary) as percent potassium oxide (K_2O), sodium oxide (Na_2O), and calculated sodium oxide (Na_2O) equivalent. Refer to relevant method for extracting alkalis from cement/binder (water soluble and acid soluble alkalis).
- Average length change in percent at each reading of the specimens.
- Any relevant information concerning the preparation of mortar samples.
- Any relevant information concerning the preparation of aggregates if the method uses extraction aggregates from cores taken from structures.
- Any significant features revealed by examination of the specimens during and after test.

10.10 Precision and Bias

Precision shall be considered satisfactory if the difference in the value for percentage expansion of any given specimen in a group from the average value for percentage expansion of the group does not exceed 15 %. Interlaboratory test data for the determination of the repeatability of the method: Repeatability shall be considered satisfactory if the percentage expansion of each specimen from the same concrete is within 20 % of the average.

References

1. Sellier, A., Bourdarot, E., Multon, S., Cyr, M., Grimal, E.: Combination of structural monitoring and laboratory tests for assessment of alkali-aggregate reaction swelling: application to gate structure dam. ACI Mater. J. **106**(3), 281–290 (2009)
2. Grimal, E., Caractérisation des effets du gonflement provoqué par la reaction alcali-silice sur le comportement mécanique d'une structure en béton. Ph.D. thesis. Toulouse 3 (2007)
3. Multon, S., Barin, F., Godart, B., Toutlemonde, F.: Estimation of the residual expansion of concrete affected by alkali silica reaction. J. Mater. Civ. Eng. **20**, 54–62 (2008)
4. Gao, X., Cyr, M., Multon, S., Sellier, A.: A three-step method for the recovery of aggregates from concrete. Constr. Build. Mater. **45**, 262–269 (2013)
5. Gao, X., Cyr, M., Multon, S., Sellier, A.: A comparison of methods for chemical assessment of reactive silica in concrete aggregates by selective dissolution. Cem. Concr. Compos. **37**, 82–94 (2013)
6. Gao, X., Multon, S., Cyr, M., Sellier, A.: Optimising an expansion test for the assessment of alkali-silica reaction in concrete structures. Mater. Struct. **44**(9), 1641–1653 (2011)

Chapter 11
Accelerated Expansion Test: UK

Jonathan Wood

11.1 Scope

This test method determines the variability of long term dimensional changes in sets of mature concrete cores as they take up moisture by capillary absorption. The dimensional changes are initially due to normal swelling (shrinkage recovery) of concrete from measured increased moisture availability. The moisture then leads to the slow swelling of any pre-existing gel from AAR and, in the longer term, to further AAR reaction with swelling of its gel. The swelling due to AAR is a measure of the "free residual expansion potential" which can arise in the cored concrete at a rate depending on the in-situ moisture and the temperature. The "free" expansion will be modified in the structure by the effects of stress and reinforcement restrain acting in three distinct directions.

The test is specific to the sample core and there is no prejudgment of relative availability of reactive silica or alkali, but this can be checked chemically or petrographically on adjacent samples from the core. Silica is not mobile. Alkali can have leached out or concentrated in the structure and also in the core in a test environment. Moisture availability, with temperature, controls the rate of reaction and consequent expansion in this test and in the structure.

Microcracking, structural cracking and expansion result from localised volumes where silica, alkali and moisture create local reaction with gel swelling and expansion. This expansion cracks the adjacent more stable concrete. It is the variability of expansion within a core, between adjacent cores and in adjacent concrete pours which causes the damage from AAR. This test varies from most procedures in that

J. Wood (✉)
Structural Studies Design, Chiddingfold, UK
e-mail: jonathan@ss-design.demon.co.uk

© RILEM 2021

V. E. Saouma (ed.), *Diagnosis & Prognosis of AAR Affected Structures*,
RILEM State-of-the-Art Reports 31,
https://doi.org/10.1007/978-3-030-44014-5_11

the objective is to quantify the damaging variability of expansion within the core and between cores from the same pour and between the pours of concrete with varied composition within real structures.

The results from a representative set of cores need to be related to their location, stress field and in situ moisture and temperature environment. Result from a few cores can never provide a reliable prognosis of potential future damage to the overall structure. Statistical analysis of the sampling and variability is necessary for major structures and this can need testing hundreds of cores. The low cost simple storage of samples in individual containers makes long term testing of large sample sets of cores both practical and economic.

Quantified cracking of the structure and the Stiffness Damage Test SDT (measuring the change in stiffness and hysteresis of the core) provides the best indications of the expansion prior to the extraction of the core. SDT can be carried out on the core before this expansion test but it is prudent and sufficient to keep the stress range between 0.5 and 5.5 MPa to prevent stress damage.

11.2 Principles and Methodology

This expansion test procedure was initially developed by Mott MacDonald SSD in the 1980s [1] for the assessment of a wide range of UK bridges and buildings suffering damage from developing AAR [2]. Over 100 initial expansion tests carried out by a range of laboratories, using a range of then recommended procedures, showed erratic results so the weighing of cores to measure moisture changes was introduced. This, with other checks, showed fundamental problems with achieving a consistent *100% Relative Humidity*. Other checks and the literature established that alkali loss from leaching from the core was often terminating expansion prematurely. Many test procedures were for a period of months and so failed to show the longer term expansions for slow late expanding aggregates being measured in structures. The AAR conference literature on expansion testing over the last 30 years has confirmed the limitations of testing at *100% RH* without resolving these problems.

Observation of the variation of AAR cracking with the availability of water in structures has shown that expansion arises when sufficient liquid water is available in the capillary pores of the concrete. It is liquid water, not humidity that is necessary for both the reaction and the expansion. The Water Supply test aims quantify the relationship between the uptake of water and the dimensional changes from the recovery of normal drying shrinkage typically less than 0.5 mm/mm and from the longer term expansions caused by AAR which can range up to 5 mm/m in very severe cases. The test is carried out in a sealed containers in which the core is in contact with a small amount of water which is topped up to maintain 5–10 g "free" water per kg of core. Moisture is absorbed by capillary action into the pores of the concrete. The small amount of "free" water limits the leaching of alkali from the core. The test should be continued for years in parallel with the monitoring of cracking and movements of the structure [3, 4].

The early tests showed a very high variation in the expansions measured on different 50 mm lengths on three axes of a core. Adjacent cores from the same pour gave a wide range of average expansions and similar variability between the nine gauge lengths. In UK cases the reactive aggregates were generally in the coarse end of the sand fraction, eg. 3–5 mm cherts, in the core. Variability in cores may well be influenced by the size and dispersion of the reactive aggregate fraction. This should be discussed with the petrographer in evaluating the results.

The Water Supply testing was included in [4] and this procedure has been progressively refined and adapted for the appraisal of over 100 of structures since then. The method should not be standardised as the core locations, size of cores and the temperature of test need to be tailored to the mix characteristics, the structural form and range of exposure conditions of the structure and the specific management questions for prognosis.

The test was developed in parallel with the field monitoring of a range of structures with AAR. It is to answer two questions about the specific structure from which cores are taken.

1. Does the availability of more water cause expansion?
2. What is the magnitude and rate of expansion in the test conditions?

The aim is moderate acceleration so that expansions will develop faster than in the field, but without distortion from very high temperatures or from the leaching or addition of Alkali.

The controlled availability of moisture accelerates relative to conditions in most structures. Using 20–25 °C "Lab Ambient" enables testing of large numbers of cores in simple individual sealed containers without the costs and complexity of ovens and humidity control that go with testing at constant 38 °C. Testing at 38 °C gives earlier expansion and may be appropriate for early diagnosis, as distinct from prognosis.

Many field structures have slow AAR expansions on site and in this test, which may develop over a period of decades [5]. The procedures set out below are for predicting future trends in field structures as a guide to their management. They are distinct from those for short term "Diagnostic" expansion tests. These laboratory core expansion tests are normally run in parallel with site monitoring of expansions and crack movements [3, 4].

11.3 Sampling and Sample Preparation

11.3.1 Core Samples

Because of the high variability of AAR expansions within and between pours of concrete in a structure, it is essential to test large representative groups of samples. The procedures for inspection and selection of locations for the coring, coordinated with any parallel diagnostic study, should follow Part 1 of [6].

Core samples should be taken clear of any reinforcement and cracks and should normally be about 70 mm diameter taken with a 75 mm core. Where reacting aggregates are >20 mm larger cores are appropriate.

The length of core for expansion testing should typically be 180–200 mm. This length of sample enables Stiffness Damage Testing (SDT) and compressive or torsional tensile strength tests to be carried out before and/or after expansion testing. The use of larger cores greatly increase the risk of cutting reinforcement which can be more damaging than AAR.

Immediately after wet coring the cores should be rinsed, surface dried, identified and then allowed to dry in sheltered, shady conditions for 1 h., before being cling-film wrapped and delivered to the laboratory. Without this drying, water absorbed during the coring process can initiate expansion before gauges are fitted. The above drying procedure gives a consistent initial moisture state and reduces the risk of expansion from coring water. Dry coring heats and dries cores and so it offers no advantage to justify extra time and expense.

Some moisture loss, which should be weight monitored, may occur during sample storage and STD testing. The condition of the cores should be checked when they are trimmed and the Demec or similar stain gauge points, are fitted. Any cracking should be noted on a developed sketch of the core showing the Demec locations and photographed. Differentiate cracking of the core from cracks in individual reacting aggregates. Any sweaty gel patches under the cling-film, a feature of UK cherts, should be noted and discussed with the petrographer. The end offcuts may be required for detailed petrographic examination and/or for chemical analysis [7]. Any cracking developing during the test should be similarly recorded.

11.4 Apparatus

11.4.1 Demec Gauge Points

The typical 70 mm diameter 180–200 mm long core, Fig. 11.1 [h!] should have 3 longitudinal rows each of three 50 mm gauge lengths fitted using the setting bar, This comes in the case with a demec Other suitable gauges can be used. The main priority is to have enough gauge lengths to establish the variability of expansion (six minimum, nine preferred). The variability of expansion between gauge lengths, Fig. 11.2, is important identifying localised expansion and in differentiating between AAR and the more uniform recovery of drying shrinkage. The Demec points should be mounted on a locally abraded area using "Plastic Padding" rigid grade or similar adhesive. "Plastic Padding" is a two part epoxy stiff enough when applied (like chewing gum) for points not to slide while it sets. It has a long term record of stability in damp conditions. There are similar products but many, including some very expensive ones, degrade in wet conditions. During the fitting of the Demec

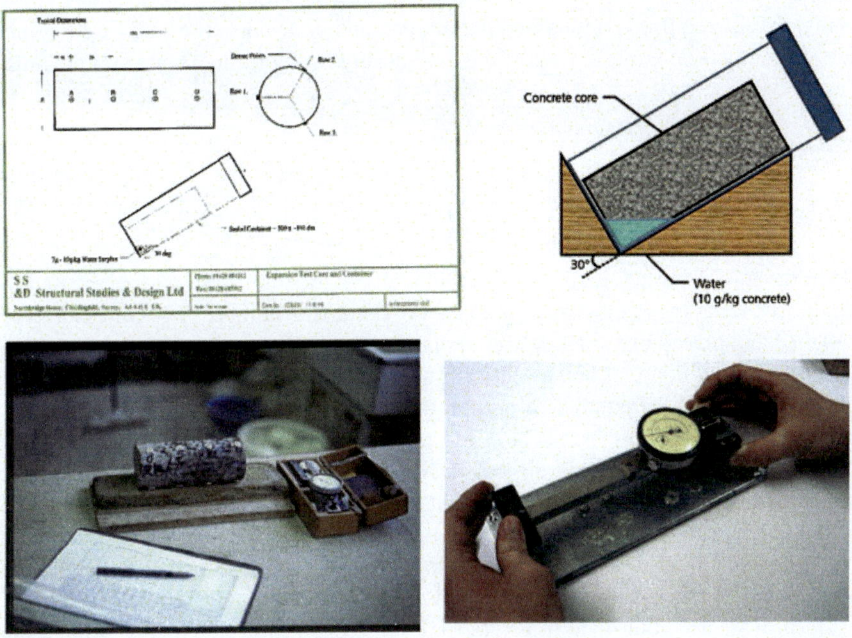

Fig. 11.1 Experimental setup

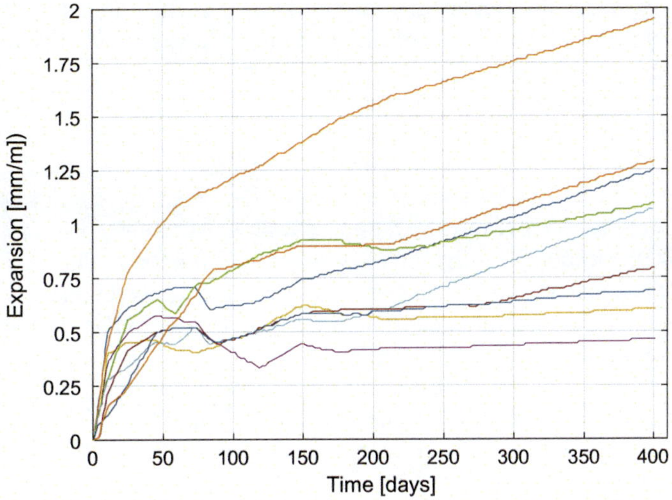

Fig. 11.2 Variation of expansion on nine 50 mm gauge lengths on one core (9 gauges lengths on core K7; water supply 20 °C, 9 50 mm gauges)

points care is required so that they are aligned and sit level to the surface so points seat correctly. Excessive drying of the core during fitting of the Demec points must be avoided.

11.4.2 *Water Supply*

Once the initial Demec readings have been taken, the cling-film wrapping used to minimise the drying of the core prior to testing should be removed. The core should be weighed and placed in a smooth walled plastic container typically 100 mm diameter, 300 mm high which can be sealed with a close fitting lid, as shown on Fig. 11.1. Plastic food storage containers are suitable for this.

The container should be set at a 30° angle to horizontal so that the core slides down into the water and can absorb it. Normally "Row 1" of Demecs should be uppermost in the container with the A end at the bottom in the water. Where the end of the core is not flat, the core alignment should be adjusted so that the projecting end is lowest and in the water, to ensure that water can be absorbed directly into the core.

To provide the additional moisture to develop expansion of existing gel and the further reaction of alkali with silica to create gel, the core needs to be maintain in moist conditions. By using the natural wick action of the concrete to carry the moisture into the core, rather than attempting to produce a consistent humidity around the core, conditions similar to that in structures and which give reproducible results can be achieved.

The records of testing many hundreds of UK cores for expansion using this procedure have shown that the water uptake during the test ranges between 1g/kg sample weight and 50 g/kg sample weight depending on the site exposure conditions.

The higher water uptakes relate to samples either from dry interior conditions or which have been excessively dried prior to testing. By maintaining a free "water surplus" of 10 g/kg sample weight of tap water (distilled water leaches faster) in the bottom of the container, consistent conditions of water supply are maintained. The free water is replenished when absorption into the sample and/or evaporation has reduced it to below 7 g/kg sample weight.

The availability of water in these conditions is greater than that in most structures and in consequence the expansion rates on test, including the effect of the higher temperature, are typically 5–10 times faster than those recorded on the large exposed masses of bridge structures in the field.

The weights of the container, free water and the core, with steel plate, are monitored to an accuracy of 0.1g on every occasion the Demecs are read, so that there is a record of the weight gain by the sample and the availability of surplus moisture. Although some alkali does leach from the sample into the free water, the 10 g/kg is sufficiently small relative to the volume of free moisture in the sample (typically 50 g/kg) to minimise loss of alkali. This can be analysed to check, but do not encourage leaching by replacing the water or distort the test by adding more alkali than was in the core.

11.4.3 Temperature

There is no advantage in maintaining a precise constant temperature. Real structures have a wide temperature range. In some instances samples in Water Supply conditions have been exposed adjacent to the structure so they had the full temperature range in parallel with tests at "Lab ambient" temperatures.

The simple low cost procedure enables a sufficient number of samples to be tested. This is particularly important because of the very high variably in expansion within cores, within sets of cores from one pour of concrete and between the wide range of pours in a real structure. It was not unusual to test 100 cores from a structure with major problems. Storing containers on shelves in normal lab conditions made this possible on the available budgets. Samples may be cost effectively stored in the laboratory, away from direct heat or sunlight, at the ambient laboratory temperature. Having cores, demecs and reference plate at the same temperature avoids the difficulties of temperature differentials and moisture loss associated with 38 °C testing.

There are problems when samples are kept in the mythical "100%" RH 38 °C conditions. As samples are cooled there is a tendency to lose moisture due evaporative cooling. When cold cores are returned to 3838 °C there is condensation, which leads to leaching, when placed in the warmer humid conditions. There can also be problems from the temperature difference between measuring gauge and the hot core. LCPC (Larive) did everything in a large 100% RH 3838 °C room at LCPC to minimise these problems from thermal variations.

Regular maximum and minimum temperature records (±0.538 °C) for the area of laboratory adjacent to the control plate and cores, should be maintained and supplied with the results. The temperature of the cores and steel control plate, when demecs are read, should be recorded to ±0.538 °C.

Because we now know that AAR damage development is slow and steady, unless more water is suddenly made available, there is no need to accelerate the test at 3838 °C to get quick but distorted results. As well as testing at Lab Ambient (20–2538 °C) the test containers can be stored on site but if the temperature drops below freezing, the near saturated microcracked samples can fragment. These site expansion tests are linked to the monitoring of overall and crack movements on the structure.

11.5 Experimental Procedure

11.5.1 Core Demec and Invar Reading

Once the adhesive has fully hardened three separate "Initial" sets of Demec readings should be taken to bed in the Demec points and to establish a reliable initial datum. Invar and the steel control plate readings should be taken at the same time and the sample and laboratory temperature should be recorded.

For simplicity and economy, the test is best carried out in "Lab Ambient" conditions, away from direct sunlight or sources of heat, with temperatures in the range 20–25 °C. With storage and reading temperatures the same there are no complications from heating and cooling cycle or temperature differences between gauge and core. The steel plate kept with the cores will show similar temperature length change to the core as their Coefficients of thermal expansion are similar. The Demec invar gives a stable length but may be subject to wear or change of instrument over a long test. The steel plate readings will show up anomalies from wear, changing gauges and reading errors and enable readings to be adjusted for changes in the Demec used or after servicing.

Demecs should be read with the core firmly located in wooden V blocks, coated to prevent absorption of moisture. The Demec readings should be made adjacent to the storage area for the cores to minimise temperature changes. Demec Gauges are designed and calibrated for one gauge to be used for a particular experiment, normally over a period of a few months at most. For long term testing with changing operatives for years additional procedures and checks are required.

The Demec bezel adjustment should be locked at the start of the test and sealed with "Snopake" or paint. 10 Demecs should always be read with the pivot point to the same side. Two separate readings should be taken 11 and recorded without averaging. To reduce wear the invar bar and steel plate should be read once only at the start and end of each of each batch of 6–12 core readings in a morning or an afternoon. These invar values should be entered on all core sheets for that batch.

11.5.2 Steel Control Plate

The steel plate with measuring points on has a triple purpose. It is kept with the cores and at their temperature and has a similar coefficient of expansion. This enables thermal dimension changes to be monitored and simply adjusted for. Importantly it also shows up any changes due to the wear or mistreatment of the measuring gauge or poor reading ability by the technician. At times results have reported "expansions" of the steel!

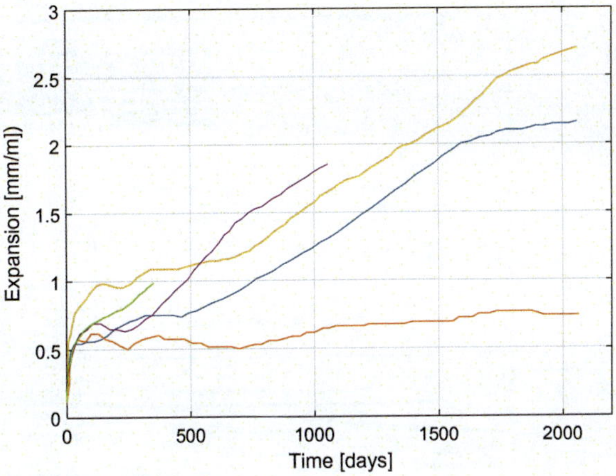

Fig. 11.3 Need for long term testing (5 Aggregates, 5 year Expansion Tests)

The Demec gauge is used for long duration test, see Fig. 11.3, and continuity of readings over periods of up to 5 years are essential. The use a steel control plate provides a check and means of recalibration for gauge wear and damage. The best practice is for a dedicated Demec gauge to be used for one group of tests. The use of a steel control plate allows cores to be moved between laboratories and read with different gauges.

The control plate should be a steel flat, as shown in Fig. 11.4, about 250 mm long, 100 mm wide and 10 mm thick to which 5 mm high studs are welded prior to galvanising. Three rows of 50 mm gauge lengths Demec studs similar to the three rows on a core should be fitted to the plate. The Demec readings for these points on the plate should be about 800 for the first length (i.e. as set with the Demec setting bar) and about 1,200 and 1,600 for the other two lengths in each row. These can be set by moving the Demec point using the gauge to set points.

The steel plate should be maintained next to the cores on expansion test so it's temperature is the same. The weight and Demec readings on the control plate should be read at the same time as those on the cores. All sets of core readings must be accompanied by the corresponding set of control plate readings. The control plate readings need to be regularly checked for any drift in the Demec strain readings so that the reading procedures and/or Demec Gauges can be corrected.

If, due to damage, a Demec gauge has to be reconditioned and recalibrated, the standard Demec calibration for both invar length and gauge constant must be checked and corrected against the steel plate readings. Gauges should not be changed, reconditioned or recalibrated during a test sequence unless this is essential due to damage. When damage to the Demec or the loss of Demec points occurs, the engineer should be informed immediately, so that appropriate remedial action can be agreed.

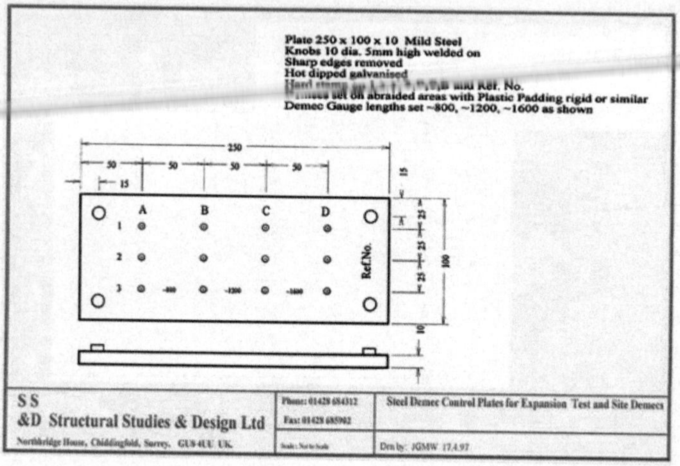

Fig. 11.4 Steel plate for demec reading checks

11.5.3 Readings

he results should be recorded in manuscript at the time of reading on the attached pro-forma. The results should not be typed up or processed from the manuscript. Transcription errors can occur. Transcription with post processing and preparation of the data should be carried out by the engineer. Copies of the results pro forma should be forwarded as soon as the 28 day readings are available and then after 3 months testing, after 6 months testing and then annually.

The initial 3 sets of readings of Demec and core weight and container weight should be taken as three separate sets, with at least one hour between each reading of a core. 10 g/kg sample weight of tap water should then be added to the container, and the core inserted. It should be sealed and placed in a rack inclined at 30°.

Before each reading the full container should be weighed, the core should be removed from the container, shaken so that loose moisture is retained in the container, and placed on its drier (D) end on the scales for weighing. The container and free water should be then be weighed, additional water added if required and reweighed. When the weight of free water in the container 24 falls below 7 g/kg, additional tap water to bring the free water up to 10 g/kg should be added.

The Demec readings should be taken immediately and the core reweighed and returned to the sealed container and the whole reweighed.

Readings should be taken at 1, 2, 3, 7, 14, 21 and 28 days, at which stage the first results should be reported. The further reading frequency is 42 days, 2–6 months then at 8, 10 and 12 months as shown on the pro-forma. It should continue 3 monthly until both the expansion and weight are stable. An Excel pro-forma for recording and processing the data is available.

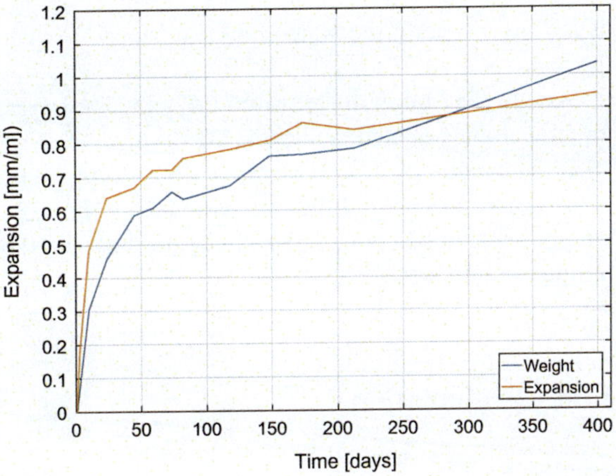

Fig. 11.5 Typical comparison of water absorbed v average expansion

The shrinkage recovery and water uptake during the early part of the test need to be clearly differentiated in analysis from long term expansions, see Fig. 11.5. Shrinkage recovery swelling is reasonably uniform on all gauge lengths and stabilises after a month. In contrast AAR gauge length expansions are highly variable, see Fig. 11.2, on the different gauge lengths. The data from copies of manuscript sheets for the cores and steel plate should be checked for anomalies by the engineer as they are entered into an Excel spreadsheet. Excel can be used to calculate expansions on each gauge length and weight changes and plot the data. Typical anomalies are weight loss and shrinkage of 2 the core due to failure to fit the container top properly and sudden shifts in expansion due to erratic reading of Demec gauge, sometimes after it has become worn or has been dropped or replaced.

Manuscript recording of data onto proformas, dated and signed by the operative, should be insisted on. This data can be then be checked for anomalies by the engineer when input into a spread sheet and again as the data is analysed.

11.5.4 Concluding Test

Normally a proportion of cores will be left on test with readings every 3 or 6 months for many years in parallel with the monitoring of structures. Some cores may be taken off test earlier.

At the conclusion of an expansion test, the “Last Expansion” double set of readings (ie core readings separated by at least one hour) should be taken. The lids of containers are then removed to allow slow drying in normal laboratory air conditions (typically 50–60% RH). Exact humidity is not critical as the aim is to slow dry down to initial weight. The surplus water in the bottom of the container will be progressively absorbed and evaporate and any alkali in it will be drawn back into the core as it dries. Some of the water can be analysed to quantify alkali leached out during the test.

Weight (including the removed lid) and Demec readings should be taken at 7 day intervals for 28 days to determine the drying shrinkage characteristics with a double set of final readings. If necessary continue drying until core weight is less than initial value. The cores should then be cling film wrapped and stored. The shrinkage and weight loss can be compared with the normal rewetting expansion of the core when initially put on test.

11.6 Preparation

The full set of recorded data of Demec lengths on cores, Invar Steel plate and weights of container, sample and free water should be maintained on an Excel spreadsheet. Weight uptake in the core and length change from day one should be calculated for all gauge lengths and those for the steel plate with temperature should be tabulated. As each set of readings is added it should checked for anomalies against the manuscript data sheet.

Typical anomalies include occasional errors in recording or transcribing data, loss of weight and shrinkage due to failure to properly close containers, change of procedure when operatives change, failure to maintain at least 7 g/kg free water in container, loss of points or erratic readings on some Demec points, dropping or changing the gauge (recalibrate and re zero using the steel plate).

The movements of the nine gauge lengths should be plotted against weight gain. No rigid rule will differentiate the normal concrete recovery of shrinkage from the start of expansions due to pre-existing gel swelling and the longer term formation of gel and its swelling. The environment of the core in the structure, the risk of accidental drying prior to testing need to be considered. Cores from dry soffit of a well waterproofed bridge deck may swell by 0.5 mm/m without any AAR and much of the swelling of pre-existing gel, if AAR is present, may occur in weeks. Cores from saturated foundations will have little recovery of shrinkage and pre-existing gel will have swollen and expansions from further AAR will only develop over months or years.

Fortunately in any batch of cores one or two will not show expansive behaviour and on any core some gauge lengths will not show any AAR expansion. These will show the magnitude and timescale for the recovery of drying shrinkage and this can be deducted from other readings to show the early and late AAR expansions. Because the timescale of gel formation is highly dependent on the form of silica and the size range and proportion of reactive aggregates the relationship of expansion, weight change and time needs to be discussed with the petrographer. Experience on a wide range of structures shows that once the risk of reaction being curtailed by loss of alkali has been eliminated expansions continue at a slow rate for many years.

The use of the simple water supply containers at Lab ambient temperatures makes it practical to continue test in parallel with monitoring on site for many years.

11.7 Examination at End of Test

After the final length measurements the characteristics of each core should be examined by a petrographer for comparison with Petrographic examination of core ends in parallel with the expansion test.

This includes:

1. Presence, location, and pattern of cracking,
2. Appearance of surfaces, surface mottling, and
3. Surficial deposits or exudations, their nature, thickness and distribution.
4. Size, distribution and density of reactive aggregate particles and their relationship to the variability of expansions within the core.

The stability of the condition and reliability of readings on Demec points should also be reported.

11.8 Report

The report on the testing of cores, together with site monitoring, petrographic and physical testing reports, should be an appendices to the engineers overall structural appraisal report with recommendations for the management of the structure.

11.9 Limitations and Applicability to Analysis

11.9.1 Limitations

The Water Supply Expansion Test was initially developed and applied by Mott MacDonald in the 1980s on over 30 structures with indications of AAR structural damage. It is part of a linked series of tests as summarized in ISE 1992, with further developments reported in papers to at ICAAR 1983–2008. The overall objective was to obtain data necessary for reliable prognosis and assessment of structural effects on the range of major structures with developing AAR.

The initial series of expansion tests was based variants of storage in "100 % RH" conditions using a range of UK laboratories. This showed up showed serious limitations particularly in long term testing. The test procedure set out above has evolved to eliminate most of these limitations and to quantify the variability of expansions within and between cores and between elements of a structure

Our unsatisfactory experience with "100% RH" testing environment in special containers mirrors that reported in the literature with alkali leaching and with actual moisture conditions being indeterminate in the range 90–98% RH. The introduction of weighing the cores, to relate water absorbed to expansion, showed weight fluctuations indicative of variable RH in tests.

Observation in the field showed that cracking from expansions developed most severely where water ponded (eg on bearing shelves) or below ground, so liquid water was absorbed. This lead to the adoption of the sealed container with a small amount of water available to be progressively absorbed by capillarity into the concrete. This represents the worst moisture conditions found so that on site expansion rates will be slower than the upper bound test value. Final total expansion should be similar.

The weight data showed that more water was taken up by capillarity in "Water Supply" conditions than in "100% RH" testing. There will be a little alkali leaching into the 10g/kg water surplus. This can be quantified by analysis, but does not seem to be significant as most long term tests show no significant slowing of expansion, even after years.

Temperature effects expansion rate and so tests were carried out with on-site exposure, in "laboratory ambient" temperature (fluctuating 18–23 °C) and at 38 °C controlled.

The simplicity and economy of "laboratory ambient" for long term testing, with core and instruments at the same temperature, avoided errors form temperature effects when storing at 38 C but measuring in laboratory temperature. This enabled large batches to be cost effectively tested over years so that variability within and between cores and long term expansions could be quantified avoiding the limitations inherent in short term and rapid testing.

Using a Demec or similar gauge in the long term there are problems with operator error, accidental resetting, damage and refurbishment etc. By monitoring for quality control with the steel plate with points like those on the cores, these errors became identifiable and correctable. Likewise errors in weighing for water absorbed could be identified by weight record of steel plate. Calculating the movements of the core relative to those on the steel plate corrected for thermal movements if lab temperatures fluctuated, as the steel plate which has similar coefficient of thermal expansion to concrete.

The six or nine gauge lengths on each core measure the variability of expansion within the core. The loss of a couple of gauge lengths does not abort the test.

The initial shrinkage recovery data needs to be checked for indications that the sample has been dried accidentally between coring and the start of test. The exposure conditions of location of coring, and the handling and core preparation procedures need to be considered in interpreting this data. The inclusion of non reactive concrete in the sets of cores helps distinguish between uniform shrinkage recovery and the variable expansions within the core from localised AAR.

11.9.2 Application to Computer Analysis

Any analysis of the concrete and its interaction with the reinforcement and restraints must start with consider the strain changes and cracking during construction and loading of the structure including thermal and shrinkage of normal concrete. The initial swelling and measured water uptake during an expansion test gives a measure of the usually uniform non AAR shrinkage recovery This needs to be differentiated from the distinctly variable AAR expansions which follow. The parallel testing of non reactive cores of similar concrete from different environments provides another check on non AAR strains. These are typically 0.3–0.6 mm/m, so of the same order as those from mild AAR.

Because of the anisotropy and high variability of AAR expansions an iterative approach to modelling its effects is appropriate. A load case based on uniform triaxial (1 mm/m, 1000 μ strain) with the reinforcement modelled in detail will show how effective reinforcement is in containing triaxial expansion. Critically for structural assessment it will show where high bond stresses, reinforcement laps and curtailed straight bars with no anchoring bends may initiate delamination failure and loss of containment. Local analysis of vulnerable details is often needed. Overall analysis it will show up where intended or accidental restraints may become distressed by AAR expansions.

Next the range of current "expansion to date", in the structure (determined from SDT testing and crack summation) prior to coring needs to be estimated. The further analysis need considering current condition and with magnitude of future expansion derived from expansion testing.

The magnitude and orientation of cracking on the structure will accurately reflect the current surface tensile strains exceeding the cracking strain 0.15–0.25 mm/m (ie. 125–250 μ strain) as well as uncracked areas where AAR is insufficient to exceed compressive strains. So the structure has done part of a finite element analysis for you! However, delamination and bond failures and fracture between more or less expansive pours or adjacent structures may be hidden. So further analysis for current and future performance of the structure is appropriate as follows:

1. Add the deadload and construction strains to the uniform triaxial 1 mm/m case. Based on the variation in "expansion to date" of different parts of the structure increase or reduce part by part the uniform tri-axial 1 mm/m case to represent current condition with reductions for compressive stress and increases for tensile stress according to orientation.
2. Re-run this and check reinforcement strains and bond and zones of compression in the structure. Some of the secondary reinforcement may have exceeded its yield stress (typically shear stirrups in beams). Revise by reducing stiffness of yielding bars to reflect their ductility and improving adjustment to triaxial expansions to reflect new stress conditions.
3. Iterate until there is a balance of expansions and constraints in 3-D.

Further variable expansions, based on expansion tests adjusted for temperature and moisture availability, can be iteratively applied to predict future structure performance.

11.10 Precision

The precision of the testing data can be determined from the reported results of the weight and length changes of the steel plate with adjustment for any changes in temperature. Variations due to unauthorised changes to gauges, scales and wear show up in steel plate data. The inclusion of non-reactive cores which should show relatively uniform response of swelling and shrinkage related to changes in weight indicating water uptake or drying. These checks on precision were progressively introduced during the development of the test following erratic and anomalous results in some test from the wide range of commercial and university laboratories used in UK AAR test programmes.

The inherent heterogeneity of site concrete which becomes greater with AAR, is such that expecting laboratories to report matching result is unrealistic.

The objective is to quantify the variability of local expansions which cause overall AAR expansion and cracking damage. In comparative trials between methods and laboratories on similar materials need to be judged on the extent to which they achieve similar water uptake and coefficient of variability of the expansion data. Measuring alkali leaching should be another criteria.

References

1. Crouch, R., Wood, J.: Damage evolution in AAR affected concretes. Eng. Fract. Mech. **35**(1), 211–218 (1990)
2. Wood, J., Johnson, R.: The appraisal and maintenance of structures with alkali-silica reaction. Struct. Eng. **71**(2) (1993)
3. Godart, B.: Méthodes de suivi dimensionnel et de suivi de la fissuration desstructures: Avec application aux structures atteintes de réaction de gonflement interne du béton: Guide technique (2009)
4. ISE.: Structural effects of alkali-silica reaction—technical guidance appraisal of existing structures. Technical Report. 11 Upper Belgrave Street, London SW1X 8BH: Institution of Structural Engineers (ISE), 1992
5. Wood, J., Angus, E.: Montrose Bridge. In: Proceedings of Structural Faults & Repair-95 Conference. London (1995)
6. Godart, B., de Rooij, M., Wood, J.: Guide to Diagnosis and Appraisal of AAR Damage to Concrete in Structures: Part 1 Diagnosis (AAR 6.1), vol. 12. Springer Science & Business Media (2013)
7. J. Wood, P. Nixon, P. Livesey. "Relating ASR structural damage to concrete composition and environment". In: International Conference on Alkali-Aggregate Reaction in Concrete, 10th: Melbourne, p. 1996. Victoria, Australia (1996)

Part III
Prognosis; Round Robin Expansion Tests

Chapter 12
Round Robin for ASR Expansion

Ammar Abd-elssamd, Sihem Le Pape, Z. John Ma, Yann le Pape, and Samuel Johnson

12.1 Introduction

Alkali-silica reactivity(ASR) has been recognized as one of the most deleterious phenomena in concrete. In fact, ASR can cause significant loss of mechanical properties and cracking in concrete structures that could lead to structural failure. The challenge now exists in evaluating the degree of the ASR damage in existing structures so that informed decisions can be made toward mitigating ASR progression and damage. Reliable methods and tools are needed in order to evaluate the condition of ASR affected structures. This evaluation includes (a) tests to determine the degree of expansion that has already occurred; (b) the degree of expansion that can be expected; and (c) the rate at which that expansion can take place.

Several test methods have been used across the world to address the residual ASR expansion in affected concrete. RILEM TC-259 "Prognosis of deterioration and loss of serviceability in structures affected by alkali-silica reactions" chaired by

A. Abd-elssamd (✉) · S. Le Pape · Z. John Ma
Department of Civil and Environmental Engineering, University of Tennessee, Knoxville, TN, USA
e-mail: Ammar-Abd-Elssamd@mocs.utc.edu

S. Le Pape
e-mail: sihem.lepape@gmail.com

Z. John Ma
e-mail: zma2@utk.edu

Y. le Pape
Ridge National Laboratory, Oak Ridge, TN, USA
e-mail: lepapeym@ornl.gov

S. Johnson
Electric Power Research Institute, Charlotte, NC 28262, USA
e-mail: sjohnson@epri.com

© RILEM 2021

V. E. Saouma (ed.), *Diagnosis & Prognosis of AAR Affected Structures*, RILEM State-of-the-Art Reports 31,
https://doi.org/10.1007/978-3-030-44014-5_12

Pr. Saouma at the University of Colorado collected laboratory test procedures for the Estimation of the Residual Expansion of Concrete in Structures Affected by Alkali Silica Reactions, including:

1. LPC N°44 (POC: R. P. Martin, B. Godart and F. Toulemonde)
2. Empa method (POC: C. Merz and A. Leemann)
3. Toulouse method (POC: S. Multon and A. Sellier)
4. Quebec method (Laval University)/U.S. Department of Transportation (DOT) (POC. M. Bérubé)
5. EPFL method (POC: L. Sofia, formerly C. Dunant)
6. UK method (POC: J. Wood)

The methods cover the laboratory determination of the swelling potential and the residual free expansion of concrete extracted from structures affected by ASR. The RILEM TC-259 supports that the next step should be a round-robin test of the different methods. Indeed, a round robin analysis for existing residual expansion techniques is needed to address the variability among these methods. To accomplish this objective, researchers from the University of Tennessee Knoxville, USA (UTK) and Oak Ridge National Laboratory (ORNL) in collaboration with the Electrical Power Research Institute (EPRI) provided cylindrical core samples extracted from an aged slab affected by ASR, and delivered to laboratories in Canada, France, Japan, Switzerland, Portugal, and the US. The participating laboratories conducted residual expansion and other ASR related assessment tests to evaluate the condition of the core samples. The current document describes the detailed information related to this analysis.

12.2 Samples, Materials, and Conditioning History

12.2.1 Samples

The core samples were extracted from an existing ASR-affected slab with the dimensions of $4 \times 3 \times 0.67$ ft ($1219 \times 914 \times 203$ mm.). The slab has been cast for another EPRI project [1] with reactive sand and coarse aggregates, and conditioned for three years prior coring for RR analysis. Figure 12.1 shows the initial condition of the slab after casting and the condition just before taking core samples for RR analysis as well as the location of core samples within the slab. The cores were extracted in the direction parallel to the slab width to avoid exposed surfaces of the slabs.

(a) Initial condition

(b) Current condition

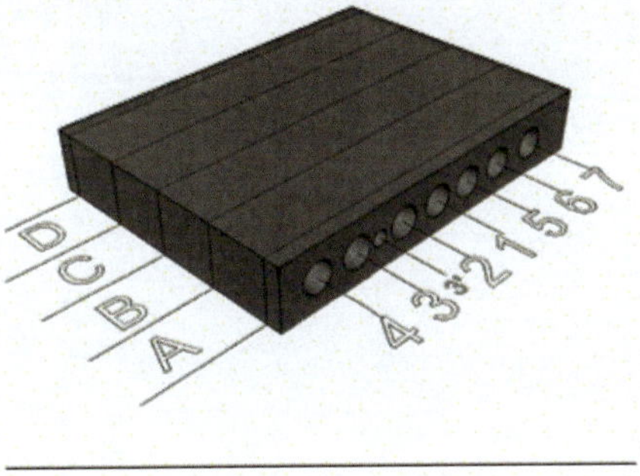

(c) Core Locations

Fig. 12.1 Photos showing the existing laboratory-aged samples [1]

12.2.2 Materials

The concrete mix for the RRA slab contained highly-reactive fine aggregate from El Paso, Texas, and two reactive coarse aggregates one from Wells, Maine and the other from Bernalillo, New Mexico. The cement used in the mix was Type I Portland cement, relatively high alkali cement. The alkali content of the concrete mix was boosted by adding sodium hydroxide (NaOH) to achieve a 1.25% equivalent alkali content, by mass of cement. The concrete mixture proportion is summarized in Table 12.1. Coarse aggregates sieve analysis can be found in Fig. 12.2 [1].

12.2.3 Conditioning History

The RRA slab was conditioned in two phases. The initial phase, immediately after casting, was an accelerated conditioning conducted in a curing room at 80.6°F (27 °C) temperature and 65% relative humidity (RH) to accelerate ASR development. Continuous moisture supply was maintained by converting the slab with a wet burlap and wrapping with plastic sheets every week. This phase lasted for 107 days, and the slab

Table 12.1 Concrete mixture proportions for RRA slab [1]

Materials	Quantity	
	lb/yd^3	kg/m^3
High alkali type I cement	700	420
Water	354	210
Coarse aggregate- Wells, Main (oven dry)	723	429
Coarse aggregate- Bernalillo, New Mexico (oven dry)	723	429
Fine aggregate- El Paso, Texas (oven dry)	1379	818
NaOH	2.74	1.63

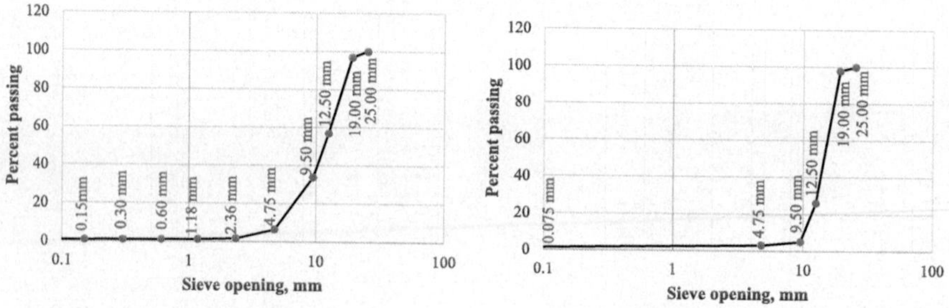

(a) Wells-Main (b) Bernalilo-New Mexico

Fig. 12.2 Aggregate sieve analysis [1]

average expansion was 0.05%. During the second phase, the slab was removed from the curing room and placed in a room with an HVAC system at temperature range between 65–75 °F (18–24 °C) and relative humidity (RH) around 40% for about three years until coring was conducted for RRA. The slab average expansion at time of coring was 0.24%. The expansion history (past expansion) for the slab is shown in Fig. 12.3.

12.3 Cores Extraction and Conditioning

The slab was placed vertically to allow coring along the slab width, as shown in Fig. 12.4. All cores were taken about 2 in. (50 mm) from the exposed surfaces using wet coring technique to avoid excessive drying of core samples. A total of 8 long cores with 36 in. (914 mm.) in length were extracted: 6 long cores with 4 in. (100 mm.) diameter, and one long core with 2.2 in. (56 mm.) diameter. The long cores were cut to provide a total of 30 cylindrical core samples for RRA. All cores were approximately

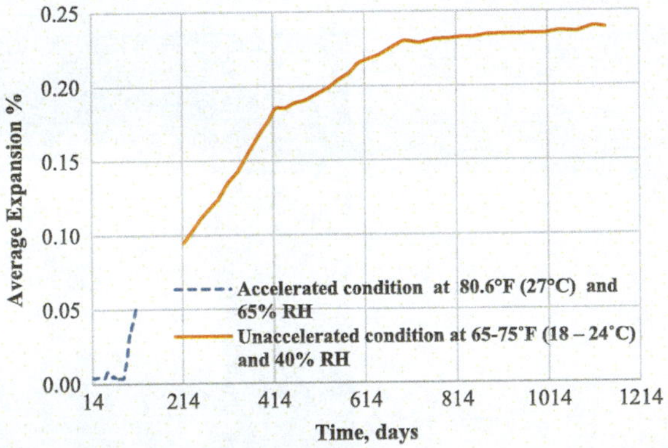

Fig. 12.3 History of ASR expansion of the slab [1]

Fig. 12.4 Slab coring [1]

4 × 8 in. (100 × 200 mm.) in size except for Japanese's methods 2.2 × 8 in. (56 × 200 mm.) cores were prepared. The samples were extracted and shipped on the same coring day to the six RRA participants following a procedure developed by Leandro Sanchez (University of Ottawa) and Cyrille Dunant (EPFL) to minimize transportation and storage effects on the moisture condition of the cores. A summary of the steps is provided below:

1. Two inches (50 mm.) were removed from top and bottom of each long core sample. These areas should be removed due to high microcracking from drying and wetting cycles, leaching of alkali, or concentration of alkali due to evaporation. The remaining core, 32 in. (813 mm.) in length, was cut into 4 small core samples with 8 in. in length (200 mm.).
2. The samples were quickly washed with clean water and immediately wiped with clean towels to remove dust and surface water. Then, the weight of each core was recorded to provide a measure for moisture condition after coring.

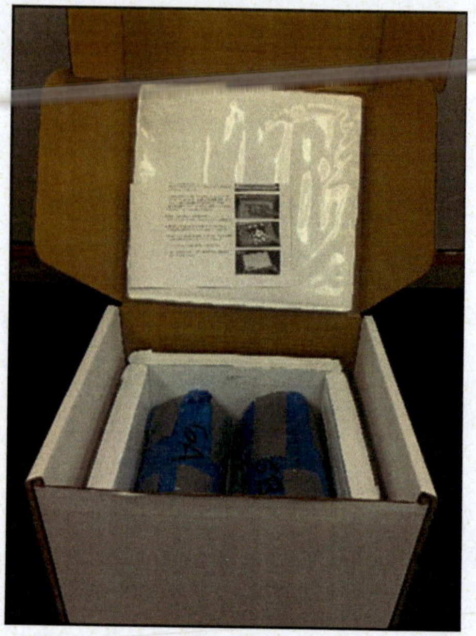

Fig. 12.5 Temperature-controlled box [1]

3. Immediately after the weight measurement completed, cores were wrapped with 20 layers of Polyurethane sheet and placed in vacuum- sealed plastic bags to avoid drying during shipping and transportation.
4. A temperature around 50°F (10 °C) was recommended during storage and transportation to minimize further progress of ASR before the residual expansion testing starts. For this purpose, core samples were placed in a temperature-controlled box (provided by FedEx) and shipped to all participants, see Fig. 12.5. The temperature-controlled box provided 41° to 46 °F (5–8 °C) environment, up to 96 h (4 days), for the core samples. The four days was enough for samples to reach all different participants at a similar condition state, see Table 12.2 for sample details.
5. Once samples were received by participants, weight measurement was conducted to assess potential drying during transportation. If the samples weight are significantly different from their original weight (immediately after coring), samples are recommended for pre-conditioning at 73 °F (23 °C) and 100% R.H before residual expansion test. The sample are pre-conditioned until either the samples reach their initial measured weight or for a maximum of 48 h.

Table 12.2 Samples details [1]

No	Participants	Core ID	Weight (grams)	Dimension (mm)
1	**Pierre Nicot & Alain Sellier** (Insa, Toulouse) **France**	4C	4249.8	100 × 200 (DXL)
		4D	MISSED	100 × 200 (DXL)
		5C	4211.4	100 × 200 (DXL)
2	**Andreas Lemann** (EMPA) **Switzerland**	2C	4287.8	100 × 200 (DXL)
		2D	4222.8	100 × 200 (DXL)
		3A	4226.6	100 × 200 (DXL)
		3B	4224.4	100 × 200 (DXL)
3	**Leandro Sanchez** (University of Ottawa) **Canada**	1A	4240.5	100 × 200 (DXL)
		1B	4261.4	100 × 200 (DXL)
		1C	4221.4	100 × 200 (DXL)
		1D	4218.9	100 × 200 (DXL)
		2A	4222.0	100 × 200 (DXL)
		2B	4242.2	100 × 200 (DXL)
4	**Lionel Sofia** (EPFL) **Switzerland**	3C	4272.7	100 × 200 (DXL)
		3D	4262.3	100 × 200 (DXL)
		4A	4223.6	100 × 200 (DXL)
		4B	4226.3	100 × 200 (DXL)
5	**Tetsuya Katayama** (Taiheiyo Consultants Co. Ltd) **Japan**	3'A	1231.1	56 × 200 (DXL)
		3'B	1223.8	56 × 200 (DXL)
		3'C	1211.8	56 × 200 (DXL)
		3'D	1217.8	56 × 200 (DXL)
		5A	4880.0	100 × 230 (DXL)
		5B	4826.3	100 × 230 (DXL)
6	**João Custódio** National Laboratory for Civil Engineering (LNEC) **Portugal**	7A	4318.1	100 × 200 (DXL)
		7B	4304.7	100 × 200 (DXL)
		7C	4193.0	100 × 200 (DXL)
		7D	4285.2	100 × 200 (DXL)
		6C	4262.2	100 × 200 (DXL)
7	**Victor Saouma*** (University of Colorado), **USA**	6A	4252.3	100 × 200 (DXL)
		6B	4234.4	100 × 200 (DXL)

*Cores were stored, cut and then shipped to David Rothstein and Tetsuya Katayama

12.4 Results

Participants submitted their results to the Round Robin organizer [1] as well as to the RILEM TC 259 ISR committee.

Each set of results will be discussed separately.

12.4.1 *Organizers Analysis and Summary*

This section will discuss the first set of results submitted to the University of Tennessee.

The round robin analysis (RRA) was conducted to assess the variability among several commonly used test methods to evaluate the condition of ASR affected concrete and the ASR residual expansion. Core samples were extracted from an aged ASR-affected slab and shipped to laboratories in Canada, France, Japan, Switzerland, Portugal, and the US to evaluate the ASR condition of the slab. All these methods are based on expansion tests conducted on cylindrical core samples except in the Toulouse method the residual expansion is evaluated on mortar bars made with aggregates extracted from cores. The laboratory from the US conducted petrographic examination assessment on core samples. Detailed reports from each participants are included in the following chapter.

The residual expansion results collected from laboratories in Canada, France, Japan, Switzerland, and Portugal are summarized Fig. 12.6. Expansion results from Toulouse method was not included in Fig. 14.6 as the expansion was measured on mortar bars made of aggregates extracted from core samples (i.e not a concrete core test). The majority of residual expansion tests were conducted at 38–40 °C except two Japanese methods were conducted at 80 and 50 °C. The core samples were mainly condition at 95–100% RH, immersed or wrapped in NaOH solution, immersed in pore solution, or immersed in NaCl.

The Japanese methods showed the highest average expansion values of 0.55% and 0.66% on cores conditioned at 50 °C–NaCl immersion and 80 °C–NaOH immersion, respectively. The lowest average expansion value of 0.12% was reported on the Empa method at 38 °C–100% RH. Although the Quebec, Empa, and LNEC methods have been conducted in a similar condition (38 °C and 95–100% RH), the ASR expansion curve and ultimate expansion values were different. This variation can be attributed to differences in pre-conditioning and conditioning regimes adapted by each method.

The duration for most residual expansion tests was either 365 days or more until expansion is stabilized or ceased. However, the duration for Empa method was less than 200 days, and the Toulouse method was the quickest, which showed results within 129 days. Although a lot of similarities between expansion methods exist, especially in conditioning regimes, the conclusion about the residual expansion and ASR condition might be different, as summarized below.

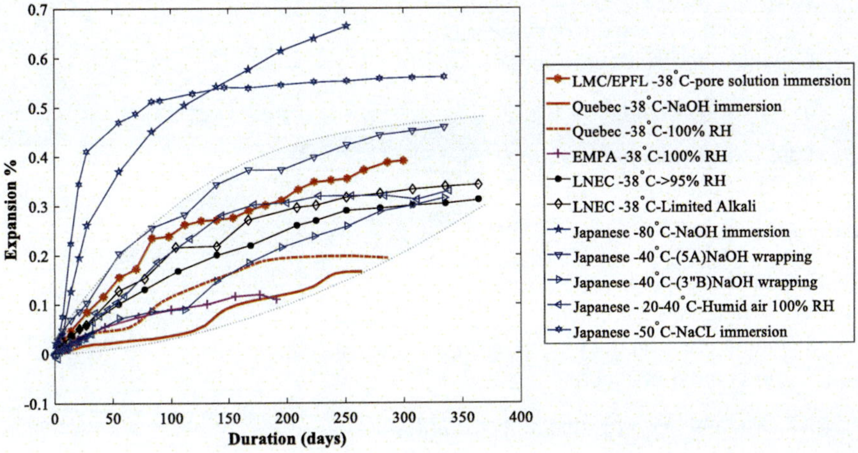

Fig. 12.6 Combined concrete core expansion tests conducted by RR participants (The gray area represents an envelope of the expansion tests at 38–40 °C temperature) [1]

12.4.1.1 Samples

- Toulouse Method:
  - 4C, 4D and 5C (100 × 200 mm)
  - Aggregates initially used for RRA slab casting.
  - Sample of cement used in slab.
- LMC or EPFL Method:
  - 3C, 4A, 4B, and 3D (100 × 200 mm)
  - Aggregates initially used for RRA slab casting.
  - Sample of cement used in slab.
- Empa Method: 3A, 3B, 2C, and 2D (100 × 200 mm)
- LNEC Method: 7A, 7B, 7C,7D, and 6D (100 × 200 mm)
- Japanese Method:
  - 3'A, 3'B,3'C, and 3'D (57.7X202mm)
  - 5A, and 5D (107.4 × 232.5 mm)
  - Aggregates initially used for RRA slab casting.
  - Sample of cement used in the slab.
- Quebec Method: 1A,1B,1C,1D,2A,2B; (100 × 200 mm).

12.4.1.2 Summary

- Toulouse Method: Mortar bars expansion, and reactive silica content for

- Aggregates extracted from core samples.
- Original aggregates used in the slab.

- LMC or EPFL Method:
  - Pore solution characterization.
  - Residual expansion test.
- Empa Method:
  - Residual expansion test.
  - Crack index measurement.
  - Scanning electron microscopy examination (SEM).
  - Alkali content in concrete.
- LNEC Method: Two residual expansion tests.
- Japanese Method:
  - Petrographic examinations:
    - Determination of cores' orientation.
    - Three-dimensional crack index (%).
    - Aggregate characteristics.
  - Four Japanese accelerated expansion tests
- Quebec Method: Two residual expansion tests.

12.4.1.3 Conditioning

- Toulouse Method: 38 °C–NaOH immersion (1 N)
- LMC or EPFL Method: 38 °C–pore solution immersion
- Empa Method: Phase 1: 20 °C, 100% RH; Phase 2 &3: 38 °C–100% RH; Phase 4: drying at 60–70% RH
- LNEC Method:
  1. 38 °C > 95% RH.
  2. 38 °C–>95% RH & limited supply of alkalis
- Japanese Method:
  1. 80 °C–NaOH immersion.
  2. 40 °C–NaOH wrapping.
  3. 50 °C–NaCl immersion.
  4. 20–40 °C 100% RH air humid
- Quebec Method:
  1. 38 °C–>95% RH
  2. 38 °C–NaOH immersion

12.4.1.4 Time to Results

- Toulouse Method: 129 days
- LMC or EPFL Method: 365 days and more
- Empa Method: 180–200 days
- LNEC Method: 365 days and more
- Japanese Method: 365 days and more
- Quebec Method: 365 days and more

12.4.1.5 Obtained Results

- Toulouse Method:

  1. Chemical ASR reaction advancement (%) = 80% for all sizes of aggregates.
  2. Chemical potential expansion (%) = 16%–20% for all sizes of aggregates.

- LMC or EPFL Method: Expansion at 300 days was 0.4%.
- Empa Method:

  1. Expansion rate of concrete = 0.22%/year
  2. Concrete has high residual expansion potential.
  3. Crack index at delivery = 0.15%.
  4. Crack index after test = 0.32%
  5. Expansion during phase 3 indicates an internal source of alkalis (presence of alkali-silica-glass in aggregates)

- LNEC Method: The average residual free expansion potential obtained at 364 days of testing was 0.32%. Whilst the minimum and maximum values were 0.29%–0.37%, respectively
- Japanese Method: When slab exposed to 100% RH:

  1. Advancement of ASR at time of coring: 39–42%.
  2. Past expansion and residual expansion: 0.22% and 0.31-0.34%, respectively.
  3. Expansion parameters usable for simulation: ε^{∞}= 0.53–0.57%, τ_L= 3.2–3.3 years, τ_C= 0.35–0.72 years.

- Quebec Method:

  1. The expansion rates obtained at 38 °C–>95% RH was 0.132%.
  2. The expansion rates obtained at 38 °C–NaOH immersion was 0.162%.

12.4.1.6 Use of Results

- Toulouse Method: The reaction advancement can be used to calibrate the kinetics parameters of modelling to assess the future behavior of damaged structures.

- LMC or EPFL Method: The test still on going to determine the kinetic of the reaction and residual expansion of concrete.
- Empa Method: (1) Results provide an indication about the expansion potential for concrete expansion. (2) The boundary conditions of a structure such as temperature, humidity and stress state must be taken into account in an overall assessment.
- LNEC Method: (1) The RRA slab is classified with a high residual expansion potential. ((2) Structural analysis is recommended to examine the consequences of such an expansion to the structure.
- Japanese Method: Overall final expansion of concrete (ε^{∞}) was estimated from four kinds of the accelerated concrete core expansion tests, aided by the thermodynamic conversions of the Larive's expansion parameters to the standard state.
- Quebec Method: The concrete from RRA slab is considered to have a high potential of residual ASR expansion.

12.4.2 RILEM TC-259 Assessment

The second set of results can be found in the next Chap. 13 for the study by Katayama (Japan), Chap. 13 for the study by Leemann (Switzerland), Chap. 14 for the LCPC (though the tested specimens did not originate from this study), 15 for the study by Rothstein (USA), Chap. 16 for the study by Sanchez (Canada), and Chap. 17 for the study by Sellier (France). It should be noted that Sofia's results (EPFL) were not incorporated in this document as the method has not been published, and the authors have not participated in the committee activities.

Finally, a thorough analysis of the above mentioned results by Andreas Leemann can be found in Chap. 20.

Acknowledgements This study is part of a project conducted by the University of Tennessee Knoxville and funded by the Electric Power Research Institute (EPRI) entitled "Aging management of ASR affected structures"- EPRI report # 3002013192.

The authors would like to express their gratitude and sincere appreciation to Maria Guimaraes (a program manager at EPRI) for supporting this study and providing an aged ASR-affected slab as well as covering the costs of cores extracting and shipping for the purpose of round robin analysis.

Reference

1. EPRI.: Evaluation of Laboratory Tests to Detect Up-to-Date Expansion and Remaining Expansion in Concrete Structures Affected by Alkali-Silica Reaction. Technical report 3002013192. Electric Power Research Institute (2018)

Chapter 13
Accelerated Expansion Test Sample Report: Japan

Tetsuya Katayama, Kozo Mukai, and Tomomi Sato

13.1 Introduction

This report presents results of a round robin test in WG-1 of RILEM TC 259-ISR (*ASR prognosis of deterioration and loss of serviceability in structures affected by alkali-silica reaction*), performed at the research laboratory of Taiheiyo Consultant (THC) in Japan. This was planned to verify the effectiveness of proposed testing methods in the RILEM WG-1 documents, with the challenges below:

Challenges Estimate the following items quantitatively, i.e. in numeric values, when a concrete core specimen of unknown expansion behavior (closed information) is received.

(1) Advancement of ASR
(2) Past expansion and residual expansion
(3) Expansion parameters usable for simulation of final free expansion.

T. Katayama (✉) · K. Mukai · T. Sato
Taiheiyo Consultant Co., Ltd., Sakura, Japan
e-mail: katayama@catv296.ne.jp

T. Sato
e-mail: Tomomi_Sato@taiheiyo-c.co.jp

© RILEM 2021

V. E. Saouma (ed.), *Diagnosis & Prognosis of AAR Affected Structures*,
RILEM State-of-the-Art Reports 31,
https://doi.org/10.1007/978-3-030-44014-5_13

13.2 Materials and Methods

13.2.1 Materials

A total of 6 core specimens from a 3-year old concrete slab (approximately 122 × 91 × 20 cm), prepared by the University of Tennessee Knoxville (UTK) in 2013 and kept at the Electric Power Research Institute (EPRI) in Charlotte, North Carolina, USA, were provided in January 2017 under the sponsorship of the latter [1], Figs. 13.1 and 13.2. Of these, 5 core specimens (two fat dia. 107 mm, and three slender dia. 57 mm) were subjected to accelerated expansion tests in various conditions (20–40, 50, 80 °C), and one remaining core (dia. 57 mm) to petrographic examination (Tables 13.1and 13.2). Three reactive aggregates from the same localities as those used in casting the concrete slab (two coarse aggregates and one fine aggregate), collected and supplied by UTK in mid-2017, were examined petrographically.

As a parallel set of concrete specimens, two fat cores (dia. 107 mm) of UTK kept by the University of Colorado Boulder (UCB) in two conditions (in refrigerat or and in 30 °C fog room for 11 months) arrived in THC in December 2017 for a comparative petrographic study (Fig. 13.1). Because they came being wrapped by a linen cloth rich in alkali-solution, they were immediately re-wrapped by a cling film and placed in a tight vinyl envelope, then further stored in refrigerator at THC (5 °C for 4 months). A comparative study started in April 2018 just before the RILEM TC259 meeting in Denver. Because of the time constraint, the present report will describe chiefly results of testing of the first 6 core specimens received directly from UTK. Summary of ongoing results of petrographic examination of two core specimens supplied by UCB will be appended in this report.

Table 13.1 History of concrete slab and core specimens received at THC from UTK for RILEM round-robin test

year		2013		2014	2015	2016	2017							2018
month/day		10/14		11/29			1/9	1/16	1/18	1/24	2/22	6/7	10/13	7/10
time (days)		0		107			1183*	1190	1192	1198	1227		1460	1719
event		casting at UTK	storage at UTK		moving to EPRI	storage at EPRI	extraction of core	ship-ping	arrival at THC	beginning of testing	storage at THC	arrival at THC	thin sectioning	end of testing
storage temperature	slab		27°C 65% RH		21°C 40%RH									
	core							6°C						
	3'D petrography									5°C			x	
	5B 100% RH									20°C	40°C			
	5A NaOH-wrapping									40°C				
	3`B NaOH-wrapping									40°C				
	3'C saturated NaCl									50°C				
	3'A 1N NaOH									80°C				
expansion (%)	slab (on-site)			0.05										
	3'D crack index (%)							0.22	←	expansion frozen to 0.22**				
aggregate	Texas sand (El Paso)											x	x	
	New Mexico gravel											x		
	Maine crushed stone											x		

**past expansion of concrete before coring (*) has been refrigerated (5 °C) until the day of thin sectioning

Table 13.2 Conditions of the concrete core expansion tests

Method		Test specimen (mm)		Storage condition		Reading
1	NaOH immersion	3′A	dia. 57.7, L202.7	80 °C, 1N (NaOH 40.0g/Lwater)	After reading, change the up-down directions for next storage	First, read the thermal expansion read the length change in hot condition
2	NaOH wrapping	3′B	dia. 57.7, L202.0	40 °C, 1N (20g solution/nonwoven fabric)** + wrapping film + vinyl envelope		
3	NaCl immersion	3′C	dia.57.3, L201.4	50 °C, saturation (NaCl 36.7 g/100 g water)		
4	NaOH wrapping	5A	dia. 107.4, L232.5	40 °C, 1N 2 × (40 g solution/nonwoven fabric) + wrapping film+ vinyl envelope		
5	Humid air (naked exposure) (modified JCI-DD2*)	5B	dia. 107.3, L230.6	20 °C RH 100% (until stabilized, e.g. 1 month) → 40 °C RH 100%		

*Original requirement of JCI-DD2 (JCI-S-011-2017): >RH 95%, read 1 day after transferring to 20 °C room

**At the next reading, the amount of alkali solution will be increased from the current pore-volumetric basis to the surface area basis (40 g solution/paper sheet)

13.2.2 Methods

A slender core (No. 3′D, dia. 57 mm), stored in a refrigerator (5 °C) for 268 days to stop expansion by ASR, was subjected to petrographic examination. Core surface was scanned (Fig. 13.3) to show a preferred orientation of coarse aggregate to estimate the original orientation (Z, X, Y) in the concrete slab. This core was dried at 40 °C for one day, impregnated with fluorescent epoxy resin, then cut into half cylinders at 45° from the estimated horizontal direction (Fig. 13.4). A half cylinder of the core, with polished surface with secondary impregnated fluorescent epoxy resin, was examined with a UV-scanner to reveal the distribution of cracks and voids in concrete (Fig. 13.5, Photo 13.1).

Another half cylinder core was cut along its longitudinal axis (X) to prepare polished thin sections (42 × 20 mm) representing the horizontal (X, Y) and vertical (Z) directions in the original slab (Fig. 13.4). Then, crack indices in three directions (Z, X, Y) were measured in thin sections along the intersecting grids at a right angle (7 lines in 42 mm, 5 lines in 20 mm) (Table 13.4). Petrographic stages of ASR defined by Katayama [2, 3] were determined under the polarizing/reflecting/fluorescence

Table 13.3 Dimensions and weight changes of core specimens upon receipt at THC from UTK

	Weight (g)			Size (mm)			Arrival from	Appearance
	Sent**	Received	Gain	Nominal	Measured			
				Length	Diameter	Length		
3′A	1231.1	1231.7	+0.6	56	57.7	202.7	UTK January 18, 2017	Core surface was dry due to vacuum treatment for sealing. Reacted aggregate had a wet and dark appearance of ASR gel. Cool box was 7 °C
3′B	1223.8	1224.2	+0.4	56	57.7	202.0		
3′C	1211.8	1212.5	+0.7	56	57.3	201.4		
3′D*	1217.8	1218.5	+0.7	56	57.3	202.6		
5A	4880.0	4882.2	+2.2	100	107.4	232.5		
5B	4826.3	4828.4	+2.1	100	107.3	230.6		

*For petrographic examination
**Informed from University of Tennessee Knoxville

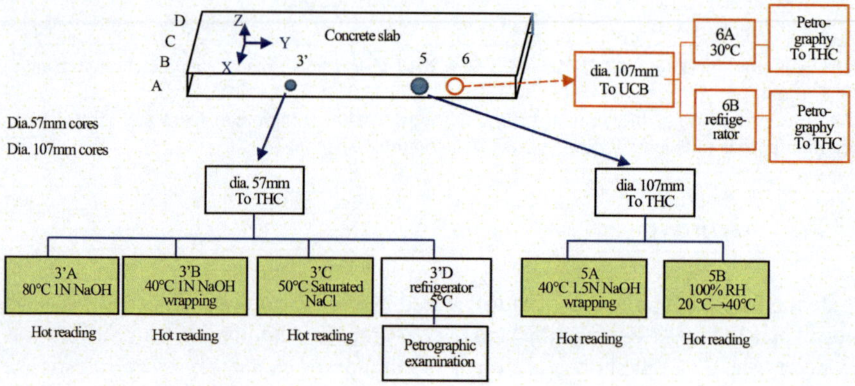

Fig. 13.1 Distribution of core specimens extracted in the horizontal direction by UTK for RILEM round-robin test

microscope in thin sections. After completing polarizing microscopy, SEM observation (JEOL JSM-IT300HR) and quantitative EDS analysis (JEOL JED2300: 15KV, 0.45nA, data acquisition 30s, ZAF correction, calibration with standards) were made of reaction products and reacting silica minerals in the same polished thin sections.

Several kinds of Japanese accelerated concrete core expansion tests (1N NaOH-immersion [4], saturated NaCl-immersion [5], 1N NaOH-wrapping [6], and 100% RH curing [7]) (Table 13.2), with slight modifications described later, were performed until 508 days at the maximum. Expansion curves were fitted to Larive equation [8]

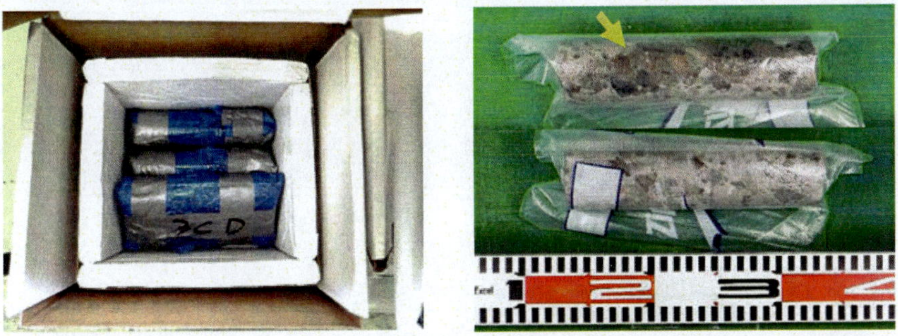

(a) Package of slender core specimens and segregation of water (wet appearance) to the bottom side of core (No.3'C)

(b) Package of fat core specimens and dry surface due to vacuum sealing, except for reacted aggregate (core No.5A)

Fig. 13.2 Core samples as received

to assess expansion parameters (ε_∞, τ_L, τ_c), then converted to a standard state (20 °C 100% RH) using Arrhenius equation and other equations [9]. Overall final expansion curve of the concrete slab was estimated by summing up the past expansion (crack index) and residual expansion that was thermodynamically corrected by re-fitting to the Larive equation.

13.3 Results

13.3.1 Materials on Arrival

All the core specimens laid horizontally in a cooling box (keeping 7 °C) arrived in a well-wrapped condition (Fig. 13.2). Before shipping, they were vacuum sealed following the protocol by Sanchez and Dunant [10] to realize a tight packing. However, all the core had a dried surface suggestive of a loss of water by the vacuum treatment. One of them presented evidence of segregation of water to the bottom side of the

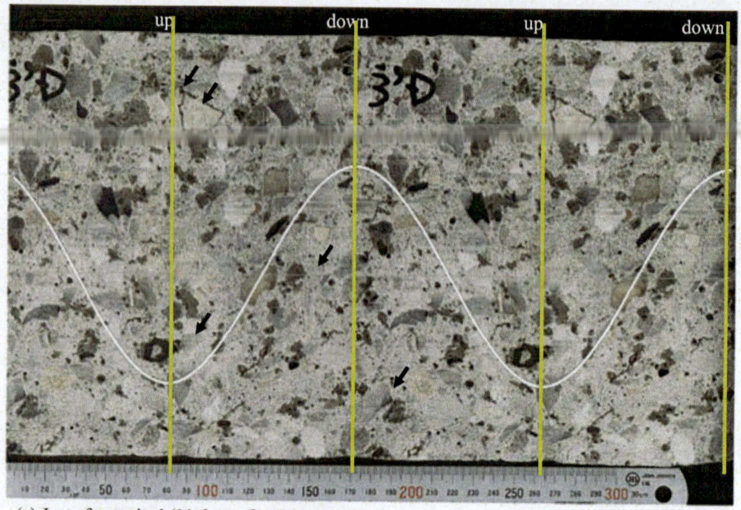

(a) Just after arrival (11 days after extraction)

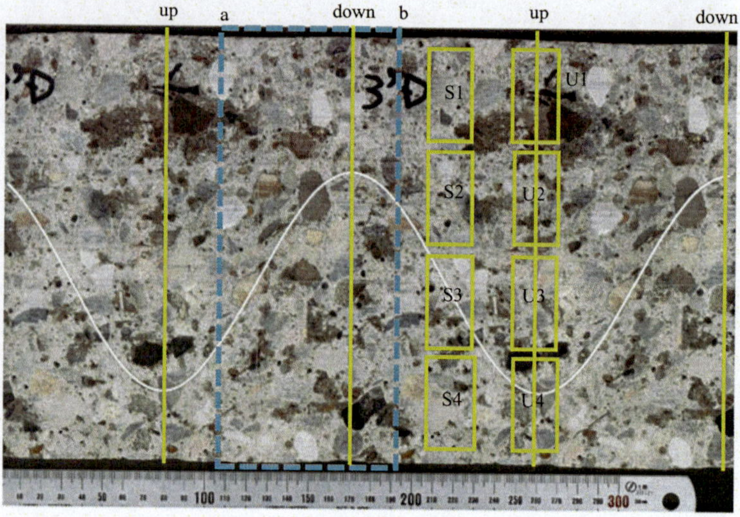

(b) After storage in refrigerator (at 6°C for 268 days after arrival)

Fig. 13.3 Scanned images of outer core surface showing a sine curve of oriented coarse aggregate, core No. 3′D

core, which was likely to occur within core specimen laid horizontally for a long time after extraction of core and during shipping (Fig. 13.2a).

Core specimens immediately after unwrapping at THC measured slightly heavier (0.03–0.06%) than those originally recorded by UTK, despite the above evidence of drying out of the core surface on arrival. One possible explanation for this apparent weight gain is that ASR gel adsorbed moisture from the atmosphere in the THC

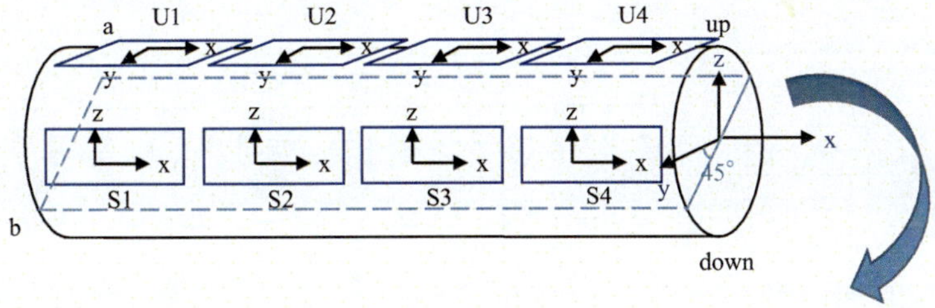

Fig. 13.4 Core cutting (dia. 5.7 cm) for thin sectioning

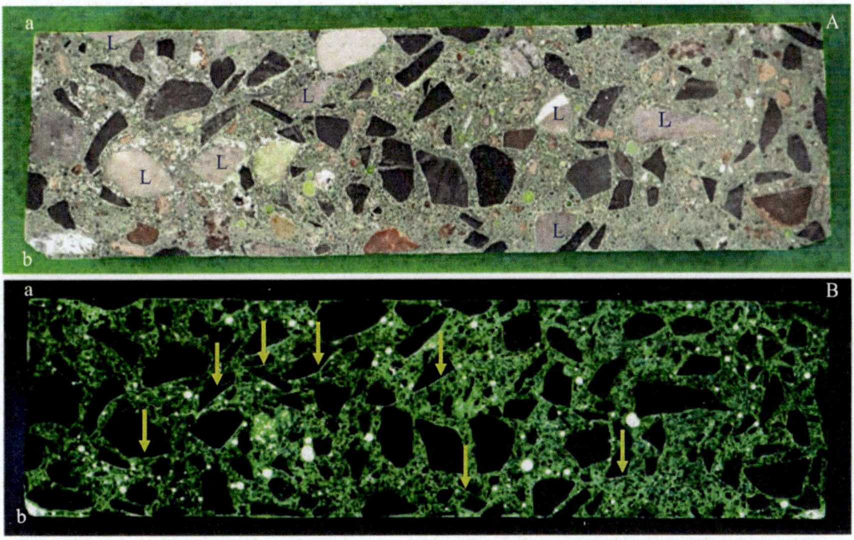

(a) Limestone (L) contained in the gravel fraction (b) Parallel cracks running through, under, or connecting the coarse aggregate, UV scanner, scale bar =1cm

Fig. 13.5 Longitudinal oblique section of concrete core, cut at 45° to horizontal direction, dia. 57 mm, core 3′D

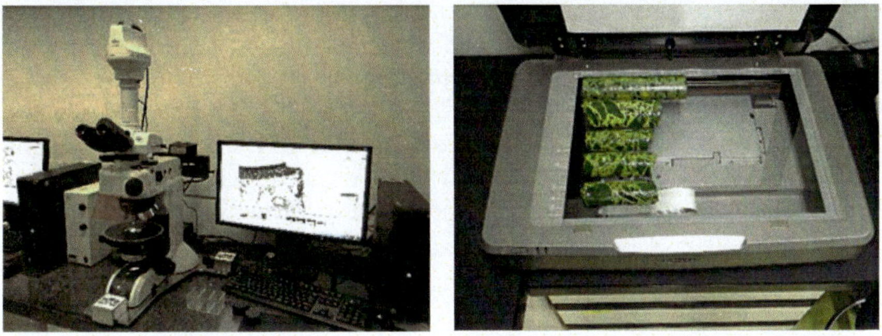

(a) Polarizing/reflecting/fluorescence microscope (b) UV scanner for half cylinder core specimens

Photo 13.1 Petrographic tools

Table 13.4 Crack index (%) of concrete in thin sections, 9 months after the storage in refrigerator (5 °C), Core 3′D

	Up					Side					Crack index (%)*
	4.2cm x 1.7cm x 4=28.6 cm					4.2cm x 1.9cm x 4=31.9 cm					
Direction	U1	U2	U3	U4	av	S1	S2	S3	S4	av	av
Z	[illegible]	[illegible]	[illegible]	[illegible]	[illegible]	[illegible]	[illegible]	[illegible]	0.14	0.28	0.28
X	0.27	0.19	0.26	0.25	0.24	0.19	0.20	0.22	0.14	0.19	0.22
Y	0.17	0.12	0.24	0.33	0.22						0.22

*To convert crack index to expansion, an addition of 0.02–0.04% may be necessary, because expansion less than this range gives no visible cracking, and is likely to be missed by the petrographic examination of the crack index

laboratory during sample handling. However, the observed difference may be due to calibration of the weighing scale (horizontal setting with a spirit level and frequency of calibration) (Table 13.3).

13.3.2 *Petrographic Examination*

13.3.2.1 Determination of Orientation

Unfortunately, orientation of the core specimen (Z, X, Y) was not recorded at time of coring (Fig. 13.1). Hence, the original directions were estimated based on the textural analysis developed by Katayama. Crest and trough of the sine curve connecting the oriented coarse aggregate represent the vertical direction (Fig. 13.3), and the down direction was determined by the presence of an interspace of breeding water and air voids trapped under the coarse aggregate (Fig. 13.5).

13.3.2.2 Three-Dimensional Crack Index (%)

Macroscopically, under the fluorescent light, parallel expansion cracks were connected under the coarse aggregate (Fig. 13.5). In thin sections (Fig. 13.4), crack index in the vertical direction (Z) varied more than three times, reflecting wide cracks within and around reacted sand particles (Table 13.4), and its average was 30% larger than that of the horizontal directions (X, Y). To convert the crack index to expansion, an addition of 0.02–0.04% may be necessary, because expansion of less than this range generally gives no visible cracking and is likely to be missed by the petrographic examination. In this report, however, averaged crack index in the horizontal directions (0.22%) was adopted as expansion of the concrete slab.

13.3.2.3 Aggregates

Rock types constituting the coarse aggregates were identified by naked-eyes aided by stereomicroscopic observation, then quantified by weighing each type after picking up (Table 13.5). Gravel from Bernalillo contained both early-expansive volcanic rocks and late-expansive quartzite. Limestone pebbles, which were contained in all concrete core specimens, were not found in the coarse aggregates received. This suggests that the gravel even from the same pits may vary considerably by time. Crushed stone from Wells was a metamorphic hornfels and typical late-expansive aggregate.

Fine aggregate from El Paso, according to point counting in thin sections under the polarizing microscope, consisted of highly reactive volcanic rocks. Vesicle-lining opal in altered hornblende dacite and interstitial cristobalite in spherulitic rhyolite are typical early-expansive pessimum-causing silica minerals, whereas vesicle-filling chalcedony in altered rhyolite and rhyolitic glass in altered rhyolitic welded tuff are moderate-expansive constituents (Table 13.6, Fig. 13.6). The occurence of these

Table 13.5 Aggregates from the same sources as used for the concrete slab

Coarse aggregate (wt %)				Fine aggregate (vol %)	
Bernalillo, New Mexico crushed gravel		Wells, Maine crushed stone		El Paso (Jobe), Texas sand	
andesite	25.6	pelitic hornfels	49.5	rhyolite	12.2
rhyolite-dacite	15.4	psammitic hornfels	34.7	glassy rhyolite	0.9
sandstone	15.1	granitic rock	10.0	spherulitic rhyolite	5.1
quartzite-chert	12.5	mafic rock	2.8	rhyolitic welded tuff	2.4
quartzite	10.3	vein quartz	2.2	altered hornblende dacite	13.1
granite	8.9	rhyolite-dacite	0.3	altered andesite	1.3
granite?	4.3	others	0.6	quartzite	1.2
chalcedony-agate	2.1			sandstone	0.1
schist?	1.8			black shale	0.3
plagioclase?	1.1			mudstone	6.1
mafic rock	0.3			granite	17.2
others	2.6			quartz	37.2
				feldspar	2.9

early-expansive   moderate-expansive   late-expansive

*Limestone was not found in the aggregate received

Table 13.6 Generalized list of reactive minerals, adapted from Katayama [11]

	Silica minerals	Pessimum proportion		Volcanic glass	Pessimum proportion
Early-expansive	opal	<5%	Moderate expansive	rhyolitic glass	100%
	cristobalite	<10%		dacitic glass	100%
	tridymite	<10%		andesitic glass	marginal
Moderate expansive	chalcedony	20%		basaltic glass	innocuous
Late-expansive	cryptocrystalline quartz	50%			
	microcrystalline quartz	>80%			

- Early-expansive: small pessimum proportion, ASR occurs even concrete alkali $Na_2Oeq < 3 kg/m^3$
- Pessimum proportion of volcanic rocks depends on the cooling rate of the magma, ranging 10% (cristobalite)–100% (glass)
- Late-expansive: cryptocrystalline quartz and microcrystalline quartz have no marked pessimum phenomenon

(a) Opal lining or filling vesicles (amygdales) in altered hornblende dacite

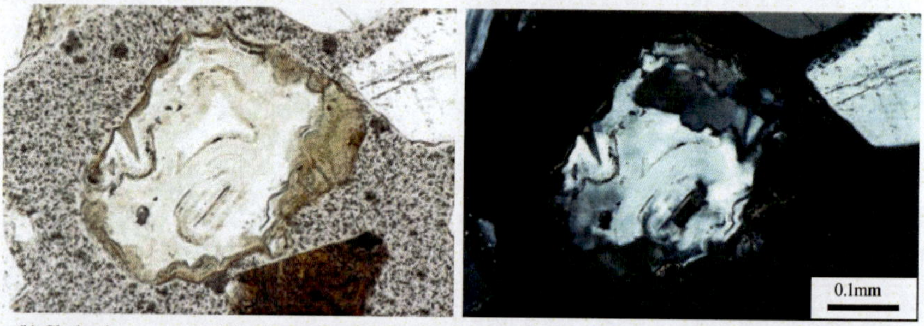

(b) Chalcedony recrystallizing into microcrystalline quartz, originally opal in the amygdale of altered rhyolite

(c) Interstitial cristobalite and anorthoclase aroundthe spherulitic texture of partly devitrified glassy rhyolite

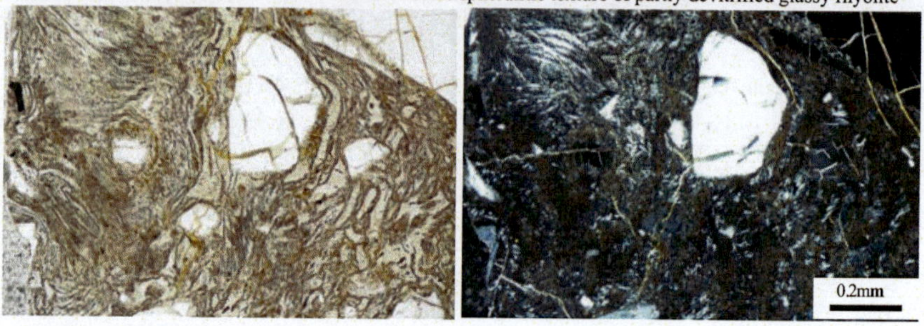

(d) Partly devitrified glass in rhyolitic welded tuff

Fig. 13.6 Alkali-reactive minerals in El Paso sand (Jobe, Texas) in plane polarized light (left) and in crossed polar silica minerals are common to fresh to moderately altered volcanic rocks in Japan [11–14], New Zealand [13], Iceland [15] and Turkey [16].

13.3.2.4 ASR in Concrete

Progress of ASR at the time of core extraction was "frozen" (estimated expansion 0.22%) by keeping the core specimen in refrigerator (9 months, 5 °C) until preparation of thin sections for petrographic examination became feasible. In thin section, sand particles of rhyolites, andesite and dacite were producing conspicuous cracks that extended into cement paste (Figs. 13.7, 13.8). This corresponds to the stage iv of ASR after Katayama [2, 3, 9] (Table 13.7). When all accelerated core expansion tests are over, their crack indices and stage of ASR will be compared with actual expansion measured.

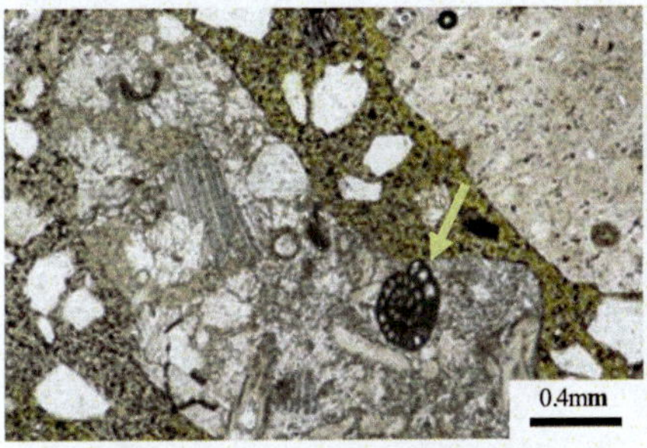

(a) Fusulinid fossil in the limestone gravel suggestive of Carboniferous formation, S3

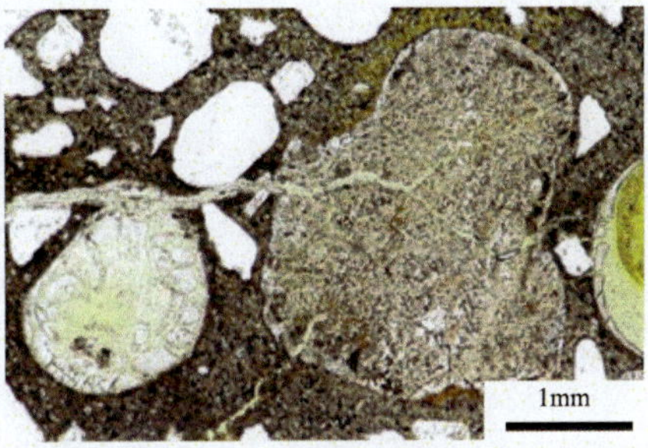

(b) Partly devitrified glassy rhyolite, S1

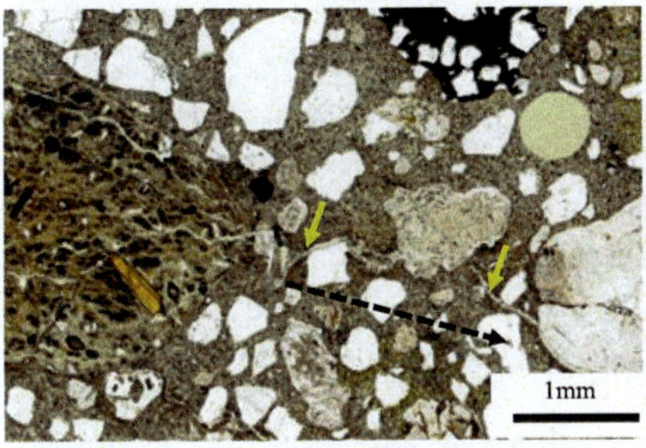

(c) Stage iv:ASR gel migrated along the crack from the reacted rhyolite into cement paste, S1

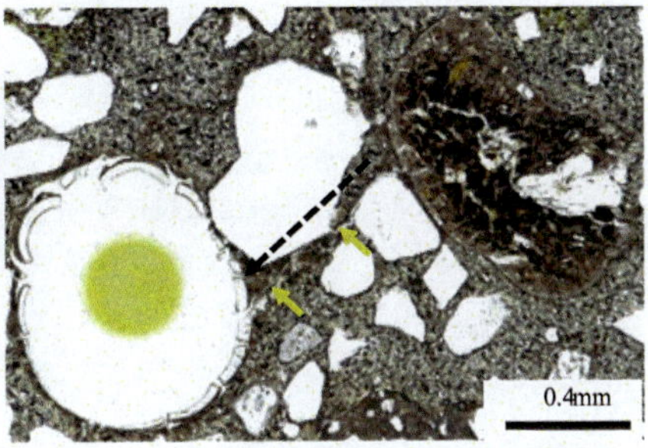

(d) Stage iv → v:ASR gel migrated along the crack into air void distant from the reacted rhyolite, U3

Fig. 13.7 Progress of ASR caused by rhyolites and migration of ASR gel. Core No. 3′D, stored in refrigerator

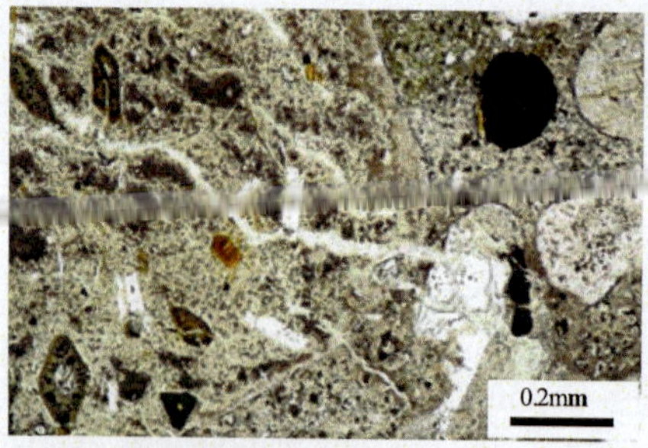

(a) Altered hornblende dacite, S1

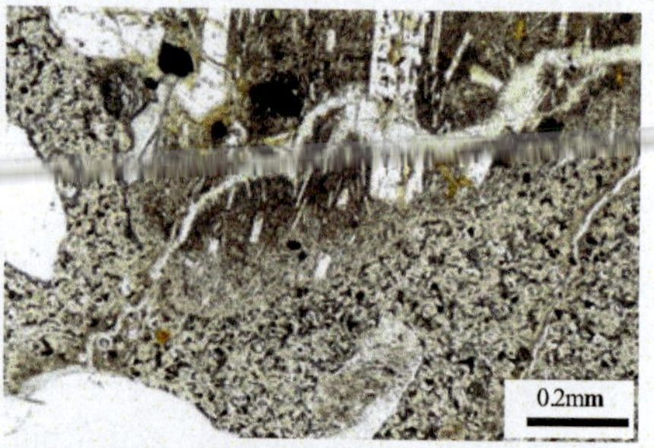

(b) Altered andesite, S3

(c) Radial expansion cracks extending from the reacted cristobalite in spherulitic rhyolite, S1

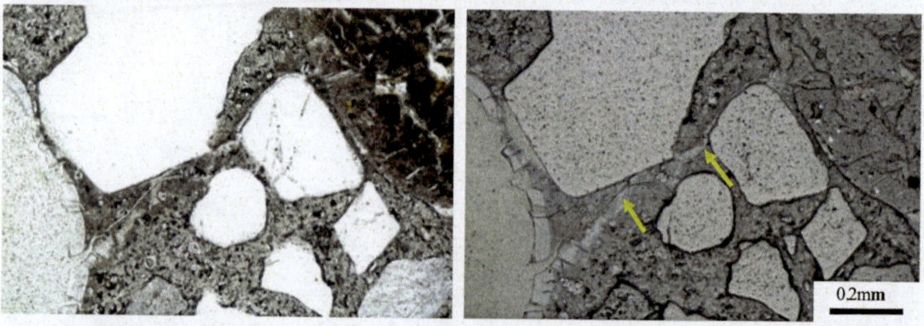

(d) ASR gel migrating alongthe crack from the reacted rhyolite, through cement paste,into air void, U3

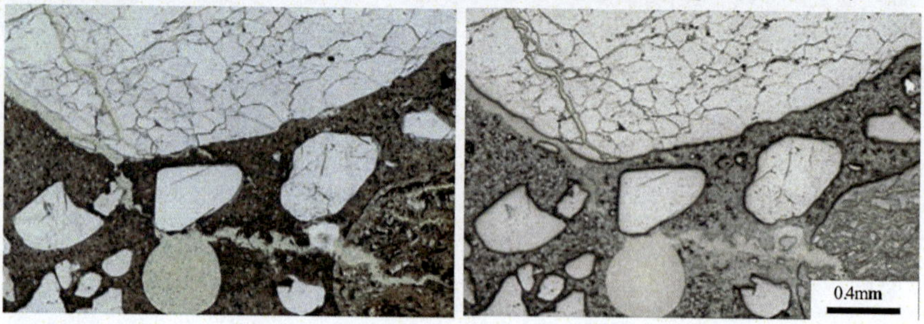

(e) Cracking within quarzite gravel and cracks in cement paste that extends from the reacted rhyolite, U3

Fig. 13.8 Reaction of sand particles and gravel in the concrete slab, Core No. 3′D, stored in refrigerator

Table 13.7 Petrographic stage of ASR in concrete, kept in refrigerator to stop ASR after extraction of core, No. 3′D

	Aggregate	Rock type	Progress of ASR →					Severity of ASR
			i	ii	iii	iv	v	
			aggregate	paste	aggregate	paste		micro-
			Reaction rim	Gel-halo	Crack-filling gel		Void-filling	scopy
Coarse	Early-expansive*	andesite		x	x			1
		rhyodacite	x	x				
	Late-expansive*	quartzite			(x)	(x)		2
	Late-expansive**	psammitic hornfels			(x)	(x)		
Fine	Early-expansive***	andesite	x		x	x		2
		rhyolite	x		xx	xx		
		glassy rhyolite			xx	xx	x	
		spherulitic rhyolite		x	x	x		
		altered andesite	x		xx	xx		2
		altered rhyolite	x		xx	xx		
		altered welded tuff	x		x			1
		altered tuff	x					
		shale			x			
	Overall assessment							2

Occurrence: xx: abundant, x common, (x) rare Severity of ASR: 1: minor; 2: moderate; 3: severe
*Bernalillo, New Mexico, **Wells, Maine, ***El Paso (Jobe), Texas

altered | Occurrence of smectite suggestive of green tuff formation of Miocene to Pliocene age

13.3.2.5 SEM Observation of ASR Gel

SEM observation on polished thin section reveals the detail of ASR gel in concrete and aggregate. ASR gel runs along crack from reacted particles of rhyolite (Fig. 13.9a, c) into cement paste (Fig. 13.9b). In this migration process, gel may enter air voids along the cracks. Where void-filling ASR gel is distant from the reacted aggregate,

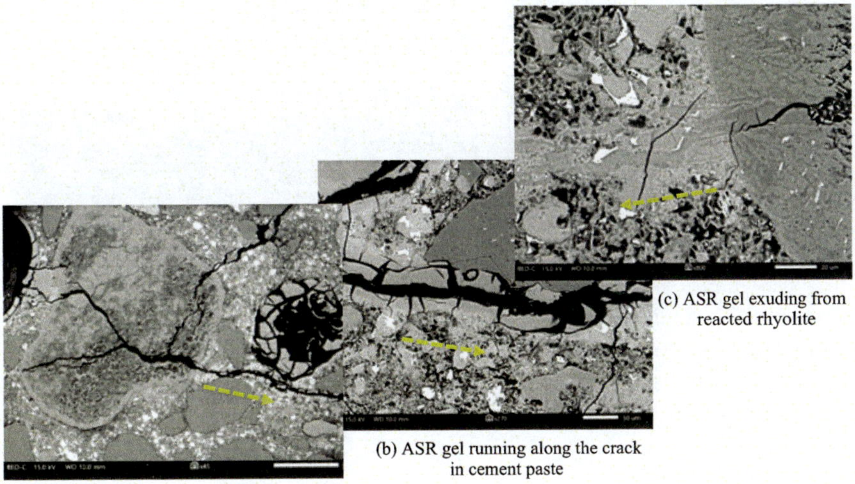

(c) ASR gel exuding from reacted rhyolite

(b) ASR gel running along the crack in cement paste

(a) ASR gel flowing into air void along the crack

Fig. 13.9 Migration of ASR gel along the crack extending from the reacted aggregate into cement paste, Core No. 3′D stored in refrigerator, section S1

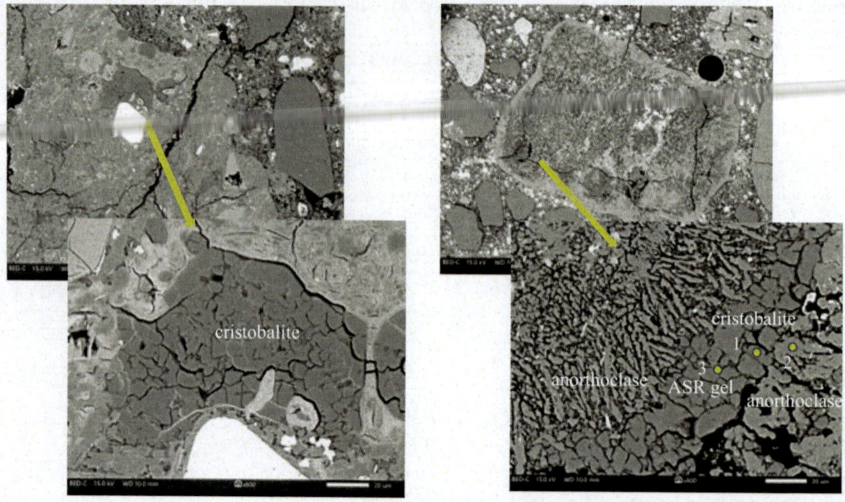

(a) Primary cristobalite patches in altered andesite, S3

(b) Secondary cristobalite forming ASR gel in the grain boudaries. Cristobalite is intermixed with elongated anorthoclase in devitrified spherulitic rhyolite, S1

Fig. 13.10 Reacting cristobalite in the volcanic aggregate, Core No. 3′D stored in refrigerator

Photo 13.2 SEM-EDS for quantitative analysis of reaction products

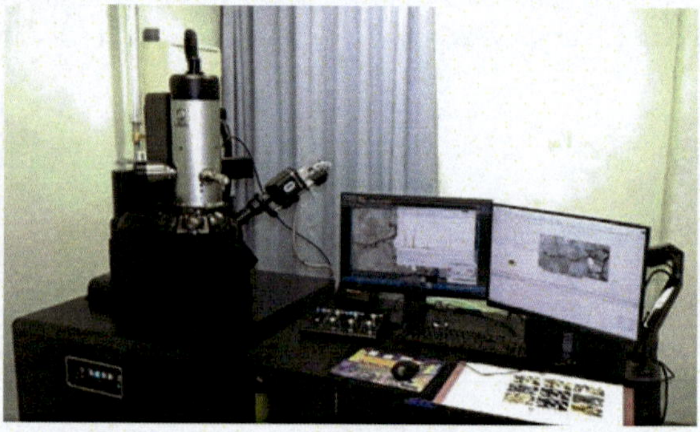

its composition tends to be rich in calcium, due to long contact with cement paste during migration from the aggregate.

Figure 13.10 show the occurrence of highly reactive cristobalite in andesite (Fig. 13.10a) and in spherulitic texture of rhyolite (Fig. 13.10b).

13.3.2.6 EDS Analysis of ASR Gel

Compositional trend of ASR gel in concrete.

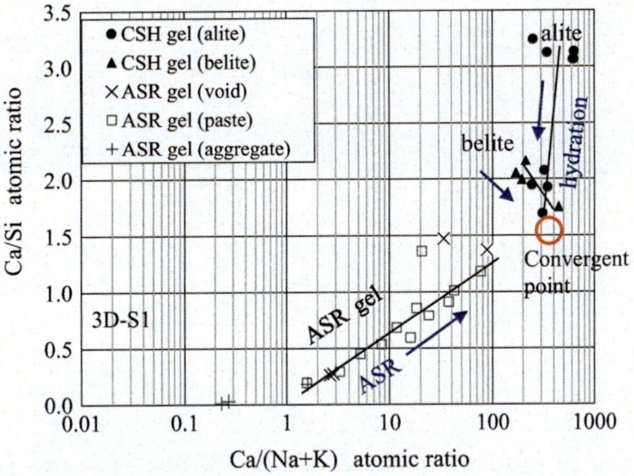

Fig. 13.11 Compositional lines of ASR gel in concrete as determined by EDS analysis on polished thin section showing the type I evolution, core No. 3′D, stored in refrigerator, thin section S1

Table 13.8 Compositions of reacting cristobalite and anorthoclase in spherulite, as determined by EDS analysis (wt%), Core No.3'D, stored in refrigerator, thin section S1

	mineral	SiO_2	TiO_2	Al_2O_3	Fe_2O_3	MnO	MgO	CaO	Na_2O	K_2O	SO_3	P_2O_5	Total
1	cristobalite	98.91	0.00	1.07	0.26	0.00	0.00	0.10	0.26	0.16	0.00	0.24	101.00
2	anorthoclase	64.32	0.00	16.68	0.19	0.00	0.04	0.41	5.25	7.05	0.00	0.05	94.01
3	ASR gel	82.69	0.10	1.20	0.70	0.06	0.18	0.85	0.64	0.53	0.00	0.05	87.02

Anorthoclase: $(Na_{0.48}, K_{0.43}, Ca_{0.02}, Fe_{0.01})_{0.94}(Al_{0.89}, Si_{3.05})_{3.98}O_8$

During migration of ASR gel along cracks from the reacted aggregate into cement paste, ASR gel gains calcium from cement paste and loses alkalis. On the Ca/Si–Ca/(Na+K) diagram, progress of ASR forms a compositional line that extends from the lower left (low Ca/Si—high alkali) to the higher right region (high Ca/Si—low alkali) heading towards the compositions of CSH gel. On the other hand, hydration of cement particles alite and belite loses calcium producing CSH gel on other two compositional lines [17, 18]. When ASR terminates, compositions of ASR gel reach the "convergent point" at which apparent chemical equilibrium is attained between ASR gel and CSH gel [17, 18]. With the core specimen studied here, compositions of ASR gel have not reached this point leaving a blanc space (Fig. 13.11). This means that ASR has not been terminated yet but is continuing. Table 13.8 identifies cristobalite and anorthoclase, ideally $(Na_{0.75}, K_{0.25})_1Al_1S_3O_8$ but with a different composition, that formed a spherulitic texture (Fig. 13.10b) in the reacted sand particle of rhyolite (Fig. 13.8c).

13.3.3 Expansion Tests

Details of the Japanese accelerated concrete core expansion tests used in this round robin test were summarized in Table 13.9. For immersion in NaOH [2, 4, 12] or in

Table 13.9 Comparison of the accelerated expansion tests performed over 50 °C

		Method	1	3
			NaOH immersion (80 °C 1N)	NaCl immersion (50 °C saturated)
		Core number	3′A	3′C
(1)	Availability of standards	Standards	Yes JCI-TC115FS proposal 5.3 [4]	Yes JCI-TC115FS proposal 5.4 [5]
			"Method of measuring ASR expansion of core samples immersed in alkali solution (Alkali-immersion method)", Japan Concrete Institute (JCI). Tokyo. 2014, committee report, pp. 316–319. (in Japanese)	"Method of measuring ASR expansion of core samples immersed in alkali solution (Saturated NaCl-immersion method)", Japan Concrete Institute (JCI). Tokyo. 2014, committee report, pp. 320–323. (in Japanese)
		Related documents	Katayama [16]: "Alkali-aggregate reaction in the vicinity of Izmir, western Turkey", *Proc. 11th ICAAR*, Quebec, Canada. pp. 365–374 Katayama et al. [2]: "Alkali-aggregate reaction under the influence of deicing salts in the Hokuriku district, Japan". *Materials Characterization*, Vol. 53 (2), pp.105-122	Nomura et al. [19]: "Case study of deterioration of concrete in road structures due to alkali-silica reaction (ASR) in Hokuriku region and its evaluation methods". *Concrete Research & Technology*, Vol. 13 (3), pp. 105–114 (in Japanese)
(2)	Applicability to large structures		RC and non-RC structures (e.g. bridges, buildings, nuclear power plants, hydraulic structures, pavements, etc.)	RC and non-RC structures undergoing ASR with supplied de-icing salt (e.g. bridge decks, abutments, piers) or sea water which may be condensed due to evaporation

(continued)

Table 13.9 (continued)

		Method	1	3
			NaOH immersion (80 °C 1N)	NaCl immersion (50 °C saturated)
		Core number	3′A	3′C
(3)	Test validation	Correlation with structure damage	Yes	Yes
		Expansion limit for field *deleteriousness*	>0.1% at 3 weeks (andesitic gravel)	>0.1% at 13 weeks (andesitic gravel)
			Katayama et al. [2]	Nomura et al. [19]
(4)	Advantages and limitations	Merit	Fast and no leaching of alkali Core diameter is small (about 5 cm): applicable to RC and PC with dense rebar configuration (e.g. girder, NPP building)	No leaching of alkali Core diameter is small (about 5 cm), applicable to RC and PC with dense rebar configuration
		Demerit	Not applicable to chert aggregate which dissolves into alkali solution and produces little expansion	Long induction period, much time is necessary to produce expansion for late-expansive aggregate (unsuitable). Replaceable by the NaOH immersion test [2], except for chert aggregate
(5)	The use of the results	Routine test of ASR potential	Many structures including bridges and nuclear power plants in Japan	Many structures under the influence of de-icing salt which produces Na concentration
		Assessment of future or maximum ASR potential	Sigmoidal expansion curve should be drawn from an extended test period (e.g. 1 year) corrected for temperature and alkali concentration (not specified in the standard)	Possible maximum expansion due to salt concentration should be accessed from a sigmoidal curve of an extended test period (e.g. 1 year) (not specified in the standard)

(continued)

Table 13.9 (continued)

		Method	1	3
			NaOH immersion (80 °C 1N)	NaCl immersion (50 °C saturated)
		Core number	3′A	3′C
(6)	Technology readiness	Applicable today at a plant?	Yes	Yes
		Performed by commercial testing laboratories?	Yes Many commercial laboratories in Japan, both private and public, have been using since around 2000	Yes Many commercial laboratories in Japan, both private and public, have been using since around 2000
		R & D needed?	Yes Determine the expansion limit for rock types other than andesite. Reaction products in high alkali and high temperature have been known similar to those in field concretes Apparent activation energy of expansion parameters (ε_∞, τc, normality of NaOH) has been determined [9]. Expansion follows the same thermo-dynamic law, even high temperature reduces viscosity of ASR gel and expansion	Yes Both ASR gel and Friedel's salt have been known to form. Effect of chloride ion (Friedel's salt and release of OH^- ion) on expansion: negligible or not Apparent activation energy of expansion parameters (ε_∞, τl, τc) should be determined
(7)	Modification used in this round-robin test			

NaCl solution [2, 5, 19], and in Table 13.10 for storage at 100% RH with [6] or without wrapping [7, 20, 21] of NaOH—soaked paper. These show the points of modification from the original standards or specifications.

Table 13.10 Comparison of accelerated tests performed at 40 °C

		Method	5	2	4
			Humid air (20, 40 °C 100% RH)	NaOH wrapping (40 °C 1N)	
		Core number	5B	3′B	5A
(1)	Availability of standards	Standards	Yes JCI-S-011-2017 “Method of measuring expansion ratio of core sampled from concrete structure”. Japan Concrete Institute (JCI). Tokyo. 2017, 10 pages. (in Japanese) [7]	No Modified from JCI-S-011-2017 Storage: core is wrapped with a non-woven paper soaked with 1N NaOH solution which is less concentrated than in JCI-TC115FS proposal 4.3 (1.5N NaOH) for concrete prism test [6]	
		Related documents	Original standard: JCI-DD2 [20] “Determination of expansion rate of AAR-damaged structure using concrete core”. Japan Concrete Institute (JCI) Tokyo. 1989, pp. 5–9. (in Japanese)	JCI-TC115FS proposal 4.3 [6] “Method of measuring ASR expansion of small concrete prism”, Japan Concrete Institute (JCI). Tokyo. 2014, committee report, pp. 272–284. (in Japanese)	
(2)	Applicability to large structures		Structures with sparse rebar configuration or mass concrete (e.g. bridges, dams and nuclear power plants)	Designed for concrete prism with total alkali Na_2Oeq 5.5 kg/m^3 after RILEM AAR-4 For structures with different alkali content, alkalinity of wrapping NaOH solution should be changed by proper calculation	
(3)	Test validation	Correlation with structure damage	Late-expansive ASR cannot be judged as deleterious with the accepted criteria	No data	

(continued)

Table 13.10 (continued)

		Method	5	2	4
			Humid air (20, 40 °C 100% RH)	NaOH wrapping (40 °C 1N)	
		Core number	5B	3′B	5A
		Expansion limit for field *deleteriousness*	>0.1% at 6 months	Not established	
			Ministry of Land, Infrastructure and Transport [21]: "Proposed guideline to repair and reinforced piers and abutments of ASR-affected road bridges". Committee on the countermeasures to ASR, 98 pages, see page 86. (in Japanese)	The use of 1.5N NaOH wrapping solution (specified in JCI-TC115FS proposal 4.3) [6] promotes expansion of aggregates but overestimates potential expansion Expansion limits of >0.1% at 6 months (core) and >0.04% at 1 year (e.g. CSA A23.2-14A, prism) are not applicable	
(4)	Advantages and limitations	Merit	Close to the natural condition (20 °C → 40 °C). For highly reactive early-expansive aggregate, liberated expansion after coring can be measured at 20 °C after coring	Close to the natural condition (40 °C), but no leaching of alkali Suitable for late-expansive ASR Fat core (5A): drying effect during measurement is small	
		Demerit	Takes time to produce expansion. Alkali-leaching and drying are likely to occur. Unsuitable for late-expansive aggregate. Large core diameter (10 cm) not applicable to RC and PC structures with dense rebar configuration	Testing a slender core (3′B) is only for comparison: gives reduced expansion due to a drying effect by evaporation, mainly during measurement	

(continued)

Table 13.10 (continued)

		Method	5	2	4
			Humid air (20, 40 °C 100% RH)	NaOH wrapping (40 °C 1N)	
		Core number	5B	3′B	5A
(5)	The use of the results	Routine test of ASR potential	Many structures including bridges and nuclear power plants in Japan to evaluate ongoing ASR expansion	Not for routine testing, no expansion limit established. Useable to evaluate ASR potential of concrete	
		Assessment of future or maximum ASR potential	Sigmoidal expansion curve should be drawn from a longer test period (e.g. 1 year), corrected for temperature, humidity and alkali concentration (not specified in the standard)	Sigmoidal expansion curve should be drawn from an extended test period (e.g. 1 year), corrected for temperature and alkali concentration (not specified in the standard)	
(6)	Technology readiness	Applicable today at a plant?	Yes	Yes	
		Performed by commercial testing laboratories?	Yes Many commercial laboratories in Japan, both private and public, have been using since before 1990 [21, 22]	Yes A few research laboratories in Japan have been using since around 2013. Confusing results (1.5N NaOH produced highly deleterious expansion of conventionally innocuous aggregates) hinders spread of this test	
		R & D needed?	Yes Expansion limit for late-expansive ASR Effect of cold-reading (dense trickling) on alkali-leaching. Apparent activation energy (ε_∞, τl, τc) of major rock type aggregates with full testing term should be determined	Yes Expansion limit of concrete cores from field structures (both deteriorated and sound) Effect of NaOH concentration (excessive dissolution of silica and ASR gel) on expansion	

(continued)

Table 13.10 (continued)

		Method	5	2	4
			Humid air (20, 40 °C 100% RH)	NaOH wrapping (40 °C 1N)	
		Core number	5B	3′B	5A
(7)	Modification used in this round-robin test		>95%RH → 100%RH Liberated expansion at 20 °C: not specified in JCI-S-011-2017 (specified in original JCI-DD2) → measured Cold reading (20 °C) → hot-reading (40 °C)	Core size: one → two for comparison Alkalinity of wrapping solution: 1.5N NaOH for prism → 1N NaOH for core Cold reading (20 °C) → hot-reading (40 °C)	

13.3.3.1 40 °C 100% RH Expansion Test

A flowchart of preconditioning and setting of zero point for 40 °C 100% RH concrete core expansion test was shown in Fig. 13.12. Judging from the weight gain of the core (Fig. 13.13b), actual equilibrium with the moisture in the humid box (20 °C 100% RH) attained 9 days after exposure. This was due to vacuum sealing of the specimen done before shipping, which produced dry surface of the core with inhomogeneous water distribution across the section.

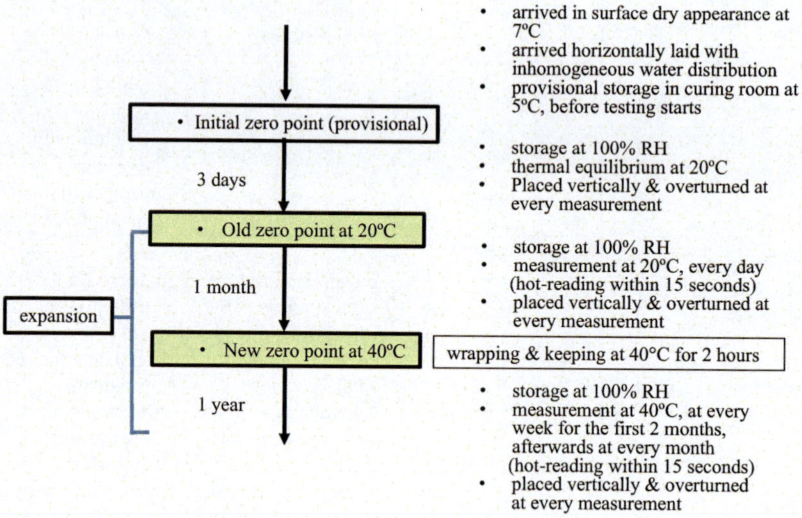

Fig. 13.12 Preconditioning and zero point for concrete core expansion test at 40 °C, 100% RH

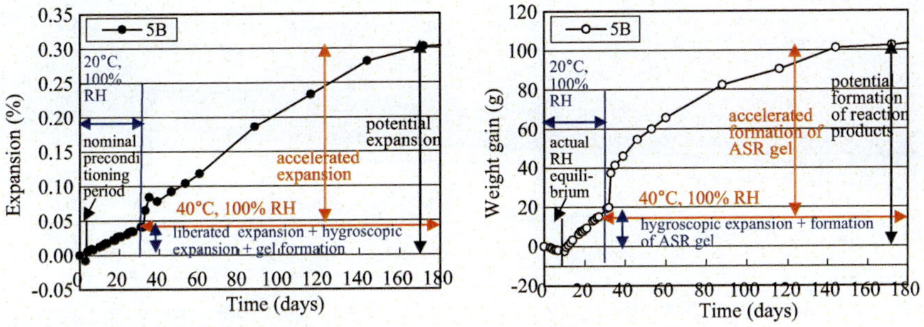

(a) Average expansion vs time
Within the first one month at 20°C, 100% RH, expansion was not stabilized but in a steady state
→ New gel formation, rather than liberated expansion or hygroscopic expansion of preexisting ASR gel

(b) Weight gain vs time
Preconditioning of 3 days is insufficient with this type of core that was vacuum sealed

Fig. 13.13 Initial stage of 40 °C, 100% RH expansion test and effect of preconditioning of core

Unlike most concrete core specimens from ASR-affected structures with highly reactive early-expansive Japanese volcanic aggregates, expansion at 20 °C 100% RH was not stabilized within the first 1 month (Fig. 13.13a). This suggests that this steady early expansion was a result of continuing gel formation, rather than liberated expansion after the extraction of core or hygroscopic expansion of preexisting ASR gel by the drilling water. After 1 month, core specimen was exposed to 40 °C 100% RH condition (Fig. 13.14e). Expansion in this test exceeded 0.3% after 200 days.

Results of the accelerated expansion tests (averaged data for each core) were shown in Fig. 13.14. Expansion curves in the axial direction (X) of core were drawn using averaged data points fitted to the sigmoidal curve of Larive equation [8] and Brunetaud equation [23].

Figure 13.15 shows raw data of expansion of core specimens measured at 10 cm spans in the axial direction (X). All the core specimens were kept vertically changing up-down directions within chambers at every measurement day, but large variations of expansion were noted within each specimen. This variation was the greatest, exceeding 5 times, with the expansion test performed in the 100% RH humid air (Fig. 13.16e).

The reason for this is (1) segregation of water within core specimens that were laid horizontally after extraction and during shipping, (2) vacuum sealing that produced intensive drying up of core surface and inhomogeneous distribution of water across the core section, and (3) inhomogeneous distribution of reactive aggregate within short measuring spans (10 cm).

13.3.3.2 80 °C 1N NaOH Immersion Test

A flowchart of preconditioning and setting of zero point for this alkali-immersion test was shown in Fig. 13.16. Core specimen in this test produced the largest expansion

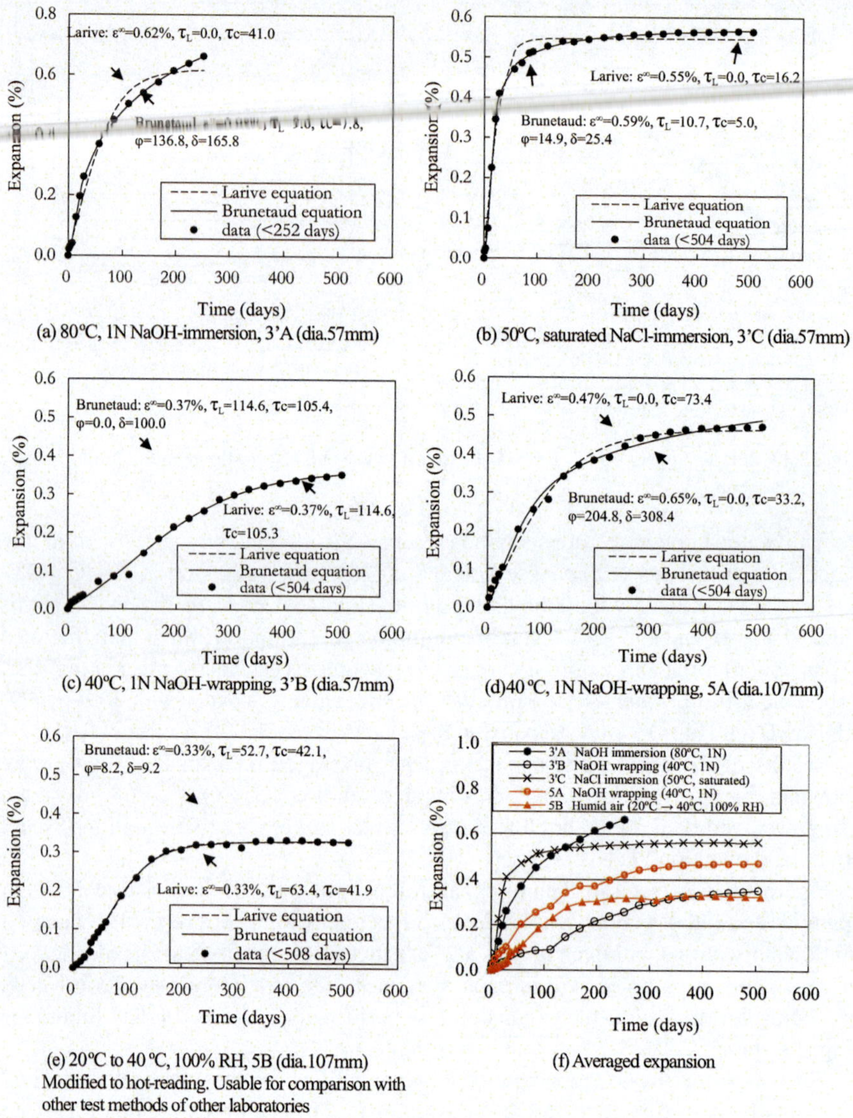

Fig. 13.14 Japanese standard expansion tests (averaged data)

among the methods tested here, exceeding 0.6% at 200 days (Fig. 13.13a). In contrast, as evident from Fig. 13.14f, the weight of the specimen was decreasing by time due to dissolution of reacted aggregate and ASR sol/gel into NaOH solution. This suggests that the concentration of 1N NaOH solution used in this test is higher than that of initial pore solution in concrete (calculated as 0.87N from Helmuth equation [24]).

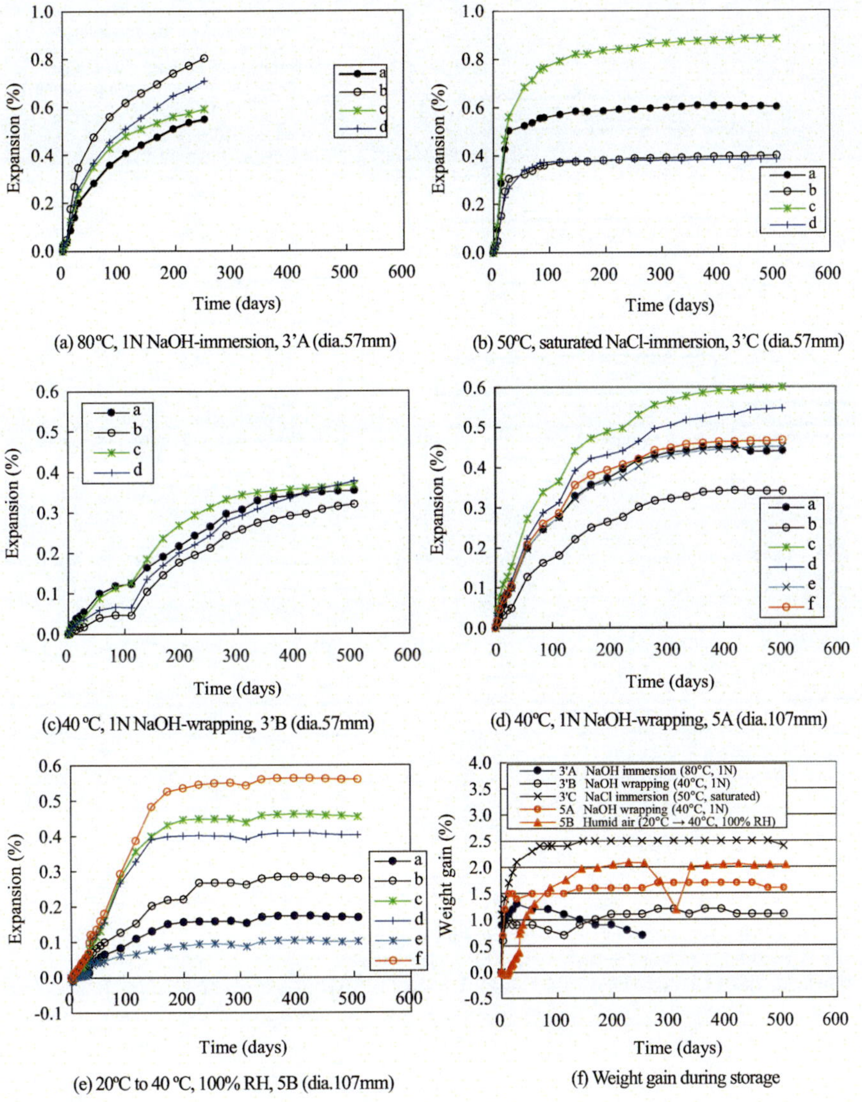

Fig. 13.15 Variations of expansion, horizontal direction (X), in 10 cm spans (raw data)

Epansion parameters obtained by fitting to Larive equation and Brunetaud expansion were shown in Table 13.11. Combining the past expansion (crack index) with the residual expansion (final expansion ε∞), overall potential expansion of concrete can be obtained by re-defining the S-shaped curves.

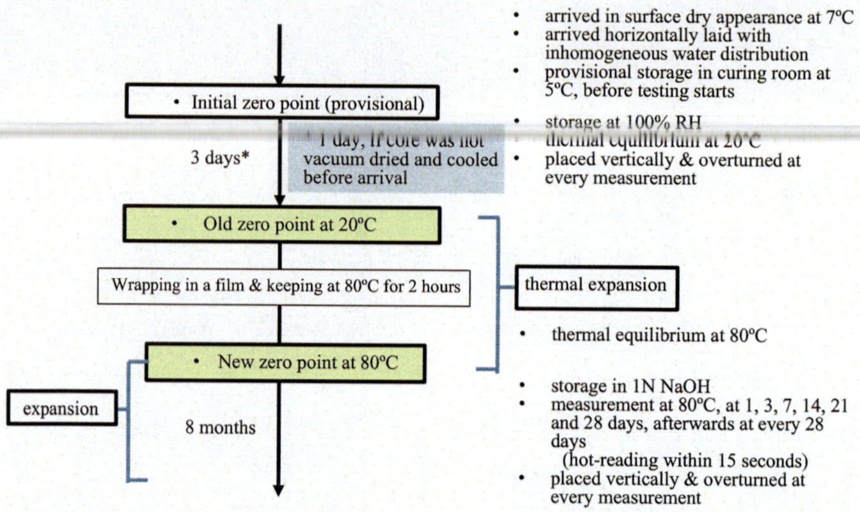

Fig. 13.16 Preconditioning and zero point for concrete core expansion test at 80 °C, 1N NaOH

Table 13.11 Expansion parameters of core and sigmoidal expansion curves fitted to two equations based on 504–508 days data (3′C, 3′B, 5A, 5B) and 252 days data (3′A)

Core specimen		Storage condition			Larive equation			Brunetaud equation				
No.	dia (cm)	T (°C)	medium	method	ε^{∞} %	τ_L days	τ_c days	ε^{∞} %	τ_L days	τ_c days	φ days	δ days
3'A	5.7	80	1N NaOH	immersion	0.62*	0.0	41.0	0.98	9.6	7.8	136.8	165.8
3'C		50	saturated NaCl		0.55	0.0	16.2	0.59	10.7	5.0	14.9	25.4
3'B		40	1N NaOH	wrapping	0.37	114.6	105.3	0.37	114.6	105.4	0.0	100.0
5A	10.7	40	1N NaOH		0.47	0.0	66.7	0.65	0.0	17.5	241.1	285.3
5B		20→40	100% RH	curing	0.33	63.4	41.9	0.33	52.7	42.1	8.2	9.2

*0.66% : processing, no less than the maximum expansion observed
Larive equation: $\varepsilon t = \varepsilon_{\infty}[1 - \exp(-t/\tau_c)]/\{1 + \exp[-(t - \tau_L)/\tau_c]\}$
Brunetaud equation: $\varepsilon t = \varepsilon_{\infty}[1 - \exp(-t/\tau_c)]\{1 - [\varphi/(t + \delta)]\}/\{1 + \exp[-(t - \tau_L)/\tau_c]\}$
Expansion parameters: ε_∞ final asymptotic expansion (%); τ_L latency time (days); τ_c characteristic time (days); φ, δ for correction (days), ($\varphi < \delta$); t time (days); ε_t expansion at time t

13.4 Simulation of Overall Expansion

Whatever the amount a concrete member had expanded in the past, overall final expansion of the concrete can be estimated based on the following simple principle: overall potential final expansion = past expansion + residual expansion.

In this report, past expansion was represented by the crack index, while residual expansion was deduced from results of the accelerated concrete core expansion tests by a thermodynamic correction. For this purpose, a set of new equations of apparent activation energy (U) was developed by Katayama [9] as a function of concentration (N) of NaOH solution, final expansion (ε_∞) and characteristic time (τ_c), based on

Table 13.12 Estimated overall expansion of concrete, based on 80 °C 1N NaOH immersion test (version 1)

Time (days)	Expansion (%)	Crack index (%)	Temperature (°C)	RH (%)	NaOH (mol/L)	Concrete	Data holder
0-107 (0.29yrs)	measured expansion $\varepsilon_0 = 0.05$*		27	65		slab	University of Tennessee
108-1183	?		21	40		slab	
0-1183 (3.24 yrs)		estimated past expansion (X,Y) $\varepsilon_1 = 0.22$	5 in refrigerator 1192-1460 days			core (dia.5.7cm) /thin section	Taiheiyo Consultant
1183-∞	Alkali-immersion test at 1N NaOH $\Delta\varepsilon = 0.66$		80		1 (immersion)	core (dia 5.7cm)	
	Conversion of expansion at 80°C, 1N NaOH → 20°C, 0.87 NaOH, $\Delta\varepsilon = 0.31$		20		0.87 (immersion)		
0-∞	Final expansion (as concrete prism) $\varepsilon_\infty = \varepsilon_1 + \Delta\varepsilon = 0.22 + 0.31 = 0.53$		independent of temperature		0.87 (pore water)	slab	
0-1183	Advancement of ASR expansion (%) (0.22/0.53)x100=42						

* Obviously, expansion $\varepsilon_0 = 0.05\%$ is higher and faster than could be expected from the exposure at 27 °C, 65% RH

expansion data from Kuroda et al. [25] with a reactive Japanese andesite (40–80 °C). With the activation energy and Arrhenius equation, main parameters of the Larive's sigmoidal expansion curve obtained at accelerated conditions were corrected to the standard state (20 °C 100% RH).

13.4.1 Storage at 100% RH

13.4.1.1 80 °C 1N NaOH Immersion Test

The effects of NaOH concentration and testing temperature on the final expansion (ε_∞) and characteristic time (τ_c, roughly reciprocal of the expansion rate) were corrected (Fig. 13.17a) by the calculation process in Table 13.13, with a result summarized in Table 13.12 as a version 1. Comparing with a steep expansion curve at 80 °C, only known expansion data at early age (0.05% at 108 days after casting the slab, storage at 27 °C, 65% RH) seemed to be higher than would be expected from the storage condition. However, overall expansion curve was drawn passing this data point (Fig. 13.17b).

- Version 1

With a steep expansion curve of the data as shown in Fig. 13.17a, Larive equation tends to underestimate final expansion. Hence, the largest expansion obtained on the last day of measurement (0.66%, 252 days) was adopted as final expansion. In this case, overall expansion of the sigmoidal curve of concrete was estimated to be 0.53% (Fig. 13.17b). For comparison, final expansion ($\Delta\varepsilon_\infty$) fitted using all expansion data to the Larive equation and overall final expansion (ε_∞) were 0.62% and 0.50%,

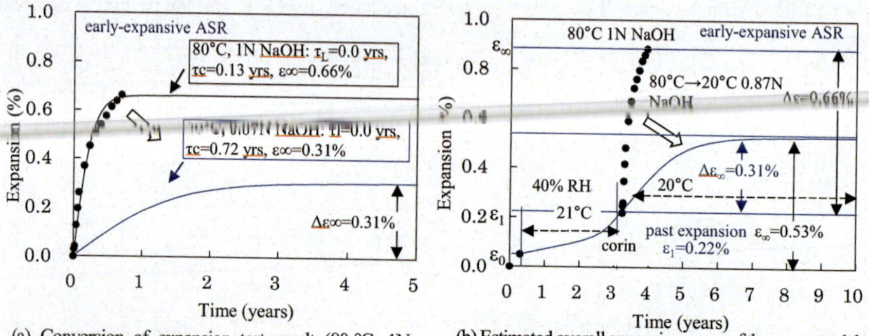

(a) Conversion of expansion test result (80 °C, 1N NaOH) to 20°C, 1N NaOH
Apparent activation energy (U) for Japanese andesite at NaOH concentration (N), calculated using data from Kuroda et al. (2008) (40°C-80°C), was used.

(b) Estimated overall expansion curve of the concrete slab
Overall expansion curve: τ_L= 3.30 years, τc= 0.72 years, $\varepsilon\infty$= 0.53%
$\Delta\varepsilon\infty$ does not change, irrespective of whether expansion curve passes ε_0 point (0.29 years, 0.05%)

Fig. 13.17 Estimated overall potential final expansion of concrete based on 80 °C 1N NaOH immersion test, assuming continuous exposure to 100% RH at 20 °C after coring Version 1

Table 13.13 Calculation process of overall expansion of concrete, based on 80 °C 1N NaOH immersion test (version 1)

1) Past expansion	ε_1 (%)	0.22	Crack index
2) Alkali-immersion test			
Final expansion at 80°C, 1N NaOH	$\Delta\varepsilon$ (%)	0.66	Fitting to Larive equation
Characteristic time	τc (years)	0.13	Fitting to Larive equation
NaOH concentration in pore water of original concrete	N (mol/L)	0.87	Helmush equation Cement Na_2Oeq=1.25%, W/C=0.5 [OH]=[NaOH]=1.25x(0.339/0.5)+0.022=0.87
Na_2O equivalent of concrete (cement Na_2Oeq1.25%, concrete 420kg/m^3)	Na_2Oeq (kg/m^3)	5.25	Helmush equation 0.87=Na_2Ox(0.339/0.5)+0.022, Na_2O=1.25%, 420x1.25/100=5.25kg/m^3
Apparent activation energy at 0.87N NaOH	U∞ (K)	1310	U∞(K) = 2950 x ln (x)+1720*, x=0.87 Based on Japanese andesite
	Uc (K)	-2940	Uc (K) = 1480 x ln (x) -2730*, x=0.87 Based on Japanese andesite
3) Conversion to 20°C, 0.87N NaOH			
Final expansion	Δε∞ (%)	0.31	Arrhenius equation Δε = 0.66x e[1310x[1/(273+80)-1/(273+20)]
Characteristic time	τc (years)	0.72	Arrhenius equation τc= 0.13x e[-2940x[1/(273+80)-1/(273+20)]
4) Overall potential final expansion of concrete at 20°C, 100% RH (wet)	ε∞ (%)	0.53	= ε_1+Δε∞ =0.22+0.31=0.53

*equations after Katayama [9]

respectively. They were essentially the same as the above estimation, hence graphs were not shown.

- Version 2

Difference in the unit alkali content in concrete corresponding to 1N NaOH pore solution and theoretical concentration of NaOH in pore solution was calculated using

Table 13.14 Calculation process of overall expansion of concrete, based on 80 °C 1N NaOH immersion test (version 2)

1) Past expansion	ε_1 (%)	0.22	Crack index
2) Alkali-immersion test			
Final expansion at 80°C, 1N NaOH	$\Delta\varepsilon$ (%)	0.62	Fitting to Larive equation
Apparent activation energy at 1N NaOH	$U\infty$ (K)	1720	Japanese andesite
Final expansion, converted to 40°C, 1N NaOH	$\Delta\varepsilon$ (%)	0.33	Arrhenius equation $\Delta\varepsilon$=0.62x e[1720x[1/(273+80)-1/(273+40)]
NaOH concentration in pore water of original concrete	N (mol/L)	0.87	Helmush equation: Cement Na_2Oeq=1.25%, W/C=0.5; [OH]=[NaOH]=1.25x(0.339/0.5)+0.022=0.87
NaOH concentration of immersion water	N (mol/L)	1.00	
3) Conversion to concrete prism			
Na_2O equivalent of concrete	Na_2Oeq (kg/m^3)	5.25	Prepared as specified (cement Na_2Oeq1.25%, concrete 420kg/m^3)
Na_2O equivalent of concrete corresponding to1N NaOH pore water	Na_2Oeq (kg/m^3)	6.05	Helmush equation 1.00=Na_2Ox(0.339/0.5)+0.022, a$_2$O=1.44%, 420x1.44/100=6.05kg/m^3
Difference in unit alkali content between original pore water and immersion water	Na_2Oeq (kg/m^3)	-0.80	5.25-6.05= -0.80
Correction of expansion for alkali difference 0.80 at 40°C (concrete prism, 100% RH)	$\Delta\infty$ (%)	-0.05	Empirical relationship: Δ=(0.063xNa_2Oeq(1)-0.16) -(0.063xNa_2Oeq(2)-0.16)=0.063x(-0.80)=-0.05
Potential expansion at 40°C (concrete prism, 100% RH)	$\Delta\varepsilon\infty$ (%)	0.28	0.33-0.05=0.28
4) Overall potential final expansion of concrete at 20°C (100% RH, independent of temperature)	$\varepsilon\infty$ (%)	0.50	=ε_1+$\Delta\varepsilon\infty$=0.22+0.28=0.50

Helmush equation [24]. Then, expansion of concrete was corrected based on empirical relationship between the difference in unit alkali content and expansion of concrete prism (Table 13.14). Because essentially the result as that from the version 1 was obtained, graphs were not illustrated here.

13.4.1.2 40 °C 1N NaOH Alkali-Wrapping Test

This test, modified from original 1.5N to 1N NaOH solution and from cold to hot-reading, presented expansion stabilizing after 400 days. Final expansion ($\Delta\varepsilon_\infty$) estimated by the Larive equation based on two testing periods (335 days Fig. 13.18a), and 504 days Fig. 13.18c) was essentially the same, resulted in almost the same final expansion (ε_∞) of 0.56–0.57% (Fig. 13.18b, d), irrespective of whether the expansion curve passes ε_0 point (0.29 years, 0.05%) (Table 13.15) or not.

13.4.1.3 40 °C 100% RH Expansion Test

With this test, expansion parameters (τ_L, τ_c) obtained at 40 °C were converted to 20 °C, using the apparent activation energy of a reactive Texas sand (El Paso) calculated here based on expansion data from Stacey et al [26] at a pessimum mix diluted by non-reactive limestone aggregate: $U_L = -7450$ K, $Uc = -5730$ K (38–60 °C). Calculation process for 509 days was given in Tables 13.16, 13.17. Because expansion of the core specimen has been stabilized after 200 days, forming a horizontal asymptotic line at final expansion ($\Delta\varepsilon$) = 0.33% (Fig. 13.19a, c), overall expansion of the concrete

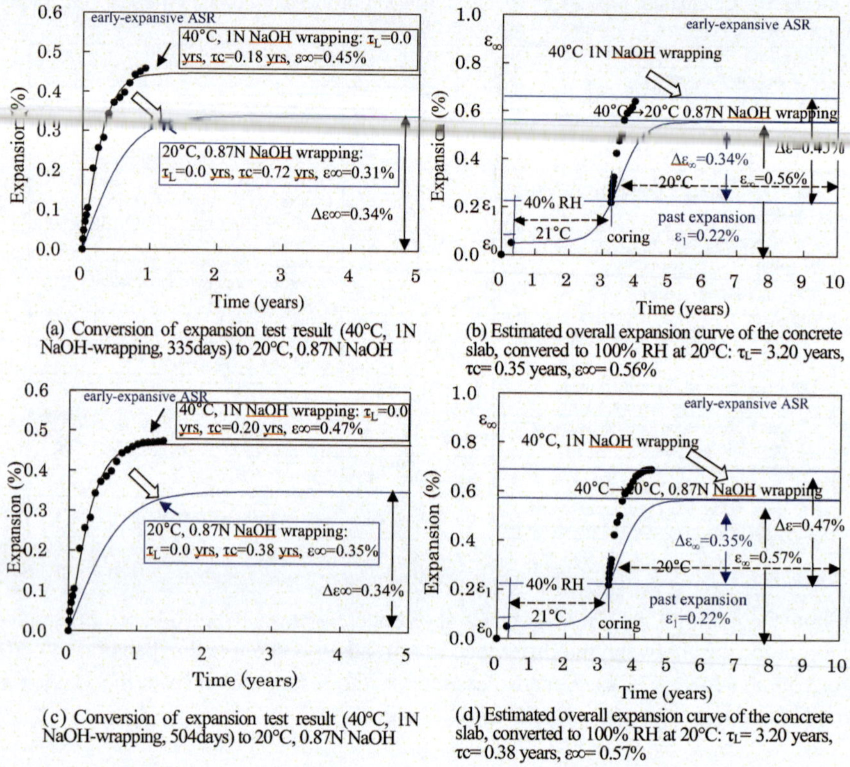

Fig. 13.18 Estimated overall potential final expansion of concrete, based on 40 °C, 1N NaOH-wrapping expansion test (dia = 10.7 cm) (a), (b) 335 days, (c), (d) 504 days, assuming continuous exposure to 20 °C 100% RH after coring

Table 13.15 Calculation of overall expansion of concrete, based on 40 °C, 1N NaOH-wrapping test (dia = 10.7 cm, 504 days)

1) Past expansion	ε_1 (%)	0.22	Crack index
2) Alkali-wrapping test			
Final expansion at 40°C, 1N NaOH-wrapping	Δε (%)	0.47	Fitting to Larive equation
Characteristic time	τc (years)	0.20	Fitting to Larive equation
NaOH concentration in pore water of original concrete	N (mol/L)	0.87	Helmush equation Cement Na_2Oeq=1.25%, W/C=0.5 [OH]=[NaOH]=1.25x(0.339/0.5)+0.022=0.87
Apparent activation energy at 0.87N NaOH	U∞ (K)	1310	U∞(K) = 2950 x ln (x)+1720*, x=0.87 Based on Japanese andesite
	Uc (K)	-2940	Uc (K) = 1480 x ln (x) -2730*, x=0.87 Based on Japanese andesite
3) Conversion to 20°C, 0.87N NaOH			
Final expansion	Δε∞ (%)	0.35	Arrhenius equation Δε = 0.47x e[1310x[1/(273+40)-1/(273+20)]
Characteristic time	τc (years)	0.38	Arrhenius equation τc= 0.20x e[-2940x[1/(273+40)-1/(273+20)]
4) Overall potential expansion of concrete at 20°C, 100% RH (wet)	ε∞ (%)	0.57	$=\varepsilon_1+\Delta\varepsilon_\infty$ =0.22+0.35=0.57

*equations after Katayama [9]

Table 13.16 Estimated overall expansion of concrete, based on 40 °C, 100% RH expansion test (508 days)

Time (days)	Expansion (%)	Crack index (%)	Temperature (°C)	RH (%)	Concrete	Data holder
0-107 (0.29yrs)	measured expansion $\varepsilon_0 = 0.05$*		27	65	slab	University of Tennessee
108-1183	?		21	40	slab	
0-1183 (3.24 yrs)		estimated past expansion (X,Y) $\varepsilon_1 = 0.22$	5 in refrigerator 1192-1460 days		core (dia.5.7cm) /thin section	Taiheiyo Consultant
1183-∞	Accelerated test $\Delta\varepsilon = 0.33$**		20→ 40	100	core (dia 10.7cm)	
0-∞	Final expansion $\varepsilon\infty = \varepsilon_1 + \Delta\varepsilon = 0.22+0.33=0.55$		independent of temperature	100	slab	
0-1183	Advancement of ASR expansion (%) (0.22/0.54)x100=41					

Table 13.17 Calculation process of overall expansion of concrete, based on 40 °C, 100% RH expansion test (508 days)

1) Past expansion	ε_1 (%)	0.22	Crack index
2) 40°C 100% RH test			
Final expansion at 40°C, 100% RH	$\Delta\varepsilon$ (%)	0.33	Fitting to Larive equation
Latency time	τ_L (years)	0.16	Fitting to Larive equation
Characteristic time	τ_C (years)	0.12	Fitting to Larive equation
Apparent activation energy at 20°C, 100% RH	U_L (K)	-7540	Based on Texas (El Paso) sand
	U_C (K)	-5730	Based on Texas (El Paso) sand
3) Conversion to 20°C, 100% RH			
Final expansion	$\Delta\varepsilon\infty$ (%)	0.33	Independent of temperature
Latency time	τ_L (years)	0.82	Arrhenius equation $\tau_C = 0.16 \times e[-7540 \times [1/(273+40)-1/(273+20)]$
Characteristic time	τ_C (years)	0.42	Arrhenius equation $\tau_C = 0.12 \times e[-5730 \times [1/(273+40)-1/(273+20)]$
4) Overall potential expansion of concrete at 20°C, 100% RH	$\varepsilon\infty$ (%)	0.55	$= \varepsilon_1 + \Delta\varepsilon\infty = 0.22+0.33=0.55$

(ε_∞) estimated from the expansion up to 339 days (Fig. 13.19b) and up to 508 days (Fig. 13.19d) was essentially the same, 0.55%.

13.4.2 Storage at 40% RH

13.4.2.1 80 °C 1N NaOH Immersion Test

Original expansion curve (Fig. 13.17) was converted to 40% RH (Fig. 13.20a). As a result, residual expansion after the extraction of core was reduced to 1/7, and overall expansion to 0.26% (Table 13.18, Fig. 13.20b).

13.4.2.2 40 °C 1N NaOH Alkali-Wrapping Test

Final expansion of concrete (ε_∞) when exposed in a dry room (40% RH, 20 °C) was calculated, based on accelerated expansion test at 40 °C 1N NaOH alkali-wrapping (Tables 13.19, 13.20, Fig. 13.21a), which resulted in 0.27% (Fig. 13.21b).

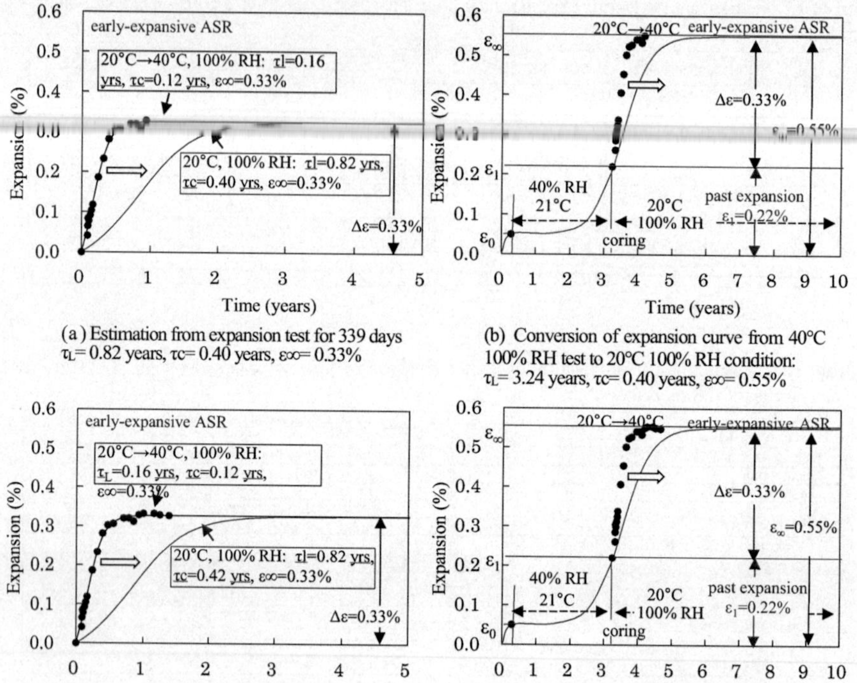

(a) Estimation from expansion test for 339 days τ_L= 0.82 years, τ_C= 0.40 years, ε_∞= 0.33%

(b) Conversion of expansion curve from 40°C 100% RH test to 20°C 100% RH condition: τ_L= 3.24 years, τ_C= 0.40 years, ε_∞= 0.55%

(c) Estimation from expansion test for 508 days τ_L= 0.82 years, τ_C= 0.42 years, ε_∞= 0.33%

(d) Conversion of expansion curve from 40°C 100% RH test to 20°C 100% RH condition: τ_L= 3.24, τ_C= 0.42 years, ε_∞= 0.55%

Fig. 13.19 Estimated overall potential final expansion of concrete based on 40 °C, 100% RH core expansion test (dia =10.7 cm): (a), (b) 339 days, (c), (d) 508 days, assuming continuous exposure to 20 °C 100% RH after coring

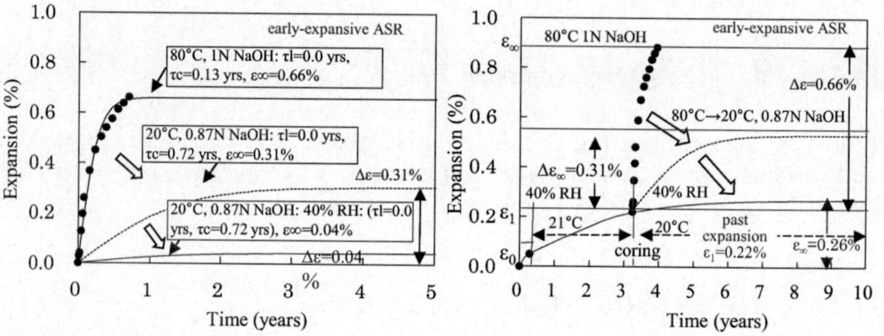

(a) Conversion of expansion curve from 80°C, 1N NaOH to 20°C, 0.87N NaOH 40% RH: test for 252 days

(b) Overall expansion curve of concrete exposed to 20°C, 40% RH: τ_L= 0.0 years, τ_C= 1.50 years, ε_∞=0.26%

Fig. 13.20 Estimated overall potential final expansion of concrete based on 80 °C, 1N NaOH-immersion test (dia = 5.7 cm): (a) 252 days, (b) assuming continuous exposure to 20 °C 40% RH after coring

Table 13.18 Calculation process of overall expansion of concrete at 40% RH, based on 80 °C, 1N NaOH expansion test

1) Past expansion	ε_1 (%)	0.22	Crack index
2) Alkali-immersion test			
Final expansion at 80°C, 1N NaOH	$\Delta\varepsilon_{\infty 1}$ (%)	0.66	Fitting to Larive equation
Characteristic time	τc (years)	0.13	Fitting to Larive equation
NaOH concentration in pore water of original concrete	N (mol/L)	0.87	Helmush equation Cement Na_2Oeq=1.25%, W/C=0.5 [OH]=[NaOH]=1.25x(0.339/0.5)+0.022=0.87
Apparent activation energy at 0.87N NaOH	$U\infty$ (K)	1310	$U\infty$(K) = 2950 x ln (x)+1720*, x=0.87 Based on Japanese andesite
	Uc (K)	-2940	Uc (K) = 1480 x ln (x) -2730*, x=0.87 Based on Japanese andesite
3) Conversion to 20°C, 0.87N NaOH			
Final expansion	$\Delta\varepsilon_{\infty 2}$ (%)	0.31	Arrhenius equation $\Delta\varepsilon_{\infty 2}$ = 0.66x e[1310x[1/(273+80)-1/(273+20)]
Characteristic time	τc (years)	0.72	Arrhenius equation τc= 0.13x e[-2940x[1/(273+80)-1/(273+20)]
4) Conversion to 40% RH			
Final expansion	$\Delta\varepsilon_{\infty 3}$ (%)	0.04	Equation by Katayama based on data by Poet et al.(2004)* $\Delta\varepsilon_{\infty 3}$=$\Delta\varepsilon_{\infty 2}$ x e[-3.28(H_1-H_2)]=0.31x e[-3.28(1-0.4)]
5) Overall potential final expansion of concrete at 20°C, 40% RH (wet)	ε_∞ (%)	0.26	= ε_1+$\Delta\varepsilon_{\infty 3}$ =0.22+0.04=0.26

Table 13.19 Estimated overall expansion of concrete, based on 40 °C, 1N NaOH-wrapping expansion test (508 days)

Time (days)	Expansion (%)	Crack index (%)	Temperature (°C)	NaOH (N)	RH (%)	Concrete	Data holder
0-107 (0.29yrs)	measured expansion ε_0 = 0.05		27		65	slab	University of Tennessee
108-1183	?		21		40	slab	
0-1183 (3.24 yrs)		estimated past expansion (X,Y) ε_1=0.22	5 in refrigerator 1192-1460 days			core (dia.5.7cm) /thin section	Taiheiyo Consultant
1183-∞	Accelerated test $\Delta\varepsilon\infty$=0.47		40		100	core (dia 10.7cm)	
	Conversion $\Delta\varepsilon\infty$=0.35		40 →20	1→0.87	100		
	Conversion $\Delta\varepsilon\infty$=0.05		20		100→40		
0-∞	Final expansion $\varepsilon\infty$ = ε_1+$\Delta\varepsilon$ =0.22+0.05=0.27		independent of temperature		40	slab	
0-1183	Advancement of ASR expansion (0.22/0.27)x100=81(%)						

13.4.2.3 40 °C 100% RH Expansion Test

Finally, overall expansion curve of the concrete slab at a continuing dry condition (40% RH) was assessed. With Larive equation containing three unknown parameters, the shape of the sigmoidal curve can be determined using two data points (ε_0, ε_1) and final expansion ($\Delta\varepsilon$). To convert the final expansion ($\Delta\varepsilon$) 0.33 at 100% RH (Fig. 13.19c, d) (Tables 13.16, 13.17) to 40% RH (Tables 13.21, 13.22), equation developed by Katayama using data from Poyet et al. [27] was used. For the latency time (τ_L) and characteristic time (τ_c), there has been no good data in the published references for formulation, but they can be ignored because only necessary data is final expansion ($\Delta\varepsilon_{\infty 2}$) at 40% RH (=0.05%) (Fig. 13.22a) in fitting a new S-curve (Fig. 13.22b).

Table 13.20 Calculation process of overall expansion of concrete at 40% RH, based on 40 °C 100% RH expansion test

1) Past expansion	ε_1 (%)	0.22	Crack index
2) Alkali-wrapping test			
Final expansion at 40°C, 1N NaOH-wrapping	$\Delta\varepsilon_{\infty 1}$ (%)	0.47	Fitting to Larive equation
NaOH concentration in pore water of original concrete	N (mol/L)	0.87	Helmush equation Cement Na_2Oeq=1.25%, W/C=0.5 [OH]=[NaOH]=1.25x(0.339/0.5)+0.022=0.87
Latency time	τ_L (years)		Fitting to Larive equation
Characteristic time	τc (years)	0.20	Fitting to Larive equation
Apparent activation energy at 0.87N NaOH	U_∞ (K)	1310	U∞(K) = 2950 x ln (x)+1720*, x=0.87 Based on Japanese andesite
	Uc (K)	-2940	Uc (K) = 1480 x ln (x) -2730*, x=0.87 Based on Japanese andesite
3) Conversion to 20°C, 0.87N NaOH			
Final expansion	$\Delta\varepsilon_{\infty 2}$ (%)	0.35	Arrhenius equation $\Delta\varepsilon_{\infty 2}$ = 0.47x e[1310x[1/(273+40)-1/(273+20)]
Characteristic time	τc (years)	0.38	Arrhenius equation τc= 0.20x e[-2940x[1/(273+40)-1/(273+20)]
4) Conversion to 20°C, 40% RH			
Final expansion	$\Delta\varepsilon_{\infty 3}$ (%)	0.05	Equation by Katayama based on data by Poet et al.(2004)* $\Delta\varepsilon_{\infty 3}=\Delta\varepsilon_{\infty 2}$ x e[-3.28(H_1-H_2)]=0.35x e[-3.28(1-0.4)]
5) Overall potential expansion of concrete at 20°C, 100% RH	ε_∞ (%)	0.27	= $\varepsilon_1+\Delta\varepsilon_{\infty 3}$ =0.22+0.05=0.27

*equations after Katayama [9]

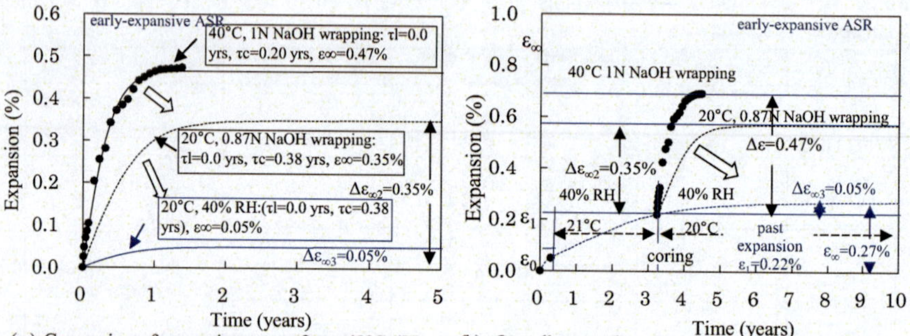

(a) Conversion of expansion curve from 40°C 1N NaOH-wrapping to 20°C, 40% RH: test for 508 days

(b) Overall expansion curve of concrete exposed to 20°C, 40% RH: τ_L= 0.0 years, τc= 1.45 years, ε_∞= 0.27%

Fig. 13.21 Estimated overall potential final expansion of concrete based on 40 °C, 1N NaOH-wrapping test (dia = 10.7 cm): (a) 504 days, (b) assuming continuous exposure to 20 °C 40% RH after coring

Table 13.21 Estimated overall expansion of concrete, based on 40 °C, 100% RH expansion test (508 days)

Time (days)	Expansion (%)	Crack index (%)	Temperature (°C)	RH (%)	Concrete	Data holder
0-107 (0.29yrs)	measured expansion ε_0=0.05		27	65	slab	University of Tennessee
108-1183	?		21	40	slab	
0-1183 (3.24 yrs)		estimated past expansion (X,Y) ε_1=0.22	5 in refrigerator 1192-1460 days		core (dia.5.7cm) /thin section	Taiheiyo Consultant
1183-∞	Accelerated test $\Delta\varepsilon_{\infty 1}$=0.33		20→ 40	100	core (dia 10.7cm)	
	Conversion $\Delta\varepsilon_{\infty 2}$=0.05		40→20	100→40		
0-∞	Final expansion ε∞ = $\varepsilon_1+\Delta\varepsilon$ =0.22+0.05=0.27		independent of temperature	40	slab	
0-1183	Advancement of ASR expansion (%) (0.22/0.27)x100=81					

Table 13.22 Calculation process of overall expansion of concrete at 40% RH, based on 40 °C 100% RH expansion test

1) Past expansion	ε_1 (%)	0.22	Crack index
2) 40°C 100% RH test			
Final expansion at 40°C, 100% RH	$\Delta\varepsilon$ (%)	0.33	Fitting to Larive equation
Latency time	τ_L (years)	0.16	Fitting to Larive equation
Characteristic time	τ_C (years)	0.12	Fitting to Larive equation
Apparent activation energy at 20°C, 100% RH	U_L (K)	-7540	Based on Texas (El Paso) sand
	U_C (K)	-5730	Based on Texas (El Paso) sand
3) Conversion to 20°C, 100% RH			
Final expansion	$\Delta\varepsilon_{\infty 1}$ (%)	0.33	Independent of temperature
Latency time	τ_L (years)	0.82	Arrhenius equation τ_C= 0.16x e[-7540x[1/(273+40)-1/(273+20)]
Characteristic time	τ_C (years)	0.42	Arrhenius equation τ_C= 0.12x e[-5730x[1/(273+40)-1/(273+20)]
4) Conversion to 20°C, 40% RH			
Final expansion	$\Delta_{\varepsilon\infty 2}$ (%)	0.05	Equation by Katayama based on data by Poet et al.(2004)* $\Delta\varepsilon_{\infty 2}$=$\Delta\varepsilon_{\infty 1}$ x e[-3.28(H_1-H_2)]=0.33x e[-3.28(1-0.4)]
5) Overall potential expansion of concrete at 20°C, 100% RH	$\varepsilon\infty$ (%)	0.27	= ε_1+$\Delta\varepsilon\infty$ =0.22+0.05=0.27

*equations after Katayama [9]

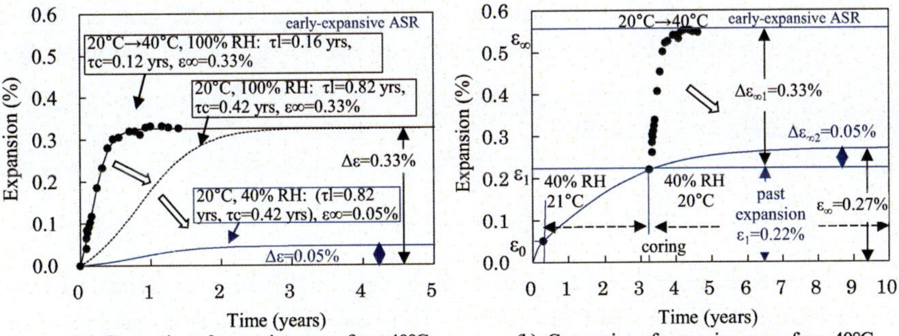

(a) Conversion of expansion curve from 40°C 100% RH test to 20°C, 40% RH condition: test for 508 days

(b) Conversion of expansion curve from 40°C 100% RH test to 20°C 40% RH condition: τ_L= 0.0 years, τ_C= 1.45 years, ε_∞= 0.27%

Fig. 13.22 Estimated overall potential final expansion of concrete based on 40 °C, 100% RH core expansion test (dia = 10.7 cm): (a) 508 days, (b) assuming continuous exposure to 20 °C, 40% RH after coring

13.5 Summary

1. Vacuum sealing and horizontal placement of core specimens before shipping resulted in dry core surface with inhomogeneous distribution of internal water, responsible for the great variations in expansion at 100% RH test.
2. The above shortcomings were not compensable by the preconditioning of core specimens within specified 3 days.
3. Storage conditions of concrete slab until extraction of the core (65% RH and mostly 40% RH by record) were not suitable for promoting expansion of ASR.
4. All the five accelerated core expansion tests produced early expansion, resembling core samples that contain highly reactive volcanic aggregates in Japan:

- dia. 107 mm (40 °C 100% RH; 40 °C 1N NaOH-wrapping),
- dia. 57 mm (40 °C 1N NaOH-wrapping; 50 °C saturated NaCl immersion; 80 °C 1N NaOH immersion).

5. Estimated past expansion before extraction of core samples (age 3.24 years), based on the crack index of the core stored in refrigerator to stop ASR (5 °C for 9 months) was
   - 0.22% in horizontal directions (X) and (Y)
   - 0.28% in vertical direction (Z), about 30% larger than those in the horizontal directions.
6. Overall final expansion of concrete (ε_∞) was estimated from three kinds of the accelerated concrete core expansion tests, aided by the thermodynamic conversions of the Larive's expansion parameters to the standard state.
7. Despite the difference in the testing conditions and correction methods, similar S-shaped expansion curves with similar final expansion were obtained. When the continuation of a humid environment of 100% RH at 20 °C is assumed:
   - 80 °C 1N NaOH-immersion test: $\varepsilon_\infty = 0.50$–53% (from 8 months data, in two correction methods)
   - 40 °C 1N NaOH-wrapping test: $\varepsilon_\infty = 0.56$–0.57% (from 11 to 16 months data)
   - 20 °C → 40 °C 100% RH test: $\varepsilon_\infty = 0.55\%$ (from 11 to 16 months data).
8. Likewise, as a more realistic condition to the EPRI laboratory, when the continuation of a dry environment of 40% RH at 20 °C is assumed, three different test methods also gave similar S-shaped curves with highly reduced expansion:
   - 80 °C 1N NaOH-immersion test: $\varepsilon_\infty = 0.26\%$ (from 8 months data)
   - 40 °C 1N NaOH-wrapping test: $\varepsilon_\infty = 0.27\%$ (from 16 months data)
   - 20 °C → 40 °C 100% RH test: $\varepsilon_\infty = 0.27\%$ (from 16 months data).
9. These suggest that the proposed thermodynamic approach was valid, even the viscosity of ASR gel would change considerably in the temperature range 20–80 °C and the solubility of ASR gel and silica minerals into NaOH solution would vary dramatically depending on the concentration of NaOH used.
10. Reacted aggregates in concrete were volcanic rocks mainly rhyolites containing highly reactive cristobalite and tridymite, indicative of marked pessimum proportion.
    - Sand from El Paso (Jobe), Texas: rhyolites, dacite and andesite contained early-expansive opal or cristobalite, and moderate-expansive chalcedony or rhyolitic glass
    - Gravel from Bernalillo, New Mexico: andesite and late-expansive quartzite with microcrystalline quartz
    - Crushed stone from Wells, Maine: hornfels containing late-expansive microcrystalline quartz.

11. Volcanic rocks in the El Paso (Jobe) sand and Bernalillo gravel were responsible for the early expansion due to pessimum phenomenon.
12. According to petrography under the polarizing microscope and SEM, the progress of ASR in the core specimen with the past expansion of 0.22% (stopped by the storage in 5 °C refrigerator) was stage iv, and not on the maximum.
13. Compositions of ASR gel in the core specimen with the past expansion of 0.22% have not reached the "convergent point" with those of CSH gel. This suggests that ASR is continuing with an ability of further expansion.
14. In terms of expansion based on three different expansion tests, advancement of ASR in concrete at the time of coring (0.22% 3.24 years) against the final expansion was
    - Humid condition (20 °C, 100% RH): 39–42%
    - Dry condition (20 °C, 40% RH): 81–84%.
15. Concrete slab has a larger thickness (20 cm) than usual core and prism specimens, hence the loss of moisture at 40% RH may be slower and residual expansion could be larger than that estimated above. Also, considering that the crack index tends to underestimate expansion (by 0.04%), overall expansion of the slab at EPRI may exceed 0.3%.

13.6 Conclusions

The answer to the questions on the ability of assessing the swelling potential and the residual free expansion of concrete extracted from ASR-affected structure is summarized in the blanket below, as well as in Tables 13.23 and 13.24.

Table 13.23 Estimated overall expansion curves fitted to Larive equation at 20 °C, assuming exposure to 100% RH

Core specimen		Japanese accelerated tests and storage condition			Expansion		Past expansion	Overall expansion parameters at 20°C			Advancement of expansion (%)	Parameters used for thermodynamic corrections (overall final expansion = past expansion + residual expansion)
No.	dia (cm)	T (°C)	medium	method	days	ε	ε_1	ε_∞	τ_L	τ_c	$\varepsilon_1/\varepsilon_\infty$	
3'A	5.7	80	1N NaOH	immersion	252	0.66	0.22*	0.53	3.30	0.72	42	activation energy (U) for ε_∞, τ_c, NaOH concentration**
						0.62		0.51	3.30	0.63	43	
3'C		50	saturated NaCl		335	0.56						to be analyzed for immersion in NaCl solution
					504	0.57						
3'B		40	1N NaOH	wrapping	335	0.32						to be analyzed for the influence of drying up of slender core
					504	0.35						
5A	10.7	40	1N NaOH		335	0.45		0.56	3.20	0.35	39	activation energy (U) for ε_∞, τ_c, NaOH concentration**
					504	0.47		0.57	3.23	0.38	39	
5B		20 → 40	100% RH	curing	339	0.33		0.55	3.24	0.40	40	activation energy (U) for τ_L, τ_c at 100% RH***
					508	0.33		0.55	3.24	0.42	40	

*crack index, **Katayama [9] for Japanese andesite, ***calculated by Katayama for Texas (El Paso) sand

Table 13.24 Estimated overall expansion curves fitted to Larive equation at 20 °C, assuming exposure to 40% RH

Core specimen		Japanese accelerated tests and storage condition			Expansion		Past expansion	Overall expansion parameters at 20°C			Advancement of expansion (%)	Parameters used for thermodynamic corrections
No.	dia (cm)	T (°C)	medium	method	days	ε	ε_1	ε_∞	τ_L	τ_c	$\varepsilon_1/\varepsilon_\infty$	
3'A	5.7	80	1N NaOH	immersion	252	0.66		0.26	0.0	1.50	85	activation energy (U) for ε_∞, τ_c, NaOH concentration
5A	10.7	40	1N NaOH	wrapping	504	0.47	0.22*	0.27	0.0	1.45	81	activation energy (U) for ε_∞, τ_c, NaOH concentration
5B		20 → 40	100% RH	curing	508	0.33		0.27	0.0	1.45	81	activation energy (U) for τ_L τ_c at 100% RH

Answer to challenges Three independent accelerated expansion tests of core samples, coupled with petrographic determination of the crack index, gave the following similar conclusions based on the thermodynamic approach.

- When exposed to 100% RH
  (1) Advancement of ASR at time of coring: 39–42%
  (2) Past expansion, residual expansion and overall expansion: 0.22%, 0.31–0.34% and 0.53–0.57%, respectively.
  (3) Expansion parameters: $\varepsilon_\infty = 0.53$–$0.57\%$, $\tau_L = 3.2$–3.3 years, $\tau_c = 0.35$–0.72 years
- When exposed to 40% RH
  (1) Advancement of ASR at time of coring: 81–85%
  (2) Past expansion, residual expansion and overall expansion: 0.22%, 0.04–0.05% and 0.26–0.27%, respectively
  (3) Expansion parameters: $\varepsilon_\infty = 0.26$–$0.27\%$, $\tau_L = 0.0$ years, $\tau_c = 1.45$–1.50 years.

Of the Larive's three expansion parameters determined here (ε_∞, τ_L, τ_c), latency time (τ_L) is most likely subject to error, because information on the expansion of the concrete slab was limited to an early age (0.29 years, 0.05%).

As a first version of the solution, continuous exposure to 100% RH was assumed. In this case, a downward convex curve was drawn, reflecting a steep expansion curve at the beginning of the accelerated testing, i.e. small characteristic time (τ_c).

However, before the extraction of core, this slab had been exposed to a dry environment (40% RH, 21 °C, 3.24 years) unfavorable to promote ASR. In general, expansion of ASR concrete at 100% RH decreases to 1/7 when exposed to 40% RH. Hence the second version assumed a dry environment in which expansion curve of the slab before coring was upward convex with gradually decreasing expansion rate. In any case, early expansion of the slab (0.29 years, 0.05%) was anomalously high with this age and temperature, hence it was used as a virtual zero point in fitting to Larive equation to draw an overall S-shaped expansion curve. This curve should be refined after knowing real early expansion data.

13.7 Recommendations and Suggestions

1. Coring and shipping: record the directions (vertical Z, horizontal X, and horizontal Y) on core specimen before it is extracted from the concrete structure. Put the core specimens vertically, not horizontally, after the extraction and in the shipping box.
2. Preconditioning: do not apply vacuum sealing. Instead, wrap the core specimen tightly with a cling film, expelling air bubbles under the film to keep a wet surface of the core. If core specimens can reach the testing laboratory within two days after coring by courier, and if testing can be started immediately, cooling process of the core specimens to stop ASR and re-conditioning of the sample temperature may not be necessary.
3. Early measurement of expansion: if the concrete contains highly reactive volcanic aggregates or opal-bearing rocks, pop-outs may occur within a few days after extraction of core. In this case, measurement of expansion can be started immediately after coring or upon receipt, so that not to miss early expansion due to accumulated expansion pressure of ASR gel within concrete. This is practiced in Japan where these aggregates are common.
4. Crack index: this should be measured in three directions (Z,X,Y) to check if there is a heterogeneous expansion in the concrete structure [3].
5. Concrete slab and crack index: check whether the estimated past expansion by the crack index in this report was correct or not by knowing the actual expansion measured by UTK. Also check the actual shape of the expansion curve of the concrete slab in the UTK data for the period 0.29–3.24 years before the extraction of core. Redraw a new S-shaped overall expansion curve using the actual expansion data obtained by UTK.
6. Expansion test and crack index: when accelerated expansion tests at THC are finished, crack index of core specimens should be measured to compare with the actual expansion data obtained in these tests.
7. Petrographic examination should be performed to check the progress of ASR in the test specimens.
    - Petrographic stage of ASR, whether it has reached the advanced stages v or vi
    - Progress of ASR of the late-expansive aggregate in concrete (quartzite, hornfels).
    - Any prominent changes in the morphology of ASR gel, e.g. crystallization into rosette.
8. Compositions of ASR gel: the following aspects should be clarified by EDS analysis.
    - Check if the compositional line of ASR gel in the core specimens tested after the long-term accelerated tests (>500 days) was closer to the "convergent point" with CSH gel than was in the core specimen examined in this report (0 day, after coring)

- Absorption of Na ions by ASR gel/rosettes. Compare the accessibility to alkali solution from the surrounding media (1N NaOH immersion test, 1N NaOH alkali-rapping tests) and any shift of the compositional line.
- Examine whether rosette crystals are products of drying up of ASR gel or not, i.e. whether they are concentrated in the dried part of concrete (surface of the alkali-wrapping slender core).

9. Alkali-budgets in concrete: check the following aspects with the core specimens after finishing the expansion tests.
   - How much water-soluble alkalis from the aggregate and leaching of alkalis from concrete core specimen affect the expansion of concrete (40 °C 100% RH test)
   - Contribution of water-soluble alkalis from 1N NaOH solution (alkali-immersion test, alkali-wrapping test).
10. Formulation of activation energy: develop new equations to estimate apparent activation energy (U) of final expansion (ε_∞), latency time (τ_L) and characteristic time (τ_c).
    - Saturated NaCl-immersion expansion test
    - Alkali-wrapping expansion test of slender core. Additionally, take the influence of drying (relative humidity or saturation of water) into consideration.

Acknowledgements The first author is grateful to Mr. Ammar Elhassan, Dr. Sihem Le Pape and Prof. John Ma of the University of Tennessee Knoxville for proving the concrete and aggregate samples. Thanks also go to Dr. Yann Le Pape, formerly at the Oak Ridge National Laboratory, and Dr. Maria Guimaraes of the Electric Power Research Institute Charlotte for arranging and backing up this round robin test. Prof. Victor Saouma of the University of Colorado Boulder, chairman of the RILEM TC 259-ISR, is greatly acknowledged for proposing this round robin test in WG-1, as well as for providing the stored core specimens for a comparative petrographic study. All the testing and research work performed at Taiheiyo Consultant Co., Ltd. in this round robin program were financially supported by this company in Sakura, Japan.

Appendix: Petrographic Examination of Accompanying Core Specimens

A1 Core Samples

Two half cut core specimens (No. 6A, No. 6B) from the concrete slab, supplied by UCB in December 2017, were subjected to petrographic examination (Fig. 13.23). One of these specimens (No.6A, stored at around 30 °C) was a longitudinal half cylinder, unfortunately cut oblique to the vertical direction (Fig. 13.24a, b) that was estimated following the Katayama's method (Fig. 13.24c). Hence, to measure the

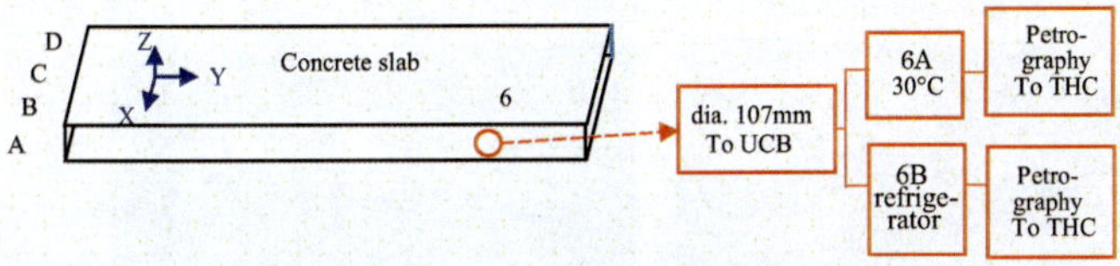

Fig. 13.23 Distribution of core specimens extracted in the horizontal direction by UTK for RILEM round-robin test

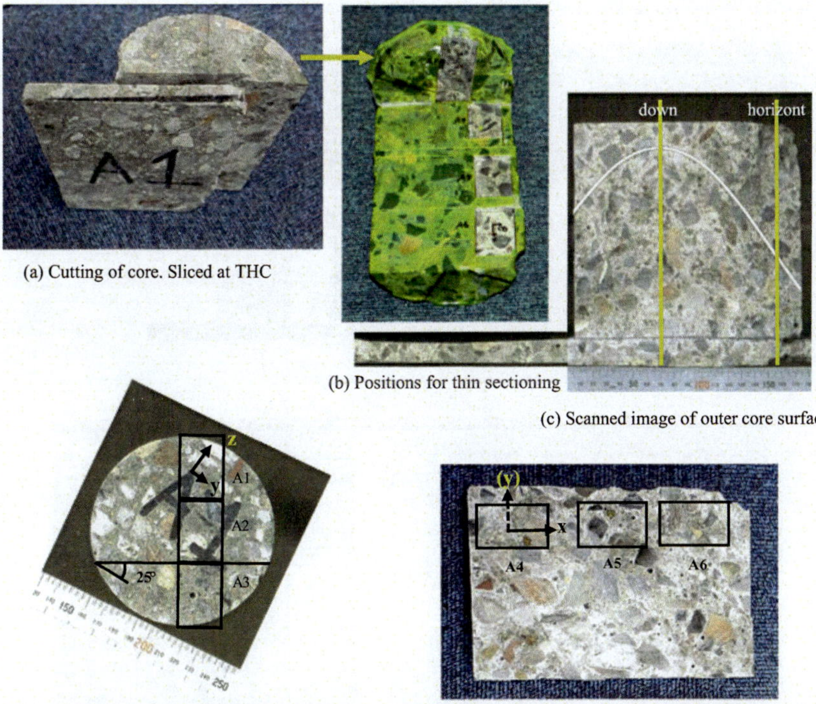

(a) Cutting of core. Sliced at THC

(b) Positions for thin sectioning

(c) Scanned image of outer core surface

(d) Thin sectioning from corner disc

(e) Thin sectioning from half cylinder: Y axis is submerged due to rotation of cutting plane

Fig. 13.24 Core sample No. 6A, stored at 30 °C, with estimated directions

crack indices in the original directions (Z, X, Y) of the slab, a rotation angle around the X axis in the polished thin sections (36 × 25mm) should be taken into account for correction (Fig. 13.24d, e).

The other core specimen (No. 6B, stored in refrigerator) was a transversally cut fat cylinder (Fig. 13.25a). This was cut parallel to the vertical direction in the original concrete slab (Fig. 13.25a, b), according to the method by Katayama (Fig. 13.2c). One half cylinder was subjected to oriented thin sectioning (Fig. 13.25d, e).

Three-dimensional crack index.

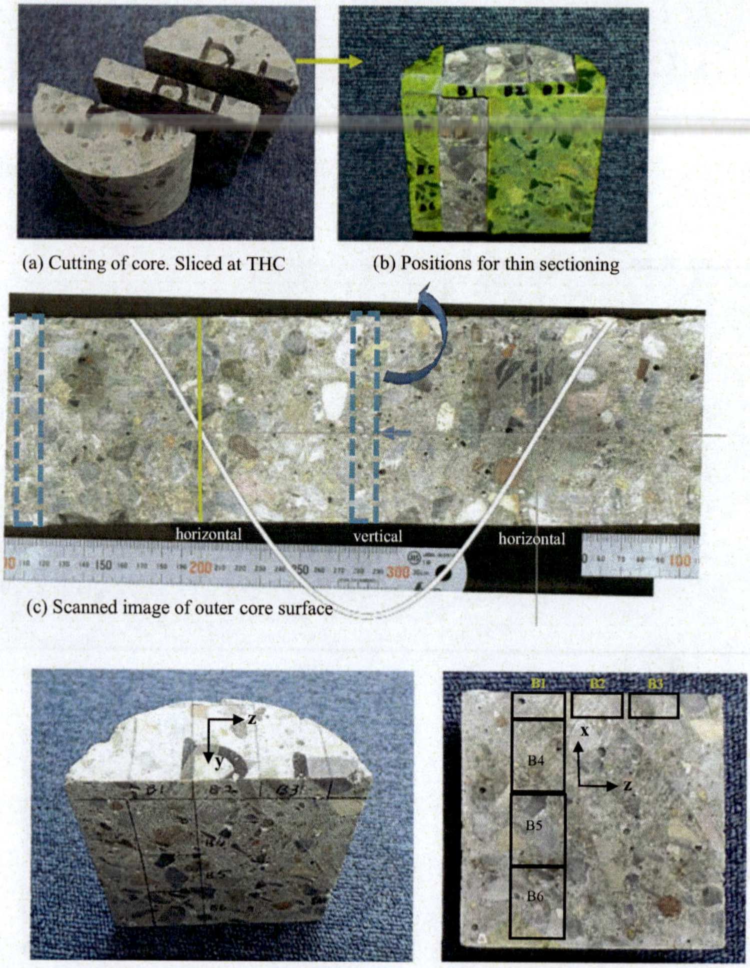

(a) Cutting of core. Sliced at THC

(b) Positions for thin sectioning

(c) Scanned image of outer core surface

(d) Thin sectioning from the corner

(e) Thin sectioning from half cylinder

Fig. 13.25 Core sample No. 6B, stored in refrigerator, with estimated directions

Crack indices of the core specimen before and after the storage at 30 °C for 11 months were shown in Tables 13.25 and 13.26, respectively. Average crack indices in the horizontal (X, Y) and vertical (Z) directions of the core stored in the refrigerator at UCB (to keep expansion at time of extraction) was 0.25% and 0.31%, respectively (Table 13.26), which were close to those obtained at THC (0.22, 0.28%) for a different core specimen (Table 13.28, main report).

On the other hand, with the core stored at 30 °C, average crack index in the horizontal direction (X, Y) was 0.33% (Table 13.25). This was even smaller than expected from the result of the 40 °C 100% RH expansion test at THC (Figs. 13.19a,

Table 13.25 Crack index (%) of concrete in thin sections, 11 months after the storage at 30 °C, Core No. 6A

	Corner disc 3.5cm x 2.5cm x 3 = 26.3 cm				Half cylinder 3.7cm x 2.5cm x 3 = 27.8cm					Crack index (%)**	
Direction	A1	A2	A3	av	A4	A5	A6	av	correction	av	
Z	0.56	0.42	0.34	0.44						0.44	
X	0.45	0.44	0.22	0.37	0.41	0.23	0.19	0.28		0.33	0.33
Y*					0.43	0.31	0.34	0.36	0.36 x cos25°	0.32	

*Longitudinally cut plane was about 25° rotated from the horizontal direction
**To convert crack index to expansion, an addition of 0.02–0.04% may be necessary, because expansion less than this range gives no visible cracking, and is likely to be missed by the petrographic examination of the crack index

Table 13.26 Crack index (%) of concrete in thin sections, 15 months after the storage in refrigerator, Core No. 6B

	Half cylinder 4.3cm x 2.2cm x 3 = 28.4 cm				Half cylinder 3.0cm x 2.6cm x 3 = 23.4cm					Crack index (%)**	
Direction	B1	B2	B3	av	B4	B5	B6	av	correction	av	
Z	0.30	0.15	0.44	0.30	0.32	0.22	0.43	0.32		0.31	
X					0.62*	0.26	0.25	0.38	*discard	0.26	0.25
Y	0.23	0.18	0.27	0.23						0.23	

*Excessive expansion due to concentrated volcanic rock particles was discarded in calculating average
**To convert crack index to expansion, an addition of 0.02–0.04% may be necessary, because expansion less than this range gives no visible cracking, and is likely to be missed by the petrographic examination of the crack index

13.22a), because conversion of the expansion parameters from 40 to 30 °C at 100% RH produces an expansion curve with overall expansion (as crack index) around 0.55% (= past expansion before core extraction 0.25% + calculated new expansion in the next 11 months 0.30%). It is therefore necessary to clarify the reason for the difference observed (e.g. by increasing the number of thin sections and checking the relative humidity and temperature history).

A2 ASR in Concrete

In terms of the petrographic stage based on the optical microscopic features (Katayama et al. [2, 3, 9], ASR in the core specimen stored at 30 °C (Table 13.27) was stage iv, not particularly advanced than in the specimen before the accelerated expansion test (Table 13.7). However, some rock types (welded tuff, spherulitic rhyolite and rhyodacite) had reacted more, and exudation of ASR sol/gel around the aggregate (gel-halo) of the stage ii became evident.

Under the microscope, both the concrete specimens (in refrigerator, and at 30 °C) were non-AE concrete scant in air voids, and exhibited a similar progress of ASR corresponding to the stage iv (Figs. 13.26, 13.27). However, in the specimen stored at 30 °C, crack width in the vertical direction (Z) became larger, producing a larger crack index (Table 13.25), and ASR gel lining air voids occurred in association

Table 13.27 Petrographic stage of ASR in concrete, kept at 30 °C for 11 months, Core No. 6A

	Aggregate	Rock type	Progress of ASR →					Severity of ASR
			i	ii	iii	iv	v	
			aggregate	paste	aggregate	paste		micro-
			Reaction rim	Gel-halo	Crack-filling gel		Void-filling	scopy
Coarse	Early-expansive*	andesite		[illegible]	[illegible]			1
	Late-expansive*	quartzite		(x)	(x)	(x)		2
	Late-expansive**	psammitic hornfels			(x)	(x)		
Fine	Early-expansive***	andesite	x		x	x		
		rhyodacite	x	x	xx	xx	x	
		rhyolite	xx	x	xx	xx		2
		glassy rhyolite		x	xx	xx	x	
		spherulitic rhyolite		x	xx	xx	x	
		altered andesite	x	x	xx	xx		
		altered rhyolite	x		xx	xx		2
		altered welded tuff	x	x	xx	xx		
		altered tuff	x	x				1
		shale	x					
	Overall assessment							2

Occurrence: xx: abundant, x common, (x) rare Severity of ASR: 1: minor; 2: moderate; 3: severe
*Bernalillo, New Mexico, **Wells, Maine, ***El Paso (Jobe), Texas
altered: Occurrence of smectite suggestive of green tuff formation of Miocene to Pliocene age

(a) Stage iv: Propagation of gel-filled crack from reacted aggregate into cement paste

(b) Stage iv→v: ASR gel migrated along the crack into air void distant from the reacted aggregate

Fig. 13.26 Progress of ASR caused by rhyolite in the sand, Core No. 6B, stored in refrigerator, thin section B1

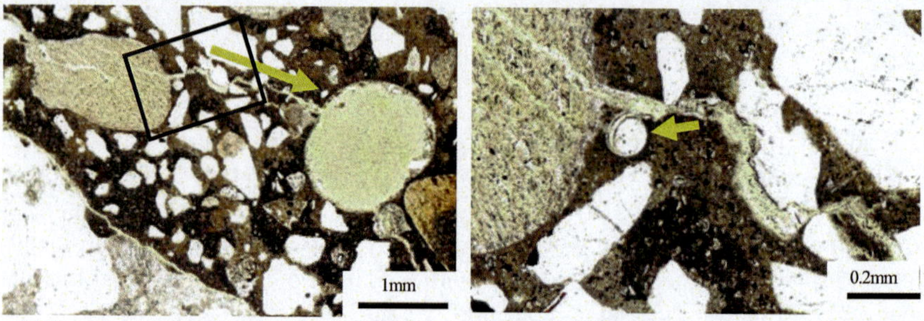

(a) Stage iv →v: Migration of ASR gel along the crack into air void distant from the reacted aggregate

(b) Air void -filling ASR gel in the stage ii adjacent to the reacted aggregate. This gel, with the gel halo surrounding the reacted aggregate, migrated through cement paste

Fig. 13.27 Progress of ASR caused by rhyolite in the sand. Core No. 6A, stored at 30 °C, thin section A1

(a) Interstitial cristobalite reacting in spherulitic rhyolite

Fig. 13.28 Pessimum-causing silica minerals, core No. 6B, stored in refrigerator, thin section B1

with a gel halo just surrounding the reacted aggregate (Fig. 13.27). This suggests an increased mobility and increased formation of ASR gel at higher temperature. Cristobalite was the main reactant in this concrete (Fig. 13.28).

- Pessimum-causing silica minerals

Under the polarizing microscope, several kinds of silica minerals in the volcanic rocks, such as highly-reactive cristobalite in andesite (Fig. 13.29b), tridymite in rhyolite (Fig. 13.29a), and chalcedony in altered andesite (Fig. 13.29c), were reacting in concrete. This reaction was more evident in the core specimen stored at 30 °C. Such an occurrence of ASR of the pessimum-causing silica minerals (Katayama [11] in the volcanic rocks was common to Japan (Katayama [14, 28], and there was nothing different from ASR in Japan.

A3 SEM-EDS Analysis

Under SEM on polished thin section of the core stored in refrigerator (core No. 6B), silica minerals cristobalite was found to have reacted to form ASR gel (Fig. 13.30a, b). Cristobalite in the spherulitic texture of rhyolite is associated with radiating fibers of an alkali-feldspar anorthoclase (Fig. 13.30b). ASR gel filling cracks in the reacted rhyolite aggregate partly converted to rosette crystals (Fig. 13.30c, d), one of which had a composition: $(Na_{1.31}, K_{0.76})_{2.07}(Ca_{1.82}, Na_{0.16}, Mg_{0.01}, Fe_{0.04}, Ti_{0.01})_{2.06}(Si_{7.83}, Al_{0.19})_{8.03}O_{19}nH_2O$ (Fig. 13.30d), Table 13.28). This broadly resembles a natural calcium alkali silicate hydrate mineral shlykovite, $(K, Na)_2.Ca_2.Si_8.O_{18}(OH)_2.6H_2O = (K, Na)_2Ca_2Si_8O_{19}7H_2O$, but it is too Na-rich to be called shlykovite. Rosette crystals resembling shlykovite with varying Na/K ratios have been identified from ASR-affected concrete structures in the world [28–30]. Some analytical results were shown in Tables 13.28. The appearance of the reacting cristobalite in the volcanic aggregates was similar to that reported [14, 28].

(a) Tridymite reacting in rhyolite

(b) Interstitial cristobalite reacting in hornblende dacite

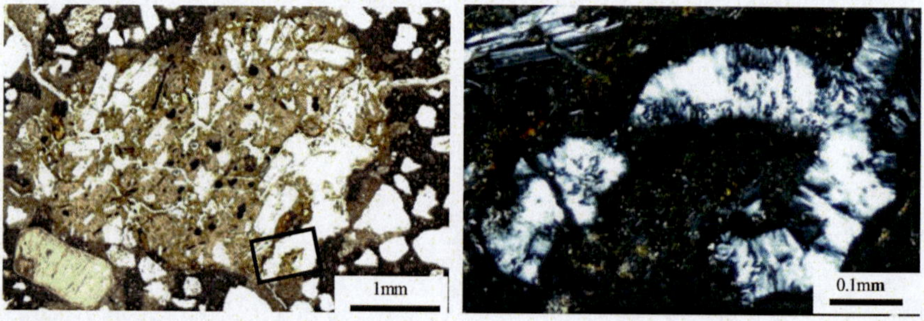

(c) Chalcedony reacting in altered andesite

Fig. 13.29 Reaction of sand particles in the concrete slab, Core No. 6A, stored at 30 °C, thin section A1

In the core specimen stored at 30 °C (core No. 6A), under SEM on polished thin section, silica minerals cristobalite (Fig. 13.31a), tridymite (Fig. 13.31b) and chalcedony (Fig. 13.31c, d) were found to have reacted to form ASR gel. Some analytical results were shown in Table 13.29. The appearance of reacting cristobalite and tridymite in the volcanic aggregates was similar to that reported [14, 28].

- Compositional trend of ASR gel in concrete

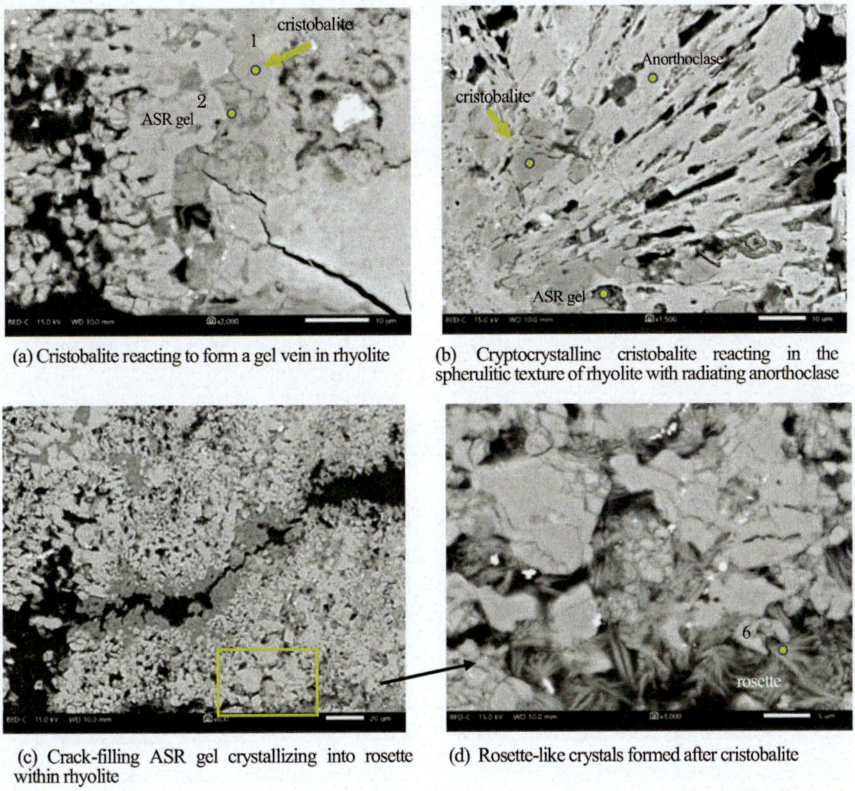

(a) Cristobalite reacting to form a gel vein in rhyolite

(b) Cryptocrystalline cristobalite reacting in the spherulitic texture of rhyolite with radiating anorthoclase

(c) Crack-filling ASR gel crystallizing into rosette within rhyolite

(d) Rosette-like crystals formed after cristobalite

Fig. 13.30 Reaction of pessimum-causing silica mineral, Core No. 6B, stored in refrigerator, thin section B1

Table 13.28 Compositions of reacting silica mineral and reaction product as determined by EDS (wt%), Core No. 6B, stored in refrigerator, thin section B1

		SiO_2	TiO_2	Al_2O_3	Fe_2O_3	MnO	MgO	CaO	Na_2O	K_2O	SO_3	P_2O_5	Total
1	cristobalite	96.40	0.00	0.38	0.52	0.10	0.00	1.00	0.46	0.10	0.00	0.17	99.12
2	ASR gel	44.89	0.11	2.50	0.39	0.18	0.05	5.91	1.72	2.10	0.07	0.05	57.98
3	cristobalite	95.57	0.17	1.28	0.06	0.24	0.01	0.06	0.15	0.42	0.00	0.00	97.97
4	ASR gel	54.68	0.15	3.07	0.98	0.06	0.65	7.40	3.52	3.24	0.33	0.03	74.11
5	anorthoclase	66.30	0.13	17.30	0.24	0.00	0.06	0.36	5.94	6.18	0.00	0.00	96.50
6	rosette	42.72	0.10	0.88	0.28	0.00	0.05	9.26	4.11	3.23	0.08	0.00	60.71

Anorthoclase: $(Na_{0.53}, K_{0.36}, Ca_{0.02}, Fe_{0.01})_{0.93}(Al_{0.94}, Si_{3.04})_{3.98}O_8$ The vacancy in the octahedral site ($0.07 = 1 - 0.93$) is possibly alkalis (Na, K) leached from the anorthoclase into pore solution within cement paste to accelerate ASR

In terms of gel compositions plotted on the Ca/Si–Ca/(Na+K) diagram, there was no essential difference between the two core specimens with (No. 6A in 30 °C, 11 months) and without the exposure to 30 °C (No. 6B in refrigerator, 11 months), plus additional 4 months in refrigerator. In both cases, ASR gel on this diagram did not

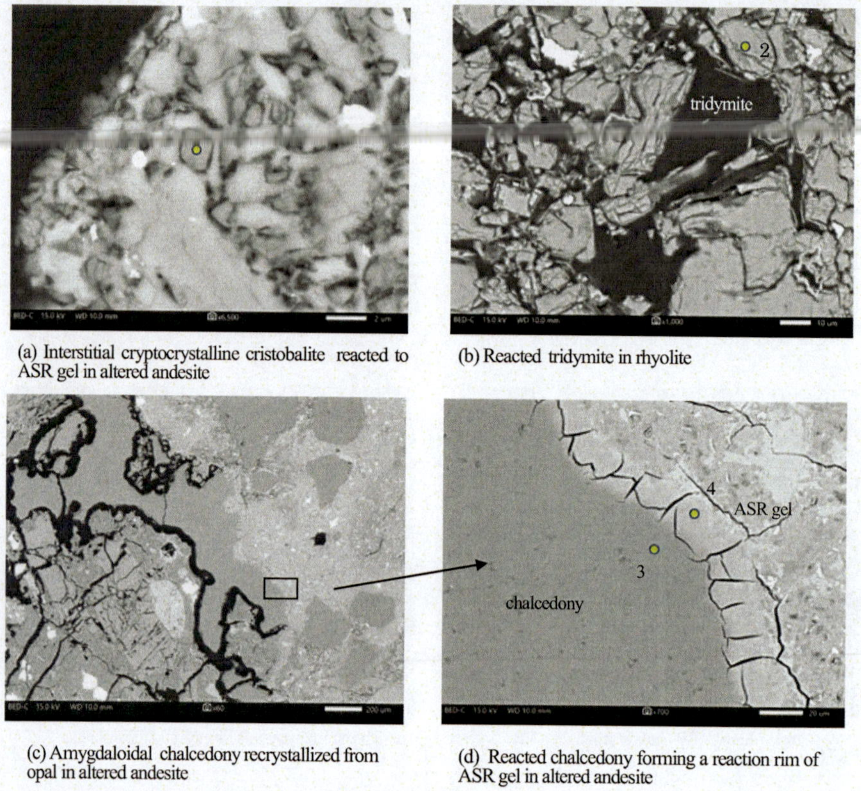

(a) Interstitial cryptocrystalline cristobalite reacted to ASR gel in altered andesite

(b) Reacted tridymite in rhyolite

(c) Amygdaloidal chalcedony recrystallized from opal in altered andesite

(d) Reacted chalcedony forming a reaction rim of ASR gel in altered andesite

Fig. 13.31 Reaction of pessimum-causing silica minerals, Core No. 6A, stored at 30 °C, thin section A1

Table 13.29 Compositions of reacting silica minerals by EDS (wt%), Core No. 6A, stored at 30 °C, thin section A1

		SiO_2	TiO_2	Al_2O_3	Fe_2O_3	MnO	MgO	CaO	Na_2O	K_2O	SO_3	P_2O_5	Total
1	ASR gel*	68.20	0.32	9.86	0.32	0.07	0.00	0.40	1.85	5.54	0.22	0.00	87.28
2	tridymite	94.04	0.30	0.60	0.00	0.00	0.08	0.00	0.31	0.20	0.00	0.00	95.64
3	chalcedony	95.79	0.00	0.08	0.08	0.24	0.00	0.17	0.00	0.25	0.02	0.18	96.91
4	ASR gel	47.62	0.07	0.42	0.84	0.14	0.10	24.92	1.34	4.46	0.09	0.00	79.99

*ASR gel with a pseudomorph after original grain of cristobalite

reach the "convergent point" [17, 18] with CSH gel (Fig. 13.32) but left a blank space before reaching this point. This means that ASR has not been terminated.

Core specimen kept in refrigerator presented a single compositional line (Fig. 13.32a), corresponding to the type I evolution after Katayama [14, 28]. On the other hand, specimen stored at 30 °C contained ASR gel with much dispersed data points, and an additional parallel dotted compositional line could be drawn above the main compositional line (Fig. 13.32b), corresponding to the type II evolution.

Fig. 13.32 Compositional lines of ASR gel in concrete as determined by EDS analysis on polished thin section

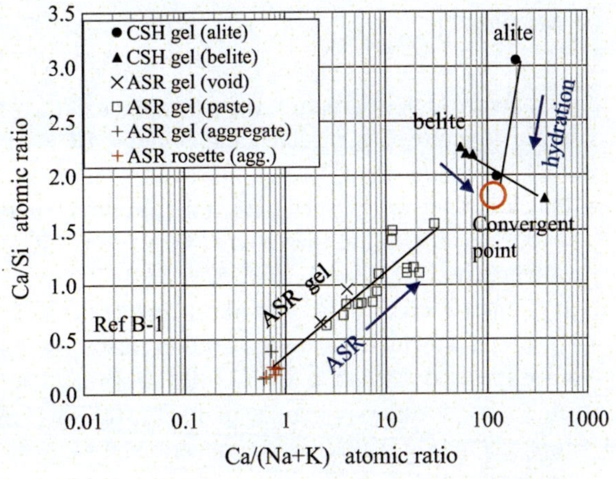

(a) Core No.6B, stored in refrigerator, thin section B1
Type I evolution of ASR

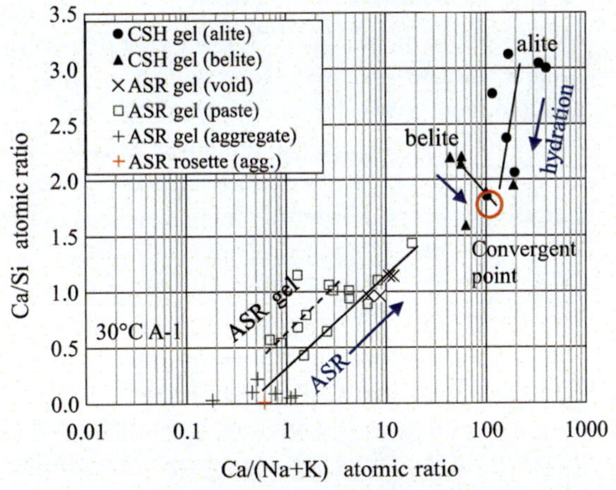

(b) Core No.6A, stored at 30°C, thin section A1
Type II evolution of ASR

Compared with the same Ca/Si ratio, ASR gel in this additional line had a lower Ca/(Na+K) ratio, i.e. higher alkali content. This aspect should be examined in detail whether the excessive alkalis were derived as water-soluble alkali from the aggregate used, or were absorbed from the linen cloth (possibly containing NaOH solution) that was wrapping the core specimen in the vinyl envelope when received.

A4 Summary

1. Comparative petrographic study was made of the core specimens from the round robin concrete slab, received and stored at UCB for 11 months then examined at THC after additional storage in 5 °C refrigerator in for 4 months.
2. Crack indices of the specimen stored at refrigerator were: vertical direction (Z) = 0.31%, horizontal directions (X, Y) = 0.25%, which were similar to those obtained at THC (Z = 0.28%, X, Y = 0.22%).
3. By contrast, crack indices of the specimens stored at 30 °C were Z = 0.44%, X, Y = 0.33%, which were even smaller than those expected (X, Y = 0.55%) for a core specimen stored at 30 °C 100% RH for 11 months, when the test result of other core specimen obtained at 40 °C 100% RH by THC was extrapolated.
4. It is a common nature to the concrete slabs cast in the horizontal direction that expansion due to ASR occurs maximum in the vertical direction (Z).
5. To examine the expansionary nature of the core specimens (including the crack index), a parallel petrographic examination will be performed by Dr. David Rothstein of DRP using another half cut core specimens.
6. The causative material of ASR in the concrete slab was mainly rhyolitic rock in the sand aggregate, and cristobalite, which is known as a typical pessimum-causing silica mineral, was the chief reactant. Beside this, tridymite and chalcedony in the volcanic aggregate also produced ASR.
7. In general, the progress of ASR in two concrete specimens was categorized as stage iv, based on the petrographic examination, but the core specimen stored at 30 °C presented a slightly advanced symptom.
8. In terms of compositions of ASR gel on the Ca/Si–Ca/(Na+K) diagram, ASR in concrete has not terminated yet, because compositions of ASR gel has not reached the convergent point with CSH gel.
9. Compositional line of ASR gel in the core specimen stored at 30 °C seems to present the type II evolution, suggestive of addition of water-soluble alkalis either from the aggregate in concrete or from the linen cloth that wrapped the core specimen. This needs confirmation by further study increasing the number of thin sections for EDS analysis.

Addendum: Apparent Activation Energy of Characteristic Time Corrected, November 2019

In general, characteristic time (τc) of concrete stored in humid air tends to be shorter when alkali content of concrete is higher. In this round robin report, an equation for apparent activation energy of characteristic time was given: Uc (K) = 1480 × ln (x) −2730. This produces a negative Uc (−2940 K) at 0.87N NaOH (mol/l) (e.g. Tables 13.13, 13.15, 13.18 and 13.20) and gently sloping expansion curve at 20 °C

0.87N NaOH with larger τc', about 5.5 times that of the original τc at 80 °C 1N NaOH ((τc'/τc) = exp{−2940 × [1/(273 + 80)] − [(1/273 + 20)]}).

By contrast, a close look at parameters calculated using data from concrete specimens (andesite) stored in alkaline solution [25], a reverse situation occurs. That is, Uc presents a positive sign. Because of this, positive value of Uc (2940 K) after equation, Uc (K)= −1480 × ln (x) +2730 (RILEM main report, Chap. 6, eq. 6.5), gives a steeper initial slope with smaller τc', 0.18 times (τc′/τc) = exp{2940 × [1/(273 + 80)] – [(1/273 + 20)]} that of the original τc at 80 °C 1N NaOH. Hence, figures of estimated expansion curves based on concrete core expansion tests with NaOH solutions should be revised. Two of them were shown in in Fig. 13.33a, b.

However, to estimate overall expansion curve and overall final expansion of concrete (ε_∞), only two past expansion data of concrete (at different times) and one final expansion ($\Delta\varepsilon_\infty$) from accelerated concrete core expansion test are necessary. By applying this method, same final expansion can be obtained (Tables 13.30, 13.23 and 13.24), irrespective of whether Uc has plus or minus sign. This means that latency time and characteristic time, often difficult to convert parameters from accelerated conditions to ambient conditions, are not necessary (Fig. 13.33a) [31].

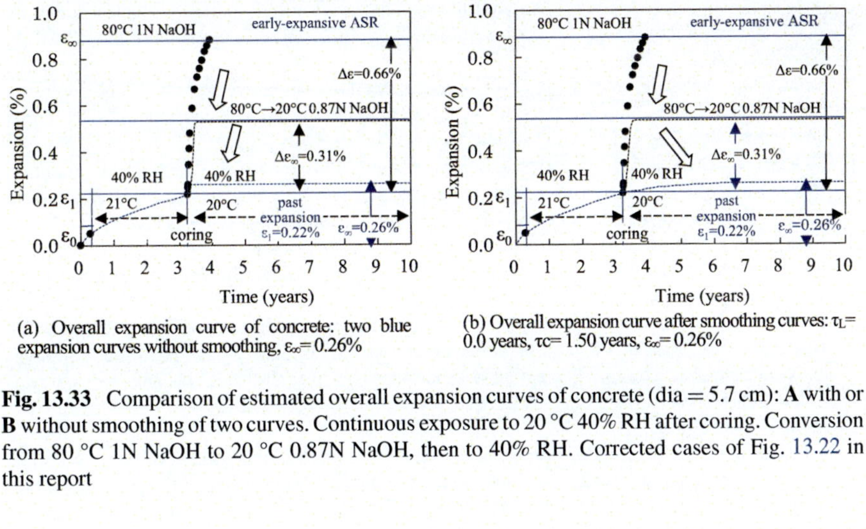

(a) Overall expansion curve of concrete: two blue expansion curves without smoothing, ε_∞= 0.26%

(b) Overall expansion curve after smoothing curves: τ_L= 0.0 years, τc= 1.50 years, ε_∞= 0.26%

Fig. 13.33 Comparison of estimated overall expansion curves of concrete (dia = 5.7 cm): **A** with or **B** without smoothing of two curves. Continuous exposure to 20 °C 40% RH after coring. Conversion from 80 °C 1N NaOH to 20 °C 0.87N NaOH, then to 40% RH. Corrected cases of Fig. 13.22 in this report

Table 13.30 Estimated overall expansion curves fitted to Larive equation at 20 °C, assuming exposure to 40% RH

Core specimen		Japanese accelerated tests and storage condition			Expansion		Past expansion	Overall expansion parameters at 20°C			Advancement of expansion	Parameters used for thermodynamic corrections
No.	dia (cm)	T (°C)	medium	method	mon	ε	ε_1	ε_∞	τ_L	τ_c	$\varepsilon_1/\varepsilon_\infty$	
3'A	5.7	80	1N NaOH	immersion	8.3	0.66	0.22*	0.26	0.0	1.50	85	activation energy (U) for ε_∞**
5A	10.7	40	1N NaOH	wrapping	16.5	0.47		0.27	0.0	1.45	81	activation energy (U) for ε_∞**
5B		20→ 40	100% RH	curing	16.7	0.33		0.27	0.0	1.45	81	activation energy (U) for ε_∞**, τ_L τ_c***

*crack index, **Katayama [4] for Japanese andesite, ***calculated for reactive volcanic sand from Texas (El Paso) using data by Stacey et al. [26]

References

1. Elhassan, A., LePape, S., Ma, Z.J., Le Pape, Y.: Round Robin Analysis, submitted to round robin participants, pp. 1–6 (2016)
2. Katayama, T., Tagami, M., Sarai, Y., Izumi, S., Hira, T.: Alkali-aggregate reaction under the influence of deicing salts in the Hokuriku district, Japan. Mater. Character. **53**(2), 105–122 (2004)
3. Katayama, T.: An attempt to estimate past expansion of concrete based on petrographic stage of alkali-silica reaction. In: Proceedings of 39th International Conference on Cement Microscopy, pp. 217–236. Toronto, Canada (2017)
4. JCI-TC115FS proposal 5.3: Method of measuring ASR expansion of core samples immersed in alkali solution (Alkali-immersion method), pp. 316–319. Japan Concrete Institute (JCI), Tokyo (2014) (in Japanese)
5. JCI-TC115FS proposal 5.4: Method of measuring ASR expansion of core samples immersed in saturated NaCl solutions (saturated NaCl immersion method), pp. 320–323. JCI, Tokyo (2014) (in Japanese)
6. JCI-TC115FS proposal 4.3: Method of measuring ASR expansion of small concrete prism, pp. 272–284. Japan Concrete Institute (JCI), Tokyo (2014) (in Japanese)
7. JCI-S-011-2017: Method of measuring expansion tat:0 of core sampled from concrete structure. Japan Concrete Institute (JCI), Tokyo, 10 pp. (in Japanese)
8. Larive, C.: Apports Combinés de l'Expérimentation et de la Modélisation à la Comprehension de l'Alcali-Réaction et de ses Effets Mécaniques. Doctoral thesis, Laboratoire Central des Ponts et Chaussées, Paris (1997). https://hal.inria.fr/docs/00/52/06/76/PDF/1997TH_LARIVE_C_NS20683.pdf
9. Katayama, T.: Petrographic stage of alkali-silica reaction in concrete and associated compositional changes of reaction products. In: 40th International Conference on Cement Microscopy. Oral Power Point presentation, Deerfield Beach, Florida, USA (2018)
10. Sanchez, L., Dunant, C.: Appendix A, Sample conditioning. In: RILEM Technical Committee 259-ISR (ASR Prognosis of deterioration and loss of serviceability in structures affected by alkali-silica reactions), pp. 7–14. WG-1 Benchmark/Round Robin Analysis (2016)
11. Katayama, T.: Petrography of alkali-aggregate reactions in concrete-reactive minerals and reaction products. In: East Asia Alkali-Aggregate Reaction Seminar, Tottori. Tottori University, Supplementary paper, Tokyo, pp. A45–A59 (1997)
12. Katayama, T., Kaneshige, Y.: Diagenetic changes in potential alkali-aggregate reactivity of volcanic rocks in Japan—a geological interpretation. In: Grattan-Bellew (ed) Proc. 7th International Conference on Alkali-Aggregate Reaction in Concrete (ICAAR), pp. 489–495. Ottawa, Canada (1986). Concrete Alkali-Aggregate Reactions. Noyes Publications, Park Ridge, New Jersey, USA (1987)
13. Katayama, T., St John, D.A., Futagawa, T.: The petrographic comparison of some volcanic rocks from Japan and New Zealand – Potential reactivity related to interstitial glass and silica minerals. In: Proceedings of 8th International Conference on Alkali-Aggregate Reaction in Concrete (ICAAR), pp. 537–542. Kyoto, Japan (1989)
14. Katayama T (2010): Diagnosis of alkali-aggregate reaction—Polarizing microscopy and SEM-EDS analysis. In: Castro-Borges, P., Moreno, E.I., Sakai, K., Gjorv, O.E., Banthia, N. (eds.) Proceedings of 6th International Conference on Concrete under Severe Conditions (CONSEC'10), Merida, Mexico, vol. 1, pp. 19–34. CRC Press, Taylor & Francis
15. Katayama, T., Helgason, T.S., Olafsson, H.: petrography and alkali-reactivity of some volcanic aggregates from Iceland. In: Proceedings of. 10th International Conference on Alkali-Aggregate Reaction in Concrete (ICAAR), pp. 377–384. Melbourne, Australia. (1996)
16. Katayama, T.: Alkali-aggregate reaction in the vicinity of Izmir, western Turkey. In: Proceedings of 11th International Conference on Alkali-Aggregate Reaction in Concrete (ICAAR), pp. 365–374. Quebec, Canada (2000)

17. Katayama, T., Oshiro, T., Sarai, Y., Zaha, K., Yamato, T.: Late-expansive ASR due to imported sand and local aggregates in Okinawa island, southwestern Japan. In: Proc. 13th International Conference on Alkali-Aggregate Reaction in Concrete (ICAAR), pp. 862–873. Trondheim, Norway (2008)
18. Katayama, T.: ASR gel in concrete subject to freeze-thaw cycles—Comparison between laboratory and field concretes from Newfoundland, Canada. In: Proceedings of 13th International Conference on Alkali-Aggregate Reaction (ICAAR), pp. 174–183. Trondheim, Norway (2008)
19. Nomura, M., Aoyama, M., Hira, T., Torii, K.: Case study of deterioration of concrete in road structures due to alkali-silica reaction (ASR) in Hokuriku region and its evaluation methods. Concr. Res. Technol. **13**(3), 105–114 (2002). (in Japanese)
20. JCI-DD2: Determination of expansion rate of AAR-damaged structure using concrete core, pp. 5–9. Japan Concrete Institute (JCI) Tokyo (1989) (in Japanese)
21. Ministry of Land, Infrastructure and Transport (2008): Proposed guideline to repair and reinforce piers and abutments of ASR-affected road bridges. Committee on the countermeasures to ASR, 98 pp., see page 86 (in Japanese)
22. General Project Method: Development of technology to improve the durability of concrete structures, vol. 2, Ministry of Construction, Japan. Core expansion method, pp. 91–92. Measurement of expansion of core, pp. 114–117 (1988) (in Japanese)
23. Brunetaud, X.: Étude de l'influence de différents paramètres et de leurs interactions sur la cinétique de l'amplitude de la réaction sulfatique interne au béton. Doctoral thesis. Châtenay-Malabry, Ecole Centrale de Paris, 2005
24. Helmuth, R.: Alkali-silica reactivity: An overview of research. SHRP-C-342, Strateg. Highw. Res. Prog. 13–16. National Research Council, Washington D.C. (1993)
25. Kuroda, T., Inoue, S., Yoshino, A., Nishibayashi, S., Miyagawa, T.: Effects of accelerated test conditions on ASR expansion of concrete core. In: Proceedings of 13th International Conference on Alkali-Aggregate Reaction (ICAAR), pp. 195–205. Trondheim, Norway (2008)
26. Stacey, S., Folliard, K., Drimalas, T., Thomas, M.: An accelerated and more accurate test method to ASTM C1293: the concrete cylinder test. In: Proceedings of the 15th International Conference on Alkali-Aggregate Reaction (ICAAR). paper No. 130, Sao Paolo, Brazil (2016)
27. Poyet, S., Sellier, A., Capra, B., Foray, G., Torrenti, J., Cognon, H., Bour-darot, E.: Modelling of alkali-silica reaction in concrete, part 2: influence of water on ASR. In: Proceedings of 12th International Conference on Alkali-Aggregate Reaction, pp. 185–190. Beijing, China (2004), pp. 185–190
28. Katayama, T.: Late-expansive ASR in a 30-year old PC structure in eastern Japan (2012b)
29. Katayama, T.: ASR gel and their crystalline phases in concrete—Universal products in alkali-silica, alkali-silicate and alkali-carbonate reactions. In: Proceedings of 14th International Conference on Alkali-Aggregate Reaction in Concrete (ICAAR). Austin, Texas, USA (2012). Paper 030411-KATA-03, 12 pp., Paper 030411-KATA-05, 10 pp.
30. Katayama, T.: Petrographic study of alkali-aggregate reactions in Concrete. Doctoral thesis. University of Tokyo. 2012
31. Katayama, T., Mukai, K., Sato, T.: Estimation of the past and residual expansion of concrete affected by alkali–silica reaction—Japanese method, RILEM TC 259-ISR WG-1, short version, Sample report, June 2019, 25 pp. (2019)

Chapter 14
Accelerated Expansion Test Sample Report: Switzerland

Andreas Leemann and Christine Merz

14.1 Introduction

The details of the Swiss method to determine the residual expansion of concrete from structures are described in detail in Chap. 9 of WP-1 in the STAR and corresponds to the test described in [1, 2]. In the following, two examples are given, how this method can be used in combination with others. The first example performed on a laboratory produced concrete element shows which techniques can be used to complement the residual expansion test. The second example is a dam where it is demonstrated how differences in expansion and damage can be identified using the residual expansion measurements and additional techniques. In both cases recommendations are given how the determined expansion rates can be used in models.

As there are confidentiality restrictions in both cases no details about the concrete component and structure are given.

A. Leemann (✉)
Empa, Swiss Federal Laboratories for Materials Testing and Research, Laboratory for Concrete and Construction Chemistry, Überlandstrasse 129, 8600 Dübendorf, Switzerland
e-mail: andreas.leemann@empa.ch

C. Merz
Merz Ingenieurberatung, Möriken, Switzerland
e-mail: merz@merz-ingenieurberatung.ch

© RILEM 2021

V. E. Saouma (ed.), *Diagnosis & Prognosis of AAR Affected Structures*,
RILEM State-of-the-Art Reports 31,
https://doi.org/10.1007/978-3-030-44014-5_14

14.2 Concrete Slab

14.2.1 General Information

A concrete slab containing reactive aggregates was stored at 27 °C and 65% RH after production for a few month. Afterwards, the temperature of the storing room varied between 18 and 24 °C at approximately 40% RH. The slab was three years old and its average expansion was 0.24% at the time of coring.

14.2.2 Materials and Methods

Four concrete cores (diameter: 100 mm, length: 200 mm) were used for the investigation. Three cores were tested according to the Swiss method.

In order to get additional information on the state of ASR in the concrete at time of delivery, one core was cut and five subsamples were prepared (~50 × 90 mm^2). They were dried in an oven at 50 °C for three days, epoxy impregnated and polished. The crack-index was determined using a Zeiss Axioplan microscope at a magnification of 100×. The samples were further documented under fluorescent light using a 24 MP camera. As an additional step, the impregnated subsamples were used to produce four samples (discs with a diameter of 50 mm) for scanning electron microscopy (SEM). They were investigated with a Nova NanoSEM 230 FEI at a pressure of 2.0–4.0 × 10^{-6} Torr at an acceleration voltage of 12 kV and a beam current of 80–88 μA. An Qxford SSD detector and INCA Energy software with ZAF correction were used for energy-dispersive X-ray spectroscopy (EDS). The concrete after the test was studied in an identical way. Four subsamples were prepared (~50 × 90 mm^2) and investigated with the optical microscope. Two samples were produced for studying the concrete with SEM.

The water-soluble alkali content was determined before and after the residual expansion test according to the procedure described in [1, 2].

14.2.3 Residual Expansion

The cores show a relatively fast expansion (Fig. 14.1). The longitudinal expansion (value used for assessment) during conditioning (20 °C, 100% RH) is above 0.03%. The irreversible longitudinal expansion at the end of the test of 0.11% indicates that about 0.02% of the swelling during the conditioning is irreversible and must be attributed to ASR. The end of the fast expansion (phase 2) has been reached very quickly after moving the samples into the reactor (temperature of 38 °C and 100%

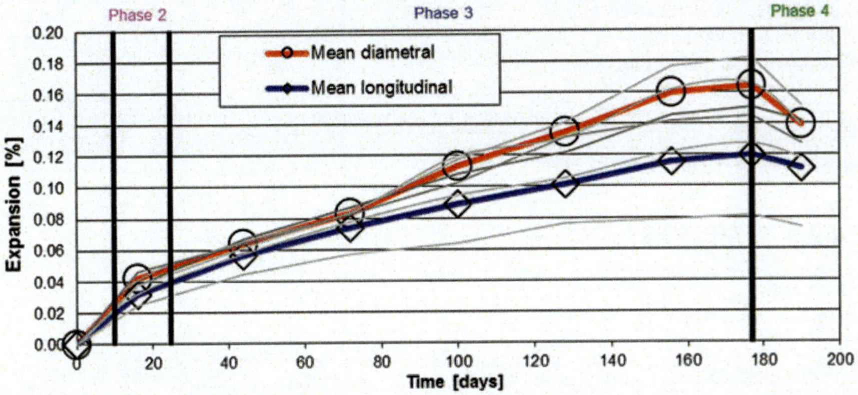

Fig. 14.1 Mean expansion values of the three measured cores in longitudinal and diametral direction

Table 14.1 Longitudinal and diametral expansion values during the different phases. The end of phase 2 has been defined at 26 days. The longitudinal expansion in phase 3 is used to assess the kinetics of expansion in concrete structures

	Expansion (%)	Expansion rate (%/year)
Total longitudinal	0.120	0.22
Total diametral	0.164	0.31
Phase 2 longitudinal	0.032	0.72
Phase 2 diametral	0.042	0.95
Phase 3 longitudinal	0.088	0.20
Phase 3 diametral	0.120	0.29
Irreversible longitudinal	0.111	0.23
Irreversible diametral	0.139	0.29

RH). Based on the current experience with the method, phase 2 was defined from 10–26 days. The cores continue to expand 0.088% until 177 days (phase 3, Table 14.1). This corresponds to an overall longitudinal expansion rate of 0.22%/year.

14.2.4 Alkali Content

The water soluble alkalis decrease during the test (Table 14.2). Alkalis can either be leached or bound in ASR products.

Table 14.2 Water soluble alkalis before and after the residual expansion test

Time of investigation	Na_2O-equivalent (%)
Before the test (delivery state)	0.179
After the test	0.109
Difference	0.070

14.2.5 Optical Microscopy

14.2.5.1 General Impression

The concrete shows already pronounced cracking in its delivery state (Fig. 14.2). The cracks are initiated in aggregates of the sand fraction. Cracks in aggregates >4 mm are practically absent.

The cracking is more pronounced after the residual expansion test. In particular, aggregates >4 mm display cracks as well, even when the cracking is much less pronounced than in the sand fraction. The majority of cracks, especially in the cement paste, are filled with ASR products after the test.

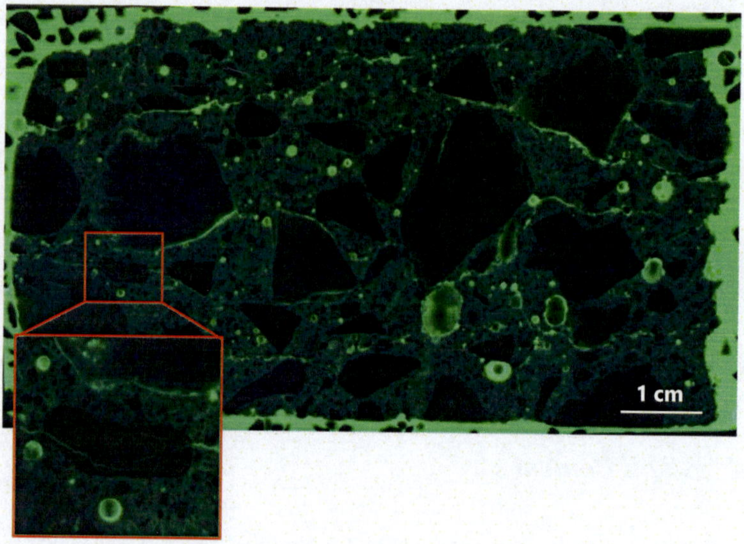

Fig. 14.2 One of the impregnated subsamples of the concrete in its delivery state under fluorescent light. The enlargement indicates the resolution of the image

Table 14.3 Crack index of the concrete before and after the residual expansion test

Time of investigation	Crack index (%)
Before the test (delivery state)	0.15
After the test	0.32
Difference	0.17

14.2.5.2 Crack-Index

The crack-index was determined both before and after the residual expansion test. The values are presented in Table 14.3. The difference between the crack-indices is 0.17%.

14.2.6 Scanning Electron Microscopy

The samples prepared for SEM (Fig. 14.3) show cracking restricted to aggregates of the sand fraction (Figs. 14.4 and 14.5) in the state of delivery, as already observed in the optical microscope. Some of the aggregates in the sand fraction contain alkali-silica-glass (determined by EDS) providing a highly reactive phase and a source of alkalis. The majority of the cracks both in aggregates and in the cement paste (width up to 30 μm) are empty or only partly filled with ASR products. There are aggregates of sedimentary, volcanic and magmatic origin, but they will not be described in detail in the following.

Cracking of the concrete is more pronounced with wider cracks up to a width of 60 μm after the residual expansion test. The majority of the cracks in the cement paste are now filled with ASR products (Fig. 14.6). The porosity of the reactive sand grains is increased by further dissolution. Aggregates >4 mm show occasional cracks containing ASR products as well (Fig. 14.7).

14.2.7 Summary and Conclusions

14.2.7.1 Test Results

The concrete expands considerably in the residual expansion test. As typical, the rate of longitudinal expansion in phase 3 is lower compared to phase 2. The irreversible expansion indicates a total longitudinal expansion during the test of 0.11%. Based on the longitudinal expansion during phase 3, an expansion rate of 0.20%/year can be calculated. This is the value to be used as a starting for assessing the expansion of the investigated concrete component. If the studied cores originated from a concrete structure, the boundary conditions like temperature, humidity and stress state would have to be taken into account in an overall assessment.

The more or less constant expansion during phase 3 indicates an internal source of alkalis, as a decrease of expansion with time is usually observed due to leaching. This seems to be confirmed by the presence of aggregate particles of volcanic origin containing alkali-silica glass.

Microscopy indicates that the very reactive aggregates are present mainly in the sand fraction. The concrete shows already obvious cracking caused by ASR in its state of delivery. The cracking is more pronounced after the test. The difference in the crack index before and after the test is 0.17%. This value should approximately correspond with concrete expansion during the test. Obviously, there is a certain overestimation of expansion by the crack-index in the current case (measured values of 0.11 and 0.14% in longitudinal and diametral direction, respectively), even if the expansion is in the same order of magnitude.

14.2.7.2 Comparison with Data Base

In comparison with expansion of cores extracted from Swiss concrete structures, the tested laboratory concrete expands fast and reaches very high expansion values (Figs. 14.8 and 14.9).

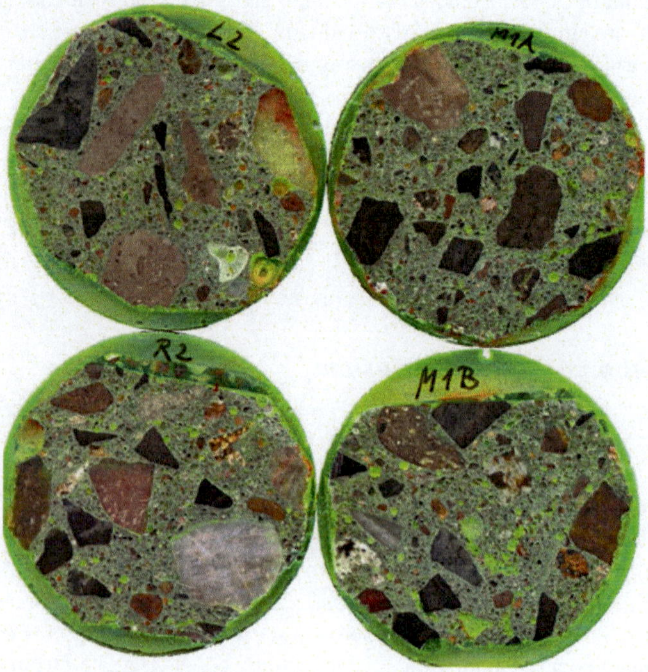

Fig. 14.3 Samples of the concrete in the state of delivery used for electron microscopy before carbon coating (sample diameter: 50 mm)

Fig. 14.4 Severely cracked sand grain in the concrete as delivered

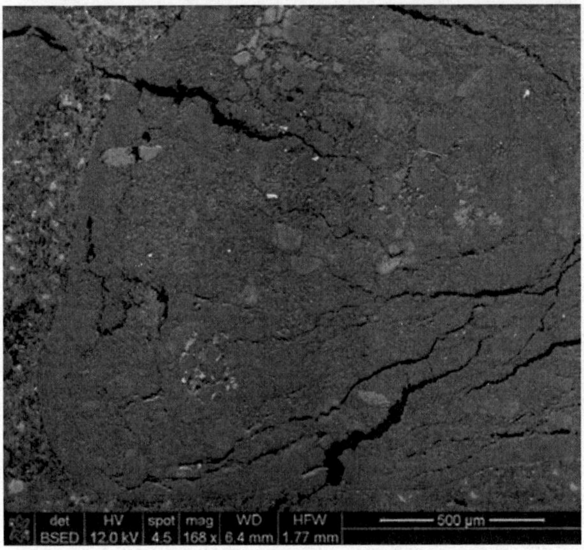

Fig. 14.5 Severely cracked sand grain in the concrete as delivered

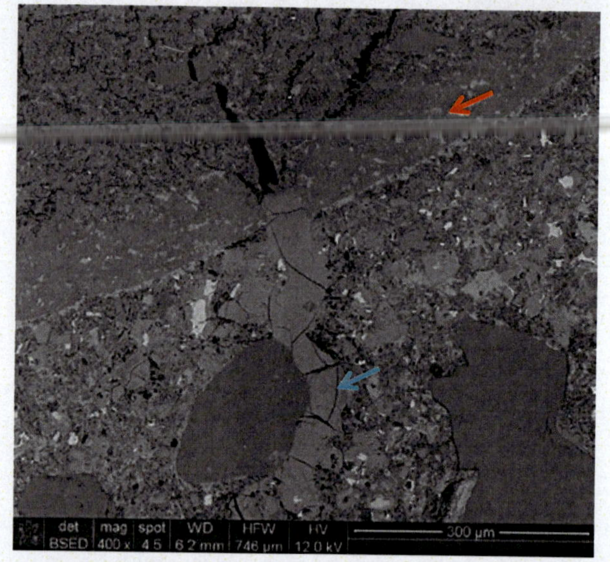

Fig. 14.6 Reactive sand grain of volcanic origin with partially dissolved matrix and denser rim (red arrow) due to deposits of ASR products and completely filled crack in cement paste (blue arrow) after the test

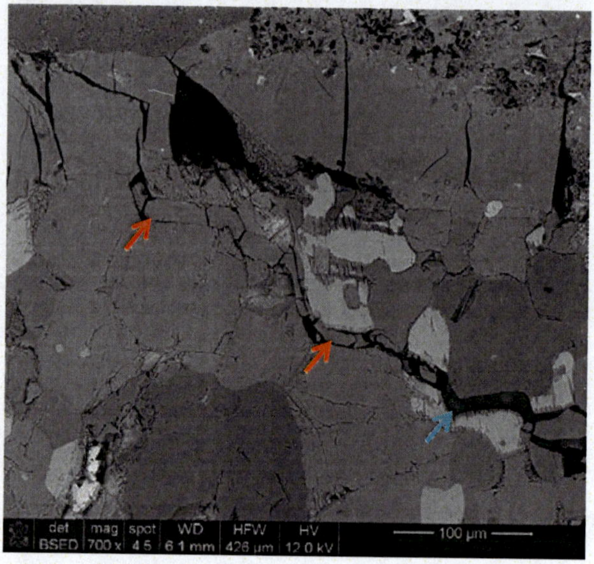

Fig. 14.7 Crack in gravel particle partially filled with crystalline (blue arrow) and amorphous (red arrows) ASR products in magmatic rock after the test

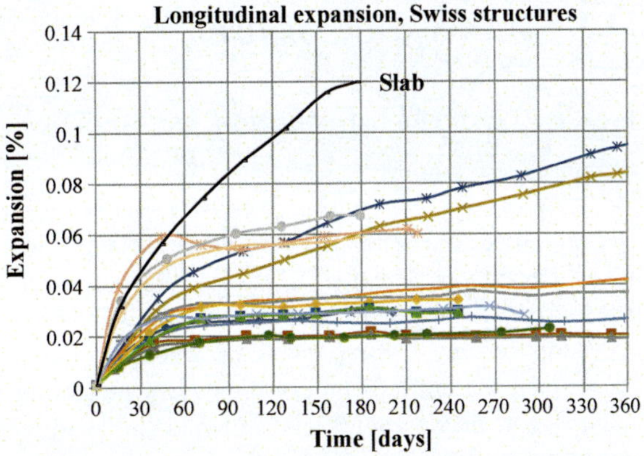

Fig. 14.8 Expansion of the concrete cores from the slab and the dam as a function of time compared to the data base of Swiss structures. Note: before test duration was defined to be six month, cores were measured up to more than a year

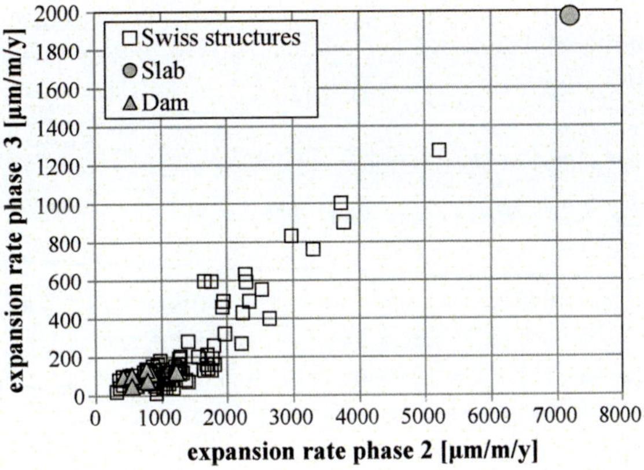

Fig. 14.9 Expansion rates during phases 2 and 3 of concrete extracted from Swiss structures (squares), from the tested slab (circle) and from the dam (triangles, see Sec. 14.3)

Based on their expansion during phase 2 and 3 in the residual expansion test, concrete from structures in Switzerland are classified as (see Chap. 7):

- concrete with a residual expansion potential zero or very low.
- concrete with a moderate residual expansion potential.
- concrete with a high residual expansion potential.

The tested concrete can be assessed as a concrete with fast kinetics and obviously a high residual expansion potential. The expansion rates during phases 2 and 3 are both higher than the ones reported from Swiss concrete structures (Fig. 14.9). Here, it has to be mentioned, that Swiss aggregates are relatively slow reacting. Volcanic rocks containing glass that are present in the studied concrete (see Sec. 14.2.6) are absent.

14.3 Gravity Concrete Dam

14.3.1 General Information

The gravity concrete dam is given as a second example for the application of the Swiss method. No details on the several decades old structure are given as it should not be identifiable. The investigation of the dam was triggered, because deformations that indicated concrete expansion were registered during monitoring.

There are only few typical AAR-cracks on the surface of the dam. However, a crack spans the entire length of the gallery located in the interior of the dam.

Only part of the analysis is shown as the example given in paragraph 2 already demonstrates, how the residual expansion measurements can be complemented with microscopy. The purpose of this example is to show the approach to assess the spatial distribution of damage and residual expansion potential in a structure. The horizontal cores taken from the dam (see the following paragraph) are used to show this approach.

14.3.2 Materials and Methods

Horizontal and vertical cores with a diameter of 100 mm were taken in three parts of the dam. The vertical and horizontal cores had a length of about 11 m and 5 m, respectively. The samples were sealed immediately after coring to avoid changes in their moisture state. Samples for microscopy were prepared from all cores at different concrete depth always using an 11 cm long section of the concrete core. This allowed to produce thin Sects. (50×90 mm^2) parallel and perpendicular to length axis of the cores (total of 54 thin sections). Impregnation followed the same protocol described in paragraph 2.2. Crack indices and percentages of aggregates exhibiting cracks typical for ASR were determined.

Data from the vertical cores, results on microstructure acquired by SEM and on water soluble alkali are not shown.

14.3.3 Results

Close to the south-ward directed surface of the dam the relative number of reacted aggregates in blocks B and C is significantly higher compared to the interior of the dam (Fig. 14.10). There is no such trend visible in block C. The same applies to the crack-indices that are clearly higher close to the surface of the dam in blocks B and C than in the interior (Fig. 14.11) confirming a more pronounced AAR in this area.

The expansion determined in the residual expansion test during phases 2 and 3 was used to calculate an expansion rate for the concrete. The highest expansion rates

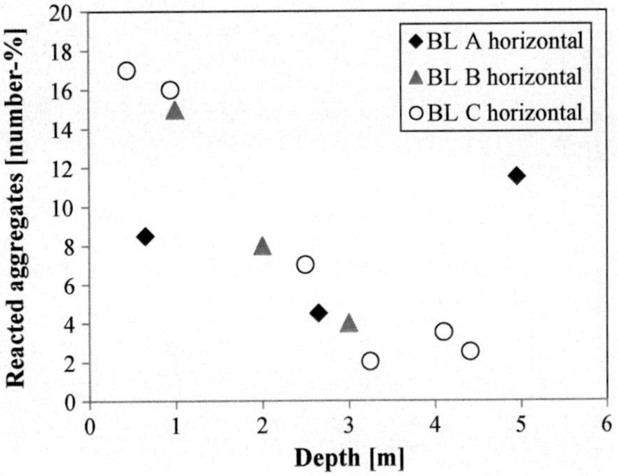

Fig. 14.10 Relative number of reacted aggregates in the horizontal cores as function of concrete depth

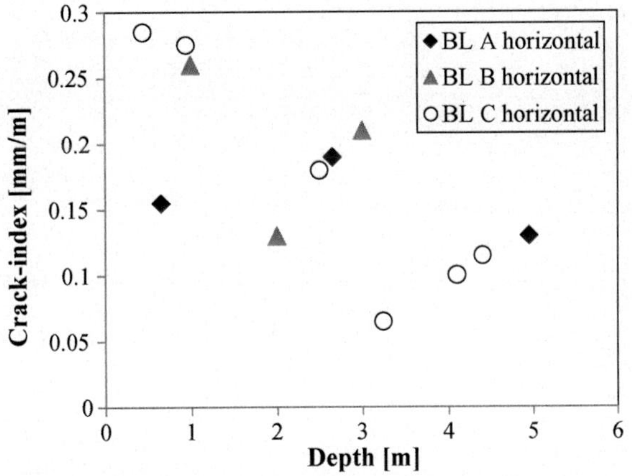

Fig. 14.11 Crack-indices of the horizontal cores as function of concrete depth

in all three horizontal cores during phase 2 occur close to the surface (Fig. 14.12). The values during phase 3 are about 10 times lower compared to phase 2 without showing a systematic trend as function of concrete depth (Fig. 14.13). The values in block C during phase 3 indicate somewhat faster kinetics compared to the other block.

14.3.4 Summary and Conclusions

The current experiences with the determination of the residual expansion measurements indicate that the expansion during phase 2 is connected with expansion of the concrete since occurrence of ASR until the time of core extraction [2, 3]. Consequently, the expansion during phase 2 indicates that the expansion in the concrete

Fig. 14.12 Expansion during phase 2 of the horizontal cores as function of concrete depth

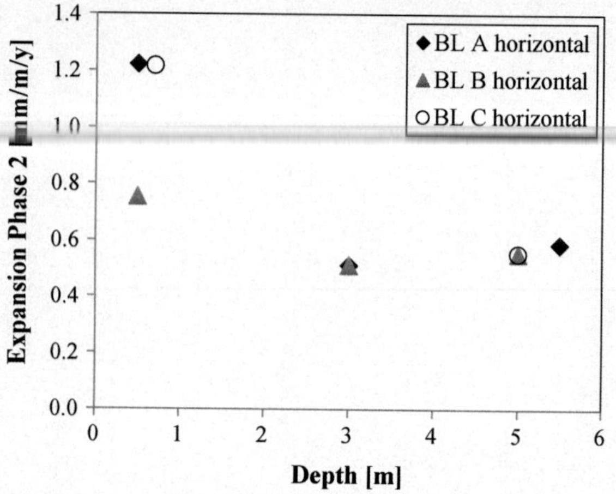

Fig. 14.13 Expansion during phase 3 of the horizontal cores as function of concrete depth

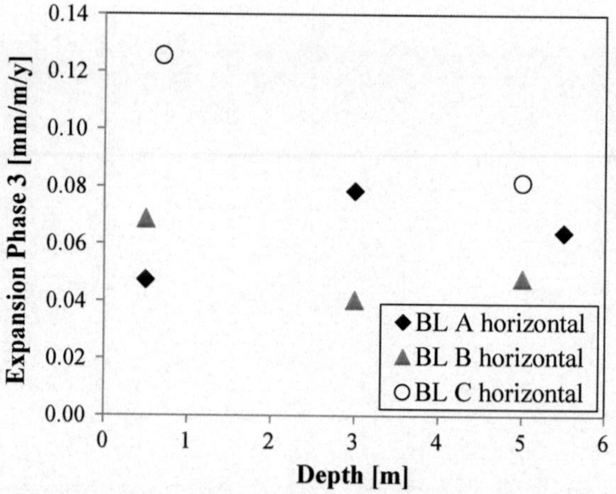

close to the surface has been larger compared to the interior of the dam. This can have two different reasons. First, the concrete in dams often contains more cement and was produced with a lower w/c resulting close to the surface resulting in a higher alkalinity of the concrete pore solution and a higher degree of ASR. Secondly, the exposure of the dam to the south (air side and location of the coring sites) leads to increased temperature during the day increasing SiO_2 solubility and ASR. However, there are no indications detected in microscopy for a significant difference in concrete composition. In any case, the higher expansion rate close to the surface goes together with a higher degree of ASR in this part of the dam, as clearly indicated by the relatively high number of reacted aggregates and the relatively high crack-indices.

The expansion during phase 3 represents the kinetics of expansion of the tested concrete [2, 3]. Because there is no clear difference in expansion as a function of concrete depth, the degree of ASR-induced damage and the higher expansion rate during phase 2 is likely attributable to the high daily temperature caused by sun radiation. A higher cement content and lower w/c should have resulted in higher expansion during phase 3 of the concrete close to the surface.

Compared to the expansion rates of concrete from other structures, the values for this specific dam are low (Fig. 14.9). This correlates well to the age of the dam and the relative low degree of cracking observed on the concrete surface. Although cracking of the concrete is only small to moderate, many of the cracks as observed in microscopy are not filled with ASR products. This is another sign for a very slow reaction. But it additionally shows that ASR has not reached a "ripe" stage (indicated by reaction products filling cracks and voids) but it still progressing.

In order to model the behavior of the dam in the future, the expansion rates during phase 3 should be used, but the likely temperature-driven higher expansion on the south side should be taken into account. Even if the expansion rates as determined during phase 3 have to be higher than in reality due to the elevated temperature of 38 °C, modelling could identify the areas of the dam where problems will likely occur. As the dam is regularly monitored, these in situ measurements could provide a valuable benchmark for the modelling.

References

1. Merz, C., Leemann, A.: Validierung der AAR-Prufüngen für Neubau und Instandsetzung. Technical Report AGB 2005/023 and AGB 2006/003. Bern, Switzerland: ASTRA-Bericht, 2012
2. Merz, C., Leemann, A.: Assessment of the residual expansion potential of concrete from structures damaged by AAR. Cem. Concr. Res. **52**, 182–189 (2013)
3. Multon, S., Barin, F., Godart, B., Toutlemonde, F.: Estimation of the residual expansion of concrete affected by alkali silica reaction. J. Mater. Civ. Eng. **20**, 54–62 (2008)

Chapter 15
Accelerated Expansion Test Sample Report: IFSTTAR

Renaud-Pierre Martin, Bruno Godart, and François Toutlemonde

15.1 Scope

This document provides an example of use of the residual expansion test described in the testing method LPC nr. 44 by [1].

Use of the residual expansion test results described in this document are part of a more complete structure management procedure proposed to deal with AAR problems in France [2, 3]. Thus, this test should not be used separately but as one of the elements of a general methodology for expert investigations; consequently, a single residual expansion measurement cannot be considered as representative enough to conclude or to take decision on the problem of an AAR-affected structure.

The experimental technique used hereafter is not a standard and has been developed as an experimental diagnosis tool by the French Public Works Research Laboratory (LCPC).

R.-P. Martin (✉)
Urban and Civil Engineering Testing and Modeling Laboratory (EMGCU), Materials and Structures Department, IFSTTAR, French Institute of Science and Technology for Transport, Development and Networks, Université Paris Est, 77447 Marne la Vall'ee, France
e-mail: renaud-pierre.martin@ifsttar.fr

B. Godart · F. Toutlemonde
Materials and Structures Department, IFSTTAR, French Institute of Science and Technology for Transport, Development and Networks, Université Paris Est, 77447 Marne la Vall'ee, France
e-mail: bruno.godart@ifsttar.fr

F. Toutlemonde
e-mail: francois.toutlemonde@ifsttar.fr

© RILEM 2021
V. E. Saouma (ed.), *Diagnosis & Prognosis of AAR Affected Structures*, RILEM State-of-the-Art Reports 31,
https://doi.org/10.1007/978-3-030-44014-5_15

15.2 Referenced Documents

Fasseu [1].
Fasseu and Mahut [2]
Godart et al. [3].

15.3 Example of Experimental Test Results

15.3.1 Introduction

During the expert investigation of a structure (not named in the following), residual expansion tests of several cored specimens have been applied to assess the risk of further expansion in the structure.

Due to the duration of this test, the corresponding results are presented in this specific report.

Among the cores taken from the structure, 10 have been tested using the methodology of the LPC 44 test method [1].

15.3.2 Principle of the Test

The objective of the test is to assess the potential of residual expansion in the investigated concrete. This is done by measuring the expansion of cores taken from the structure in affected or potentially affected locations.

15.3.3 Coring

10 cores have been drilled out from the structure for residual expansion testing. Their characteristics are provided in Table 15.1.

15.3.4 Instrumentation of the Cores

The cores are equipped with steel studs for the use of an extensometer. Their positions is described in Fig. 15.1 from the LPC 44 method [1].

The steel studs are placed every 120° on the lateral sides of the cylinders. The gauge length is 10 cm. Each measurement basis is equidistant from the top and bottom faces. When two measurement levels are chosen, two measurement bases are

Table 15.1 Position of the cores and test program

Name	Position**	Direction	Picture*	Test on core #1	Test on core #2
Core #1	Position 1	Sub-horizontal		Porosity	Mechanical behavior – compression
Core #2	Position 2	Horizontal		Mechanical behavior – compression	Residual expansion + expansive products identification
Core #3	Position 3	Vertical		Porosity	Mechanical behavior – compression
Core #4	Position 4	Sub-horizontal		Mechanical behavior – compression	Residual expansion + expansive products identification
Core #5	Position 5	Horizontal		Mechanical behavior – compression	Residual expansion
Core #6	Position 6	Vertical		Mechanical behavior – traction	Residual expansion
Core #7	Position 7	Horizontal		Mechanical behavior – compression	Residual expansion + expansive products identification
Core #8	Position 8	Vertical		Mechanical behavior – traction	Residual expansion + expansive products identification
Core #9	Position 9	Horizontal		Mechanical behavior – traction	Residual expansion + expansive products identification
Core #10	Position 10	Horizontal		Mechanical behavior – traction	Residual expansion + expansive products identification
Core #11	Position 11	Vertical		Mechanical behavior – compression	Residual expansion

(continued)

Table 15.1 (continued)

Name	Position**	Direction	Picture*	Test on core #1	Test on core #2
Core #12	Position 12	Horizontal		Porosity	Mechanical behavior – compression
Core #13	Position 13	Horizontal		Porosity	Mechanical behavior – compression
Core #14	Position 14	Horizontal		Mechanical behavior – traction	Residual expansion

*Pictures normally provided in an appendix
**Positions have been anonymized

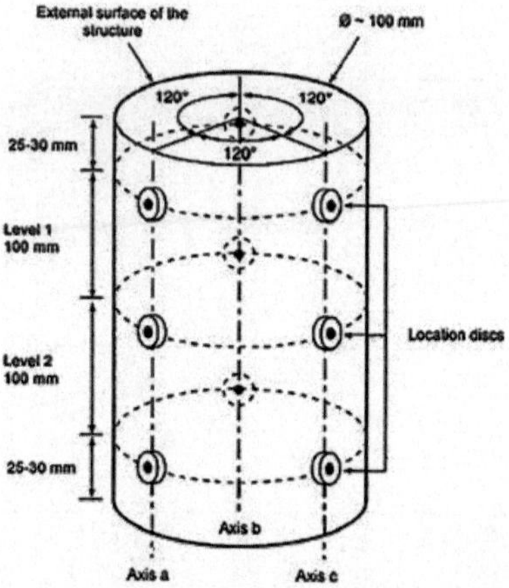

Fig. 15.1 Position of the location discs (steel studs)

placed, once again equidistant from the end faces. The studs are glued with an epoxy resin after drying of the surface with compressed air. The specimen is then stored in a laboratory at ambient temperature during 24 h.

15.3.5 Cores Storage Condition: Wrapping of the Cores

Firstly, every core is weighed. It is then wrapped in absorbent paper moistened with distilled water. A polyethylene sheet with identical dimensions is then placed with openings to have an access to the studs. This system is fixed with an adhesive tape. The end faces remain unprotected; the name of the specimen is written on them (Fig. 15.2).

The initial reading is made after wrapping. The specimen is then placed in a container (on grid above the water level). The container is stored in a reactor with

Measurement after instrumentation Wrapping of the cores

Fig. 15.2 Preparation of the specimens

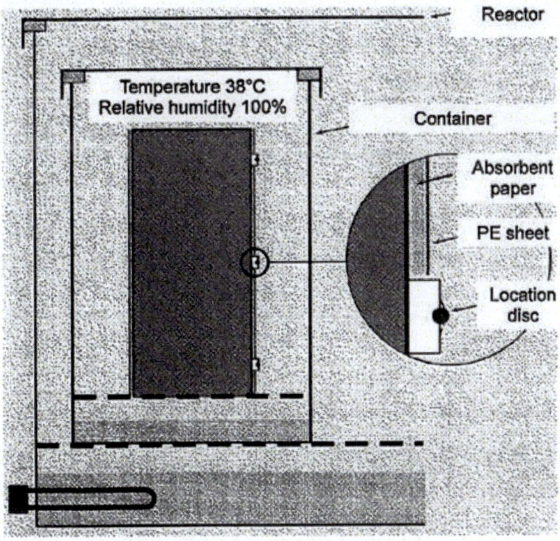

Fig. 15.3 Storage of a specimen in the reactor based on the LPC 44 method [1] (note: the initial publication of the LPC nr. 44 method required a reactor and a container conform to the standard P18-587 presently substituted by NF P18-594:2015) (See NF P 18-454:2004)

a controlled temperature (38 ± 2 °C) according to Fig. 15.3. The bottom of the container is filled with water.

15.3.6 Results

Measurements have been made after 2, 4, 6, 8, 12, 16, 20, 24, 28, 32, 36, 40, 44, 48 and 52 weeks. Table 15.2 displays the results. Figure 15.4 illustrates the evolution of residual expansion as a function of time.

Table 15.2 Results obtained for the 10 cores monitored (in μm/m)

Time (weeks)	Core #4	Core #5	Core #6	Core #7	Core #8	Core #9	Core #10	Core #2	Core #11	Core #14
0	0	0	0	0	0	0	0	0	0	0
2	0	19	213	58	19	78	78	116	0	78
4	116	23	213	155	39	116	155	369	19	136
6	175	388	369	252	58	175	252	349	58	175
8	466	446	446	524	116	369	466	679	155	388
12	679	563	543	679	127	524	679	873	233	485
16	854	621	582	815	155	660	795	951	272	563
20	970	679	601	892	175	737	873	1009	291	621
24	1067	698	640	970	194	834	951	1028	330	660
28	1125	698	640	1009	213	873	970	1048	349	679
32	1183	698	698	1048	233	931	989	1067	369	718
36	1319	698	698	1067	233	989	1028	1106	369	737
40	1339	698	718	1086	252	1028	1028	1106	388	757
44	1397	718	718	1106	252	1048	1048	1125	388	776
48	1436	718	737	1106	252	1086	1048	1145	388	776
52	1494	718	757	1125	252	1145	1086	1183	388	795

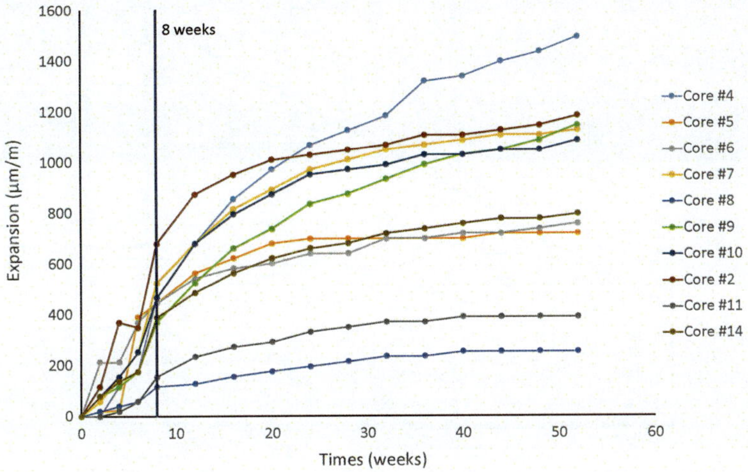

Fig. 15.4 Residual expansion versus time

15.3.7 Data Analysis

The analysis of the results is based on the feedback obtained during elaboration of the LPC 44 method [1].

The ranking of the different residual expansions is based on the LPC 44 method [1]. This method recommends the use of the expansions measured between 8 and 52 weeks (results below the double line in Table 15.2) to evaluate an expansion rate per year (values multiplied by 1.2) and propose a semi-quantitative ranking.

Indeed, during the first 8 weeks of the test, several processes are under progress, including a water uptake of the core, a re-hydration of the AAR gels already existing and the formation of new gels while from 8 to 52 weeks, expansion is believed to be mainly due to the formation of new gels.

The qualification criteria of the LPC 44 method [1] are provided in Table 15.3.

Figures 15.5, 15.6, 15.7, 15.8, 15.9, 15.10, 15.11, 15.12, 15.13 and 15.14 provide the linear regression slopes for the expansions measured between 8 and 52 weeks; each line is then extended to its intersection with the y-axis. The equations of the lines are provided with each figure.

Table 15.3 Residual expansion qualification criteria

Value	Ranking of the residual expansion
<100 µm/m/year	Negligible
From 100 to 500 µm/m/year	Moderate
>500 µm/m/year	High

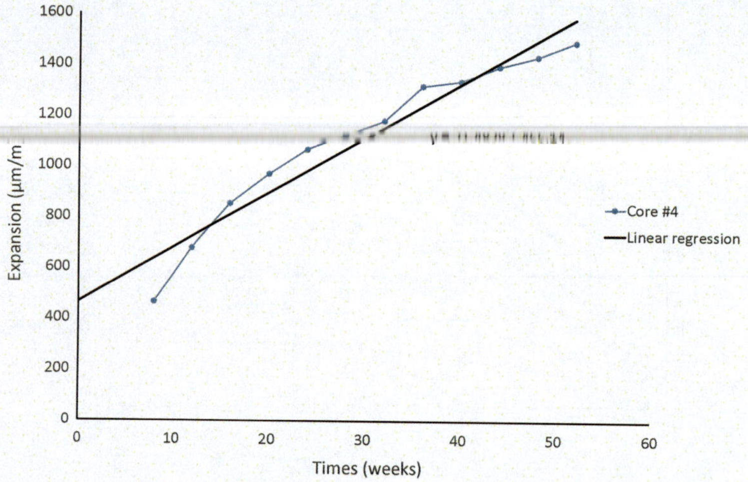

Fig. 15.5 Mean slope of expansion between 8 and 52 weeks for core #4

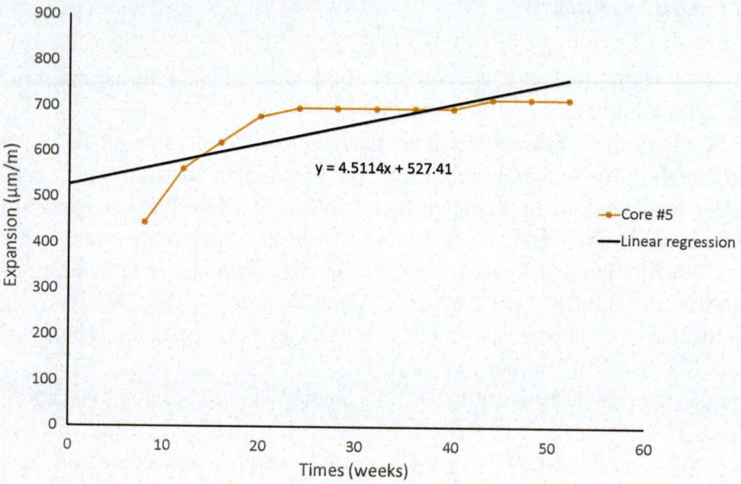

Fig. 15.6 Mean slope of expansion between 8 and 52 weeks for core #5

15.3.8 Conclusions

Based on the present interpretation that can be considered as conventional, our feedback leads us to conclude that the residual expansions measured on the cores drilled out from the structure are **moderate to high** as presented in Table 15.4.

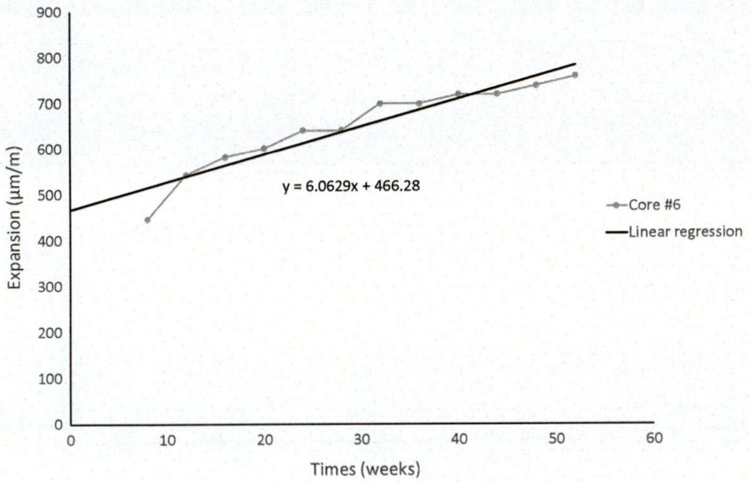

Fig. 15.7 Mean slope of expansion between 8 and 52 weeks for core #6

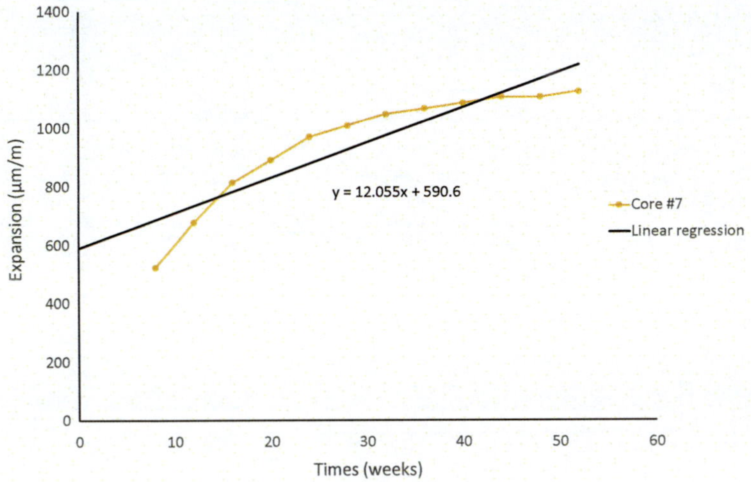

Fig. 15.8 Mean slope of expansion between 8 and 52 weeks for core #7

The concrete expansions are heterogeneous not only in the structure (they depend on the structural part considered), but also inside a given part of the structure. This fact is not specific to this structure.

Finally, one must keep in mind that expansions measured are obtained on cores under free expansion (while they were previously under a given stress state inside the structure) in environmental conditions that accelerate the reaction.

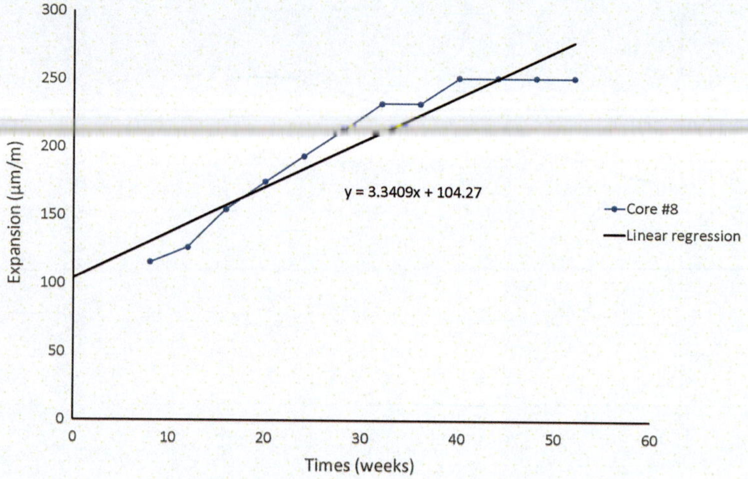

Fig. 15.9 Mean slope of expansion between 8 and 52 weeks for core #8

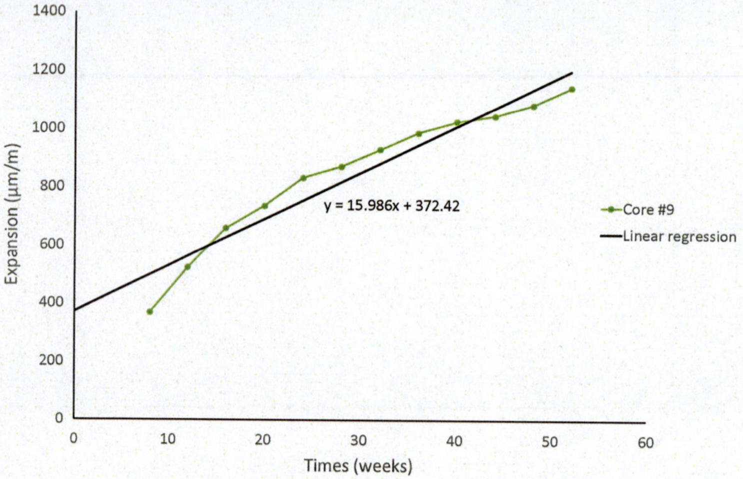

Fig. 15.10 Mean slope of expansion between 8 and 52 weeks for core #9

For a more detailed analysis, it is thus necessary to consider the stress state of the cores in the structure and their true environmental conditions (temperature and humidity).

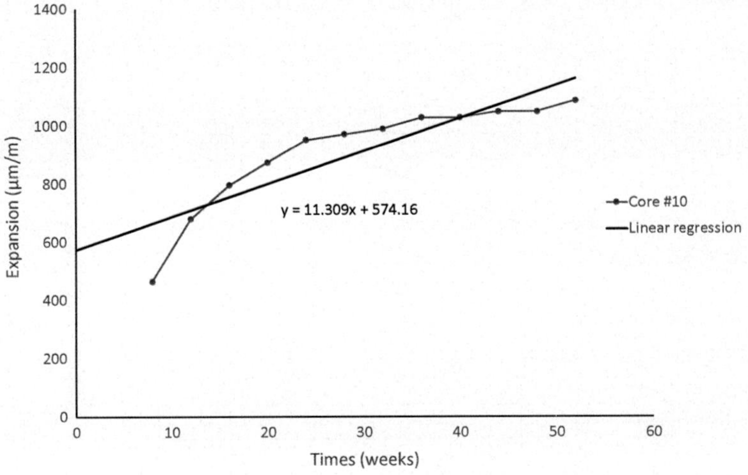

Fig. 15.11 Mean slope of expansion between 8 and 52 weeks for core #10

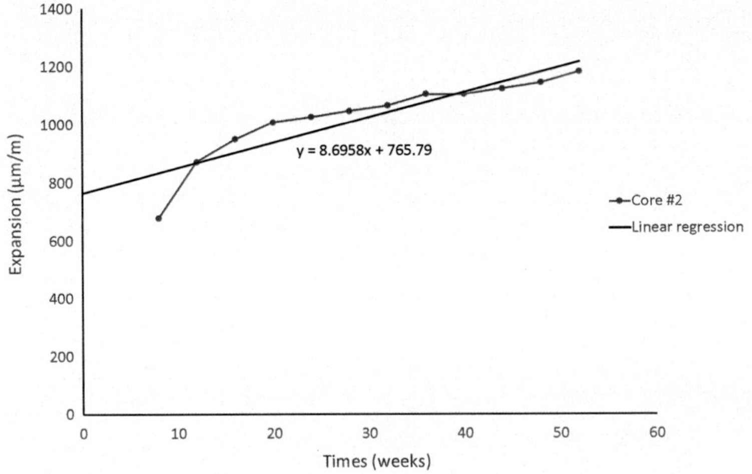

Fig. 15.12 Mean slope of expansion between 8 and 52 weeks for core #2

15.3.9 Perspectives

The present study could be continued by applying the same mechanical tests on the core after expansion testing as the mechanical tests described in the report on the structure, where a compressive test allowed to assess the compressive strength, the elasticiy modulus and the strain at failure on cores not subjected to expansion test.

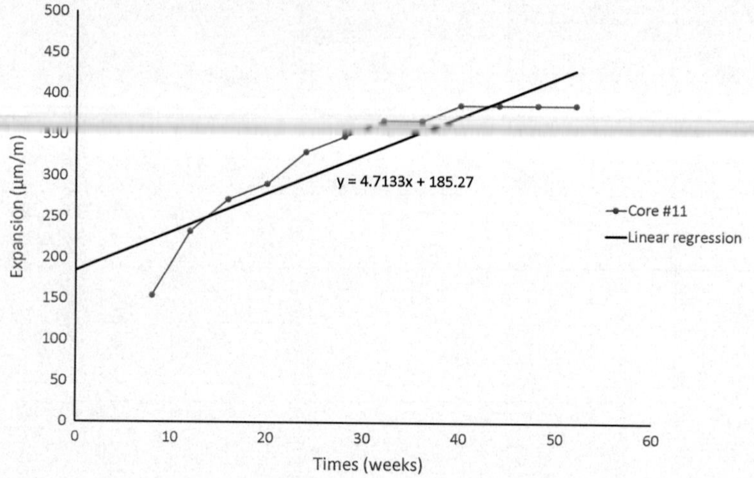

Fig. 15.13 Mean slope of expansion between 8 and 52 weeks for core #11

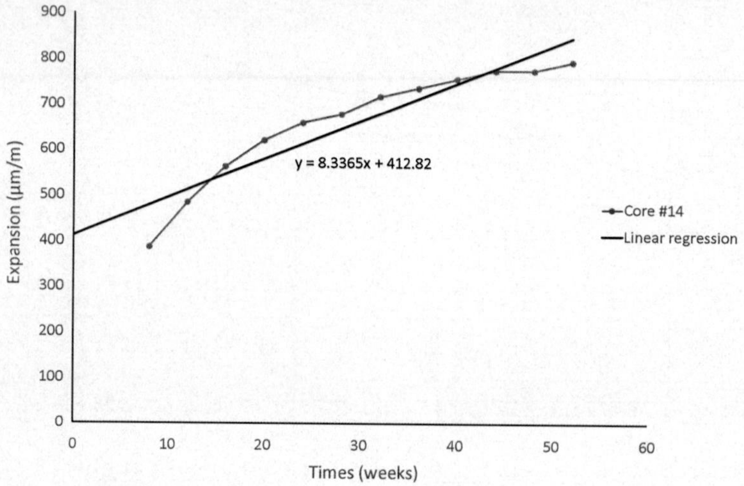

Fig. 15.14 Mean slope of expansion between 8 and 52 weeks for core #14

The values obtained with the present expanded cores could thus be compared to those from the report on the structure to illustrate the effect of the accelerated ageing procedure (38 °C and humidity ~100% RH) on the concrete.

Moreover, after the mechanical test applied on cores subjected to expansion testing, microscopic observations (scanning electronic microscope) would provide information about the nature and the quantity of the gels formed during the expansion test.

Table 15.4 Summary of the residual expansion characteristics of the cores investigated

	Linear regressions	Residual expansion (μm/m/year)	Ranking of the residual expansion
Core #4	y = 21.487x + 466.14	1117	High
Core #5	y = 4.5114x + 527.41	235	Moderate
Core #6	y = 6.0629x + 466.28	315	Moderate
Core #7	y = 12.055x + 590.6	627	High
Core #8	y = 3.3409x + 104.27	174	Moderate
Core #9	y = 15.986x + 372.42	831	High
Core #10	y = 11.309x + 574.16	588	High
Core #2	y = 8.6958x + 765.79	452	Moderate
Core #11	y = 4.7133x + 185.27	245	Moderate
Core #14	y = 8.3365x + 412.82	433	Moderate

References

1. Fasseu, P.: Alcali-Reaction Du Beton-Essai D'Expansion Residuelle Sur Beton Durci-Projet De Methode D'Essai. Technical Report No. 44. Laboratoire Central des Ponts et Chaussées (LPC), 1997. http://www.ifsttar.fr/fileadmin/user_upload/editions/lcpc/MethodeDEssai/MethodeDEssai-LCPC-ME44.pdf
2. Fasseu, P., Mahut, B.: Aide a la gestion des ouvrages atteints de reactions de gonflement interne. Technical Report 66 pp. Paris, France: Laboratoire Central des Ponts et Chaussees, 2003. http://www.ifsttar.fr/fileadmin/user_upload/editions/lcpc/GuideTechnique/GuideTechnique-LCPC-G0NFLIN.pdf
3. Godart, B., Mahut, B., Fasseu, P., Michel, M.: The guide for aiding to the management of structures damaged by concrete expansion in Franc. In: Proceedings of the 12th ICAAR, pp. 1219–1228, Bejing, China (2004)

Chapter 16
Accelerated Expansion Test Sample Report: DRP

David Rothstein and Chunyu Qiao

16.1 Introduction

This report describes the results of a petrographic investigation aimed at quantifying the degree of damage associated with alkali-silica reaction (ASR). The samples consisted of two cores extracted from the same concrete slab. One of the cores was sawn on both ends (Core 6A) and the other core (Core 6B) is a partial length core that was sawn on both ends and then sawn in half longitudinally (Fig. 16.1).

16.2 Materials and Methods

A concrete slab was prepared with reactive aggregates. The slab was stored at a high relative humidity and a few degrees above room temperature for approximately three years. Cores were then extracted and transported to the University of Colorado (Boulder). Core 6A was stored at 38 °C temperature (T) and 100% relative humidity (RH) for one year while Core 6B was sealed in a ziplock bag and stored in a refrigerator. Both cores were then delivered to DRP where they were kept in a refrigerator until they were studied.

The cores were sectioned longitudinally to provide a full length polished slab (Fig. 16.2). For Core 6B the slab was cut parallel to the received saw cut surface. The slabs were lapped and polished using standard procedures and without epoxy impregnation. Sets of petrographic thin sections were cut along the x-y and x-z

D. Rothstein (✉) · C. Qiao
DRP, A Twining Company, Boulder, CO, US
e-mail: petro@drpcinc.com

C. Qiao
e-mail: joe@drpcinc.com

© RILEM 2021

V. E. Saouma (ed.), *Diagnosis & Prognosis of AAR Affected Structures*,
RILEM State-of-the-Art Reports 31,
https://doi.org/10.1007/978-3-030-44014-5_16

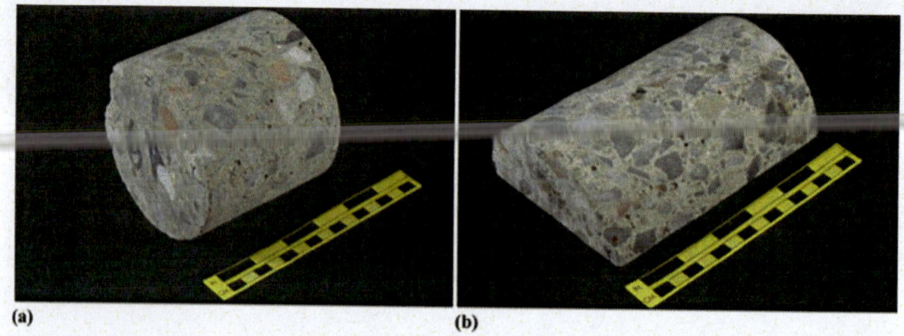

Fig. 16.1 Photographs of **a** Core 6A and **b** Core 6B in as-received condition

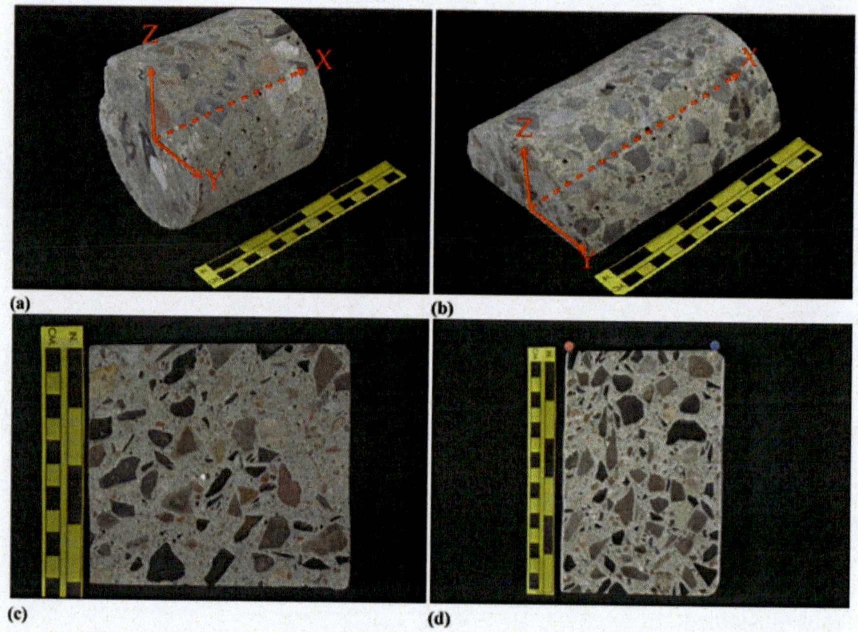

Fig. 16.2 Photographs of **a** Core 6A and **b** Core 6B showing orientation of saw cuts used to prepare lapped and polished slabs, thin sections and SEM sections. **c** and **d** show polished slabs from Core 6A and Core 6B, respectively

plane. The sections were impregnated with fluorescent epoxy prior to preparation. Six sections were prepared from Core 6A (three in the x-y plane and three in the x-z plane) and four sections were prepared for Core 6B (two in the x-y plane and two in the x-z plane). SEM sections were prepared from the same impregnated billets following standard procedures.

Three different methods were used to measure damage associated with ASR. These include (1) the Damage Rating Index (DRI) using weighing factors described by Sanchez et al. [1], (2) image analysis on transmitted fluorescent light photomicrographs (FLIA) and (3) image analysis on backscatter electron micrographs (BSEIA).

For the DRI analyses the samples were examined in reflected light at 15× magnification using a Nikon SMZ-25 stereomicroscope. General petrographic features were observed and documented. The entire area of the polished slab was examined and results were normalized to a 100 cm^2 area.

For the FLIA thin sections impregnated with fluorescent epoxy were examined using a Nikon E-Pol 600 petrographic microscope. For each thin section nine-areas were imaged at 50× and analyzed using *Image J* freeware. The *Ridges* plugin was used to identify and measure microcracks in each image. This provides information regarding the length, width and orientation of the microcracks. The amount of microcracking is then calculated as an area percentage for each image and microcrack density, which indicates the length of microcracks in a given area, is expressed in $\mu m/\mu m^2$.

For the BSEIA micrographs were obtained with a FEI Quanta 250 SEM. The instrument was run at low pressure (~80 Pa) with an accelerating voltage of 15 kV and a spot size of 5. For each section 25 images were collected at 100× magnification. These images were stitched together using the *Stitcher* plugin and then analyzed using the *Ridges* plugin. The amount of microcracking is also calculated as an area percentage and microcrack density is expressed in $\mu m/\mu m^2$.

16.3 Results

16.3.1 General Observations

Petrographic observations obtained via stereomicroscope and petrographic microscope indicate that siliceous volcanic rocks (rhyolite, andesite and dacite which commonly contain cristobalite and tridymite) in the fine aggregate are the reactive components. ASR was not observed in the coarse aggregate. Petrographic observations indicate Stage IV ASR using the classification scheme described by Katayama in other sections of the TC report. Figure 16.3 shows some of the typical reaction textures observed in the study.

16.3.2 Damage Rating Index

Table 16.1 and Fig. 16.4 show the results of the DRI analyses, which indicate more damage in the exposed core (Core 6A) than in the unexposed core (Core 6B). The

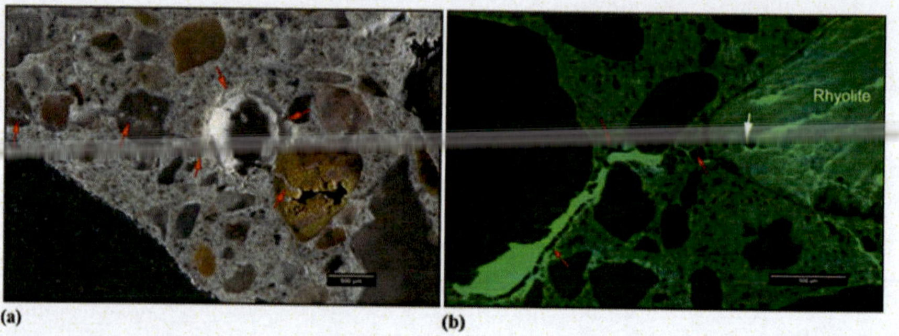

Fig. 16.3 **a** Reflected light photomicrograph of the polished surface of Core 6A showing microcracks filled with ASR gel and gel lining a void (red arrows) that are associated with reactive rhyolite particles in the fine aggregate. **b** Transmitted fluorescent light photomicrograph of thin section from Core 6B showing deposits of ASR gel (red arrows) in cracks that cut into the paste from a reactive rhyolite particle in the fine aggregate. The white arrow indicates a plug of gel near the edge of the reactive particle. Note that the rhyolite particle has elevated porosity

Table 16.1 Summary of DRI scores

Petrographic feature	Exposed (6A)	Unexposed (6B)
Crack in aggregate (CCA)	10	10
Open crack in aggregate (OCA)	51	81
Crack with ASR gel in aggregate (OCAG)	29	1
Aggregate debonded (CAD)	0	4
Crack in cement paste (CCP)	110	176
Crack with ASR gel in cement paste (CCPG)	190	68
Disaggregate/corroded aggregate (DAP)	60	40
Total score	450	380

Fig. 16.4 Summary of DRI scores shown in Table 16.1

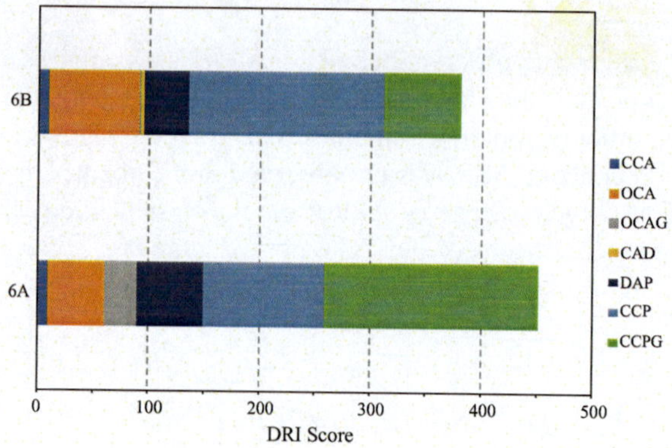

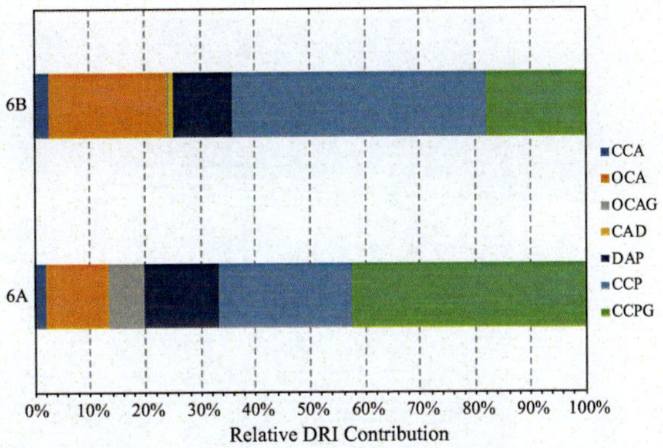

Fig. 16.5 Relative DRI contribution from different features. Core 6A shows higher percentage of the total DRI score from gel in cracks that cut through the paste (green; CCPG) and in cracks in aggregate particles (gray; OCAG) compared to Core 6B

DRI analyses include the fine aggregate because it contains the reactive components associated with ASR.

Figure 16.5 shows the results of the analyses in terms of the percent contribution from each feature to the total DRI score. The data indicate that there is a marked increase in gel in microcracks that cut through the paste and in microcracks within aggregate particles, after exposure to elevated T/RH conditions.

16.3.3 Fluorescent Transmitted Light Microscopy Image Analysis (FLIA)

The orientation, width and length of microcracks was measured using image analysis methods. The microcracks do not show a preferred orientation. Figure 16.6 shows an example of microcracks detected in an image obtained from Core 6B (not exposed) and Fig. 16.7 shows an example of microcracks detected in an image obtained from Core 6A (exposed).

Table 16.2 and Fig. 16.8 show the results of the FLIA, which indicate that Core 6A shows more microcracking than Core 6B. The FLIA does not show a significant difference between the amount of microcracking observed in the XY orientation or the XZ orientation in either sample. Core 6A shows higher microcrack densities than Core 6B.

(a)
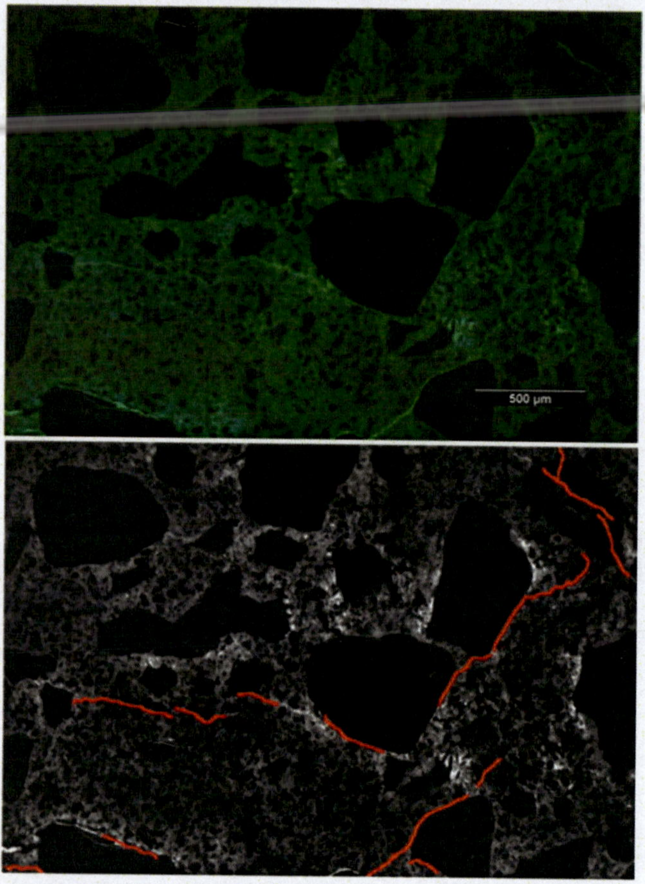

(b)

Fig. 16.6 **a** Transmitted fluorescent light image from Core 6B showing microcracking associated with ASR. **b** Grayscale image from **a** showing individual detected microcrack segments in red

16.3.4 Backscatter Electron Microscopy Image Analysis (BSEIA)

Figure 16.9 shows microcracks detected in the stitched composite image obtained from Core 6B and Fig. 16.10 shows microcracks detected in the stitched composite image obtained from Core 6A.

Table 16.3 and Fig. 16.11 show the results of the BSEIA, which indicate Core 6A shows more cracking than Core 6B. The BSEIA shows a significant difference between the amount of microcracking observed in the XY orientation or the XZ orientation.

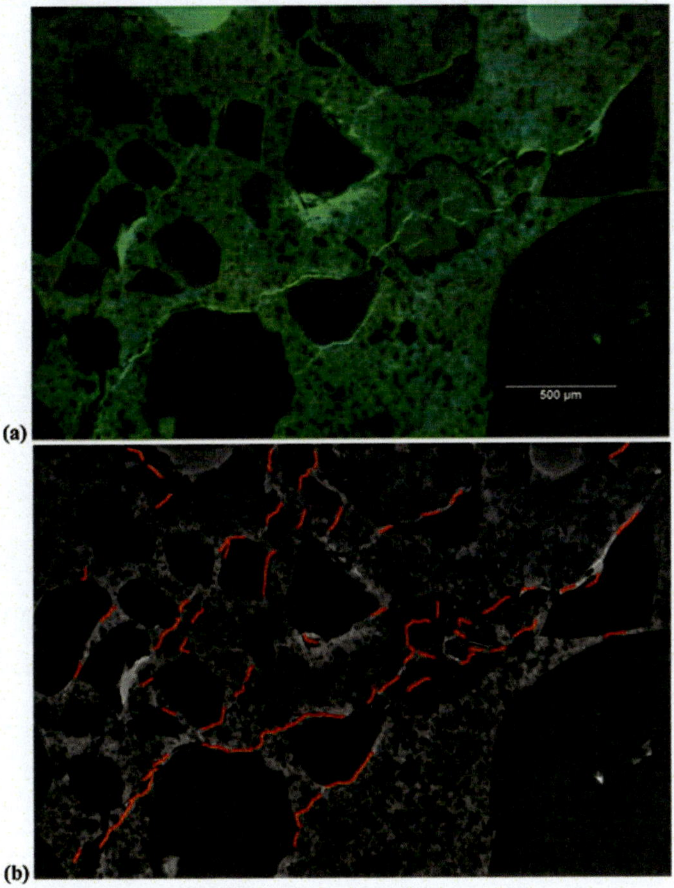

Fig. 16.7 **a** Transmitted fluorescent light image from Core 6A showing microcracking associated with ASR after exposure to elevated T/RH conditions. **b** Grayscale image from **a** showing individual detected microcrack segments in red

Table 16.2 Summary of microcracking measurements by FLIA. Core 6A was exposed to elevated T/RH conditions while Core 6B was refrigerated

	Core 6A	Core 6B
Area of microcracking (X-Y)	1.495%	0.634%
Microcracking density (X-Y)	0.0011 $\mu m/\mu m^2$	0.0005 $\mu m/\mu m^2$
Area of microcracking (X-Z)	1.441%	0.620%
Microcracking density (X-Z)	0.0013 $\mu m/\mu m^2$	0.0008 $\mu m/\mu m^2$

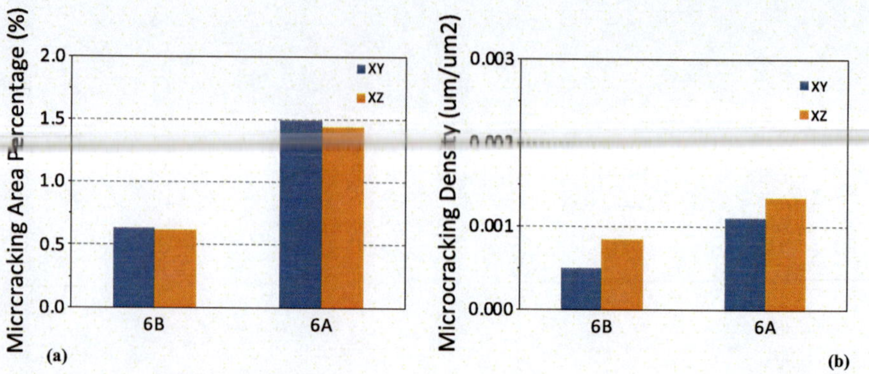

Fig. 16.8 Bar charts summarizing data shown in Table 16.2. Note that there is little difference between the area of microcracking observed in the XY and XZ direction between the samples

16.4 Summary and Conclusions

16.4.1 Summary

Table 16.4 and Fig. 16.12 summarize differences observed in the amount of damage measured by each method. The BSEIA results are relatively consistent with the DRI in the XY plane while the FLIA shows significantly greater increases in damage. The cause of the disparity between the percentage increase in damage found by FLIA and BSEIA is not clear. One potential factor may involve interference from porous aggregate in the fluorescent images, which complicates the image analysis. Other factors may include a larger interaction volume between transmitted light microscopy and scanning electron microscopy.

16.4.2 Conclusions

The findings described above demonstrate that the exposure conditions generated a measurable increase in damage related to ASR. The DRI method shows an increase in overall damage with increased amounts of gel observed in cracks that cut through the paste and in cracks within aggregate particles. The FLIA and BSEIA methods show increased damage in the exposed sample as well, with the area of microcracking increasing after exposure. The DRI and BSEIA methods showed comparable increases in the amount of damage measured after exposure.

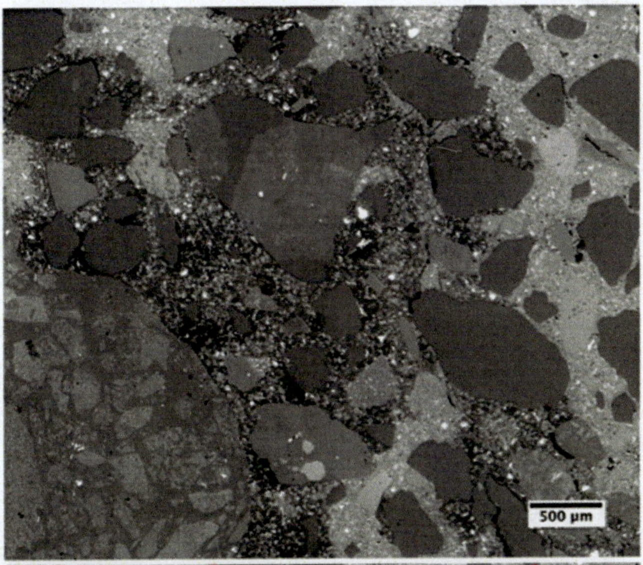

(a)

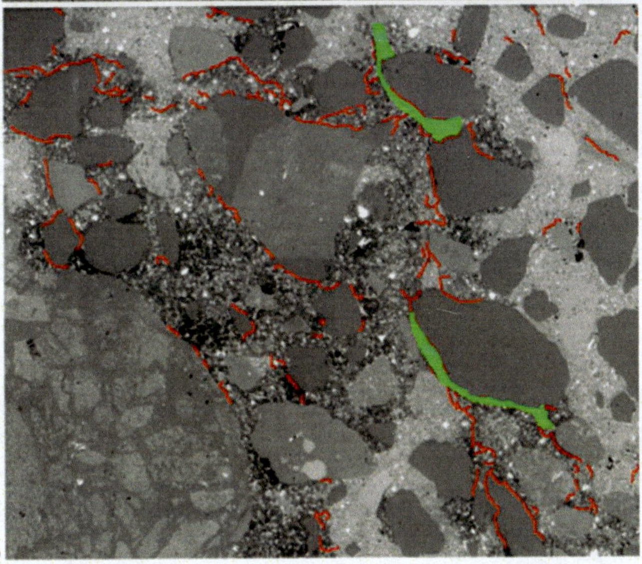

(b)

Fig. 16.9 **a** BSE montage of 25 images showing microstructure of Core 6B. **b** BSE montage shown in **a** with crack segments shown in red; the green segment indicates a crack filled with ASR gel

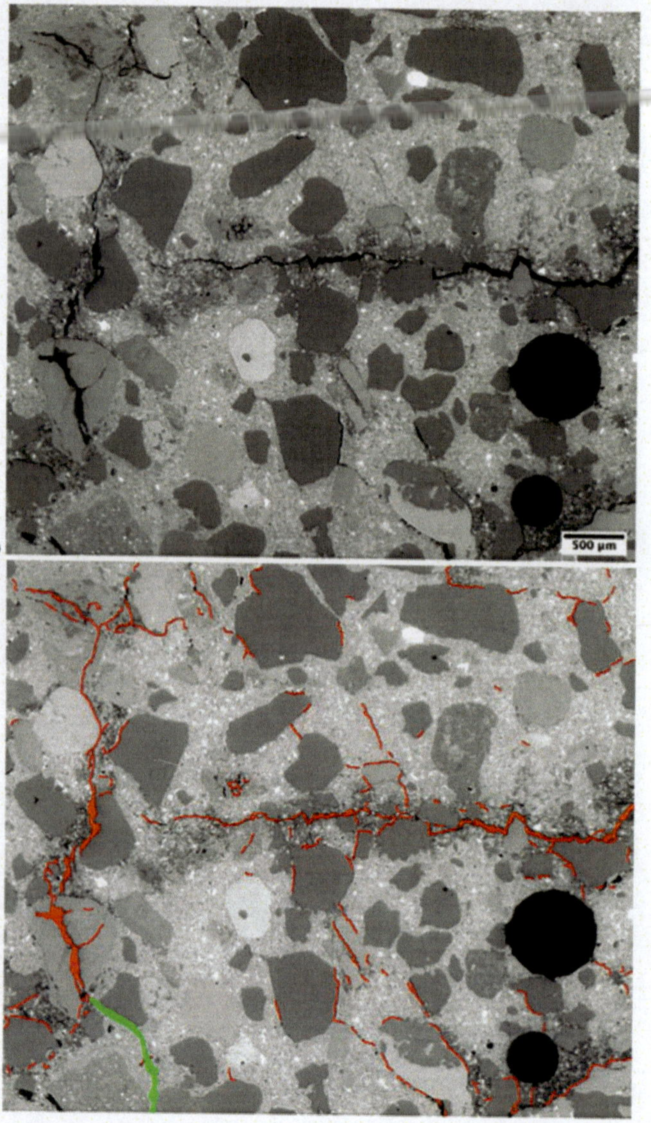

(a)

(b)

Fig. 16.10 **a** BSE montage of 25 images showing microstructure of Core 6A. **b** BSE montage shown in **a** with crack segments shown in red; the green segment indicates a crack filled with ASR gel

Table 16.3 Summary of microcracking measurements by BSEIA

	Exposed (Core 6A)	Unexposed (Core 6B)
Area of microcracking (X-Y)	2.429%	2.033%
Microcracking density (X-Y)	0.0011 μm/μm^2	0.0017 μm/μm^2
Area of microcracking (X-Z)	4.432%	3.650%
Microcracking density (X-Z)	0.0013 μm/μm^2	0.0008 μm/μm^2

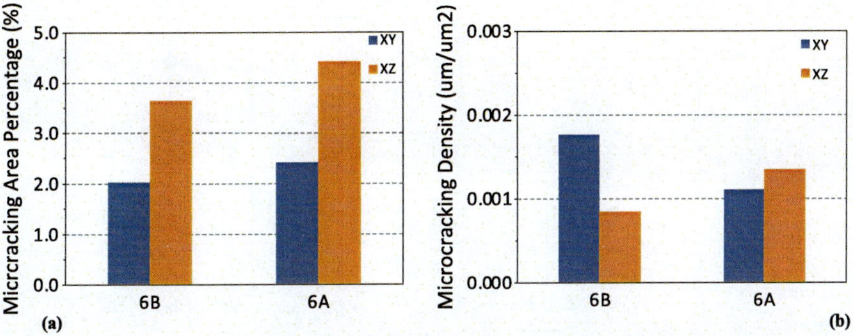

Fig. 16.11 Bar charts summarizing data shown in Table 16.3. Note that there is a significant difference in the area of microcracking observed in the XY and XZ directions, with more cracking in the XZ direction

Table 16.4 Summary of damage measurements in Core 6A (exposed to elevated T/RH) and Core 6B (refrigerated)

	Core 6B	Core 6A	Delta
Damage rating index (XY)	380	450	18.4%
FLIA cracking area (XY)	0.634%	1.495%	134.2%
BSEIA cracking area (XY)	2.033%	2.429%	20.7%
FLIA cracking area (XZ)	0.620%	1.441%	132.4%
BSEIA cracking area (XZ)	3.650%	4.432%	21.4%

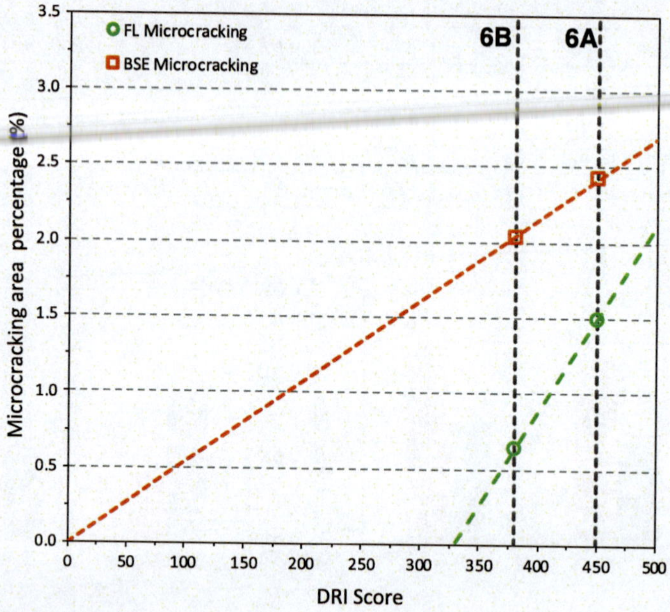

Fig. 16.12 Summary plot of damage measurements in the XY plane. The DRI method and BSEIA show a comparable increase (~20%) in damage whereas the FLIA shows a much higher increase (~130%)

Reference

1. Sanchez, L.F.M., Fournier, B., Jolin, M., Bedoya, M.A.B., Bastien, J., Duchesne, J.: Use of damage rating index to quantify alkali-silica reaction damage in concrete: fine versus coarse aggregate. ACI Mater. J. **113**(4), 395–407 (2016)

Chapter 17
Accelerated Expansion Test Sample Report: Laval

Leandro Sanchez and Diego Jesus de Souza

17.1 Introduction

Two test procedures are recommended as per *Laval University Method* [1], in which concrete cores are: (1), tested in air at >95% R.H. and 38 °C (100 °F), and (2), tested in 1N NaOH solution at 38 °C (100 °F). The first test is considered the most realistic for evaluating the "residual expansion" and the potential for further expansion of ASR-affected concrete whereas the second approach accounts for the "maximum expansion potential" of the concrete [2]. This document describes the principles of the practical use of the above two techniques.

17.2 Samples Preparation

AAR-affected samples (e.g. 100 by 200 mm in size) cored from damaged structures/structural members under analysis are first superficially dried and wrapped in plastic film in order to minimize the effects of drying shrinkage and/or further expansion due to rewetting. Then, the specimens are kept in the laboratory at room temperature (i.e. 23 ± 2 °C) for at least five days (not more than two weeks) to reach a relative equilibrium with respect to the new stress and environmental (temperature and moisture) conditions to which the concrete is now exposed. After this period, small holes (e.g. 8 mm diameter by 20 mm deep) are drilled at the ends of the concrete specimens and stainless steel bolts (e.g. 3 mm diameter by 13 mm in length, presenting "demec points") are installed with a fast-setting (shrinkage-free) cementitious slurry.

L. Sanchez (✉) · D. J. de Souza
University of Ottawa, Ottawa, Canada
e-mail: Leandro.Sanchez@uottawa.ca

© RILEM 2021

V. E. Saouma (ed.), *Diagnosis & Prognosis of AAR Affected Structures*, RILEM State-of-the-Art Reports 31,
https://doi.org/10.1007/978-3-030-44014-5_17

17.3 Experimental Procedure and Results

Expansion in air at >95% R.H. and 38 °C: The initial mass and dimensions (i.e. length, diameter) of the AAR-affected core samples are first measured; then, they are placed vertically above water in sealed plastic buckets presenting a water film at the bottom and wicks lining the inside walls, to keep the R.H. >95%. The buckets are then stored in a room and maintained at 38 ± 2 °C, as per the *Concrete Prism Test* (*CPT, ASTM C 1293*) and the samples are monitored over time. Whenever the samples are measured, the sealed buckets are taken from the storage room and kept for 16 h at 23 ± 2 °C. Afterwards, the specimens are removed from the buckets individually and their longitudinal (i.e. axial) deformation and mass are recorded. Table 17.1 shows the average length and mass variation over time from three AAR-affected specimens exposed to >95% R.H. and 38 °C for 504 days.

Expansion soaked in 1N NaOH R.H. and 38 °C: The initial mass and dimensions (i.e. length, diameter) of the AAR-affected core samples are first measured; then, they are placed vertically in sealed plastic buckets, soaked into a 1N NaOH solution and stored in a room and maintained at 38 ± 2 °C. The measurement procedure is exactly the same as the method at >95% R.H. and 38 °C. Table 17.1 shows the average length and mass variation over time from three AAR-affected specimens soaked in 1N NaOH solution.

Figure 17.1 gives a plot from the data (expansion vs time) displayed in Tables 17.1 and 17.2 for both test procedures. Analyzing the plot below, one verifies that the core samples tested at both procedures yielded high expansion levels after 504 days, and

Table 17.1 Average length and mass variation from three specimens exposed to >95% R.H. and 38 °C

Time (days)	Expansion (%)	Mass (%)
0	0.000	0.000
7	0.003	0.044
14	0.009	0.069
28	0.020	0.158
42	0.022	0.287
70	0.029	0.356
85	0.032	0.510
122	0.051	0.728
150	0.096	0.688
213	0.130	0.708
241	0.164	0.732
301	0.163	0.748
362	0.169	0.748
424	0.176	0.764
452	0.177	0.777
487	0.177	0.797

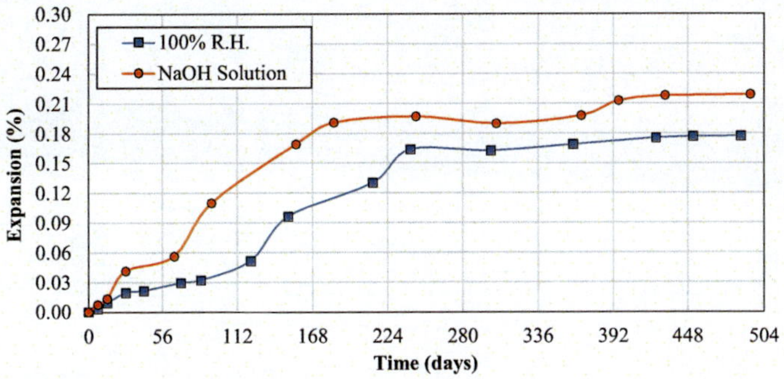

Fig. 17.1 Alkali-Aggregate Reaction (AAR)-kinetics and induced expansion over time for the concrete core samples tested at >95% R.H. and soaked in 1N NaOH R.H. at 38 °C

Table 17.2 Average length and mass variation from three specimens soaked in 1N NaOH solution and 38 °C

Time (days)	Expansion (%)	Mass (%)
0	0.000	0.000
7	0.008	0.238
14	0.014	0.277
28	0.041	0.420
65	0.056	0.372
93	0.110	0.531
156	0.169	0.654
184	0.191	0.654
245	0.197	0.733
305	0.190	0.745
368	0.198	0.745
396	0.213	0.745
431	0.218	0.757
494	0.219	0.776

thus may be classified as having a high potential of residual (or further) induced expansion as per Bérubé et al. [3]. Furthermore, two interesting points might be noticed: (1) the ultimate expansions gathered in both procedures were quite similar after 500 days (i.e. 0.177% for >95% R.H. and 0.219% for soaked in 1N NaOH). It is worth noting that after about 400 days, all samples from the two procedures presented a levelling off trend, and reached their plateaus and, (2) the expansion rates (i.e. expansion/year) obtained for the >95% R.H. procedure and 1N NaOH solution procedures were 0.132% (i.e. 0.177% over 1.33 years, which corresponds to 487 days) and 0.162% (i.e. 0.219% over 1.35 years or 494 days), respectively.

The features aforementioned indicate that the "residual expansion" and the "maximum expansion potential" of the AAR-affected core samples were quite similar; in

other words the reactive aggregates were able to display all their potential for reacting with the alkalis from the cement in either test procedure performed, which also means that important leaching has not been observed in the test conducted at >95% R.H. Yet, it is clear that AAR-kinetics and induced expansion was significantly faster at the beginning of the test (i.e. about 180 days or six months) in the 1N NaOH solution method. The latter is likely due to the fact of the significantly higher amount of alkalis supply found in this test procedure.

References

1. Bérubé, M., Tremblay, C., Fournier, B., Thomas, M., Stokes, D.: Influence of lithium-based products proposed for counteracting ASR on the chemistry of pore solution and cement hydrates. Cem. Concr. Res. **34**(9), 1645–1660 (2004). H. F. W. Taylor Commemorative Issue. ISSN: 0008-8846. https://doi.org/10.1016/j.cemconres.2004.03.025. http://www.sciencedirect.com/science/article/pii/S0008884604001528
2. Bérubé, M., Frenette, J., Pedneault, A., Rivest, M.: Laboratory assessment of the potential rate of ASR expansion of field concrete. Cem. Concr. Aggreg. **24**(1), 13–19 (2002)
3. Bérubé, M., Frenette, J., Landry, M., McPhedran, D., Pedneault, A., Ouellet, S.: Evaluation du potentiel résiduel de réaction et d'éxpansion du béton en service atteint de réactivité alcalis-silice. Technical Report Final report No. CO-93–04 to Hydro-Quebece1 (1994)

Chapter 18
Accelerated Expansion Test Sample Report: Toulouse

Alain Sellier and Stéphane Multon

18.1 Purpose

This study, realized for EPRI in order to complete round-robin tests, aims at characterizing concrete cores of a concrete and the initial aggregates used for the casting.

The chemical potential of aggregates will be evaluated on mortar made with the extracted aggregates of concrete cores. For one concrete, sand (0/5 mm), fine gravel (5/14 mm) and coarse gravel (>14 mm) will be extracted depending on the aggregates used. Thus, the aggregate sensitivity at the alkali-silica reaction will be characterized on the 3 aggregate sizes of the tested concrete and original aggregates (3 sizes). The advancement of the reaction will be deduced from both measures (in the case of structure, potential residual expansion of aggregate and total potential expansion of aggregate are respectively given by the measurements on the under-water location and on the open-air location—unaltered concrete).

18.2 Extracting Aggregates Method

18.2.1 Summarize of the Method

The extracting method consists in a chemical attack coupled with cooling and warming cycles.

A. Sellier (✉) · S. Multon
Université de Toulouse, Toulouse, France
e-mail: alain.sellier@insa-toulouse.fr

S. Multon
e-mail: stephane.multon@insa-toulouse.fr

© RILEM 2021

V. E. Saouma (ed.), *Diagnosis & Prognosis of AAR Affected Structures*,
RILEM State-of-the-Art Reports 31,
https://doi.org/10.1007/978-3-030-44014-5_18

For each type of concrete, sand (0/5 mm), fine gravel (5/14 mm) and coarse gravel (>14 mm) will be extracted. The characterization of the alkali-silica reaction sensitivity will be carried out on 3 sizes of aggregates per concrete and original aggregates (aggregates initially used for the concrete casting) i.e. 6 tests.

The reactive silica content of each size of aggregate is quantified in order to determine the part of non-reactive aggregates in mortar mix-design (quantification by a cold HF attack method). The non-reactive aggregates come from St Béat in France. The cement used is a Lafarge cement from Le Teil factory (CEM I 52.5 R CE CP2 NF). NaOH is added in the mix to obtain the same alkali content for all the mortars. Demineralized water is used for the casting.

Mortars are cast with each size of aggregate (with a mix of non-reactive and reactive aggregates in order to limit the sample damage and thus measure the expansion during all the test duration). The extracting aggregates are previously crushed and sieved in order to reconstitute a granular curve for the recomposed mortars casting. Finally, the samples are placed in a NaOH solution (1 mol/l) at 38 °C during 4 months (the test can be longer because it depends on the nature of the aggregates).

18.2.2 *Expansion of Mortars Made with Extracted Aggregates, Expansion Test at 38 °C*

18.2.2.1 Extracting

Extracting method consists in a sodium sulfate attacks combined with heat-cold cycles. Then, the aggregates are then rinsed with hot water [1]. The locations of the drilled cores are given in the Appendices.

After sieving, for each zone, it recovers:

- a sand (0.063/5 mm);
- a gravel (5/14 mm);
- a coarse gravel (>14 mm).

18.2.2.2 Distribution Size of Aggregate

Table 18.1 shows the distribution size by weight of extracted aggregates.

Table 18.1 Size distribution of extracted aggregates

Référence	RILEM (extracted aggregates)		
Size (mm)	>14	5/14	<5
% (by weight)	12	30	58

Table 18.2 Reactive silica contents

		Reactive silica content (% by weight)
Initial aggregates	<5 mm	8.4
	5/14 mm	8.9
	>14 mm	8.7
Extracted aggregates	<5 mm	7.5
	5/14 mm	7.0
	>14 mm	5.3

18.2.2.3 Reactive Silica Content

Operating Mode

To determine the reactive silica content, a chemical attack with a cold fluorhydric acid is applied of each size of extracted or initial aggregates.

Results

Table 18.2 presents the results obtained for each size of initial and extracted aggregates. The reactive silica contents of extracted aggregates are lower than the ones of the initial aggregates. It confirms the fact that ASR reactions have consumed a part of the reactive silica of all the aggregates.

18.2.2.4 ASR Characterization of Aggregates

The test is performed with mortars composed by 50% of non-reactive aggregates, 50% of supposed reactive aggregates (studied aggregates) and an addition of alkalis. These proportions are given by the chemical reactive silica content obtained in order to limit the expansion, avoid the cracking of the samples during the test and limit the gel exudation. In order to compare the results for initial and damaged aggregates, the same content of reactive aggregate (50%) was chosen for all the aggregates.

Mortars Components

The different components of mortar are:

- non-reactive aggregates: marble from St. Béat (31, France);
- reactive aggregates: extracted from cores;
- cement: CEM I 52.5 R CE CP2 NF LAFARGE Le Teil factory;
- alkalis: NaOH in water;

Table 18.3 Recomposed mortar formulations

Components	Locations	Quantités	
		St. Béat	RILEM
Non-reactive sand	St. Béat	1,497 kg	748.5 kg
Reactive sand	RILEM	–	748.5 kg
Cement CEM I 52.5 R CE CP2 NF	LAFARGE—Martres	499 kg	499 kg
Alkalis	NaOH	6.56 kg	6.56 kg
Total water	Demineralized water	300 l	300 l

– demineralized water.

Mortar Formulation

The formulations are given in the Table 18.3. The content of aggregates, alkalis and cement are in kg by dried powder. The water content is in l/m^3.

Mortar References

The mortars are cast with the 3 sizes of extracted aggregates or initial aggregates (crushed and sieved) in order to obtain 160–3150 μm sand. Three $10 \times 10 \times 160\ mm^3$ samples by size of aggregate are prepared (18 samples in total). Samples are kept at 38 °C in a NaOH solution (1 mol/l).

The mortar references are:

– Initial <5 (mm), Initial 5/14 (mm), Initial >14 (mm);
– Extracted <5 (mm), Extracted 5/14 (mm), Extracted >14 (mm).

Results

Figures 18.1, 18.2, 18.3 and 18.4 show the results for the initial aggregate mortars and the extracted aggregate mortars. All results and the standard errors are given in the Appendices.

For the extracted aggregates, the expansions are less important than the extracted aggregates (Figs. 18.3 and 18.4). These expansions are higher than the one of the non-reactive reference of the study ("St Béat").

Figure 18.5 gives the expansion of reactive and non-reactive aggregates. For the initial aggregates (quartzite from Fig. 18.5), this expansion is around 0.3% (3000 μm/m) at 130 days with a temperature of conservation about 60 °C [1].

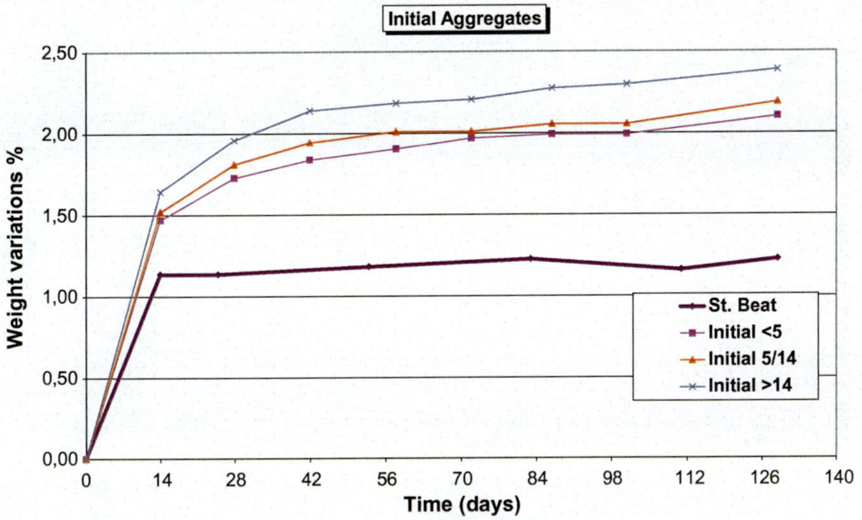

Fig. 18.1 Weight variations of mortar made with initial aggregates

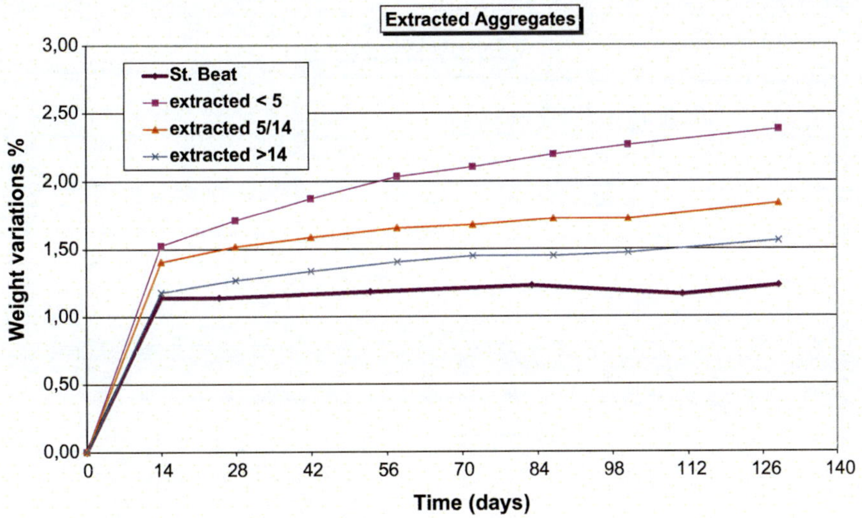

Fig. 18.2 Weight variations of mortar made with extracted aggregates

18.3 Conclusions

Table 18.4 and Fig. 18.6 allow the comparison of the results obtained in this study. Table 18.4 gives the progress of ASR reaction in the studied concrete at 129 days. This advancement is around 80% for all sizes of aggregates. The reactive silica

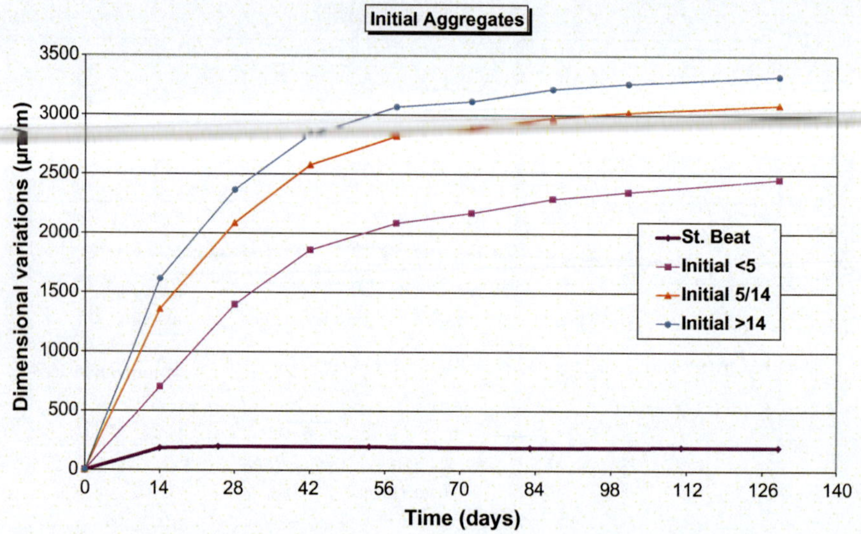

Fig. 18.3 Dimensional variations of mortar made with initial aggregates

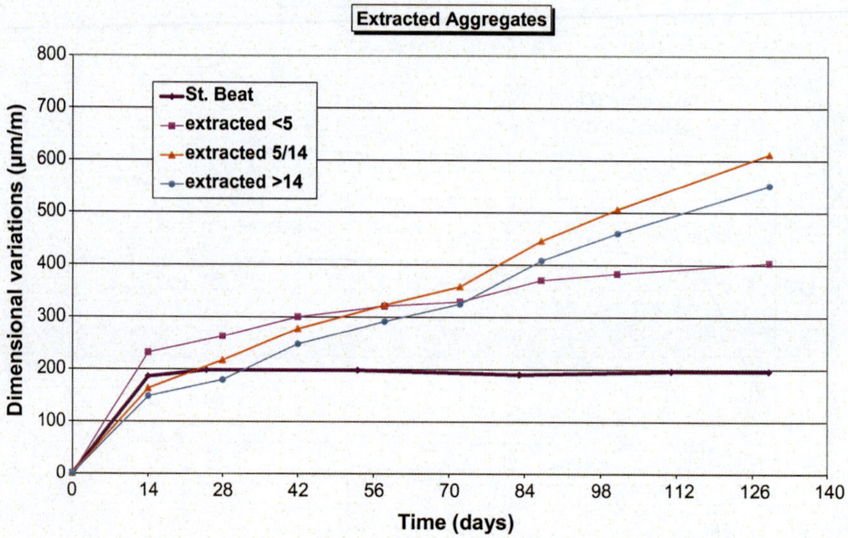

Fig. 18.4 Dimensional variations of mortar made with extracted aggregates

content of the sand is maybe overestimated because of a rest of cement on the sand and/or its petrographic nature. On structures, in field conditions, the progress of the reaction of sand is often higher than the advancement of gravels and coarse aggregates. It is not the case in the present study. This difference could be explained

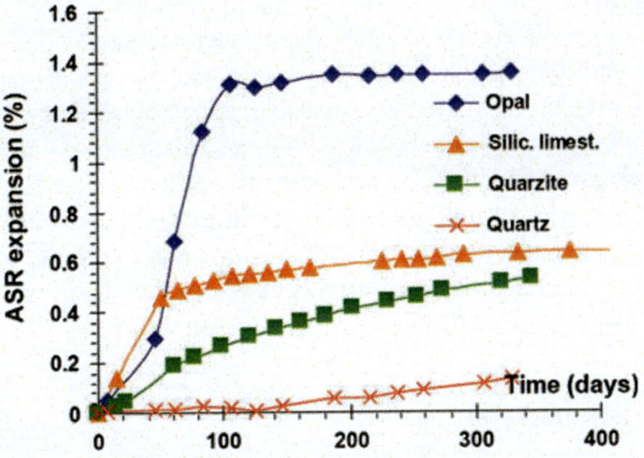

Fig. 18.5 Expansion of some aggregates [1] (0.2% correspond to 2,000 μm/m)

Table 18.4 Synthesis of the results

	<5 mm	5/14 mm	>14 mm
Proportion of extracted aggregates (% by weight)	58	30	12
Initial expansion (μm/m)/reactive silica content (%)	2,460/8.4	3,088/8.9	3,329/8.7
Extracted expansion (μm/m)/reactive silica content (%)	404/7.5	613/7.0	554/5.3
Chemical ASR reaction advancement (%)	84	80	84
Chemical potential expansion (%)	16	20	16

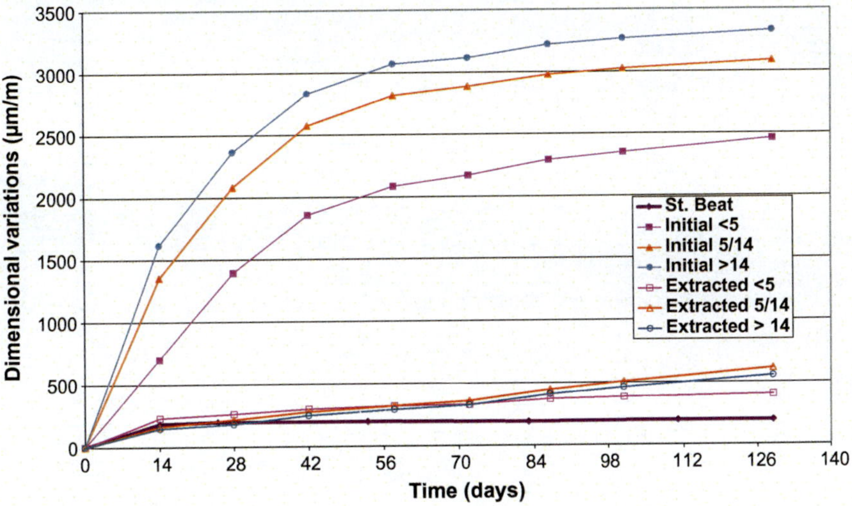

Fig. 18.6 Dimensional variations of mortar made with initial and extracted aggregates

by the difference between the accelerated conditions of expansion test and in-field conditions where the diffusion and reactive mechanisms kinetics are different. The nature of the aggregate can also impact this behavior. Moreover, we can observe that the expansions of recomposed mortar with gravels and coarse aggregates are still in progress which will decrease the final chemical ASR reaction advancement once the expansion of mortars will be stabilized (Figs. 18.4 and 18.6).

This advancement can be used to calibrate the kinetics parameters of modelling in order to assess the future behavior of damaged structures.

Appendices

The following tables give the results of dimensional variations of recomposed mortars.

Time (days)	Initial <5 mm		Initial 5/14 mm		Initial >14 mm	
	Average	SE	Average	SE	Average	SE
0	0	0	0	0	0	0
14	704	73	1356	117	1617	37
28	1396	122	2085	169	2367	91
42	1860	157	2577	207	2833	70
58	2088	191	2819	209	3071	67
72	2173	166	2888	200	3117	74
87	2294	169	2979	245	3221	78
101	2352	172	3025	246	3267	85
129	2460	176	3088	250	3329	92

Dimensional variations of mortars composed by initial aggregates (μm/m; SE: Standard Error)

Time (days)	Extracted <5 mm		Extracted 5/14 mm		Extracted >14 mm	
	Average	SE	Average	SE	Average	SE
0	0	0	0	0	0	0
14	231	6	163	0	148	7
28	263	6	217	10	179	42
42	300	13	277	10	248	4
58	321	20	323	13	292	4
72	329	24	358	10	325	0

(continued)

(continued)

Time (days)	Extracted <5 mm		Extracted 5/14 mm		Extracted >14 mm	
	Average	SE	Average	SE	Average	SE
87	371	31	446	14	408	4
101	383	32	506	17	460	4
129	404	34	613	17	552	18

Dimensional variations of mortars composed by extracted aggregates (μm/m; SE: Standard Error)

Reference

1. Gao, X.: Contribution to the requalification of Alkali Silica Reaction (ASR) damaged structures: assessment of the ASR advancement in aggregates. Thèse de l'Université de Toulouse, Laboratoire Matériaux durabilité des Constructions, 155 pp. (2010)

Chapter 19
Accelerated Expansion Test Sample Report: LNEC/Portugal

João Custódio

19.1 Introduction

In this report, the results obtained from the laboratory tests conducted at LNEC, between February 2017 and October 2019, on concrete samples extracted from an existing laboratory-aged slab cast in October 2013, at the Electric Power Research Institute laboratory (EPRI) in Charlotte (North Carolina, USA), with the intent of determining the concrete residual free expansion potential, are presented and analysed.

The cores were transported to LNEC in a temperature-controlled box, to ensure temperature was below 12 °C, provided by FedEx. This type of box is capable of maintaining temperature in the range 2–8 °C up to 96 h (4 days) [1].

The temperature-controlled box, with five cores (7A, 7B, 7C, 7D, 6C), was sent to LNEC on January 16th 2017 and has arrived at LNEC on February 1st 2017. When the box was opened, on the day of arrival at LNEC, the temperature inside the box was 12 °C. The bagged cores were immediately transferred to an environment at a temperature of 6 °C until the moment of testing. The samples were taken out of the bag and unwrapped on February 7th 2017, when the preparations for testing have begun, and were then weighted and measured immediately after. The details are presented in Table 19.1. It can be seen that there was no significant variation of the samples' mass during transportation.

J. Custódio (✉)
LNEC - National Laboratory for Civil Engineering, Av. do Brasil, 101, 1700-066 Lisbon, Portugal
e-mail: jcustodio@lnec.pt

© RILEM 2021

V. E. Saouma (ed.), *Diagnosis & Prognosis of AAR Affected Structures*,
RILEM State-of-the-Art Reports 31,
https://doi.org/10.1007/978-3-030-44014-5_19

Table 19.1 Details of the concrete samples tested at LNEC

Test specimen	Mass (g)			Dimensions (mm)	
	16/01/2017	07/02/2017	Difference	Diameter	Height
7A	4318.1	4318.8	+0.7 (0.02 %)	107	200
7B	4304.7	4306.6	+1.9 (0.04 %)	107	200
7C	4193.0	4194.8	+1.8 (0.04 %)	107	200
7D	4285.2	4287.1	+1.9 (0.04 %)	107	200
6C	4262.2	4264.1	+1.9 (0.04 %)	107	200

19.2 Test Methods

The method used at LNEC to determine the potential for further expansion of the concrete samples due to ASR was the following:

- Day -7 (01/02/2017)
    - The samples were taken out of the cold shipping package and placed in a low temperature environment (6 °C).
    - Two airtight containers (A and B) were placed in the measurements room, kept at 20 ± 2 °C and 65 ± 5 % RH. The containers had 25 mm of water on the bottom, a perforated rack with a height of 50 mm, and an absorbent material, placed around the wall of the container extending into the water. The container had a size sufficient to store up to three specimens vertically and without being in contact with each other or the sides of the container.
- Day -1 (07/02/2017)
    - The specimens were taken out of the low temperature environment and moved to the measurements room, kept at 20 ± 2 °C and 65 ± 5 % RH;
    - The specimens were unwrapped;
    - The specimens were weighed, to the nearest ±0.1 g;
    - The specimens were measured (diameter and height);
    - Three pairs of strain gauge pre-drilled stainless steel location discs, separated by 120° and using a strain gauge length of 100 mm, were bonded to the core side, using an epoxy adhesive;
    - The adhesive was left to cure during roughly three hours;
    - The specimens were then placed in the abovementioned airtight containers (specimens 7C, 7D and 6C in container A; specimens 7A and 7B in container B).
- Day 0 (08/02/2017)
    - The first readings, mass and length between strain gauge points, were made to all five specimens (the procedure followed was that defined in RILEM Recommended Test Method AAR-3 [2]);

- Specimens 7C, 7D and 6C were placed in the container A (testing regime A);
- Specimens 7A and 7B were placed in the container B, and then 11 ml of a 1.5 M NaOH solution were added to the top of each specimen (testing regime B). A rubber band was used to keep the solution in place until complete absorption occurred;
- The sealed containers were moved from the measurements room, kept at 20 ± 2 °C and 65 ± 5 % RH, and placed in a test room, maintained at a temperature of 38 °C.

Prior to each following measurement, the specimens were cooled in the sealed containers for 24 ± 2 h in the measurements room, maintained at 20 ± 2 °C and at a relative humidity of 65 ± 5 %. The procedure followed for taking readings was that defined in RILEM Recommended Test Method AAR-3 [2]. In the specimens subjected to testing regime A, after the readings made at 6 and 16 months of testing, 11 ml of deionised water were added to the ends of each specimen; therefore, in total, each specimen had 11 ml of deionised water added to its bottom and top surface. The deionised water was added with the intent of ensuring that a sufficient amount of moisture is available during the test for alkali-silica reaction to develop. In the specimens subjected to testing regime B, after the readings made at 3, 6, 9, 16 and 24 months of testing, 11 ml of a 1.5 M NaOH solution was added to the ends of each specimen; thus, in total, each specimen had 33 ml of a 1.5 M NaOH solution added to its bottom and top surface. The sodium hydroxide solution was added with the aim of compensating the eventual leaching of alkalis from the concrete specimens that may occur during the residual expansion test.

19.3 Test Results

The results obtained are presented in Figs. 19.1, 19.2 and 19.3 and in Table 19.2. The average residual free expansion potential obtained at 364 days of testing was 0.32 %, whilst the minimum and maximum values were 0.29 % and 0.37 %, respectively (Table 19.3). The cores subjected to testing regime A (with no alkalis added, *i.e.* specimens 7C, 7D and 6C) had residual expansion values ranging from 0.29 to 0.35 %, with an average of 0.31 % (Table 19.4). The cores exposed to testing regime B (with alkalis added, i.e. specimens 7A and 7B) had residual expansion values ranging from 0.31 to 0.37 %, with an average of 0.34 % (Table 19.4). When comparing the results obtained with test regimes A and B, at 364 days of testing, for specimens 7A, 7B, 7C and 7D, specimens obtained from the same core, it is seen that the minimum, maximum and average values are slightly higher for the specimens subjected to test regime B (Table 19.5).

Since the expansion did not stabilize in one year of testing, the test duration was increased and was only terminated when the expansion levelled off for all specimens, which has occurred after almost two years of testing. Analysing the data for 902 days of testing it can be observed that the general trend, reported in the previous paragraph,

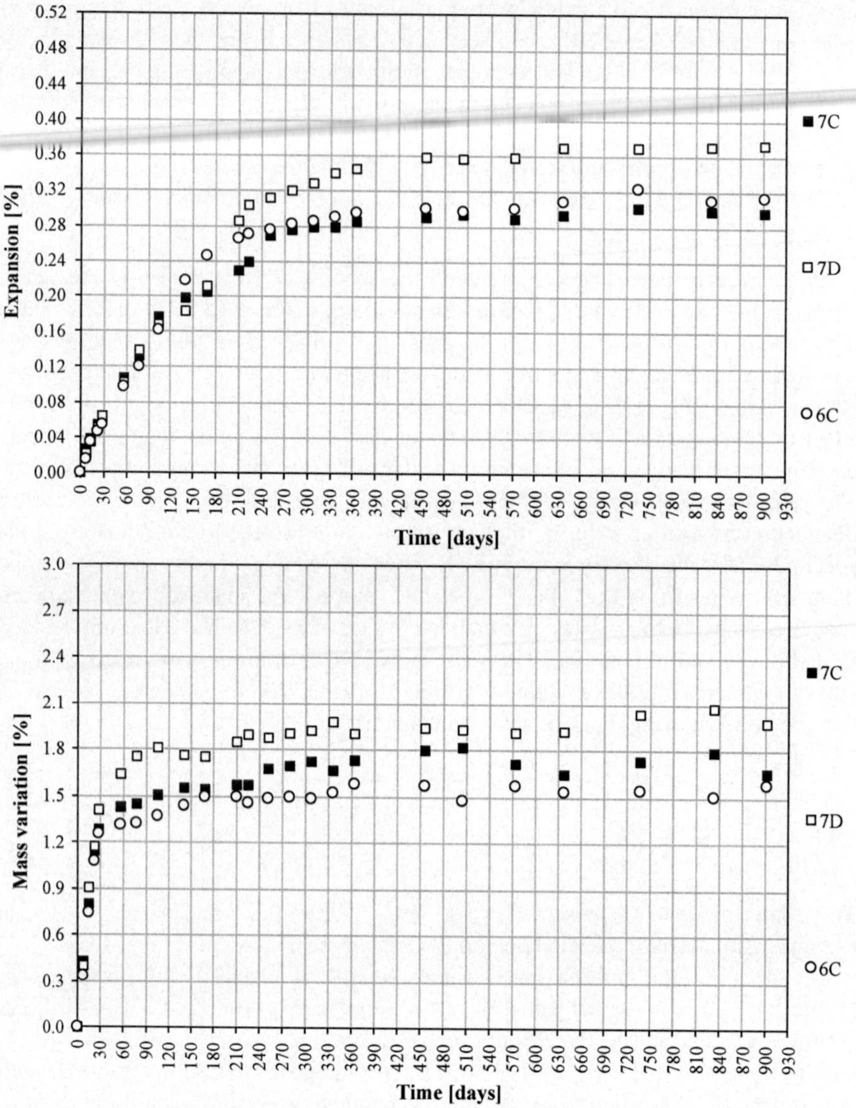

Fig. 19.1 Results obtained with test regime A

is maintained, i.e. the expansion values are slightly higher for the specimens on test regime B (Table 19.4).

Looking into the relative position that the samples occupied in the slab, it is seen that the outer specimens (7A and 7D) exhibited a residual expansion slightly higher than the inner specimens (7B and 7C), throughout the entire test period (Table 19.5). The expansion attained by specimens 6C and 7C, specimens obtained from different

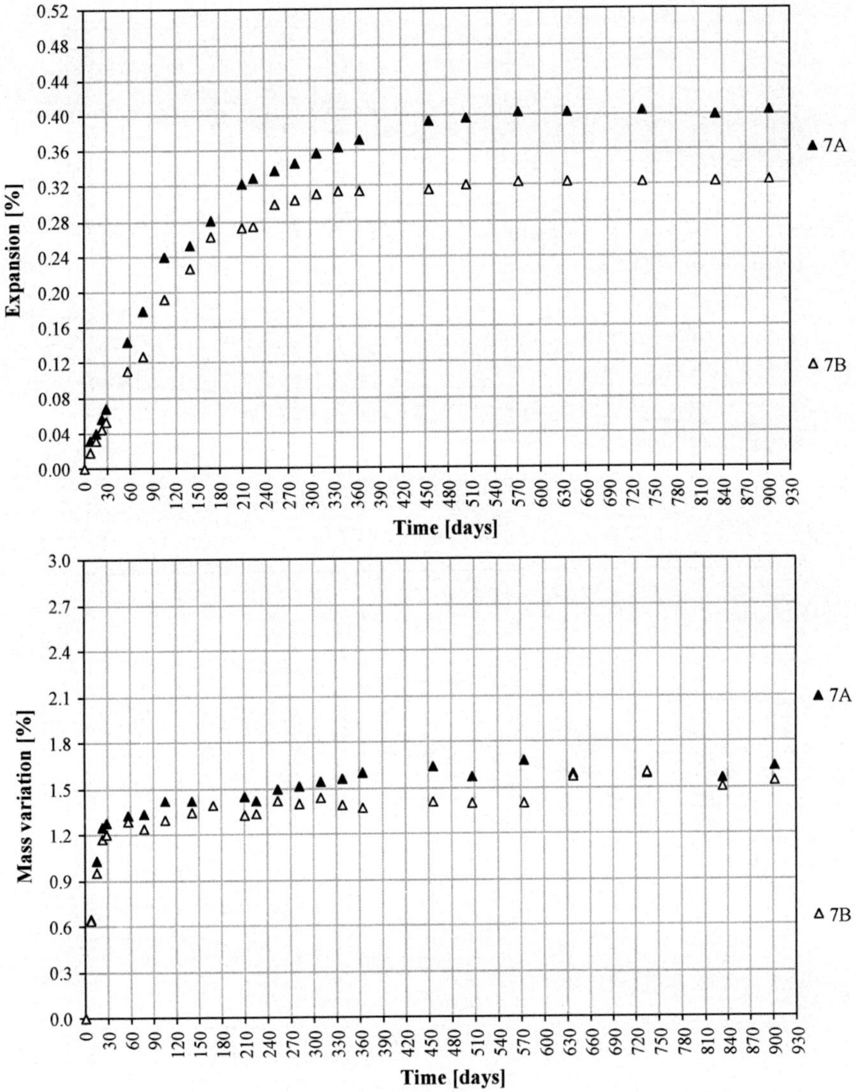

Fig. 19.2 Results obtained with test regime B

cores but from an adjacent position in the slab, was very similar (Table 19.2 and Fig. 19.3).

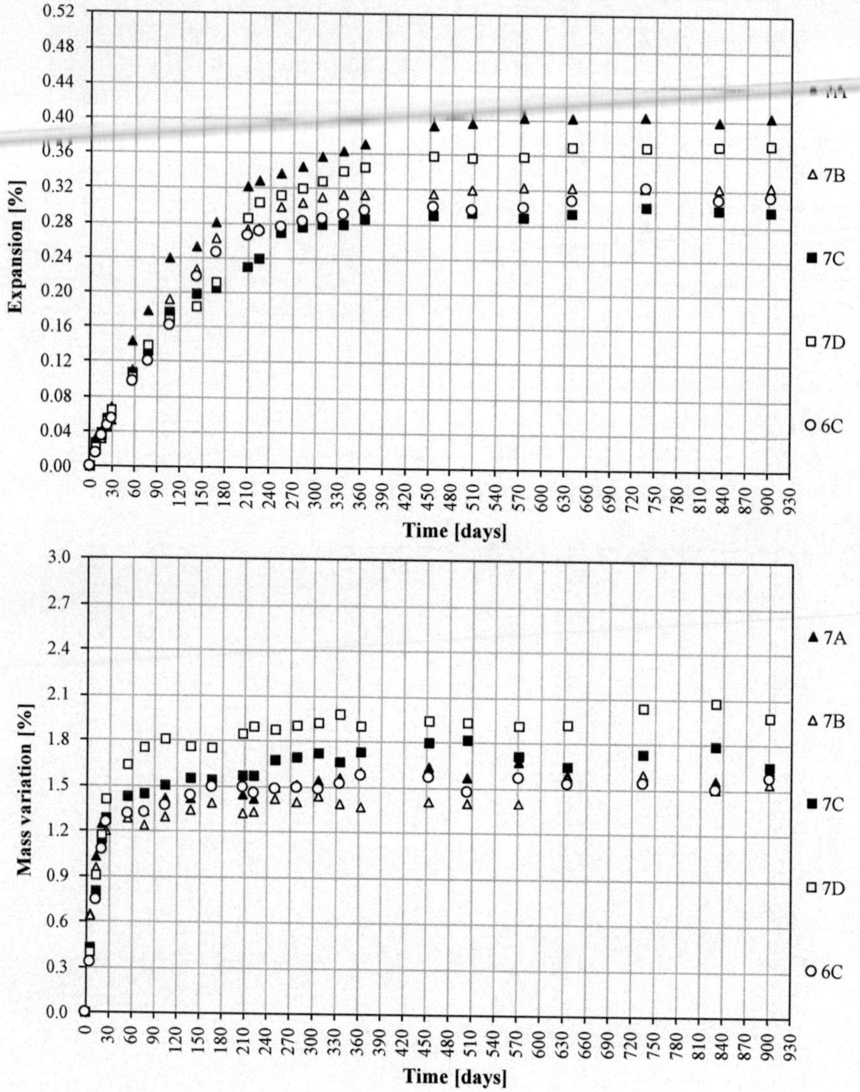

Fig. 19.3 Results obtained with test regimes A and B

19.4 Conclusions

Due to the magnitude of the residual expansion values attained in the tests, the slab could be classified as having a high residual expansion potential.

In the scenario that the assessed cores did not came from a laboratory cast slab but from an actual field structure, due to the fact that the residual expansion values obtained with the two methods are very similar, suggesting that either no significant

Table 19.2 Results obtained with test regimes A and B

Test specimen	Expansion (%)							
	105 days	252 days	364 days	505 days	637 days	735 days	832 days	902 days
7A	0.240	0.337	0.373	0.396	0.404	0.405	0.400	0.404
7B	0.192	0.299	0.314	0.321	0.323	0.323	0.324	0.326
7C	0.177	0.269	0.286	0.294	0.295	0.302	0.299	0.298
7D	0.169	0.312	0.345	0.356	0.370	0.370	0.372	0.373
6C	0.162	0.276	0.296	0.298	0.309	0.324	0.310	0.313

Table 19.3 Overall results obtained at 364 and 902 days of testing

Test specimen	Expansion, 364 days (%)			Expansion, 902 days (%)		
	Minimum	Maximum	Average	Minimum	Maximum	Average
7A	0.29	0.37	0.32	0.30	0.40	0.34
7B						
7C						
7D						
6C						

Table 19.4 Results obtained with test regimes A and B at 364 and 902 days of testing

Test specimen	Expansion, 364 days (%)			Expansion, 902 days (%)		
	Minimum	Maximum	Average	Minimum	Maximum	Average
7A	0.31	0.37	0.34	0.33	0.40	0.37
7B						
7C	0.29	0.35	0.31	0.30	0.37	0.33
7D						
6C						

Table 19.5 Results obtained with test regimes A and B at 364 and 902 days of testing for specimens 7A, 7B, 7C and 7D

Test specimen	Expansion, 364 days (%)			Expansion, 902 days (%)		
	Minimum	Maximum	Average	Minimum	Maximum	Average
7A	0.31	0.37	0.34	0.33	0.40	0.37
7B						
7C	0.29	0.35	0.32	0.30	0.37	0.34
7D						

alkali leaching occurred during the test or that if it has indeed occurred it was not sufficient to decrease the alkali loading below the critical value for the aggregate combination used in the concrete, and assuming that the leaching that might actually occur in service will not be higher than that observed in the laboratory, the expansion value that would be taken into account, for issuing recommendations on how to deal with the ASR in the affected structure, would be the maximum value obtained in the residual expansion test carried out with the supply of alkalis (i.e., with testing regime B).

Considering the maximum value obtained in the residual expansion test, 0.40 %, and considering also the 0.05 % expansion, attained in the first conditioning period, the actions that would be recommended to the structure owner depend upon the type, age and remaining service life of the structure (dam, bridge or viaduct), and also on its current condition and on the environmental service conditions to which the structure is exposed. In Portugal, such fast rate and highly reactive aggregates are very uncommon, but if this expansion behaviour is encountered then a structural analysis would be recommend to see which would be the consequences of such an expansion to the structure, and from there, evaluate the possibility of adopting mitigation measures or rehabilitation/strengthening works.

Acknowledgements João Custódio carried out the work in the scope of the FCT Investigator Project IF/00595/2015/CP1326/CT0001, financed by the Fundação para a Ciência e a Tecnologia, FCT (Portugal); the author wishes to acknowledge this financial support.

References

1. Elhassan, A., et al.: Round Robin Analysis. University of Tennessee & Oak Ridge National Laboratory, 18 Nov 2016
2. RILEM.: RILEM recommended test method: AAR-3 detection of potential alkali-reactivity—38 °C. Test method for aggregate combinations using concrete prisms. In: Nixon, P.J., Sims, I. (eds.) RILEM Recommendations for the Prevention of Damage by Alkali-Aggregate Reactions in New Concrete Structures (State-of-the-Art Report of the RILEM Technical Committee 219-ACS), Springer: Dordrecht (2016)

Chapter 20
Assessment of Round Robin Accelerated Expansion Tests

Andreas Leemann

20.1 Summary

A wide range of approaches is used to determine kinetics of expansion and the residual expansion potential of concrete specimens extracted from structures.

The Laval, IFSTTAR, LNEC and Swiss methods and one of the five different approaches used in Japan are very similar. All of them are performed with similar storage conditions (i.e. 38/40 °C and 100% RH). The IFSTTAR and the Swiss approaches rely solely on this method. The main difference between these two methods is the data's analysis and use. While the relatively fast expansion during the first phase of storage at accelerating conditions and the subsequent phase of slower expansion are considered separately in the Swiss method, a linear regression is adjusted in the IFSTTAR method on the slower expansion phase between 8 and 52 weeks. The results of both methods are used to give a rate of expansion. The IFSTTAR method may use the residual expansion potential as well in its re-assessment procedure. In the Laval method an additional set of cores is stored in 1 N NaOH at 38 °C. The expansions reached after a storage of 365 days (often the test is performed for a longer period) is assigned as "residual expansion" (storage at 100% RH) and "maximum expansion potential" (storage in 1N NaOH). In the LNEC method a small amount of water is added to one set of the cores and a small amount of 1.5 M NaOH to the other with the intent to compensate for alkali leaching. Both sets are stored at 38 °C.

The Japanese method applies a 40 °C and 100% RH storage to cores with a nominal diameter of 100 mm (received core diameter 107 mm) as well as a wrapped storage in 1N NaOH at 40 °C. This is broadly similar to the Laval method. However, additional cores with a diameter of 57 mm are stored in 1 N NaOH at 80 °C, in

---

A. Leemann (✉)
Empa, Swiss Federal Laboratories for Materials Testing and Research, Laboratory for Concrete and Construction Chemistry, Überlandstrasse 129, 8600 Dübendorf, Switzerland
e-mail: andreas.leemann@empa.ch

© RILEM 2021

V. E. Saouma (ed.), *Diagnosis & Prognosis of AAR Affected Structures*,
RILEM State-of-the-Art Reports 31,
https://doi.org/10.1007/978-3-030-44014-5_20

saturated NaCl solution at 50 °C and 1N Na OH at 40 °C. The resulting expansion is not directly used to assess an expansion rate or a residual expansion potential like in the IFSTTAR, Swiss and Laval methods. The values are used as a base to calculate the residual expansion potential using Larive's equation, Helmuth equation for estimating alkali concentration of pore water, and activation energy for reactive silica and Arrhenius law.

The Toulouse method differs significantly from the other methods, as new mortar samples are fabricated with crushed aggregates extracted from affected concrete cores along with crushed virgin aggregates corresponding to the same aggregates used in the concrete mix from the structure under analysis (or aggregates retrieved from a weakly damaged location of the ASR-affected structure). In total, seven sets of mortars are produced: three with crushed aggregates obtained from three different grain size fractions (small, medium and large aggregates) of both the extracted and the virgin aggregates and one with an innocuous aggregate as a control. The seven mortars have the same aggregate size distribution (even if the original aggregates had different sizes before crushing). These seven sets of mortar bars are exposed to a temperature of 38 °C and immersed in 1 N NaOH. The expansion of the mortars produced with the virgin aggregates is regarded as representative of the total content of reactive silica and the expansion of the mortars produced with the extracted aggregates as the residual content of reactive silica. The normalized difference between the two represents the consumption of reactive silica of the concrete between construction and the time of core extraction, also called ASR advancement (between 0 and 100%). This advancement divided by the age of the structure displays the kinetics of the reaction but not the expansion potential. The residual expansion must be assessed coupling this information with the displacement measured on the structure to assess the real swelling potential of the concrete in the structure.

In addition to these test results, a petrographic analysis of cores stored at the University of Colorado for one year either in a fridge or 38 °C and 100% RH, respectively, are provided in the chapters from Japan and DRP methods.

Table 20.1 illustrates a comparison among the results obtained from the distinct methods. The IFSTTAR method is not included in this comparison, as cores from another source were used. In spite of this, the results of the IFSTTAR method impressively illustrate, how strongly expansion can vary on cores obtained at different locations from the same structure. The Toulouse method is presented separately (Table 20.2) since it does not provide the expansion potential but only the advancement of consumption of the reactive silica per aggregate size.

The expansion of the concrete at the time of coring was assessed in the Japanese and the Swiss methods based on petrographic analysis. The values obtained are within the same expansion range (0.15 to 0.22%). These values are lower then the reported average expansion of the slab at the time of coring (0.24%).

As the duration of storage at 38/40 °C and 10% RH varies in the different methods and the expansion rate changes during the storage, the expansions obtained at the end of storage cannot be compared directly. The values range from 0.12% (167 days of storage) to 0.36% (546 days of storage). Extracted alkalis in the Swiss method

Table 20.1 Summary of the expansion potential obtained with the different methods

Method	Exposure	Past expansion [%]	Measured expansion [% / days]	Expansion rate % / year]	Residual [%]
Japan	80 °C, 1N NaOH[1]	0.22	0.66 / 250	-	0.314
	50 °C, sat. NaCl[1]		0.57 / 500		-
	40 °C, 1N NaOH[1]		0.35 / 500		-
	40 °C, 1N NaOH[2]		0.47 / 500		0.354
	20/40 °C, 100 RH[2]		0.33 / 500		0.334
Switzerland	38 °C, 100% RH[3]	0.15	0.12 / 167	0.2	-
Laval	38°C, 100% RH[3]	-	0.18 / 487	-	0.18
	38°C, 1 N NaOH[3]		0.20 / 487		0.2
LNEC	38 °, 100% RH[2]	-	0.33/546	(0.31)	-
	38 °, 1N 100% RH[2,4]		0.36/546	(0.34)	-

1 core diameter of 57 mm
2 core diameter of 107 mm
3 core diameter of 100 mm
4 calculated for 20° and 100% RH

Table 20.2 Summary of the reactive silica consumption obtained with the Toulouse method and expected structural effects amplification until the end of the reaction

Aggregate sizes extracted from concrete	< 5 mm	5–14 mm	>14 mm
Fraction of aggregate in concrete	58%	30%	12%
Chemical advancement of ASR	84%	80%	84%
Expected structural effects until the end of the reaction (displacement, strain, stresses...)	16%	20%	16%

indicated that alkalis were leached during the test at the conditions used possibly affecting expansion. Because the expansion did not stop after one year, the tests were prolonged to 546 days in the case of LNEC. The expansion was continuing during the prolongation making it not possible to determine the residual expansion potential.

The cores with diameters of 57, 100 and 107 mm stored in 1N NaOH at 38 °C during approximately 500 days reached expansion levels ranging from 0.20 to 0.47%.

The calculated values for residual expansion obtained for three of the five setups according to the Japanese method are very consistent (i.e. 0.31–0.35%). The residual expansion according to the Laval method is 0.18–0.20%

The petrographic analysis of the particular core stored at University of Colorado indicates an increase of expansion and damage during storage by a factor of 0.44 (Japan) and depending on the method used (DRP) of 0.18 (damage rating index), 1.33 (optical microscopy on thin sections) and 0.21 (scanning electron microscopy on polished sections).

20.2 Conclusions

Each of the presented methods has advantages and disadvantages. While the storage at 100 % RH may lead to alkali leaching and thus affect the resulting expansion, the immersion of specimens in alkaline solution neglects that the depletion of alkalis may limit expansion in real structures. The Japanese and the Laval methods take this into account and use storage at 100 % RH and in alkaline solutions and the a small amount of 1.5 M NaOH is added to one set of cores in the LNEC method. The Japanese method additionally uses different temperatures and solutions for immersion. The Toulouse method represents a different approach when compared to the other methods. Its main advantage is the determination of ASR advancement independently of the aggregate size in the field environmental conditions. Its disadvantage may be that it does not directly yield a residual expansion, and needs a computational inverse analysis coupled with the assessed advancement to determine it. Therefore, the method may not be performed independently of the structural analysis and of in-situ measurements. It has to be pointed out that the different aforementioned methods were conducted under different boundary conditions such as aggregate features (i.e. nationally available materials) and concrete practices. Nevertheless, the development of a unique method accounting for these different boundary conditions would be desirable. It could enable the development of a large database which would increase the reliability of two decisive parameters for ASR-prognosis that are very challenging to determine: reaction kinetics (or expansion rate) and residual expansion potential.

Part IV
Prognosis; Benchmark Numerical Studies

Chapter 21
Benchmark Problems for AAR FEA Code Validation

Victor Saouma, Alain Sellier, Stéphane Multon, and Yann Le Pape

21.1 Introduction

A number of structures worldwide are known to (or will) suffer from chemically induced expansion of the concrete. This includes not only the traditional alkali aggregate reaction (also known as alkali silica reaction) but increasingly delayed ettringite formation (DEF).[1]

There are three components to the investigation of structures suffering from such an internal deterioration: (a) Chemo-physical characterization focusing primarily on the material; (b) Computational modeling of the evolution of damage and assessing the structural response of the structure; and (c) managing the structure [1].

When focusing on the second aspect, the ultimate objective is to make a predictive assessment of the structural condition and its significance under accidental or extreme scenarios, raising numerous considerations: (a) Would future operation and serviceability be affected?, (b) Would safety be compromised at some point in time?,

[1] It is well known that DEF is often associated with AAR, however it is increasingly observed that it can occur by itself in massive concrete structure subjected to early age high temperature and under high relative humidity (above 95%).

V. Saouma (✉)
University of Colorado, Boulder, USA
e-mail: saouma@colorado.edu

A. Sellier · S. Multon
Université de Toulouse, Toulouse, France
e-mail: alain.sellier@insa-toulouse.fr

S. Multon
e-mail: stephane.multon@insa-toulouse.fr

Y. Le Pape
Oak Ridge National Laboratory, Oak-Ridge, USA
e-mail: lepapeym@ornl.gov

© RILEM 2021
V. E. Saouma (ed.), *Diagnosis & Prognosis of AAR Affected Structures*,
RILEM State-of-the-Art Reports 31,
https://doi.org/10.1007/978-3-030-44014-5_21

and (c) How will degradation and structural significance evolved over time. Answering those questions require predictive capabilities that are best addressed through numerical simulation (usually finite element analysis) accounting for the structure's inherent complexities. Assessing the capabilities of current finite element models to perform reliable are predictive structural assessment of ASR-affected concrete structure is the subject of the benchmark proposed in this chapter.

The assessment of finite element codes has been partially performed within the ICOLD International Benchmark Workshops on Numerical Analysis of Dams, and only limited discussion about AAR within the European project Integrity Assessment of Large Concrete Dams,[2] NW-IALAD were conducted. Nevertheless a rigorous and rational assessment of existing codes capabilities remain to be conducted. This observation was recently strengthened by a benchmark study about shear walls subjected to AAR before being tested under reverse cyclic loading. For the sake of practice, models calibration of large structures should ideally be based on the inherently limited past inspection data, including permanent deformations of a dam's top of the spillway, or surface cracking maps for reinforced concrete. In the field of science and engineering, sound extrapolation and prediction of future degraded states rely on the validation of sophisticated numerical tools and softwares.

To date, finite element models of ASR-affected concrete structures are yet to be validated within a formal and rather systematic framework. The objective of the proposed benchmark is perceived by its authors as the initial step toward developing a formal approach recognized by the profession.

Although the authors are fully cognizant of the complexity of reactive transport phenomenons occurring in alkali-silica concrete, and their effects on damage development and residual expansion, connecting microstructural evolution and structural significance is considered beyond the scope of this study. The proposed study is focused on the interactions of temperature, relative humidity, chemically induced expansion, and mechanical loading.

The proposed benchmark includes two sets of problems, the first on material-scale concrete specimens, and the second, at the structural scale. The material-scale problems have been conceived to test the specific capabilities (strengths and deficiencies) of the benchmarked models to capture the effects of environmental factors and loading, individually or concurrently.

The description of the tests and the mandatory format for participant to reports their results are described hereafter. Test problems are presented with increasing complexity and difficulty with only a limited number of output parameters (generally only one). It is believed such a gradual validation of the constitutive models is needed and provides adequate validation to complex simulation of large-scale aging structures such as hydro-electric concrete dams and nuclear power plants's concrete containment subjected to either static and dynamic loading.

[2]The sixth (Salzburg) and the eighth (Wuhan) benchmarks invited participants to analyze Pian Telessio and Poglia dams respectively. There was no submission to the former, and only two for the second.

21.2 Objectives

This document is submitted by the authors to the Civil Engineering community for the assessment of finite element codes which can perform a "modern" simulation of reactive concrete-induced expansion.

The study includes two parts, the first addresses material modeling, and the second structure modeling. For the material modeling each study is split in two parts: (a) parameter identification for the constitutive model (through calibration of the model with provided laboratory test results); and (b) predictive capabilities.

21.3 Important Factors in Reactive Concrete

Assuming that the final residual swelling of the reactive concrete is known, and based on experimental and field observations, indications are that the following factors[3] should be considered in the finite element analysis of a structure:

1. Environmental Conditions of the concrete
   a. Temperature
   b. Humidity
2. Constitutive models
   a. Solid concrete (tension, compression, creep, shrinkage)
   b. Cracks/joints/interfaces.
3. Load history
4. Mechanical Boundary Conditions
   a. Structural Arrangement
   b. Reinforcement
   c. Anchorage.

21.4 Test Problems

Table 21.1 describes the 11 problems defined. Participants do not have to consider all of them and may select only those appropriate for their finite element code.

[3]There is no general agreement on the importance of all these parameters, the list is intended to be inclusive of all those perceived by researchers to be worth examining.

Table 21.1 List of benchmark problems

No.	Description
P0	Textual description of finite element code/models
Material response	
P1	Constitutive model
P2	Capturing drying and shrinkage
P3	Capturing creep
P4	Effect of temperature
P5	Effect of RH
P6	Effect of confinement
Structural response	
P7	Internal reinforcement
P8	Reinforced concrete beam
P9	Dam (simplified)
P10	Reinforced concrete panel expansion
P11	Nuclear containment vessel (simplified)

21.4.1 Units

For all problems use: m, s, MN, and MPa.

21.4.2 P0: Finite Element Model Description

This very first section should include up to five pages of description of the model adopted in this particular order:

Constitutive Model

1. Basic principles of the model and its implementation.
2. Nonlinear constitutive model of sound or damaged concrete (clarify)
    a. Instantaneous response (elasticity, damage, plasticity, fracture and others)
    b. Delayed response (creep and shrinkage)
3. Effect on the chemically induced expansion by
    a. Moisture
    b. Temperature
    c. Stress confinement
4. Effect on the mechanical properties of concrete by
    a. Expansion
    b. Shrinkage and creep

Finite Element Code Features

1. Gap Element
2. Coupled hydro-thermo-mechanical
3. Others

21.4.3 *Materials*

In light of the preceding list of factors influencing AAR, the following test problems are proposed.

21.4.3.1 P1: Constitutive Models

At the heart of each code is the constitutive model of concrete. This problem will assess the code capabilities to capture the nonlinear response in both tension and compression.

It should be noted that in some codes [2], the constitutive model is tightly coupled (in parallel) with the AAR expansion one (modeled as an internal pressure), in other [3], it is more loosely coupled (in series) with the AAR (modeled as an additional strain).

Constitutive Model Calibration

Perform a finite element analysis of a 16 by 32 cm concrete cylinder with f'_c, f'_t and E equal to 38.4 MPa, 3.5 MPa and 37.3 GPa respectively.[4] Traction is applied on the top surface, and a frictionless base is assumed. Make and state any appropriate assumption necessary, use the following imposed strain histogram:

$$0 \rightarrow 1.5\frac{f'_t}{E} \rightarrow 0 \rightarrow 3\frac{f'_t}{E} \rightarrow 1.5\varepsilon_c \rightarrow 0 \rightarrow 3\varepsilon_c \tag{21.1}$$

where $\varepsilon_c = -0.002$. If needed, the fracture energy G_F in tension and compression are equal to 100 Nm/m^2 and 10,000 Nm/m^2 respectively.

Prediction

Repeat the previous analysis following an AAR induced expansion of 0.5%, you may use the experimentally obtained degradation curve, by [4] and published by [5], Fig. 21.1.

[4]These parameters should be used in all subsequent test problems.

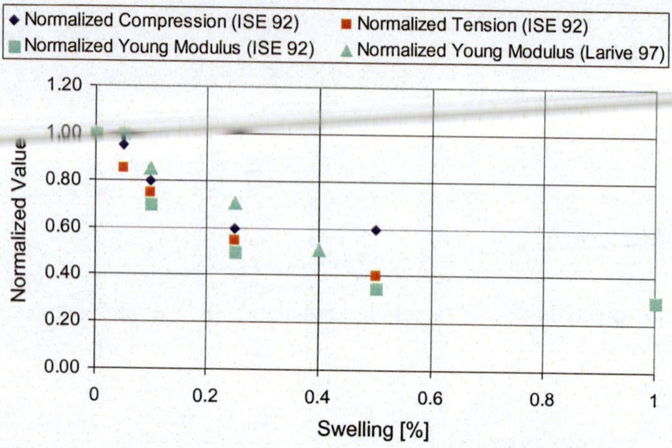

Fig. 21.1 Deterioration of AAR affected concrete Capra and Sellier [5]

Prediction should highlight concrete mechanical properties degradation in terms of ASR evolution; in particular: Young modulus, tensile and compressive strengths.

21.4.3.2 P2: Drying and Shrinkage

For some structures not necessarily under water (such as bridges or certain hydraulic structures), drying shrinkage strains may be of similar order of magnitude as the AAR induced ones. As shown in Fig. 21.2 one must consider various cases of drying and shrinkage, reactive and non reactive concrete, and at relative humidities ranging from a low 30% to a fully saturated environment, and sealed or not. There are a total of 6 potential cases of interest:

a. Non reactive concrete at 30% RH
b. Reactive concrete at 30% humidity
c. Non Reactive concrete sealed specimen
d. Non Reactive concrete under water.

Fig. 21.2 Drying and shrinkage test cases

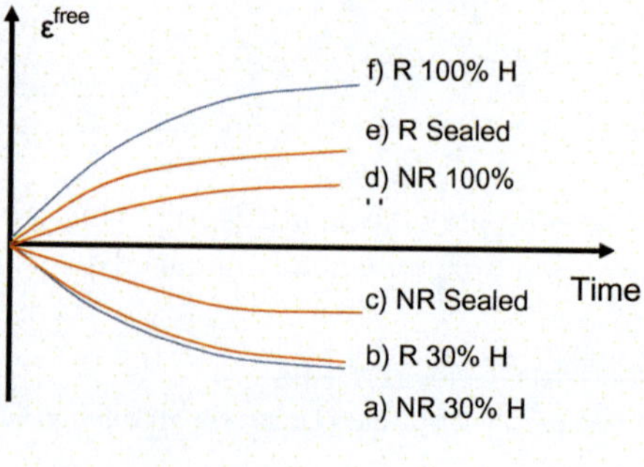

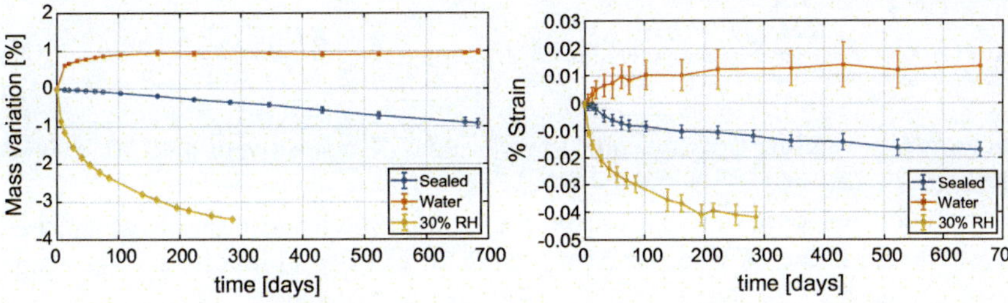

(a) Mass variations (b) Strain Variation

Fig. 21.3 Non reactive concrete under various RH conditions [6]

e. Reactive Concrete, sealed cylinder.
f. Reactive concrete under water. Note that this is not identical to a 100% RH if leaching is to be accounted for.

which will be analyzed in P2 and P5

Constitutive Model Calibration

For calibration purposes, the parameters can be fitted using a 16 by 32 cm cylinder to perform the following analyses: a, c, and d with respect to the temporal variation of mass (Fig. 21.3a) and longitudinal strain (Fig. 21.3b). For initial condition, assume an initial saturation of 0.85, and $T(t = 0) = 38\,^{\circ}\mathrm{C}$

Prediction

Using the parameter determined from the previous section, repeat the same analysis with the temporal annual variation of external RH for the cylinder shown in Fig. 21.4.

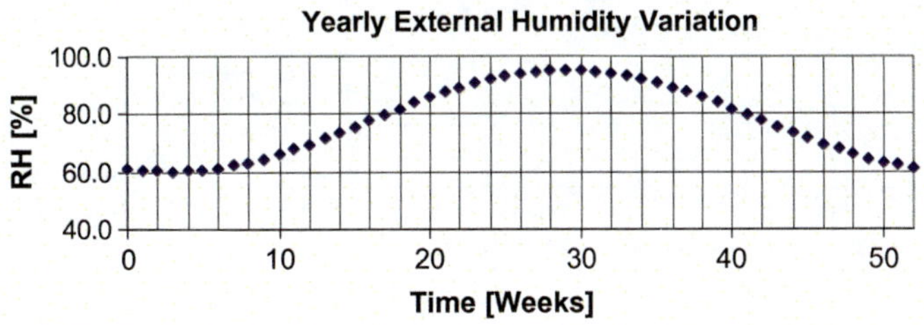

Fig. 21.4 Humidity variation

$$RH(\text{week}) = \frac{RH_{\text{max}} - RH_{\text{min}}}{2} \sin\left(2\pi \frac{t-16}{52}\right) + \frac{RH_{\text{max}} + RH_{\text{min}}}{2} \quad (21.2)$$

where RH_{max} and RH_{min} are equal to 95% and 60% respectively, and $T(t = 0) = 20\,^{\circ}\text{C}$.

The model response should first exhibit a negative strain due to shrinkage and then a positive strain due to water absorption until 30 weeks and finally new shrinkage until 52 weeks.

21.4.3.3 P3: Basic Creep

There is strong experimental and field indications that creep plays a dominant role in the irreversible long term deformation concrete subjected to constant load. Its effect must be accounted for to properly extract the AAR expansion. This may be explained through biaxially or triaxially loaded elements where swelling is restricted in one direction while free to occur on the other(s). Therefore, in the AAR constrained direction creep deformation will be predominant. This is more likely to occur in arch dams.

Constitutive Model Calibration

For a 13 by 24 cm cylinder subjected to 10 and 20 MPa axial compression, plot the longitudinal and radial displacements. You may calibrate your model on the experimental curve shown in Fig. 21.5.

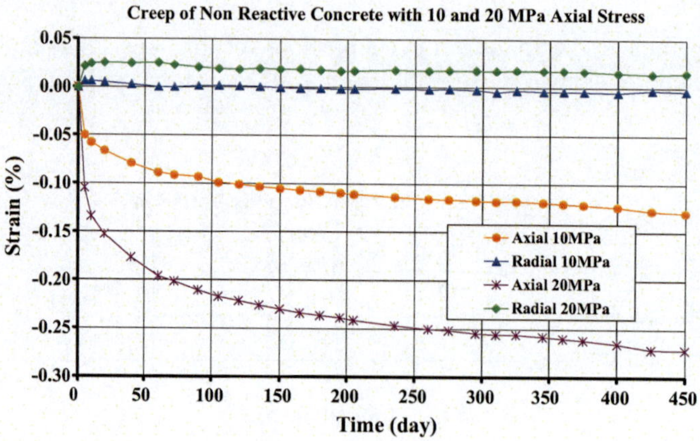

Fig. 21.5 Creep in non-reactive concrete under sealed condition for different axial stress [6]

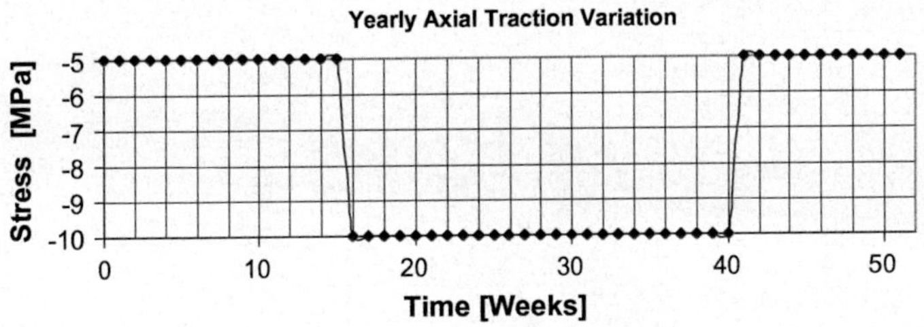

Fig. 21.6 Stress variation

Prediction

Using the previously determined parameters, repeat the same analysis for the axial load history shown in Fig. 21.6.

During the first 16 first weeks, the model should exhibit negative strain due to creep. The load increase at the 16th week should imply an instantaneous strain followed by creep until the 40th week. Then, partial creep recovery should be observed during the first days following unloading.

21.4.3.4 P4: AAR Expansion; Temperature Effect

All chemical reactions are thermodynamically driven. Reactive concrete expansion varies widely with temperature ranges usually encountered in the field or laboratories. Hence, it is of paramount importance that the kinetics of the reaction captures this dependency.

Constitutive Model Calibration

Perform the finite element analysis of a 13 by 24 cm cylinder under water, free to deform at the base and undergoing a free expansion, and for $T = 23$ and 38 °C. Fit the appropriate parameters of your model with Fig. 21.7 obtained by [7].

Prediction

Repeat the previous analysis using the variable internal annual temperature variation

$$T(\text{week}) = \frac{T_{\text{max}} - T_{\text{min}}}{2} \sin\left(2\pi \frac{t-16}{52}\right) + \frac{T_{\text{max}} + T_{\text{min}}}{2} \tag{21.3}$$

where T_{max} and T_{min} are equal to 25 °C and 0 °C respectively, as shown in Fig. 21.8. Use $RH(t = 0) = 100\%$ and $T(t = 0) = 10\,°\text{C}$.

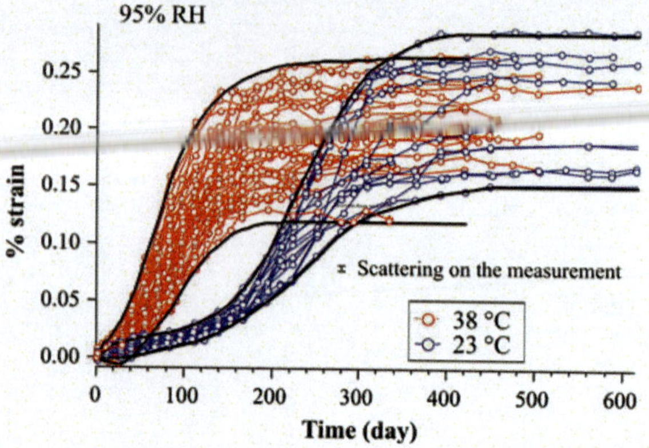

Fig. 21.7 Free expansion from Larive's tests [7]

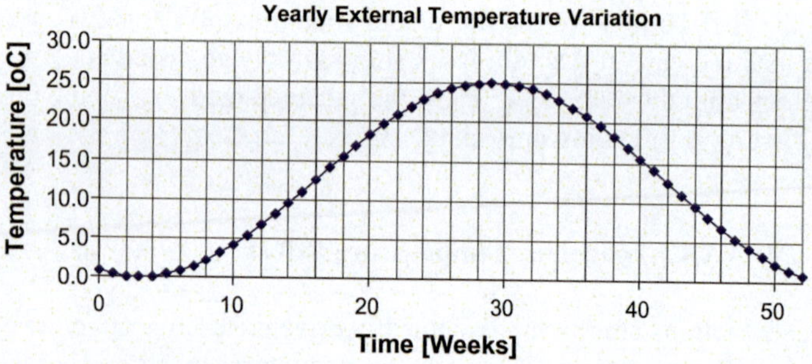

Fig. 21.8 Temperature variation

As the dependence of ASR characteristic times to temperature is exponential, the predictions should be highly non linear. ASR rate should be very slow down during cold period and accelerate during hot period without in a nonlinear response.

21.4.3.5 P5: Free AAR Expansion; Effect of RH

Relative humidity plays a critical role in the expansion of AAR affected concrete. It is well established [8] that expansion will start for a RH at least equal to 80%, and will then increase with RH (RH^8 is a widely accepted formula). For external bridge structures and some dams this can be critical.

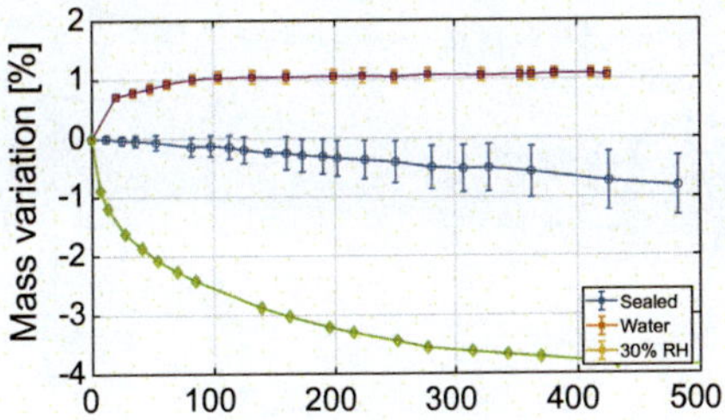

(a) Mass variation

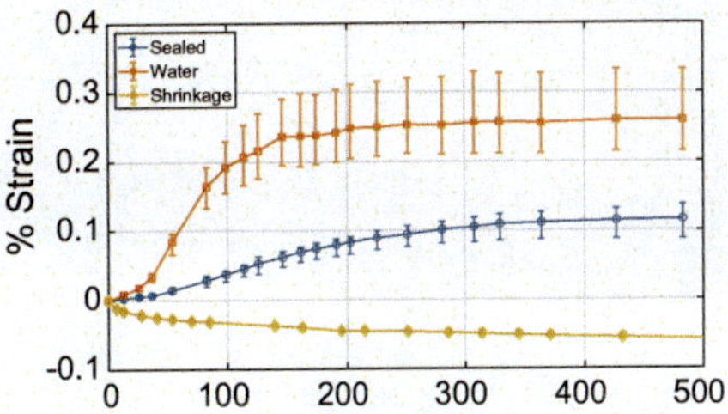

(b) Strain Variation

Fig. 21.9 Reactive concrete under various RH conditions [9]

Constitutive Model Calibration

Using a 16 by 32 cm cylinder, and assuming a temperature of 38 °C, fit the appropriate parameters for mass and vertical strain variation of reactive concrete as shown in Fig. 21.9a and b respectively. Use $RH(t = 0) = 85\%$.

Prediction

Repeat previous analysis using the RH variation shown in Fig. 21.4.

ASR rate should be increased during high saturation and decreased during dry periods.

21.4.3.6 P6: AAR Expansion; Effect of Confinement

It has long been recognized that confinement inhibits reactive concrete expansion, [10, 11] and most recently [6]. This test series seeks to ensure that this is properly captured by the numerical model.

Constitutive Model Calibration

For a 13 by 24 cm cylinder, and assuming a temperature of 38 °C, analyze the following test cases (all of which consist of sealed specimens):

P6-a. No vertical stress, no confinement (Free swelling), Fig. 21.10a.

P6-b. Vertical stress of 10 MPa, no confinement, Fig. 21.10b.

P6-c. No vertical stress, concrete cast in a 5 mm thick steel container, Fig. 21.10c.

P6-d. Vertical stress of 10 MPa and concrete cast in a 5 mm thick steel container, Fig. 21.10d.

In all cases, plot both the axial and radial strains.

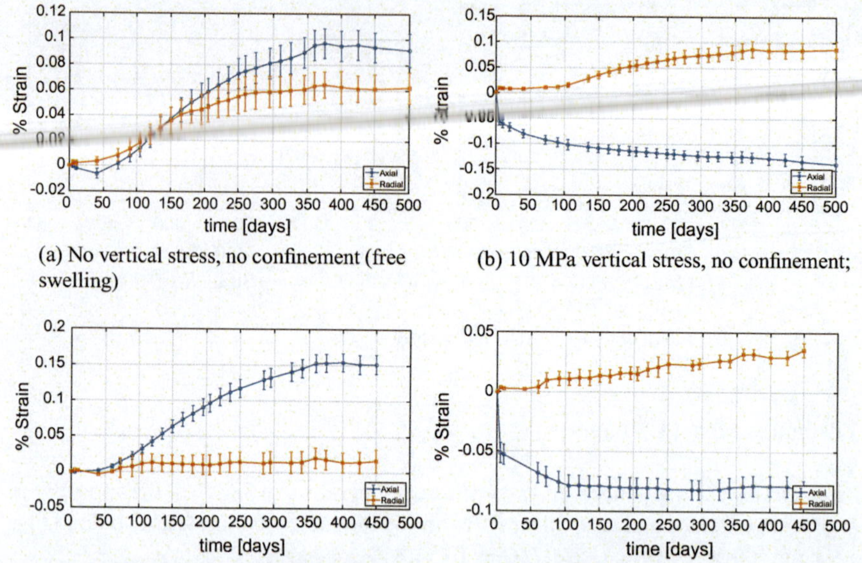

(a) No vertical stress, no confinement (free swelling)

(b) 10 MPa vertical stress, no confinement;

(c) Vertical stress of 10 MPa and concrete cast in a 5 mm thick steel container;

(d) Vertical stress of 10 MPa and concrete cast in a 5 mm steel container;

Fig. 21.10 Expansions in terms of confinements [6]

Prediction

Repeat the analysis with the vertically imposed stress histogram shown in Fig. 21.6.

With such compressive loading, ASR expansion should not be observed in the axial direction. However, creep should be the main cause of negative strain. In radial direction, ASR expansion should be higher than for stress-free expansion.

21.4.4 *Structures*

21.4.4.1 P7: Effect of Internal Reinforcement

Description

Internal reinforcement inhibits expansion and AAR induced cracking would then align themselves with the direction of reinforcement as opposed to the traditional "map cracking". This test problem seeks to determine how the numerical model accounts for this, especially when cracking (thus a nonlinear analysis is needed) occurs.

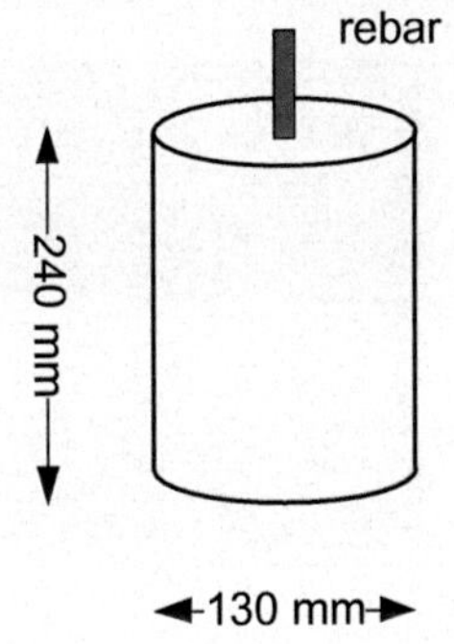

Fig. 21.11 Concrete prism with internal reinforcement

Analyze the cylinder shown in Fig. 21.11 under the same condition (free expansion, 38 °C, 100% RH), for the same duration with a single internal reinforcing bar of diameter 12 mm in the center, and $E = 200{,}000$ MPa and $f_y = 500$ MPa.

Prediction

Determine longitudinal strain in the rebar and the longitudinal and radial strains on the surface of the concrete cylinder. In both cases values are to be determined at mid-height.

Expansion should be reduced along the reinforcement and compressive stresses should develop orthogonally. Small modification of expansion should be observed in the directions perpendicular to steel bar. If evaluated, cracking should be parallel to the reinforcement.

21.4.4.2 P8: Reinforced Concrete Beams

Description

The mechanical behavior of two concrete beams, studied by S. Multon during his Ph.D. works at LCPC, is proposed. One beam is damaged by ASR during two years of exposure in a 38 °C environment and differential water supply, leading to differential ASR expansion within the structures [12]. The other made with non-reactive aggregates was stored in similar conditions. Namely, the effects of the ASR development have been quantified in a 4-points bending test of the beams, resulting in a lot of data among which the residual stiffness and the flexural strength of both reactive and non-reactive beams. The objective is to simulate the evolution of the two beams during the two years of tests, and to finish by a simulation of beam failure in four points bending, Fig. 21.12.

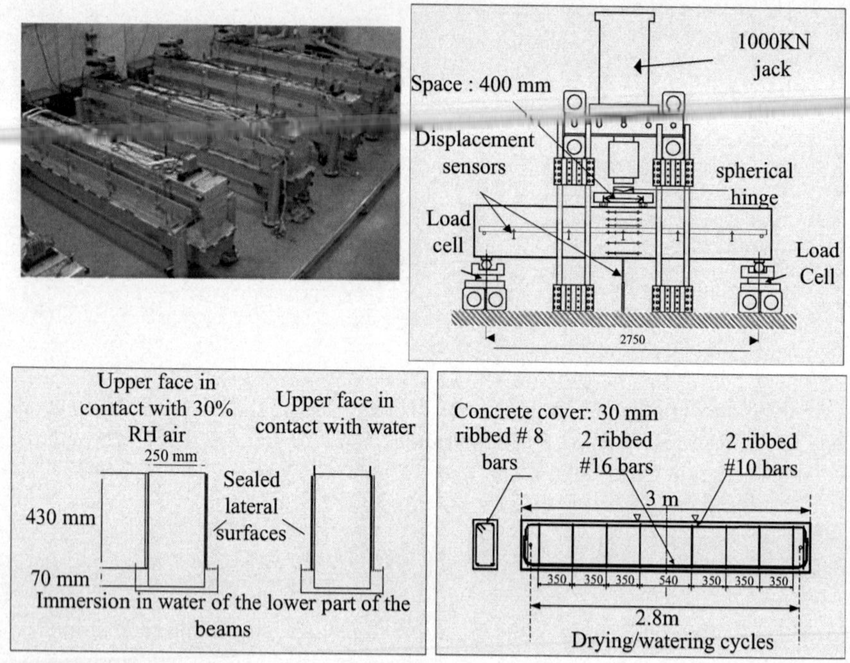

Fig. 21.12 Multon's beams

Table 21.2 Reinforced concrete beam mechanical properties

	28 days	180 days	2 years	
Reactive concrete				
E	37,300	30,100	34,600	MPa
f_c	38.4	41.2	43	MPa
f_t	3.5	3.4	3.8	MPa
Non-reactive concrete				
E	38,700	37,800	38,700	MPa
f_c	35.5	40.4	43	MPa
f_t	3.6	3.2	3.7	MPa

Material characteristic are the same then in tests P1 to P6, therefore, the LCPC performed tests at several dates since the fabrication. Results are given in Table 21.2. During the 2-years aging phase, beams were placed on simple bearings along the geometrical mid-height (span of 2.8 m): steel bars were embedded at mid-height of the structure. During the 4-point bending test, beams were simply supported on the lower face (span of 2.75 m).

In the present benchmark only beams P4 and P6, Fig. 21.12, have to be simulated.

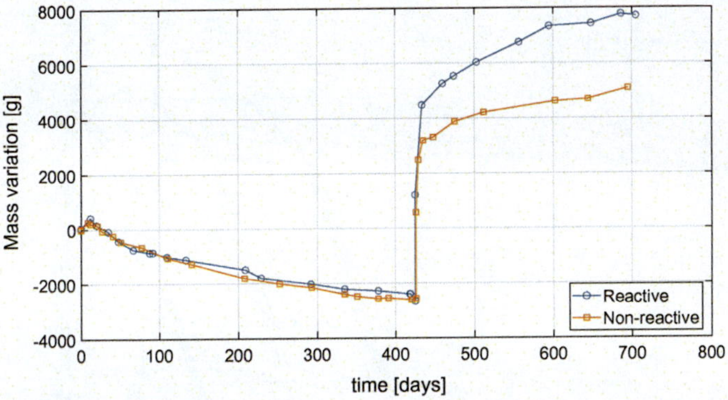

Fig. 21.13 Mass variation of the beams

As AAR depends on humidity, a humidity profile must be fitted, in order to consider effect of saturation on the reaction. In order to fit the drying-humidification cycle, the mass evolutions of the beams are given below.

Initial saturation of the beam is 0.85, temperature is constant and equal to 38 °C. The concrete porosity is around 16% (15% at the bottom and 17% at the top of the beam).

Prediction

- The first objective is to find a realistic humidity profile compatible with the mass variation history given in Fig. 21.13.
- The second objective is to predict the deflection of each beam, at mid span, versus time
- The third objective is the evolution of stress versus time, in the bottom longitudinal reinforcement #16, at mid span.
- The last stage consists in simulating for the two beams a four point bending test schematized in Fig. 21.12. Participants have to provide the Force-deflection curve until failure of each beam.

Numerical results can be compared to experimental results presented in [12] to assess precision of the model.

21.4.4.3 P9: AAR Expansion; Idealized Dam

Description

This next test problem assesses the various coupling amongst various parameters as well as the finite element code and its ability to simulate closure of joint. A common

Fig. 21.14 Idealized dam

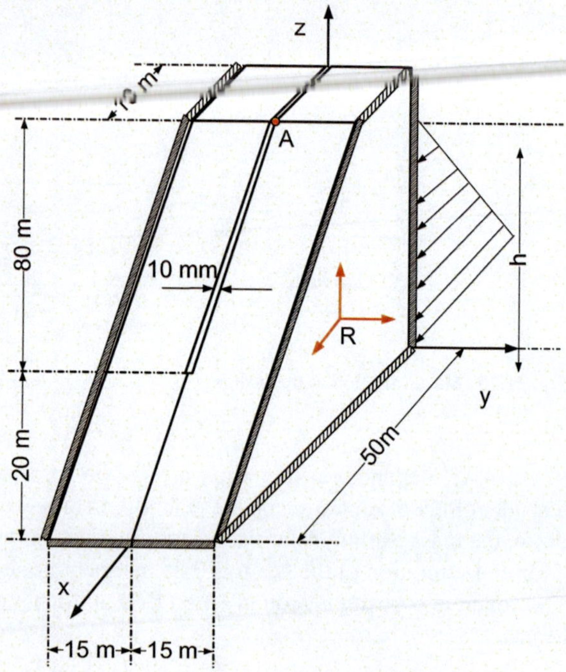

remedy for AAR induced damage in dams is to cut a slot in the structure [13–16]. This will relieve the state of stress, and allow the concrete to expand freely. However, at some point concrete swelling will result in a contact between the two sides of the slot. Hence, this problem will test the model ability to capture this important simulation aspect as well.

Consider the reduced dam model shown in Fig. 21.14 with the following conditions: (a) lateral and bottom faces are all fully restrained; (b) front back and top faces are free; (c) slot cut at time zero, total thickness 10 mm; (d) concrete on the right is reactive, and concrete block on the left is not reactive; (e) hydrostatic pressure is applied only on the right block.

Prediction

Using the fitting data of P6, and an friction angle of 50° for concrete against concrete, and zero cohesion, consider two cases:

- Homogeneous field of internal temperature (20 °C), relative humidity (100%), and an empty reservoir.
- Transient field of external temperature Fig. 21.8, relative external humidity Fig. 21.4, and pool elevation variation Fig. 21.15 given by where ELmax and ELmin are equal to 95 and 60 respectively.

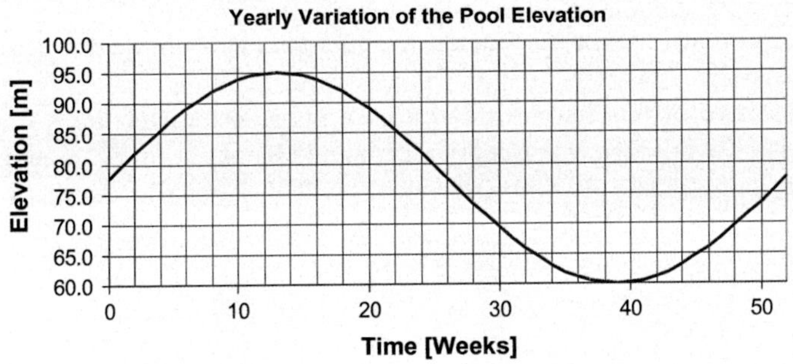

Fig. 21.15 Yearly variation of pool elevation

For both analysis, the specified temperature and relative humidity is the one of the concrete surface. Zero flux condition between dam and foundation. Reference base temperature of the dam is 20 °C.

- x, y, z displacements of point A.
- Fx, Fy and Fz resultant forces on the fixed lateral face versus time (25 years). Assume the typical yearly variations of external air temperature and pool elevation shown in Figs. 21.8 and 21.15 respectively.

This model seeks to capture: (a) general finite element program capabilities in modeling the joint response; (b) ease (or difficulty) in preparing the input data file for a realistic problem; and (c) coupling of the various parameters.

$$EL(\text{week}) = \frac{EL\text{max} - EL_{\text{min}}}{2} \sin\left(2\pi \frac{t}{52}\right) + \frac{EL\text{max} - EL_{\text{min}}}{2} \qquad (21.4)$$

where $EL\text{max}$ and EL_{min} are equal to 95 and 60 respectively.

The slot cutting and subsequent joint closure due to ASR expansion reflect the high nonlinearity of the FE calculation. In this case, modelling should capture stress release caused by cutting and subsequent contact of the surfaces without numerical convergence problems. Of primordial importance will be the stress redistribution in the dam through the various phases.

21.4.4.4 P10: Expansion of RC Panel with or Without Lateral Confinement

This section has been prepared with the assistance of Nolan Hayes, Ammar Abd-Elssamd and Qiang Gui from the University of Tennessee, Knoxville.

The University of Tennessee, Knoxville (UTK), under U.S. Department of Energy (DOE) subcontract managed by Oak Ridge National Laboratory (ORNL), have been performing large scale laboratory testing of confined and unconfined concrete blocks (simulating a typical reinforced concrete member found in light water reactor nuclear power plants).

The objective of this benchmark test case is to perform predictive numerical simulations of two large-scale reinforced concrete blocks (with different boundary conditions) and compare the simulation results with the already collected monitoring data.

Description

Geometry The laterally-confined reinforced concrete reactive specimen, referred to as *CASR* ("C" for confined), is cast inside a rigid steel frame while a similar reinforced concrete reactive specimen, referred to as *UASR* ('U', for unconfined) is allowed to expand without lateral restraints. A third specimen, non-reactive, referred to as *CTRL*, for control, is also not subjected to lateral restraints [17]. See summary in Table 21.3 All three specimens of dimensions, $136'' \times 116'' \times 40''$ (length, width and height; x-y-z axis), i.e., 3.453 m $\times$ 2.946 $\times$ 1.016 m, Fig. 21.16 are reinforced near the top and the bottom faces by two welded layers of orthogonal rebars: (22) #11 bars (1.41″ nominal diameter, cross section area: 1006 mm^2), (10) in one direction and (12) in the perpendicular direction, placed in horizontal planes—See Fig. 21.16d for layout. Rebars are made of standard carbon steel. Square plate heads (4" $\times$ 4" $\times$ 1", i.e., 10.16 cm $\times$ 10.16 cm $\times$ 2.54 cm) are welded to the rebar extremities. The concrete cover, in the least distance to the concrete outer surface, is 3″ (7.62 cm). There is no reinforcement in the third, i.e., vertical, direction, to the exception of (6) #11 debonded rebar spacers placed inside of pipes to allow free vertical expansion during the test.

Steel Confinement Frame The steel plate girder frame was designed with the primary goal of maximizing stiffness in bending. In order to achieve this goal, 3″ thick plates, height 3′4″, were chosen as flanges to the plate girder. These flanges are connected by three 2″ thick web plates, length 2′10″S. All steel was manufactured from A572 Grade 50 steel plate. Design of splice plate connection not provided here is available upon request. In order to reduce frictional effects between the steel frame and the concrete specimen, a single layer (thickness: 1.5 mm) of high-density polyethylene (HDPE) was introduced at the interface, providing a low steel-HDPE friction coefficient estimated by the vendor around 0.3 and unilateral contact conditions.

Table 21.3 Characteristics of the three specimens

ID	Label	Confined	Reactive
1	CASR	Yes	Yes
2	UASR	No	Yes
3	CTRL	No	No

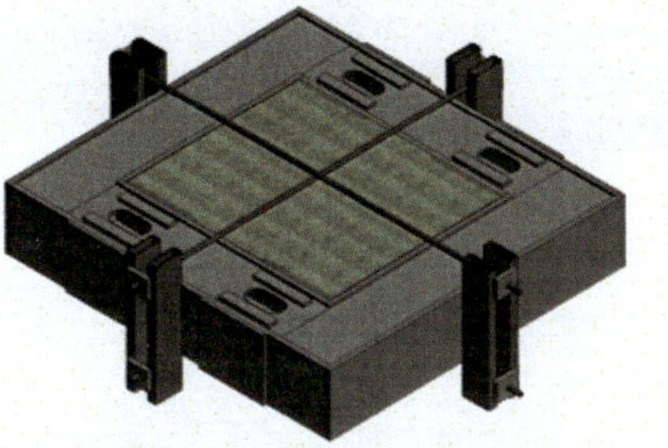

(a) Computer rendering of the confined specimen (CASR)

(b) Form construction in UTK Civ. Env. Eng. high bay

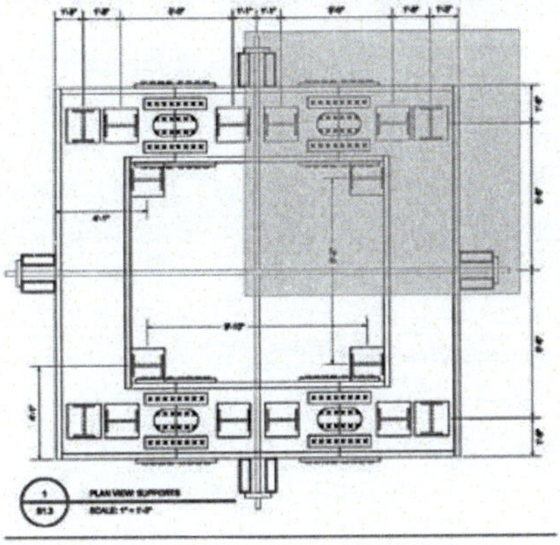

(c) Confined specimen layout – Top view. (Shaded area indicating symmetries)

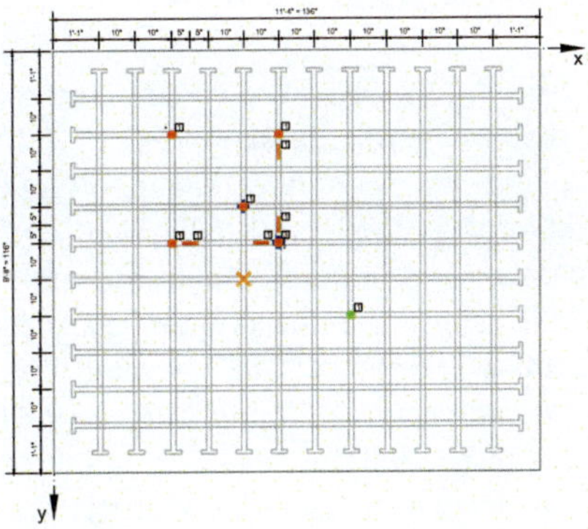

(d) Reinforcement layout – Top view. Bottom and top reinforcement layout are identical.

Fig. 21.16 Reinforced concrete panel test problem [17]

Additional Post-tensioned system Four threadbar post-tension bars (2 in each direction, 2.5 in. dia.) manufactured by DYWIDAG-Systems International (DSI) were installed in September 2016, in order to increase the confining force, if necessary. It is initially just slightly tightened to avoid "slack", and has remained, as of today.

Casting and Curing Casting took place July 23rd 2016. In an attempt to mitigate potential crack sources other than ASR, the formworks were insulated by placing rigid foam sheathing insulation with an R-value of three around the side and on top of the specimens, shortly after pouring. The insulation was placed with edges overlapping and secured in place with tape and plastic wrap. All formworks were removed on August 4, 2016. Each large specimen and concrete cylinder, for further materials testing, was covered with wet burlap to prevent moisture loss. The burlap was periodically moistened as required to keep the concrete surfaces

Table 21.4 Target mix design. Aggregate quantities are for oven-dry material. Water quantities assume aggregates in saturated-surface dry (SSD) condition. (*) To limit the early-age temperature below ≈65 °C, about 70% of the water was added to the mix as ice cubes

Materials	Quantity, $kg\,m^{-3}$ ($lb\,yd^{-3}$)	
	Reactive	Control
Coarse Aggregate	1180 (1988.8)	1180 (1988.8)
Fine Aggregate	728 (1226.6)	728 (1226.6)
Cement	350 (590)	350 (590)
Water(*)	175 (295)	175 (295)
w/c	0.5	0.5
NaOH solution	9.8 (16.6)	–
LiNO3 solution	–	11.9 (20.03)

wet. A few days after casting, the bottom support is removed, and the concrete block is vertically supported by four 18″ × 18″ (45.7 cm × 45.7 cm) corner plates. Plates are directly supporting the specimens on the concrete surface. The estimated steel-concrete friction coefficient is ≈0.6.

Operation A modular environmental chamber was designed by Norlake Scientific with the initial primary goals for temperature and humidity control being 100 °F ± 2 °F (38 °C ± 1 °C) and 95% ± 5%. The chamber was initialized for full operation early morning August 19, 2016. The chamber is periodically shutdown for inspection on a average frequency of 2 days per month. During shutdowns, the average temperature and RH are about ≈77 °F (25 °C) and 60% (transient of about 4 h). After the shutdown period, the chamber is restarted and and the temperature and humidity return to the original set points within 6 h.

Target mix design The mix design has been extensively investigated at the University of Alabama, and the one retained, including a reactive and a control mix, is shown in Table 21.4 with 1″ (25 mm) maximum size aggregate (MSA) composed of Green schist—muscovite, chlorite, quartz, Na-feldspar, K-feldspar, calcite, and, cristobalite. In this mix, only the coarse aggregate is reactive. A 50% sodium hydroxide solution (NaOH) is used to increase the alkali loading of the reactive mix to 5.25 $kg\,m^{-3}$, and a 30% lithium nitrate solution (LiNO3) is used at 150% of the manufacturerś recommended dosage to mitigate ASR for the control mix.

Mechanical properties 28 days mechanical properties compressive and tensile strengths, and the elastic modulus are shown in Tables 21.5, 21.6 and 21.7, respectively along with their mean and standard deviations. A representative 28 days stress-strain curve is shown in Fig. 21.17.

Shrinkage Shrinkage has been measured in the CTRL specimen. The datapoints for the shrinkage curve are shown in Table 21.8.

Expansion curves obtained from earlier material testing Expansion curves were obtained by Pr. E. Giannini, at the University of Alabama (UA), while testing different aggregates-forming concrete. The concrete blocks, 300 × 300 × 600

Table 21.5 Reported 28 days compressive strengths f_c' (MPa)

Specimen type	AVG	STD
CASR	22.2	2.07
UASR	20.7	1.17

CASR Confined reactive specimen
UASR Unconfined reactive specimen

Table 21.6 Reported 28 days tensile strengths f_t' (MPa)

Specimen type	AVG	STD
CASR	2.70	0.215
UASR	2.13	0.044

CASR Confined reactive specimen
UASR Unconfined reactive specimen

Table 21.7 Reported 28 days elastic modulus E_c (GPa)

Specimen type	AVG	STD
CASR	34.5	3.03
UASR	34.4	2.22

CASR Confined reactive specimen
UASR Unconfined reactive specimen

mm, are stored in UA climate chamber at 38 °C and 95%RH, shown in Fig. 21.18, and their expansion was periodically monitored using DEMEC points.

Data are tabulated in Table 21.9 and shown in Fig. 21.19a where the vertical expansions were taken over a 150 mm gauge length, and longitudinal expansions (same direction as longitudinal) were taken over a 500 mm gauge length. It should be noted that the reported mean (or average) corresponds to the average of all the experimental values.

Recorders/sensors location Recorder[5] locations are shown as follows:

Embedded KM strain transducer (KM-100B), referred as *strain gauges*, gauge length 100 mm, in Fig. 21.19 and Table 21.10.

Resistive strain gauges General purpose resistive strain gauges (gauge length: 1.52 mm) were attached to the reinforcing bars in the specimens. These sensors are attached to the top and bottom of the rebar in the select locations to measure rebar strain. The location of resistive strain gauges of interest are shown in Fig. 21.19 and Table 21.10.

Long gauges fiber-optics-based deformation sensors (SofO, gauge length ≈1.0–1.5 m with location) measure (1) the vertical deformation between the bottom and top rebars layers, and, (2) horizontal deformation at the bottom surface as illustrated and tabulated in Fig. 21.20 and Table 21.11.

[5]In a finite element analysis, point from which we determine computed values are commonly referred to as "recorders".

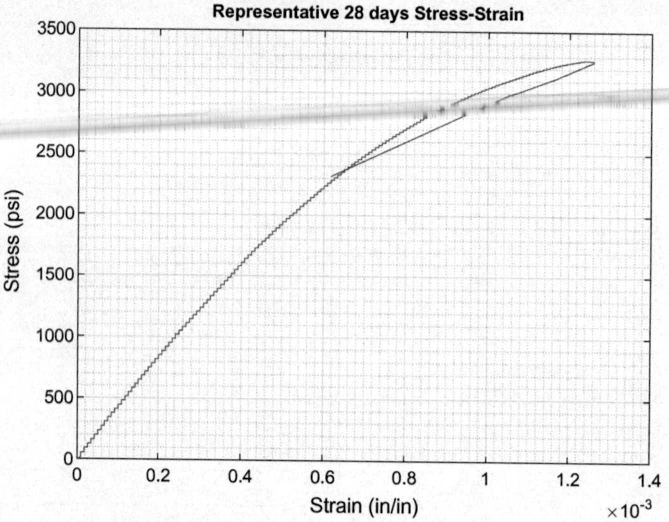

Fig. 21.17 Stress strain curve (28 days)

Table 21.8 Provided shrinkage curve data

Measured shrinkage	
Age (Days)	Shrinkage (%)
5	−0.0031
10	−0.0104
20	0.0162
30	−0.0178
40	−0.0185
50	−0.0190
60	−0.0194
100	−0.0214
200	−0.0245
300	−0.0275

Test duration Casting occurred July 23rd 2016. Assuming testing will end April 19, 2019, it is requested to model a total duration of 1,000-days.

Predictions

Plot for both specimens, CASR and UASR, as a function of time (increments of one month) the following model outputs:

Fig. 21.18 Concrete expansion block tested by Prof. E. Giannini

Table 21.9 Provided expansion curve data

Calculated expansions		
Age (Days)	Average exp. (%)	STD (%)
6	0.000	0.0000
40	−0.004	0.0045
68	0.000	0.0031
87	0.012	0.0081
103	0.020	0.0091
117	0.028	0.0103
138	0.045	0.0193
152	0.057	0.0250
170	0.070	0.0307
190	0.088	0.0382
220	0.103	0.0440
304	0.146	0.0634
312	0.157	0.0733
350	0.165	0.0729
371	0.174	0.0782
459	0.192	0.0885
504	0.197	0.0903

1. Vertical displacements at D1.
2. Concrete strain at S1, S2, S3, S4 and S5.
3. Reinforcement strains at R1 and R2.

Numerical results can be compared to experimental data published in [17].

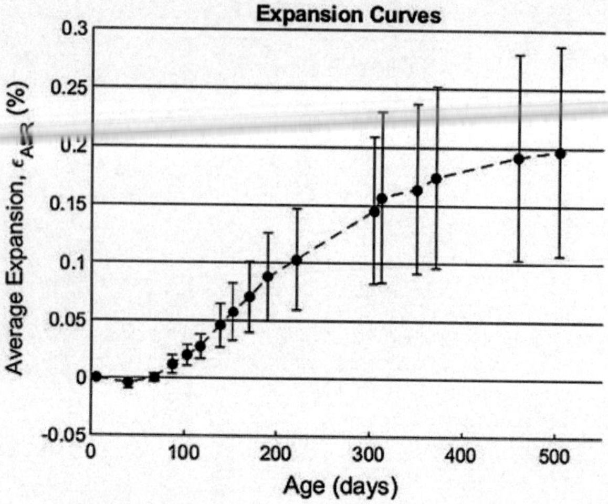

(a) Laboratory measured expansion. Error bars: standard-deviation.

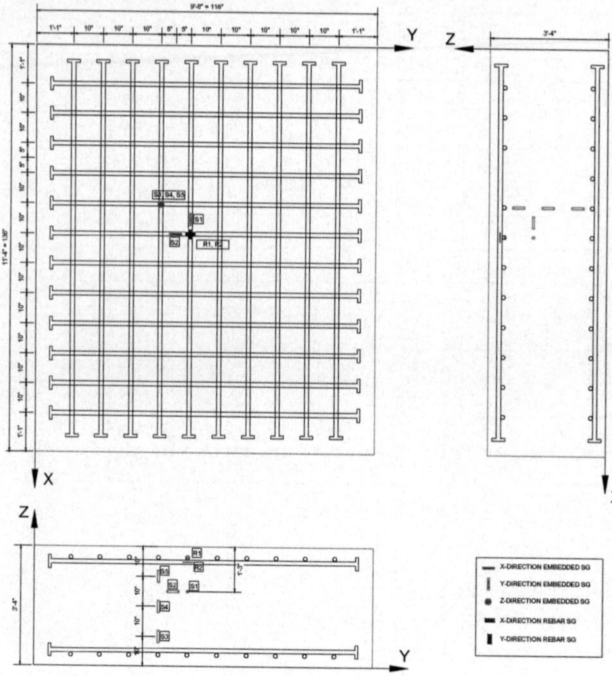

(b) Location of internal concrete gauges

Fig. 21.19 Recorded expansion

Table 21.10 Strain gauges location points. 'S' refers to KM embedded sensors, while 'R' refers to resistive strain gauges placed directly on the rebars

ID	Dof	Coord. (in.)			Coord. (m)		
		x	y	z	x	y	z
S1	1	58	53	25	1.473	1.346	0.635
S2	2	63	48	25	1.600	1.219	0.635
S3	3	53	43	10	1.346	1.092	0.254
S4	3	53	43	20	1.346	1.092	0.508
S5	3	53	43	30	1.346	1.092	0.762
R1	1	63	53	36.375	1.600	1.346	0.924
R2	2	63	53	34.875	1.600	1.346	0.886

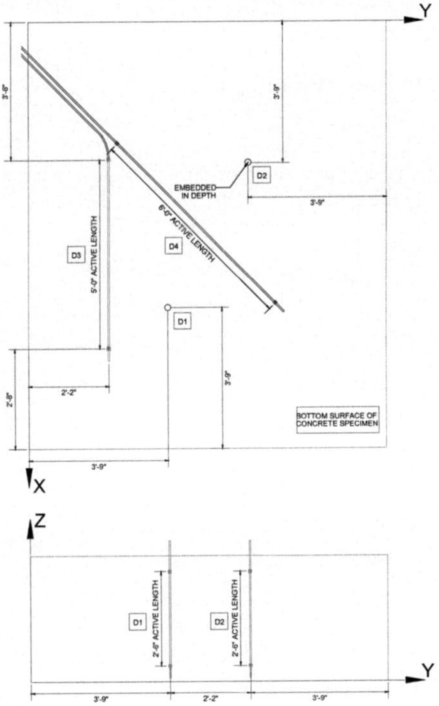

Fig. 21.20 Location of deformation sensors

21.4.4.5 P11: AAR Expansion of Nuclear Containment Vessel Followed by Earthquake

Description

Ultimately, codes should be able to analyze nuclear containment vessel structures suffering from AAR under dynamic excitation.

Accordingly, a much simplified geometry, inspired by [18], is adopted. Fig. 21.21a shows the dimensions as well as the key material parameters. Note that the mat foundation and the walls only are subjected to AAR, the dome is not.

Total reinforcement is 1% vertically, and 0.5% circumferentially. Reinforcement in each direction is to be split in two layers, each 10 cm from the wall. Ignore reinforcement of the dome, however triple the elastic modulus of the concrete. Steel elastic modulus is 200 GPa, and yield stress 250 MPa.

For added clarity, the boundary conditions, and the expansion curve is shown in Fig. 21.21b. Only gravity and AAR loads are first considered. Note that the AAR expansion is assumed to follow Larive's curve Larive's curve [19]

$$\varepsilon(t) = \varepsilon^{\infty} \frac{1 - \exp(-\frac{t}{\tau_c})}{1 + \exp(-\frac{(t-\tau_l)}{\tau_c})} \tag{21.5}$$

21.4.4.6 Prediction

Two sets of analyses are required.

Static

Though an axisymmetric analysis is possible, it is highly recommended that a 3D one (using 180° segment) be performed. Plot

1. Horizontal displacement of point A (Δ_x) versus time (increments of one month).
2. Maximum (positive) principal stress ($\sigma_{(1)}$) in the wall versus time.
3. Crack profiles at $t = [5, 10, 20, 30]$ years.

Dynamic

Perform a 3D dynamic analysis, for a harmonic intensifying dynamic excitation, shown in Fig. 21.21c, assumed to occur at age $t = 20$ years. Assume a 5% Rayleigh damping. Report the following:

1. Time of failure (may be defined when the analysis failed to converge).
2. Time displacement curves for point A starting with the AAR displacement that occurred at time 20 years, until failure (as defined by the user) occurs.
3. Maximum (positive) principal stress ($\sigma_{(1)}$) in the wall versus time.
4. Deformed shapes and crack profiles at 1 sec. increment (starting with $t = 0$) until reported failure.

Results should capture the expansion trends with particular emphasis on the confining effects of the base-mat and dome. The dynamic analysis should capture crack localization,localized failures. Indicate to which extent soil-structure interaction have been

Table 21.11 Deformation sensor location points

ID	Dof	Start Coord. (in.)			End Coord. (in.)			Start Coord. (m)			End Coord. (m)		
		x	y	z	x	y	z	x	y	z	x	y	z
		Coord. (in.)			Coord. (m)								
D1	3	91	45	4.25	91	45	35.75	2.311	1.143	0.108	2.311	1.143	0.908
D2	3	45	71	4.25	45	71	35.75	2.311	1.143	0.108	2.311	1.143	0.908
D3	1	45	26	0	104	26	0	1.143	0.660	0	2.642	0.660	0
D4	1–2	38.75	28.75	0	89.75	79.75	0	0.984	0.730	0	2.280	2.026	0

accounted for, and whether rocking has been prevented. Dynamic analysis should be a restart from the static analysis resulting in already damaged structure.

21.5 Conclusion

Given that ASR is already a prevalent problem worldwide and that even more are likely to be identified in the near-distant future, it is of the utmost importance that proper numerical tools are available to offer a diagnosis and prognosis.

For a credible prediction, those tools ought to be first validated through the analysis of simple experimental tests to determine if separate and identifiable phenomena can indeed be captured.

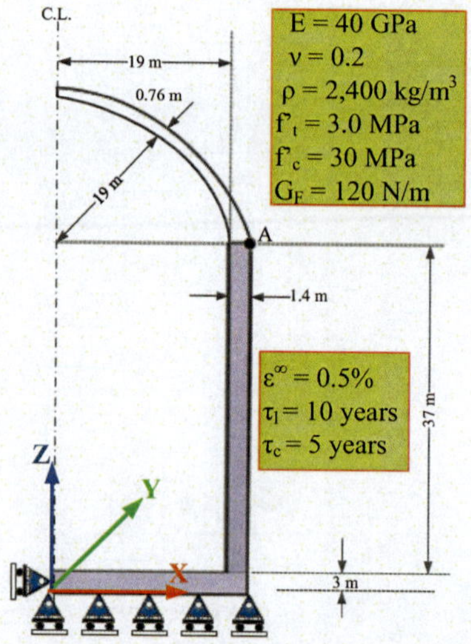

(a) Geometery and Material Properties

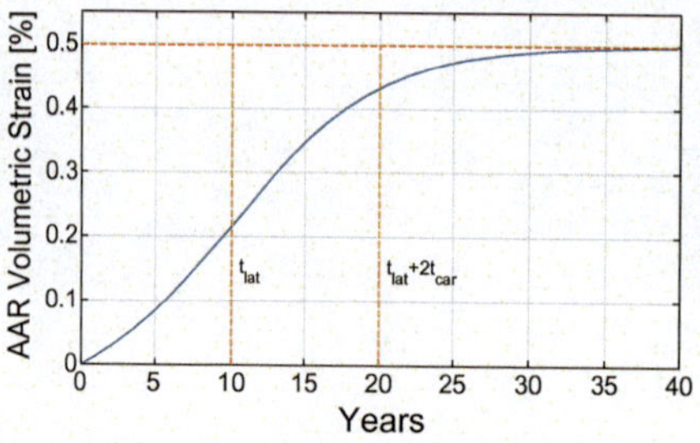

(b) Expansion Curve

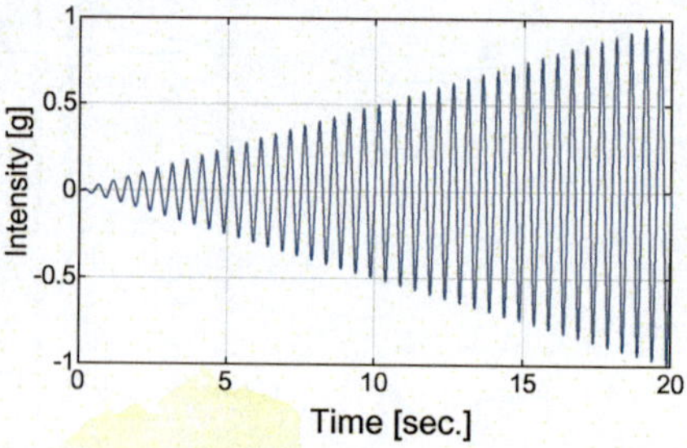

(c) Expansion Curve

Fig. 21.21 Characteristics of the NCVS

Only, once these tools have been validated through material testing, then they ought to be assessed through the analysis of structural components where many separate phenomenon interplay.

Then, and only then, should those tools be deemed (to various degrees) capable of providing engineers with credible set of predictions.

Analysts are encouraged to summarize their findings, determined where the model succeeded and where it filed to capture the intended response and possibly provide an explanation.

Finally, it should be noted that the authors have placed a very high bar to reach in these benchmark problems. Hence, it is supposed that probably not a single finite element code can successfully analyze all problems hereby presented, however strength and limitation should be always identifiable.

References

1. Fasseu, P., Mahut, B.: Aide à la gestion des ouvrages atteints de reactions de gonflement interne. Technical report, 66 pp. Laboratoire Central des Ponts et Chaussées, Paris, France (2003). http://www.ifsttar.fr/fileadmin/user_upload/editions/lcpc/GuideTechnique/GuideTechnique-LCPC-GONFLIN.pdf
2. Sellier, A., Bourdarot, E., Multon, S., Cyr, M., Grimal, E.: Combination of structural monitoring and laboratory tests for assessment of alkali-aggregate reaction swelling: application to gate structure dam. ACI Mater. J. **106**(3), 281–290 (2009)
3. Saouma, V., Perotti, L.: Constitutive model for alkali aggregate reactions. ACI Mater. J. **103**(3), 194–202 (2006)
4. Institution of Structural Engineers: Structural effects of alkali-silica reaction. Technical guidance on the appraisal of existing structures. Technical report, Report of an ISE task group (1992)
5. Capra, B., Sellier, A.: Orthotropic modeling of alkali-aggregate reaction in concrete structures: numerical simulations. Mech. Mater. **35**, 817–830 (2003)
6. Multon, S., Toutlemonde, F.: Effect of applied stresses on alkali-silica reaction induced expansions. Cem. Concr. Res. **36**(5), 912–920 (2006)
7. Larive, C.: Apports combinés de l'expérimentation et de la modélisation à la compréhension de l'alcali-réaction et de ses effets mécaniques. Technical report OA28, 404 pp. Laboratoire Central des Ponts et Chaussées (1998)
8. Poole, A.: Introduction to alkali-aggregate reaction in concrete. In: Swamy, R. (ed.) The Alkali-Silica Reaction in Concrete, pp. 1–28. Van Nostrand Reinhold, New York (1992)
9. Multon, S., Seignol, J., Toutlemonde, F.: Structural behavior of concrete beams affected by alkali-silica reaction. ACI Mater. J. **102**(2), 67 (2005)
10. Charlwood, R.G., Steele, R., Solymar, S.V., Curtis, D.D.: A review of alkali aggregate reactions in hydroelectric plants and dams. In: CEA and CANCOLD (eds.) Proceedings of the International Conference of Alkali-Aggregate Reactions in Hydroelectric Plants and Dams, pp. 1–29. Fredericton, Canada (1992)
11. Léger, P., Côte, P., Tinawi, R.: Finite element analysis of concrete swelling due to alkali-aggregate reactions in dams. Comput. Struct. **60**(4), 601–611 (1996)
12. Multon, S.: Evaluation expérimentale et théorique des effets mécaniques de l'alcali-réaction sur des structures modéles. Technical report. Etudes et recherches des Laboratoires des ponts et chaussées, Série Ouvrages d'art OA46 (2004)
13. Newell, V., Wagner, C.: Fontana dam: a crack in the curve. In: Waterpower' 99: Hydro's Future: Technology, Markets, and Policy, pp. 1–10 (1999)

14. Caron, P., Léger, P., Tinawi, R., Veilleux, M.: Slot cutting of concrete dams: field observations and complementary experimental studies. Struct. J. **100**(4), 430–439 (2003)
15. Gilks, P., Curtis, D.: Dealing with the effects of AAR on the water retaining structures at Mactaquac GS. In: Proceedings of the 21st Congress on Large Dams, Montreal, Canada, pp. 681–703 (2003)
16. Metalssi, O., Seignol, J., Rigobert, S., Toutlemonde, F.: Modeling the cracks opening-closing and possible remedial sawing operation of AAR affected dams. Eng. Fail. Anal. **36**, 199–214 (2014)
17. Hayes, N., Gui, Q., Abd-Elssamd, A., Le Pape, Y., Giorla, A., Le Pape, S., Giannini, E.R., Ma, Z.: Monitoring alkali-silica reaction significance in nuclear concrete structural members. J. Adv. Concr. Technol. **16**(4), 179–190 (2018)
18. NUREG/CR-6706: NUREG/CR-6706: Capacity of Steel and Concrete Containment Vessels with Corrosion Damage (2001). http://pbadupws.nrc.gov/docs/ML0110/ML011070123.pdf
19. Larive, C.: Apports Combinés de l'Experimentation et de la Modélisation à la Comprehension del'Alcali-Réaction et de ses Effets Mécaniques. Ph.D. thesis. Laboratoire Central des Ponts et Chaussées, Paris (1998). http://hal.inria.fr/docs/00/52/06/76/PDF/1997THLARIVECNS20683.pdf

Chapter 22
Benchmark Study Results: EdF/LMDC

Pierre Morenon, Alain Sellier, Stéphane Multon, Etienne Grimal, and Philippe Kolmayer

22.1 P0: Finite Element Model Description;Constitutive Model

The model presented here and used for the numerical benchmark has been under development by the LMDC and EDF for fifteen years and is implemented in the finite element code, Code_Aster.

22.1.1 Basic Principles of the Model and Its Implementation

The concrete behavior law is established in the poro-mechanical framework. The rheological model is built to accept the two kinds of cracking (Fig. 22.1) that are encountered in structures affected by ASR. The first one is ASR diffuse cracking,

P. Morenon · A. Sellier (✉) · S. Multon
Civil Engineering, Paul Sabatier University, and Construction and Durability Materials Laboratory of Toulouse, Toulouse, France
e-mail: alain.sellier@insa-toulouse.fr

P. Morenon
e-mail: morenon@insa-toulouse.fr

S. Multon
e-mail: stephane.multon@insa-toulouse.fr

E. Grimal · P. Kolmayer
Electricité de France, Centre d'Ingénierie Hydraulique, EDF-CIH Technolac, 73373 Le Bourget du Lac Cedex, France
e-mail: etienne.grimal@edf.fr

P. Kolmayer
e-mail: philippe.kolmayer@edf.fr

© RILEM 2021

V. E. Saouma (ed.), *Diagnosis & Prognosis of AAR Affected Structures*, RILEM State-of-the-Art Reports 31,
https://doi.org/10.1007/978-3-030-44014-5_22

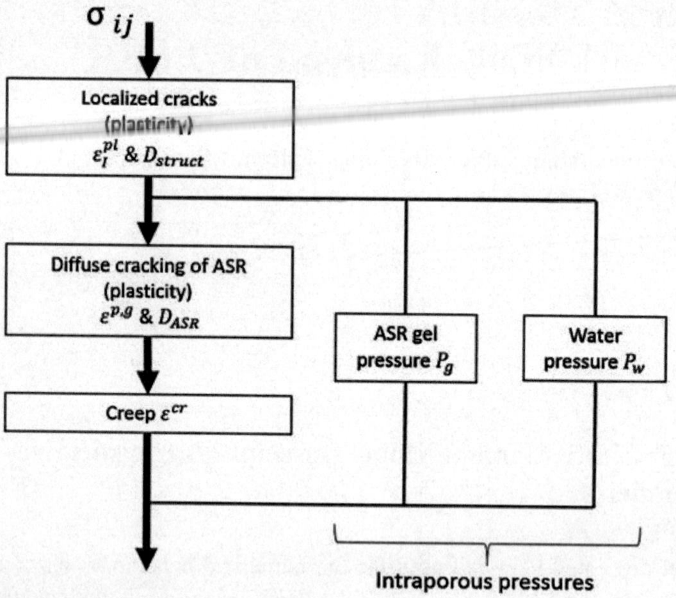

Fig. 22.1 Rheological model

which is a direct consequence of the ASR gel pressure inside the material (Fig. 22.1) at the microscopic scale. The structural localized cracks induced by external loads or expansion gradients are evaluated separately (Fig. 22.1). The creep is also taken into account by Sellier's model, which contains a Kelvin-Voigt (visco-elastic) and an anisotropic non-linear Maxwell level in order to reproduce multi-axial delayed strains due to loading [1]. The model takes creep reversibility and irreversibility, nonlinearity with stress, anisotropy, and dependence on moisture and temperature into account so as to be as realistic as possible and usable in structural calculations [1]. The rheological model uses a poro-mechanical scheme to take ASR gel pressure, water pressure (shrinkage), and external stress into account in the creep calculation. This global scheme directly couples the concrete swelling and the tensile creep produced around the reactive sites [2, 3].

22.1.2 *Nonlinear Constitutive Model of Sound or Damaged Concrete*

22.1.2.1 Instantaneous Response

Details of the numerical implementation algorithm used to combine poro-mechanics, creep, and damage are available in [4]. Only the main features of this implementation are presented here. In Fig. 22.1, the total stress $\tilde{\sigma}_{ij}$ is calculated from the damage D_{ASR}

(due to ASR) and D_{struct} due to structural cracks), and from the effective stress $\tilde{\sigma}_{ij}$ (Eq. 22.1). In the context of damage theory, the effective stress $\tilde{\sigma}_{ij}$ corresponds to the stress in the undamaged zone of the material. The effective stress $\tilde{\sigma}_{ij}$ can also be defined in the framework of poro-mechanics theory [5] by Equation 22.3 [1], with $\tilde{\sigma}'_{ij}$ the stress part in the solid and $(-b_g P_g - b_w P_w)$ the contribution of the interstitial phase pressures (P_g for the pressure due to ASR and P_w for the pressure due to shrinkage induced by water). Each interstitial pressure is affected by the corresponding Biot coefficient (respectively b_g and b_w [6]). δ_{ij} is the Kronecker symbol, which is equal to 1 only if $i = j$. The effective poro-mechanical stress increment $\dot{\tilde{\sigma}}_{ij}$ is calculated from the stiffness matrix S_0 and the elastic strain. The elastic strains can be obtained by subtracting the non-elastic strain increments ($\dot{\tilde{\varepsilon}}_{plkl}$ for the plastic strain, $\dot{\tilde{\varepsilon}}_{crkl}$ for the creep strain and $\dot{\tilde{\varepsilon}}_{thkl}$ for the thermal strain) from the total strain increment $\dot{\tilde{\varepsilon}}_{kl}$ according to Eq. 22.3.

$$\sigma_{ij} = (1 - D_{ASR})(1 - D_{struct})\tilde{\sigma}_{ij} \tag{22.1}$$

$$\tilde{\sigma}_{ij} = \tilde{\sigma}'_{ij} - \delta_{ij} b_g P_g - \delta_{ij} b_w P_w \tag{22.2}$$

$$\dot{\tilde{\sigma}}_{ij} = S_0(\dot{\epsilon}_{kl} - \dot{\epsilon}_{plkl} - \dot{\epsilon}_{crkl} - \dot{\epsilon}_{thkl}) \tag{22.3}$$

22.1.2.2 Delayed Response (Creep and Shrinkage)

Creep and shrinkage are modeled using the poro-mechanical framework as the water pressure induces creep (Fig. 22.1). The shrinkage is reproduced through the Van Genuchten law (Eq. 22.4 [7]). Two material parameters are necessary: M_{sh} (usually fitted on a shrinkage test) and b (usually fitted on a sorption-desorption test).

$$P_w = M_{sh}\left(1 - S_r^{-\frac{1}{b}}\right)^{(1-b)} \tag{22.4}$$

The creep model is based on the consolidation theory. It is represented by a Burger chain divided into three anisotropic stages (Fig. 22.2): elastic, visco-elastic (Kelvin-Voigt), and plastic (Maxwell) [1].

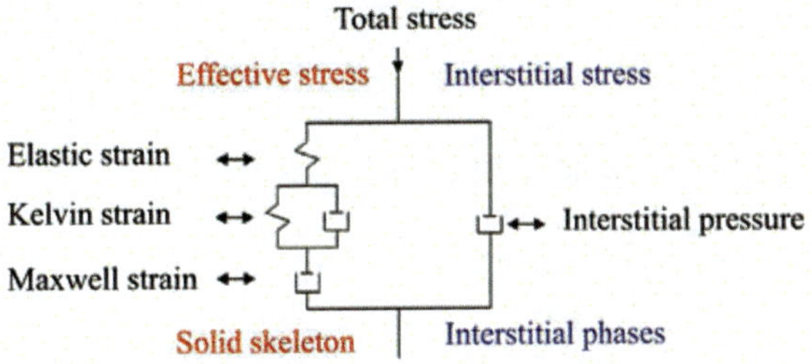

Fig. 22.2 Idealized rheological scheme for poro-mechanical creep model [1]

22.1.3 Gel Pressure

The gel pressure P_g is managed by Eq. 22.5:

$$P_g = M_g < \phi_g - \left(\phi_g^v \left(\frac{P_g}{\tilde{R}_I^t} \right) + b_g \mathrm{Tr}\,(\varepsilon^e + \varepsilon^{cr}) + \mathrm{Tr}\,(\varepsilon^{p.g}) \right) >^+ \tag{22.5}$$

where:

- M_g is the interaction modulus between the gel and the matrix $\phi_g = \phi_g^\infty . A^{asr}$ is the volume of gel produced by ASR, which comes from the chemical advancement A^{asr} and the final gel volume ϕ_g^∞.
- ϕ_g^v is the volume reachable without gel pressure around the reaction site.
- $\tilde{R}_I^t$ is the tensile strength.
- b_g is the Biot coefficient, which comes from the poro-mechanics.
- ε^e is the elastic strain.
- ε^{cr} is the creep strain.
- $\varepsilon^{p.g}$ is the ASR plastic strain, which represents the opening of ASR micro-cracks.

The gel pressure P_g represents the pressure due to the difference between the gel volume ϕ_g and the volumes reachable around the reaction site:

- created by mechanical loading $b_g.\mathrm{Tr}\,(\varepsilon^e + \varepsilon^{cr})$,
- due to connected porosity ϕ_g^v,
- due to ASR microcracking $\mathrm{Tr}\,(\varepsilon^{p.g})$.

22.1.4 ASR Chemical Advancement

The kinetics of the ASR advancement (Eq. 22.6) depends on the temperature and the moisture content around the reaction site. The characteristic time τ_{ref}^{asr} is a parameter to fit the free swelling kinetics.

$$\frac{\delta A^{asr}}{\delta t} = \frac{1}{\tau_{ref}^{asr}} C^{W,asr} < Sr - A^{asr} >^+ C^{T,asr} \tag{22.6}$$

22.1.4.1 Impact of the Water Content

The advancement law proposed in [2] is based on Poyet's law [8] (Eq. 22.7). It uses a minimum threshold to initiate the reaction $S_r^{th,asr}$ (here 0.1) and evolves non-linearly in order to strongly accelerate the reaction kinetics with the water saturation of the material.

$$C^{W,asr} = \begin{cases} \left(\frac{S_r - S_r^{th,asr}}{1 - S_r^{th,asr}}\right)^2 & \text{if } S_r > S_r^{th,asr} \\ 0 & \text{if } S_r \leq S_r^{th,asr} \end{cases} \tag{22.7}$$

22.1.4.2 Impact of the Temperature

An Arrhenius law manages the impact of the temperature T on the ASR advancement kinetics (Eq. 22.8)

$$C^{T,asr} = \exp\left(-\frac{E^{asr}}{R}\left(\frac{1}{T} - \frac{1}{T_{ref}}\right)\right) \tag{22.8}$$

E^{asr} is the thermal activation energy (~40,000 J Mol^{-1} [9]), R is the perfect gas constant (8.3145 J mol^{-1} K^{-1}), and T_{ref} is the temperature corresponding to the τ_{ref}^{asr} calibration).

22.1.5 *Effect of ASR on Concrete Cracking*

In the model, two kinds of cracking are distinguished at two different scales [3]. The first one is the microcracking directly due to the gel pressure, which leads to a decrease of mechanical characteristics such as Young's modulus. The second one is the macro structural cracking due to mechanical loading and swelling gradients inside a structure. It leads to tensile macrocracking, which can be a serious problem for the monolithic nature of the structure or the durability of the steel bars in reinforced concrete.

22.1.5.1 Effect of the ASR Expansion on the Mechanical Properties of Concrete

The ASR microcracking is managed by an anisotropic Rankine criterion (Eqs. 22.9, 22.10).

$$f_I^{tASR} = \tilde{\sigma}_I' - R_{tmicro}; \qquad I \in [I, II, III] \tag{22.9}$$

$$\tilde{\sigma}_I' = k.P_g + \min(\tilde{\sigma}_I; 0); \qquad I \in [I, II, III] \tag{22.10}$$

In cases of external compressive loading $\tilde{\sigma}_I$, the microcracking is delayed or eliminated [2]. The value of k is 1. The ASR plastic strain $\varepsilon_I^{p,g}$ is then calculated from a plastic flow. The corresponding ASR damage is finally deduced from Equation 22.11 [10]:

$$D_I^{t,g} = \frac{\varepsilon_I^{p,g}}{\varepsilon_I^{p,g} + \varepsilon^{k,g}} \tag{22.11}$$

$\varepsilon^{k,g}$ is a parameter fixed at 0.3% for ASR.

22.1.5.2 Effect of the ASR Swelling Gradients on Macrocracking

Swelling gradients due to water transfer or alkali leaching can lead to structural cracks. They are managed by an anisotropic Rankine plastic criterion.

$$f_I^{tstruct} = \tilde{\sigma}_I - \tilde{R}_I^t \tag{22.12}$$

A plastic strain ε_I^{pl} is then obtained and a crack opening $w_I^{pl,max}$ is calculated [1]. The structural damage D_I^{struct} is deduced (Eq. 22.13). $_I^k$ is a characteristic crack opening (calculated to dissipate the fracture energy Gft). An energy regularization is carried out based on the Hillerborg method [11].

$$D_I^{struct} = 1 - \left(\frac{w_I^k}{w_I^l + w_I^{pl,max}} \right) \tag{22.13}$$

22.2 P2: Drying and Shrinkage

The model's ability to reproduce drying and shrinkage is evaluated on samples (cylinders 16 cm by 32 cm) subjected to three environmental conditions:

- Immersed.
- Under aluminum layers.
- Under 30% of relative humidity.

The mass variations are fitted on the experimental data [12] (Fig. 22.3a). The immersed sample is quickly saturated (100 days), the sample under aluminum layers loses water linearly, and the sample under 30% of relative humidity (RH) loses a lot of water during the first days.

Two parameters (M_{sh} = 25 MPa and b = 0.42 from Eq. 22.4) are fitted to obtain the strains (Fig. 22.3a). The model is able to reproduce strains due to gain or loss of water without ASR.

22.3 P3: Basic Creep

The numerical basic creep test consists in modelling two samples (cylinders 13 by 24 cm) under uniaxial loadings (10 and 20 MPa) [13].

From the fitted mass loss of the samples under aluminum layers (Fig. 22.3b), the hydric conditions and the strains are faithfully reproduced by calibrating the two coefficients of the law of Van Genuchten ($M_{sh} = 25$ MPa and $b = 0.42$). Then, the creep tests at 10 and 20 MPa make it possible to simulate the behavior of the material under long-term loading. Four parameters are necessary for the creep calibration: two

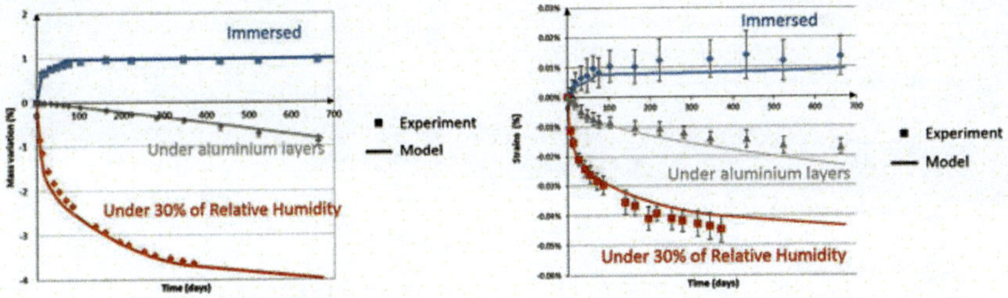

(a) Mass variations for non-reactive samples. Experiment from [285]

(b) Strains for non-reactive samples Experiment from [285]

Fig. 22.3 EdF/LMDC; P2: drying shrinkage

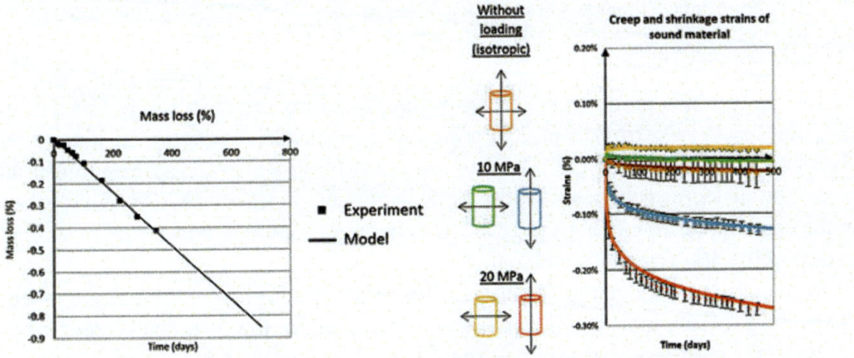

(a) Mass variations for non-reactive sample. Experiment from [284]

(b) Shrinkage and creep (uniaxial loadings 10 MPa and 20 MPa): comparison between numerical modeling [272] and experiment results [284]

Fig. 22.4 EdF/LMDC; P3: basic creep

characteristic times for reversible creep (0.7 days) and irreversible creep (10 days), one for the modulus of the reversible part (4.65 times the elastic modulus) and one for the deformation characteristic of the irreversible part (0.94×10^{-4}) [1] (Fig. 22.4b).

22.4 P5: Free AAR Expansion; Effect of RH

The effect of relative humidity (RH) on swelling modeling is analyzed for the same environmental conditions [12] as in the Drying and Shrinkage part. Thus, the shrinkage parameters have already been calibrated.

The mass variations are fitted with respect to the experimental results [12] (Fig. 22.5a). They are close to the mass variations of the non-reactive samples (Fig. 22.3a).

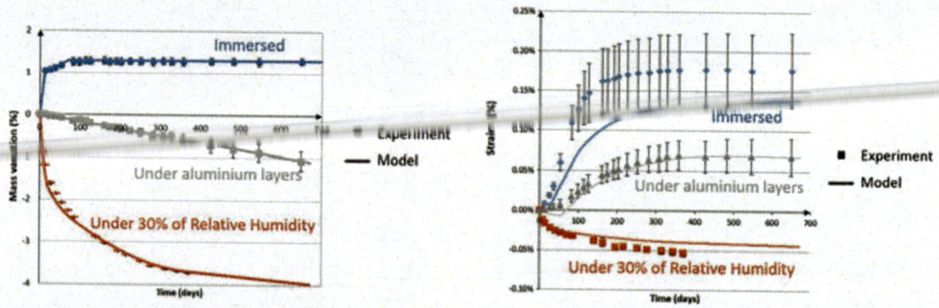

(a) Mass variations for reactive samples [271]. Experiment from [285]

(b) Strains for reactive samples [271]. Experiment from [285]

Fig. 22.5 EdF/LMDC; P5: free Asr expansion; effect of RH

The ASR model parameters ($\tau_{ref}^{asr} = 60$ days from Eq. (22.6, $\phi_g^{\infty} = 0.54\%$ in Eq. (22.7), $\phi_g^{v} = 0.13\%$ from Eq. 22.8) are calibrated on the sample under aluminum layers and then applied to the other humidity conditions (Fig. 22.5b). The threshold of saturation degree necessary to observe a significant swelling of AAR is not reached in the case of specimens kept at 30% RH. The deformation is thus almost identical to that of the non-reactive specimen.

In the case of endogenous swellings (under aluminum layers), the curve reproduces the experimental results correctly, since these are the data that were used to calibrate the parameters of the swelling model. In the case of specimens immersed in water, the kinetics is well reproduced. The final expansion seems small compared to the mean experimental value, but expansion is within the interval of measurement. These tests made it possible to determine the exponent "2" of the law of dependence of the kinetics of advancement of the AAR as a function of the degree of saturation [2] to obtain a better reproduction of the kinetics in the immersed case. The model generally well reproduces the behavior of test pieces under different water conditions: a minimum saturation threshold to observe a swelling, a period of latency before a larger swelling and an asymptotic difference between the immersed specimens and the sealed specimens.

22.5 P6: AAR Expansion; Effect of Confinement

In this experiment [13], cylindrical specimens (13 × 24 cm) are loaded in uniaxial compression. In some tests, the radial displacement is restrained by steel rings of different thicknesses, which makes it possible to obtain multiaxial stress states. The steel rings, which prevent the concrete from swelling radially, put the material in a state of multi-axial stress. The numerical concrete parameters are strictly the same as in the previous part on RH impact.

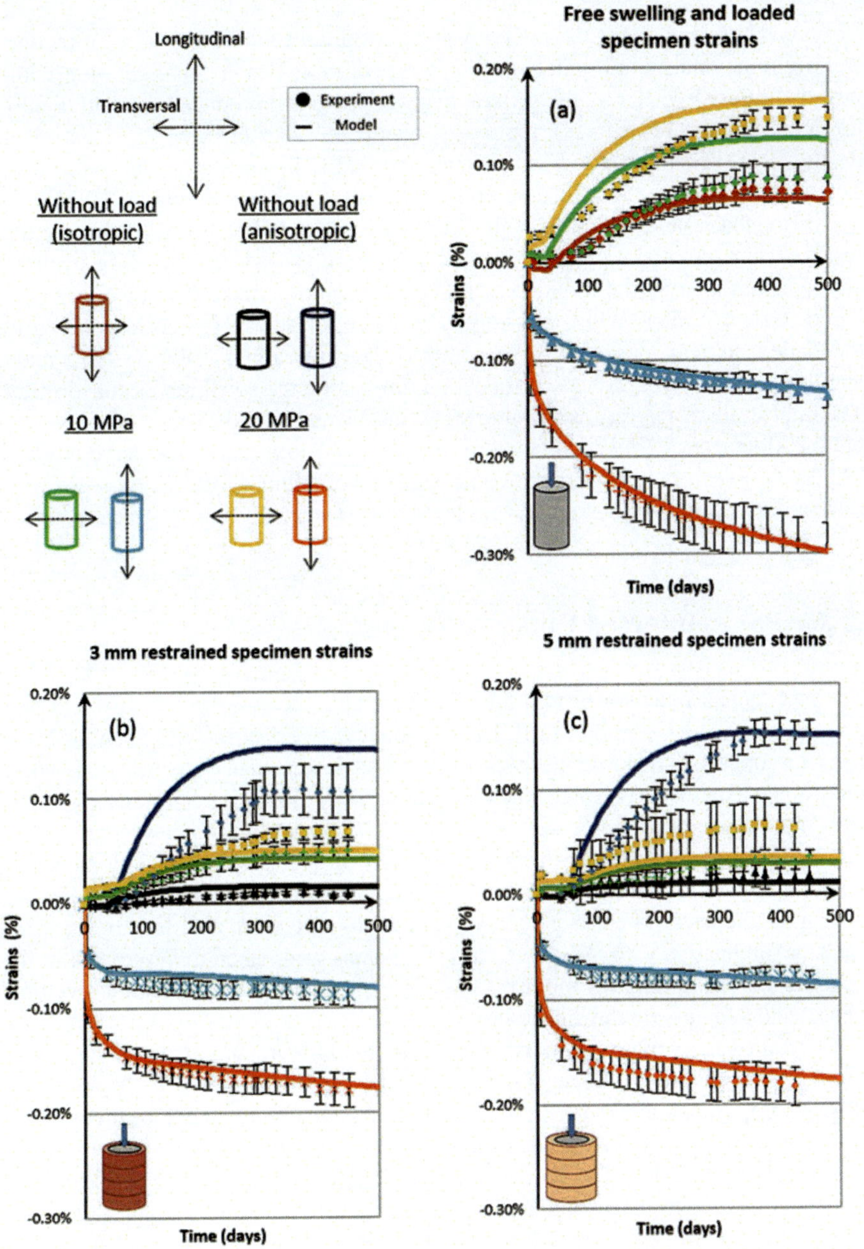

Fig. 22.6 Multon's tests [13]: **a** Swelling tests with or without uniaxial loading, **b** samples with 3 mm steel rings, **c** samples with 5 mm steel rings. Numerical results from Morenon et al. [2]

Figure 22.6 shows that the model is able to reproduce the expansions of moderate swellings under multi-axial stresses and restraints. For the configuration without steel rings (Fig. 22.6a), the general behavior is well reproduced. Longitudinally (brown, blue and red curves), the model results are very close to the experiment results. Transversally (yellow and green curves), the kinetics are slightly too fast. The asymptote of the loaded case at 20 MPa is correct but the one at 10 MPa is overestimated. The difference obtained on the case loaded at 10 MPa could be the result of experimental scatter because the points are very close to the free swelling case whereas a larger deformation could be expected due to the Poisson effect.

For tests restrained with steel rings (Fig. 22.6b, c), the curves corresponding to the loaded cases are successfully reproduced. Transversally, swelling without axial stresses (black curves) is satisfactory. Longitudinally, the swellings occur too fast. The final strain of the case with 5 mm steel rings is realistic (dark blue curve on (c)) but the kinetics is slightly too fast with the rings of 3 mm (dark blue curve on (b)). The two curves obtained for the 3 and 5 mm rings without axial stress were very different experimentally whereas, for all the other tests with rings, they were close.

22.6 P8: Reinforced Concrete Beams

The mechanical behavior of two reinforced concrete beams, one reactive and one non-reactive, has been studied [15]. Their geometry was 3.0 × 0.5 × 0.25 m for 2.8 m between supports. The lower part of the beams was immersed in water to 70 mm while their upper face was subjected to air drying at 30% of relative humidity (RH) for 428 days. The upper faces were wet at this date. The rewetting phase consisted in putting the upper face of the beams in contact with water while keeping the same boundary conditions in the lower part. The temperature was maintained at 38 °C during all the tests. These beams were simply supported on steel bars 0.1 m from each end and halfway up. After 700 days, the beams were taken to failure in 4-point bending tests with a slightly modified span (2.75 m against 2.8 m during the swelling phase) and supports on the underside.

Several experimental data make it possible to calibrate and evaluate the representativeness of the modeling. First, the beams were weighed at different times (Fig. 22.7a). The mass variations were calibrated on the non-reactive beam case. To simulate drying (upper part) and capillary rise (lower part), a model of water diffusion is used (Eq. 22.11). The water diffusion coefficient D (Eq. 22.15) takes account of the permeation transfers and the dependence of diffusion on the water saturation degree, W, according to Mensi's model [16]. This model is simple but it gives a realistic representation of moisture gradients in the beams, which is an important point for their structural analysis.

The calibration is based on the data provided by the experimental program [17] (Fig. 22.7a). For the coefficient a (Eq. 22.15, the values vary between 1.2 10–13 m^2 s^{-}1 (upper part) and 5.8 10–12 m^2 s^{-}1 (lower part). The coefficient b is taken as 0.051 for the zone in imbibition and 0.06 for the zone under drying

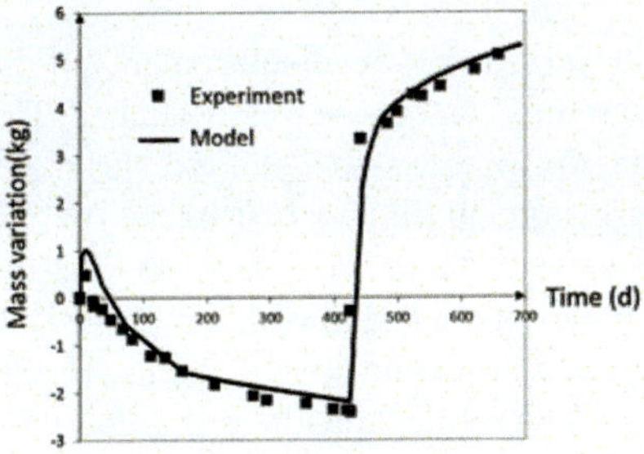

(a) Global mass variation of the non-reactive beam [273]; [283]

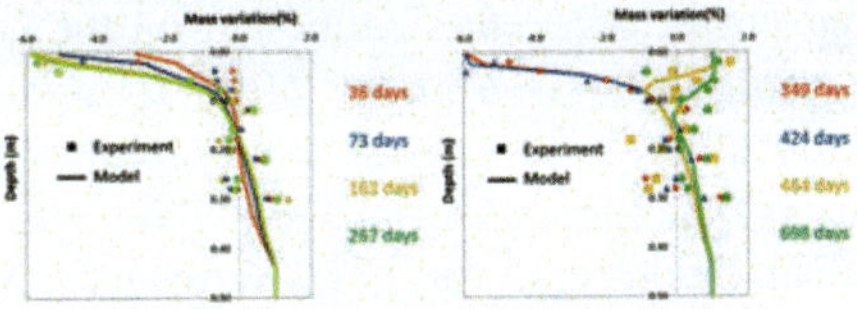

(b) Local mass variation of the non-reactive beam [273]; [283]

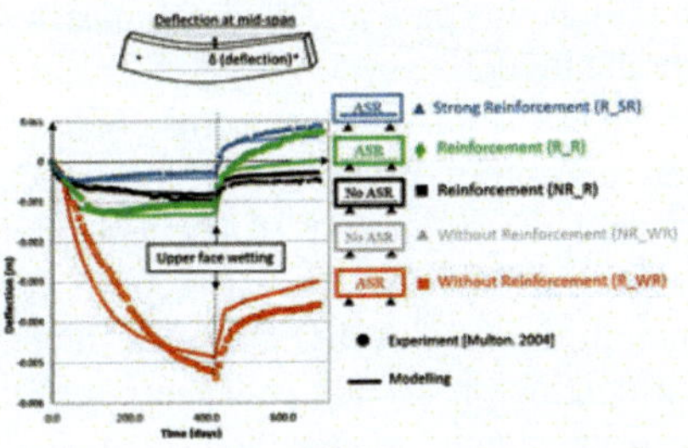

(c) Comparison between deflections obtained in the experimental program presented in [285] and by modeling [273]

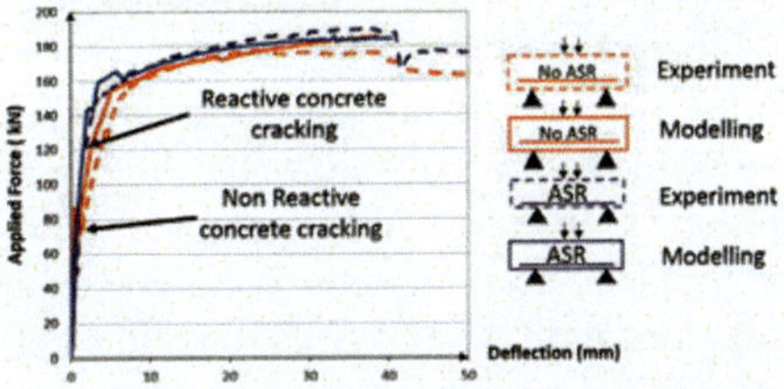

(d) Force-Deflection curves of the failure test: model [273] and experiment [278] of the normally reinforced beams

Fig. 22.7 EdF/LMDC; P8 reinforced concrete beam

[14, 18]).

$$\frac{\delta w}{\delta t} = Div[D(W)\nabla W] \tag{22.14}$$

$$D(W) = a.e^{b.w} \tag{22.15}$$

The modeled mass loss is consistent with the experimental data (Fig. 22.7a). Mass recovery from rewetting (428 days) is also correctly reproduced. Elements of local mass losses are provided by the authors to model the drying and imbibition in the height of the beam (Fig. 22.7b). Despite the scatter of the experimental results, the loss of mass in the height of the beam over time is well reproduced by the modeling (Fig. 22.7b, [19]).

The experimental data [12] present the aging phase of five beams (Fig. 22.7c). The benchmark focuses on the non-reactive reinforced beam (NR_R) and the reactive reinforced beam (R_R) in Fig. 22.7c.

All the model parameters (shrinkage, creep, ASR swelling) come directly from Multon's calibration of the samples (see Figs. 22.4b, 22.6). The results from the modeling of ASR beams are very satisfactory (black and grey curves in Fig. 22.7c.

On the basis of the calibration obtained on the specimens, the displacements of the reactive or non-reactive beams with or without reinforcement are well evaluated. The modeling of the non-reactive beams validates the water profile and the shrinkage and creep parameters calibrated on the specimens. During the first phase (before rewetting), the movements of the reactive beams are found accurately. At first, the deflections become negative because the lower part of the beam, which is immersed, reacts quickly. The shrinkage in the upper part and the expansions in the lower part are at the origin of this negative displacement. The chemical prestress is then set up in the direction of the reinforcements. It has the effect of changing the direction of the deflections of the reinforced beams when the swelling of the median part becomes predominant over that of the rest of the beam. After rewetting, the upper part of the beams is saturated. So, the concrete swells quickly in this part, and creates a positive deflection. Trends are correctly reproduced by the model.

Figure 22.7d compares model responses to experimental measurements obtained during the four-point bending tests [17]. For the non-reactive beam (blue lines), the experimental data show the first flexural tensile cracking at 75 kN and steel plasticization at 150 kN. The model reproduces this behavior correctly (60 kN for the first tensile crack in the concrete) even though it seems to slightly overestimate the initial rigidity. The cracking and steel plasticizing phases are well reproduced. The differences in rigidity after the first cracking can come from the displacement of the sensor used to measure the deflection. As cracking developed around the sensor, the measurement could overestimate the displacement as discussed in [17].

The reactive beam (red curves) follows the same trend with a later start of cracking (120 kN) and a start of steel plasticization equivalent to the non-reactive beam. The model reproduces the experimental behavior well throughout the test (first crack at 140 kN) and thus allows correct prediction of the future behavior of the beam. The reactive beam shows a flexural cracking delay with respect to the non-reactive beam. This is due to the chemical prestressing developed in the concrete. Reinforcement restrains the expansion of the concrete, which is thus subjected to compressive stress, responsible for this prestressing. For the reactive beam, the applied force is first used to decompress the concrete of the lower part before putting it under tension and cracking it. This difference is well reproduced by the model.

22.7 P10: Expansion of RC Panel with or Without Lateral Confinement

The exercise consists in modeling Reinforced Concrete (RC) panels subjected to ASR with or without lateral confinement [20] exerted by external steel frames (Fig. 22.8a). The panels contain two steel lattices (Fig. 22.8b) orthogonal to the vertical direction Z.

Shrinkage parameters ($M_{sh} = 40$ MPa and $b = 0.45$ from Eq. 22.4) and ASR swelling parameters (τ_{ref}^{asr} =140 days from Eq. 22.6, $\phi_g^{\infty} = 1.93$ % in Eq. 22.7), $\phi_g^{v} =$ 0.73 % from Eq. 22.8, $R_t = 4.5$ MPa) are calibrated on samples (Fig. 22.9a). The

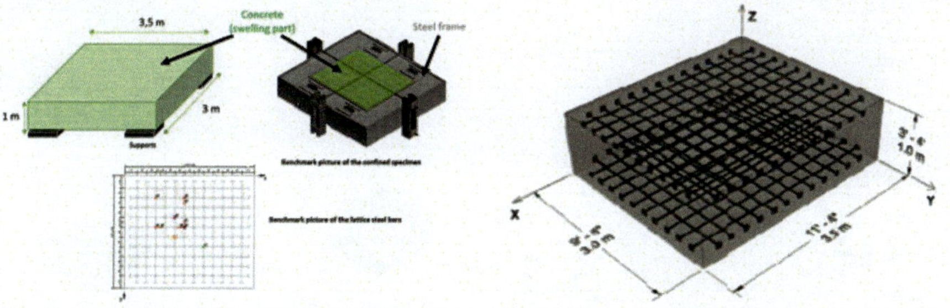

(a) Presentation of the reinforced concrete panels Hayes, Gui, Abd-Elssamd, Le Pape, Giorla, Le Pape, Giannini, and Ma [151]

(b) Dimensions and coordinate axes of concrete specimens [151]

Fig. 22.8 EdF/LMDC; P10 RC panel, problem statement

other parameters are the same as on the ASR beams and ASR samples. Three meshes are created: the concrete mesh, the steel bar mesh and the external frame mesh (Fig. 22.9b). Two limit conditions are used: the free RC panel and the restrained RC panel (Fig. 22.9c).

For the free RC panel, the strains in both directions are faithfully reproduced (Fig. 22.9d). However, two structural macrocracks appear along the steel bar lattice. This cracking does not seem realistic. This phenomenon is due to the simplicity in the representation of the interface between steel and concrete in the modeling. In reinforced concrete modeling, an interface layer is usually added to manage the slip between concrete and steel subjected to high strains [21]. Without this layer, the strain gradient between swelling concrete and passive steel induces such cracking. Further research on the properties of the interface between ASR swelling concrete and steel reinforcement is needed to obtain a more realistic evaluation.

For the RC panel restrained by a steel frame, the strains obtained by the model are still very close to the experiment results (Fig. 22.9e) in both directions. Figure 22.9 highlights the impacts of the steel bars effect and the frame effect on the strains in the panels. The steel bars induce a swelling anisotropy: the swelling is reduced in the lattice direction X and there is a swelling increase in the orthogonal direction Z. This phenomenon is accentuated with the external steel frame, which restrains the swelling in the X direction.

22.8 Conclusion

The numerical benchmark consists of reproducing the mechanical behavior, from samples to structures, of ASR affected structures. To obtain a faithful structural assessment, it is necessary to:

- combine expansion, damage and creep,

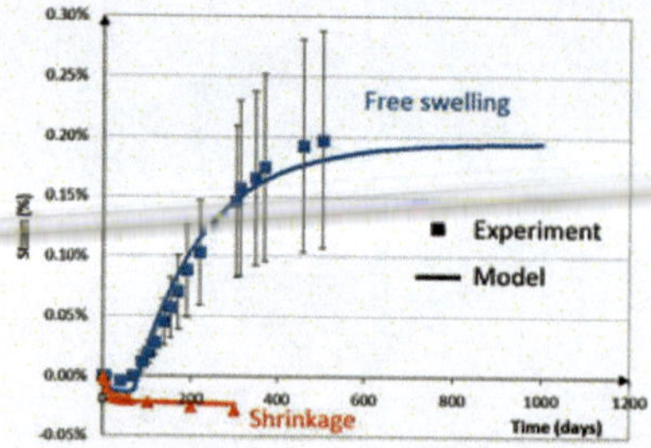

(a) Free swelling and shrinkage fitting on experimental data provided in the benchmark

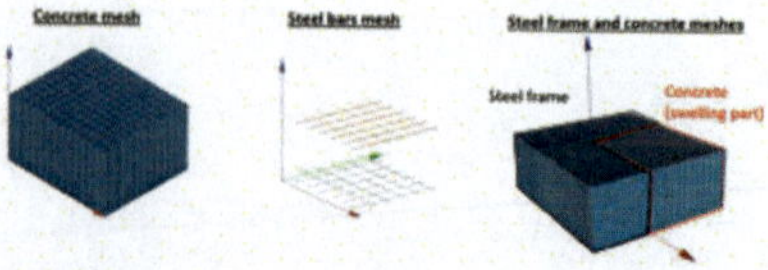

(b) Concrete, steel bar and external steel frame meshes used

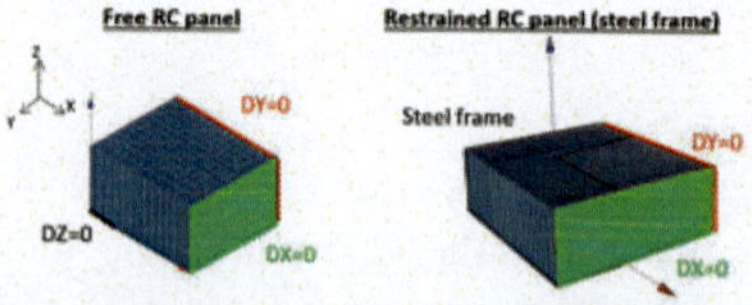

(c) Limit conditions for the reinforced concrete (RC) panels

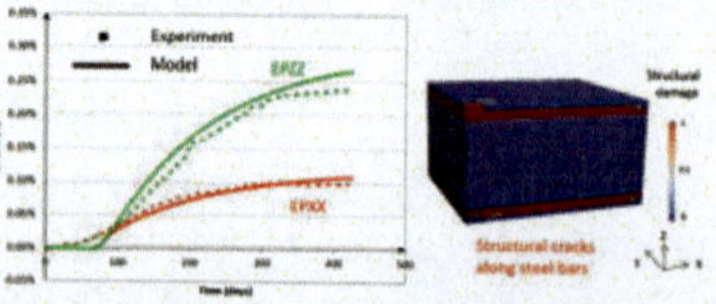

(d) Strains and damage obtained for the free RC panel

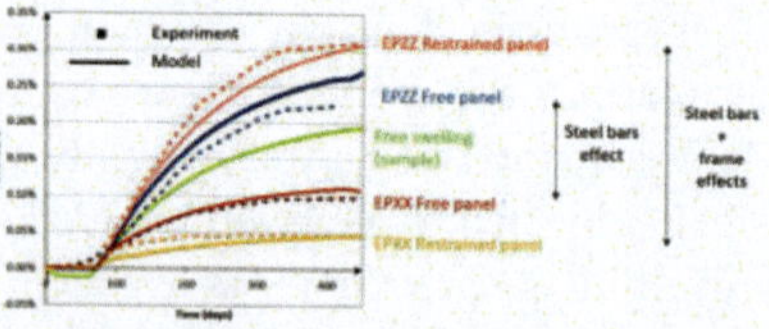

(e) Strains obtained for the RC panels with and without external steel frame (restrained panels)

Fig. 22.9 EdF/LMDC; P10 RC panel, results

- distinguish the two types of cracking (ASR microcracking and structural macrocracks),
- model anisotropic ASR damage depending on the multi-axial stress state.

The model validation methodology consists of calibrating the model parameters on samples and predicting the behavior of laboratory structures.

These numerical benchmark tests have been performed with the poro-mechanical model developed by the LMDC and EDF and implemented in Code_Aster. It is used daily by EDF engineers.

References

1. Sellier, A., Multon, S., Buffo-Lacarrière, L., Vidal, T., Bourbon, X., Camps, G.: Concrete creep modelling for structural applications: non-linearity, multi-axiality, hydration, temperature and drying effects. Cement Concrete Res. **79**, 301–315 (2016)
2. Morenon, P., Multon, S., Sellier, A., Grimal, E., Hamon, F., Bourdarot, E.: Impact of stresses and restraints on ASR expansion. Construct. Build. Mater. **140**, 58–74 (2017)
3. Morenon, P., Multon,S., Sellier, A., Grimal, E., Hamon, F., Kolmayer, P.: Flexural performance of reinforced concrete beams damaged by Alkali-Silica Reaction. Cement Concrete Compos. **104**, 103412 (2019)
4. Sellier, A. : Anisotropic damage and Visco-Elasto-plasticity applied to multiphasic materials. Tech. rep. LMDC-Laboratoire Matériaux et Durabilité des Constructions de Toulouse; Université de Toulouse III-Paul Sabatier; INSA de Toulouse (2018)
5. Coussy, O.: Mécanique des milieux poreux. Editions Technip (1991)
6. Biot, M.A.: General theory of three-dimensional consolidation. J. Appl. Phys. **12**(2), 155–164 (1941)
7. van Genuchten, M.T.: A closed form equation for predicting the hydraulic conductivity of unsaturated soils. Soil Sci. Soc. Am. J. **44**(5), 892–898 (1980)
8. Poyet, S., Sellier, A., Capra, B., Thèvenin-Foray, G., Torrenti, J.-M., Tournier-Cognon, H., Bourdarot, E.: Influence of water on alkali-silica reaction: experimental study and numerical simulations. J. Mater. Civil Eng. **18**(4), 588–596 (2006)
9. Larive, C.: Apports Combinés de l'Experimentation et de la Modélisation à la Comprehension del'Alcali-Réaction et de ses Effets Mécaniques. Ph.D. thesis. Laboratoire Central des Ponts et Chaussées, Paris (1998). https://hal.inria.fr/docs/00/52/06/76/PDF/1997TH_LARIVE_C_NS20683.pdf
10. Capra, B., Sellier, A.: Orthotropic modeling of Alkali-aggregate reaction in concrete structures: numerical simulations. Mech. Mater. **35**, 817–830 (2003)
11. Hillerborg, A., Modéer, M., Petersson, P.-E.: Analysis of crack formation and crack growth in concrete by means of fracture mechanics and finite elements. Cement Concrete Res. **6**(6), 773–781 (1976)
12. Multon, S., Toutlemonde, F.: Effect of moisture conditions and transfers on alkali silica reaction damaged structures. Cement Concrete Res. **40**(6), 924–934 (2010)
13. Multon, S., Toutlemonde, F.: Effect of applied stresses on alkali-silica reaction-induced expansions. Cement Concrete Res. **36**(5), 912–920 (2006)
14. Morenon, P.: Modélisation des réactions de gonflement interne des bétonsavec prise en compte des couplages poro-mécaniques et chimiques. Ph.D. thesis. LMDC-Laboratoire Matériaux et Durabilité des constructions (2017)
15. Multon, S., Seignol, J., Toutlemonde, F.: Structural behavior of concrete beams affected by alkali-silica reaction. ACI Mater. J. **102**(2), 67 (2005)
16. Mensi, R., Acker, P., Attolou, A.: Séchage du béton: analyse et modélisation. Mater. Struct. **21**(1), 3–12 (1988)
17. Multon, S.: Evaluation expérimentale et théorique des effets mécaniques de l'alcali-réaction sur des structures modèles. Ph.D. thesis. Université de Marne-la-Vallée (2004)
18. Grimal, E., Sellier, A., Multon, S., Le Pape, Y., Bourdarot, E.: Concrete modelling for expertise of structures affected by alkali aggregate reaction. Cement Concrete Res. **40**(4), 502–507 (2010)
19. Multon, S., Merliot, E., Joly, M., Toutlemonde, F.: Water distribution in beams damaged by Alkali-Silica Reaction: global weighing and local gamma densitometry. Mater. Struct. **37**(5), 282 (2004)
20. Hayes, N., Gui, Q., Abd-Elssamd, A., Le Pape, Y., Giorla, A., Le Pape, S., Giannini, E.R., Ma, Z.: Monitoring alkali-silica reaction significance in nuclear concrete structural members. J. Adv. Concrete Technol. **16**(4), 179–190 (2018)
21. Handika, N., Casaux-Ginestet, G., Sellier, A.: Influence of interface zone behaviour in reinforced concrete under tension loading: an analysis based on modelling and digital image correlation. In: COMPLAS XIII: Proceedings of the XIII International Conference on Computational Plasticity: fundamentals and applications, pp. 122–133. CIMNE (2015)

Chapter 23
Benchmark Study Results: IFSTTAR

Boumediene Nedjar, Claude Rospars, Renaud-Pierre Martin, and François Toutlemonde

23.1 P0: Finite Element Model Description

23.1.1 Constitutive Model

Within the continuum, the kinematical choice is based on an additive split of the *total* strain, denoted as usual by the tensor $\boldsymbol{\varepsilon}$, into an elastic part $\boldsymbol{\varepsilon}^{\mathrm{e}}$ and complementary parts, each one corresponding to a phenomenon. We write,

$$\boldsymbol{\varepsilon} = \boldsymbol{\varepsilon}^{\mathrm{e}} + \boldsymbol{\varepsilon}_{\mathrm{th}} + \boldsymbol{\varepsilon}_{\mathrm{hyd}} + \boldsymbol{\varepsilon}^{\mathrm{cr}} + \boldsymbol{\varepsilon}_{\chi} \ldots, \tag{23.1}$$

where $\boldsymbol{\varepsilon}_{\mathrm{th}}$ and $\boldsymbol{\varepsilon}_{\mathrm{hyd}}$ are respectively the thermal and hydric expansion tensors, $\boldsymbol{\varepsilon}^{\mathrm{cr}}$ is the basic creep strain, and $\boldsymbol{\varepsilon}_{\chi}$ is the chemical expansion tensor. The dots in this expression mean that other standard and/or non stantard phenomena can be integrated within

B. Nedjar (✉) · C. Rospars · R.-P. Martin
Materials and Structures Department, Urban and Civil Engineering Testing and Modeling Laboratory (EMGCU), IFSTTAR, French institute of science and technology for transport, development and networks, Université Paris Est, 77447 Marne-la-Vall'ee, France
e-mail: boumediene.nedjar@ifsttar.fr

C. Rospars
e-mail: claude.rospars@ifsttar.fr

R.-P. Martin
e-mail: renaud-pierre.martin@ifsttar.fr

F. Toutlemonde
Materials and Structures Department, IFSTTAR, French institute of science and technology for transport, development and networks, Université Paris Est, 77447 Marne-la-Vall'ee, France
e-mail: francois.toutlemonde@ifsttar.fr

© RILEM 2021

V. E. Saouma (ed.), *Diagnosis & Prognosis of AAR Affected Structures*, RILEM State-of-the-Art Reports 31,
https://doi.org/10.1007/978-3-030-44014-5_23

this formalism whenever needed, for instance, plastic strain $\boldsymbol{\varepsilon}^{\mathrm{p}}$, shrinkage strain $\boldsymbol{\varepsilon}_{\mathrm{sk}}$, and so on. For the thermal and hydric parts, we can use the classical relations,

$$\boldsymbol{\varepsilon}_{\mathrm{th}} = \alpha\,(T - T_0)\,\mathbf{1}\,, \qquad \boldsymbol{\varepsilon}_{\mathrm{hyd}} = \beta\,(S_r - S_{r0})\,\mathbf{1}\,, \tag{23.2}$$

where α and β are respectively the thermal and hydric expansion coefficients, and $\mathbf{1}$ denotes the second-order identity tensor. T and S_r are the temperature and saturation, respectively, while T_0 and S_{r0} are their respective initial values. Notice that a form based on the relative humidity can equivalently be used instead of the above one based on saturation. Now if we consider a purely volumetric chemical expansion, we then write:

$$\boldsymbol{\varepsilon}_\chi = \varepsilon_\chi \mathbf{1}, \tag{23.3}$$

where the scalar functional $\varepsilon_\chi \equiv \varepsilon_\chi(S_r, T, t\ldots)$ is the so-called chemical expansion. In a fully saturated condition, i.e. $S_r = 1$ for all $t \in [0, +\infty[$, one has the nowadays well known Larive's law given by, see for instance [1, 2],

$$\varepsilon_\chi = \varepsilon^0_\infty, \frac{1 - e^{\left(-\frac{t}{\tau_c}\right)}}{1 + e^{\left(-\frac{t-\tau_l}{\tau_c}\right)}}, \tag{23.4}$$

where ε^0_∞ is the potential chemical strain that constitues the amplitude of full expansion, and τ_c and τ_l are respectively the characteristic and latent times that characterize the kinetics of expansion. However, for the case of variable hydric and thermal conditions, this latter expression must be adapted, see below.

23.1.1.1 Instantaneous Response

The above kinematic decomposition must now be embedded into a constitutive relation. The simplest choice is to consider an elastically reversible behavior as

$$\boldsymbol{\sigma} = \mathbb{C} : \boldsymbol{\varepsilon}^{\mathrm{e}}, \tag{23.5}$$

where $\boldsymbol{\sigma}$ is the stress tensor, and $\mathbb{C}$ is the fourth-order elasticity tensor that can in turn be affected by chemical damage as

$$\mathbb{C} = (1 - d_\chi)\mathbb{C}_0, \tag{23.6}$$

where $\mathbb{C}_0$ is the elastic modulus for the undamaged concrete and d_χ is a damage variable in the sense of continuum damage mechanics. Intuitively, this latter can naturally be driven by the chemical expansion itself, then explicitly given as a function of the quantity ε_χ. For instance, the following form is adopted:

$$d_\chi = 1 - e^{\left[-\omega\left\langle\varepsilon_\chi - \varepsilon_{\text{trs}}\right\rangle_+\right]}, \tag{23.7}$$

where ε_{trs} is the strain-like chemical damage threshold, and $\omega \geq 0$ a convenient parameter, i.e. no chemical damage takes place if we set $\omega = 0$.

23.1.1.2 Creep Response

The creep strain $\boldsymbol{\varepsilon}^{\text{cr}}$ is here treated as an internal variable that moreover can in turn be the sum of as many as necessary internal contributions $\boldsymbol{\varepsilon}_i^{\text{cr}}$:

$$\boldsymbol{\varepsilon}^{\text{cr}} = \sum_{i=1}^{\ell} \boldsymbol{\varepsilon}_i^{\text{cr}}, \tag{23.8}$$

where the $i = 1, \dots \ell$ hidden variables $\boldsymbol{\varepsilon}_i^{\text{cr}}$ characterize viscoelastic processes with corresponding relaxation times $\tau_i \in [0, +\infty[, i = 1, \dots \ell$. The way all these internal variables evolve is motivated by the generalized Kelvin-Voigt rheological model sketched in Fig. 23.1. In this case, the complementary evolution equations that govern the creep strain components are given by, see for example [3],

$$\dot{\boldsymbol{\varepsilon}}_i^{\text{cr}} + \frac{1+\omega_i}{\tau_i}, \mathbb{N} : \boldsymbol{\varepsilon}_i^{\text{cr}} + \frac{\omega_i}{\tau_i}, \sum_{j=1, j\neq i}^{\ell} \mathbb{N} : \boldsymbol{\varepsilon}_j^{\text{cr}}, =, \frac{\omega_i}{\tau_i}, \mathbb{N} : \left(\boldsymbol{\varepsilon} - \boldsymbol{\varepsilon}_{\text{th}} - \boldsymbol{\varepsilon}_{\text{hyd}} - \boldsymbol{\varepsilon}_\chi\right), \qquad i = 1, \dots \ell \tag{23.9}$$

where the dimensionless factors ω_i, $i = 1, \dots \ell$ are material parameters. Here $\mathbb{N}$ is the fourth-order tensor which depends solely on the Poisson's ratio ν. In Voigt

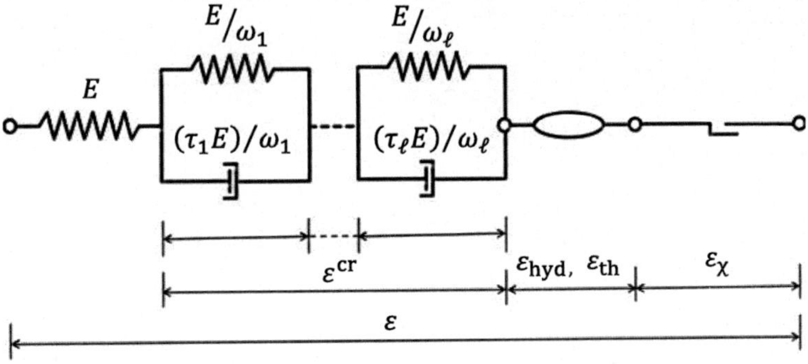

Fig. 23.1 One-dimensional viscoelastic rheological model

engineering notation, it is given by,

$$\tilde{\mathbb{N}} = \frac{1}{(1+\nu)(1-2\nu)} \begin{bmatrix} 1-\nu & \nu & \nu & & & \\ \nu & 1-\nu & \nu & & & \\ \nu & \nu & 1-\nu & & & \\ & & & 1-2\nu & & \\ & & & & 1-2\nu & \\ & & & & & 1-2\nu \end{bmatrix}. \tag{23.10}$$

23.1.1.3 Effect of Variable Thermal Conditions on the Chemical Expansion

It has been shown experimentally that the ambient temperature has an influence on the kinetics of expansion. Among other choices, we can consider that the two characteristic times in Eq. (23.4) be thermo-activated with the forms

$$\tau_l = \overline{\tau}_l \, e^{\left[\frac{U_l}{R}\left(\frac{1}{T} - \frac{1}{\overline{T}}\right)\right]}, \qquad \tau_c = \overline{\tau}_c \, e^{\left[\frac{U_c}{R}\left(\frac{1}{T} - \frac{1}{\overline{T}}\right)\right]}, \tag{23.11}$$

as suggested for example in Larive's work [1]. Here $\overline{\tau}_l$ and $\overline{\tau}_c$ are reference characteristic times for a reference temperature $\overline{T}$, and U_l and U_c are activation energies.

23.1.1.4 Effect of Variable Hydric Conditions on the Chemical Expansion

Experimental evidences have also shown that the expansion stops below a certain threshold of humidity. Larive's relation must then be adapted to take into account this strong dependency. In terms of saturation, denoting the threshold below which the expansion reaction stops by $\overline{S}_r$, we can formally write, see [4]:

$$\begin{cases} \dot{\varepsilon}_\chi = 0, & \text{if} S_r \leq \overline{S}_r, \\ \dot{\varepsilon}_\chi > 0, & \text{if} S_r > \overline{S}_r, \\ \text{and } \varepsilon_\chi \text{ is given by Larive's law}, & \text{if } S_r = 1. \end{cases} \tag{23.12}$$

Now for $\overline{S}_r < S_r < 1$, we need to establish a *continuous* link. We introduce for this an *effective* time that we denote by $\tilde{t}$ and such that $\dot{\tilde{t}} \equiv \dot{\tilde{t}}(S_r, t, \ldots) \in [0, 1]$. A possible choice would be:

$$\dot{\tilde{t}} = \left(\frac{\langle S_r - \overline{S}_r \rangle_+}{1 - \overline{S}_r}\right)^{\overline{m}} \quad \Longrightarrow \quad \tilde{t} = \int_0^t \left(\frac{\langle S_r - \overline{S}_r \rangle_+}{1 - \overline{S}_r}\right)^{\overline{m}} \mathrm{d}t, \tag{23.13}$$

where the Macauley bracket $\langle . \rangle_+$ denotes the positive part function, and the exponent parameter most probably depends on the saturation, i.e. $\overline{m} \equiv \overline{m}(S_r)$. Hence, by replacing the real time t by the effective time $\widetilde{t}$ into Larive's expression (23.4), we obtain a free expansion law that covers the above requirements for variable humidity.

23.1.1.5 Evolution Equation for Chemical Expansion

For the well posedness of the coupled boundary-value problem at hand, the internal variables must be set in the form of evolution equations. So far, this is the case for the treatment of basic creep as given above by Eq. (23.9) and, even not developed here, also for plasticity. Now it remains to give the evolution equation for the chemical expansion that matches Larive's law. From the thermodynamic developments, for instance see [1, 2], we deduce the following form:

$$\tau_c\, \varepsilon^0_\infty \left(1 + e^{\frac{\tau_l}{\tau_c}}\right) \dot{\varepsilon}_\chi + \left(e^{\frac{\tau_l}{\tau_c}}\right) \varepsilon^2_\chi + \varepsilon^0_\infty \left(1 - e^{\frac{\tau_l}{\tau_c}}\right) \varepsilon_\chi = \left(\varepsilon^0_\infty\right)^2, \quad (23.14)$$

which exactly gives Larive's relation (23.4) for the case of fully saturated conditions at constant temperature. However, the form (23.14) is now well suited for variable thermal and hydric conditions: for the former by making the characteristic times τ_l and τ_c temterature-dependent as given above in (23.11), while for the latter by replacing the real time derivative by the derivative with respect to the effective time $\widetilde{t}$ as,

$$\dot{\varepsilon}_\chi \equiv \frac{\partial \varepsilon_\chi}{\partial \widetilde{t}}. \quad (23.15)$$

23.1.2 Outlines of the F.E. Approximation

In the finite element context, the above developments are classical. However, care must be taken in evaluating the chemical expansion. Indeed, this latter is now computed with the effective time $\widetilde{t}$ that must in turn be stored locally at the integration points level since each material point has its own hydric history. Consequently, the effective time is treated as an internal field variable ; $\widetilde{t} \equiv \widetilde{t}(\boldsymbol{x}, t)$.

23.2 P1: Constitutive Model

23.2.1 *Constitutive Model*

For sound concrete, the constitutive model model we use is a classical elastoplastic model where the plastic flow is driven by a Drucker-like associated rule. Mechanical damage has not been introduced so far. Hence, short-terms softening responses of concrete are not taken into account in the following tests.

23.3 P2: Drying and Shrinkage for Non-reactive Concrete

23.3.1 *Constitutive Model Calibration*

Figure 23.2a gives the mass variations for the three cases of: (a) immersed in water, (b) sealed under aluminium, and (c) exposed to 30% HR. Notice that from our modeling point of view, immersion in water is considered here as equivalent to a 100% HR. Furthermore, for the modeling of the sealed sample, exchange boundary conditions of the Fourier-type have been used.

The corresponding longitudinal strains are plotted in Fig. 23.2b.

23.3.2 *Prediction*

For the given humidity variation, the predicted longitudinal strain evolution is plot on Fig. 23.3a. For the sake of comparison, this latter is superposed with those of

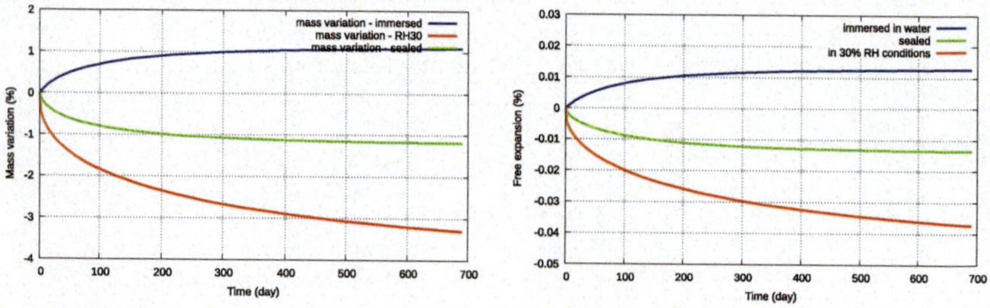

(a) Mass variations for non-reactive concrete: immersed (in blue), sealed (in green), and at 30 % RH (in red)

(b) Longitudinal strains for non-reactive concrete: immersed (in blue), sealed (in green), and at 30 % RH (in red)

Fig. 23.2 Constitutive model calibration

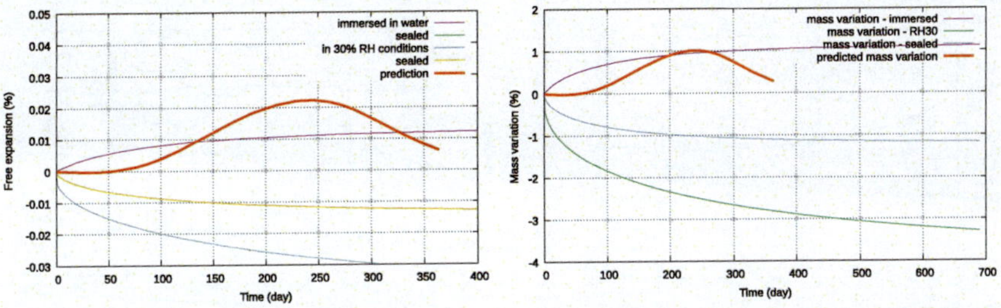

(a) Predicted longitudinal strains for non-reactive concrete (red curve). Superposition with the computed curves of Figure 23.2(b)

(b) Predicted mass variation for non-reactive concrete (red curve). Superposition with the computed curves of Figure 23.2(a)

Fig. 23.3 Constitutive model prediction

Fig. 23.2b. Figure 23.3b shows the predicted mass variation that is superposed with those of Fig. 23.2a.

23.4 P3: Basic Creep for Non-reactive Concrete

23.4.1 Constitutive Model Calibration

Figure 23.4a shows the longitudinal and radial displacements under 10 and 20 MPa compression tests.

23.4.2 Prediction

For the given axial stress variation, Fig. 23.4b shows the computed prediction for longitudinal strain.

23.5 P4: AAR Expansion; Temperature Effect

23.5.1 Calibration and Prediction

Figure 23.5 shows the calibration of the free expansion for temperatures $T = 23\,°C$ (blue curve), $T = 38\,°C$ (red curve), and the prediction result for varying sinusoidal temperature (green curve).

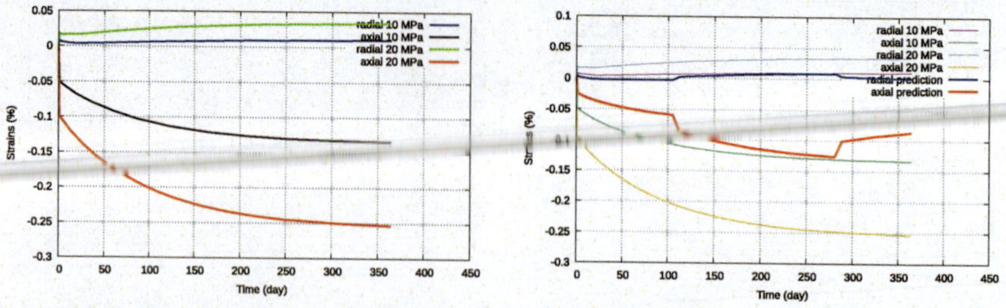

(a) Creep in non-reactive concrete under sealed condition for different axial stress

(b) Predicted creep under the given axial loading. Superposition with the computed curves of Figure 23.4(a)

Fig. 23.4 P3: Basic creep for non-reactive concrete

23.6 P5: Free AAR Expansion; Effect of RH

23.6.1 Constitutive Model Calibration

Assuming a temperature of 38 °C, Fig. 23.6a shows the longitudinal expansions for the three cases: (a) immersed in water, (b) sealed under aluminium, and (c) exposed to 30% HR.

23.6.2 Prediction

For the given humidity variation, the predicted longitudinal strain evolution is plot on Fig. 23.6b. For the sake of comparison, this latter is superposed with those of Fig. 23.6a.

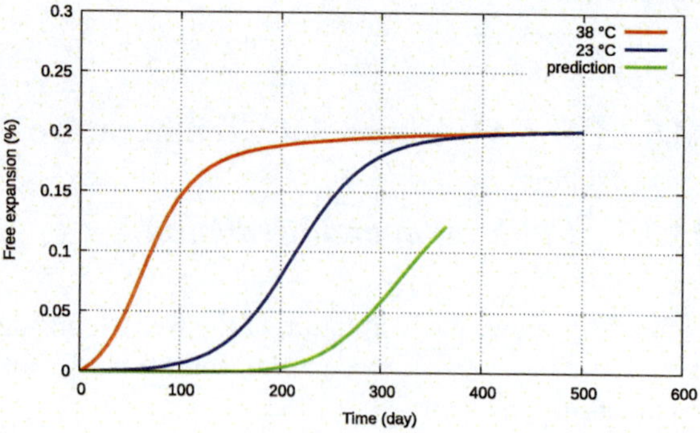

Fig. 23.5 Free expansion under different temperatures

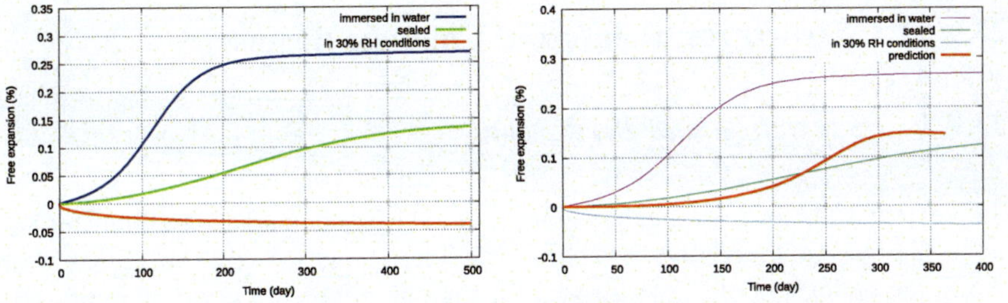

(a) Longitudinal strains for reactive concrete: immersed (in blue), sealed (in green), and at 30 % RH (in red)

(b) Predicted longitudinal strains for reactive concrete (red curve). Superposition with the computed curves of Figure 23.6(a)

Fig. 23.6 P5: free AAR expansion; effect of RH

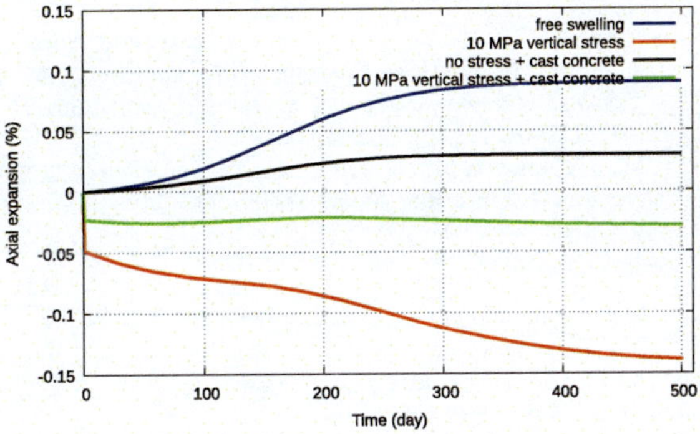

Fig. 23.7 Longitudinal strains for reactive concrete under different confinements

23.7 P6: AAR Expansion; Effect of Confinement

23.7.1 Calibration and Prediction

Figure 23.7 shows the calibration at temperatures $T = 38\,°C$ (blue curve) for different load levels and confinements.

23.8 P7: Structures; Effect of Internal Reinforcement

23.8.1 Description and Prediction

The problem has been discretized by using an axisymmetric analysis. Figure 23.8 shows the evolutions of the longitudinal and radial strain on the midsurface of the concrete, together with the longitudinal strain of the rebar.

23.9 P8: Structures; Reinforced Concrete Beams

23.9.1 *Description and Prediction*

For obvious symmetry reasons, one fourth of the beam is modeled in three dimensions. The result of the relative fluid mass variation is shown in Fig. 23.9a. Figure 23.9b shows the deflexions of the mid-spans for the unreinforced and rebars reinforced beams.

References

1. Larive, C.: Apports Combinés del'Experimentation et de la Modélisation à la Comprehensiondel'Alcali-Réaction et de ses Effets Mécaniques. Ph.D. thesis. Paris: Laboratoire Central des Ponts et Chaussées, (1998). https://hal.inria.fr/docs/00/52/06/76/PDF/1997TH_LARIVE_C_NS20683.pdf
2. Ulm, F., Coussy, O., Li, K., Larive, C.: Thermo-chemo-mechanics of ASR expansion in concrete structures. ASCE J. Eng. Mech. **126**(3), 233–242 (2000)

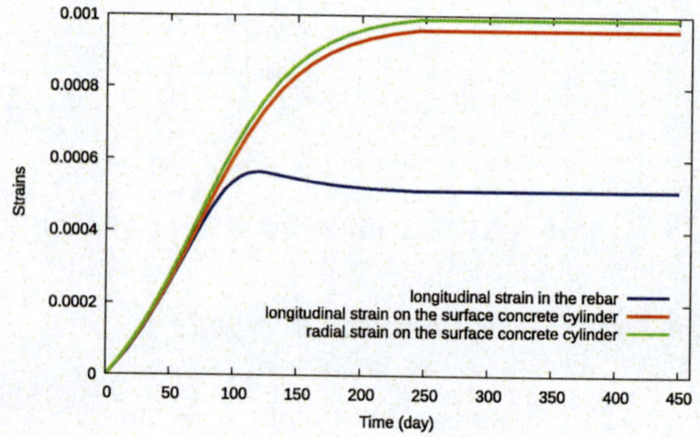

Fig. 23.8 Longitudinal strains on the rebar (blue) and concrete (red), and radial strain of the concrete (green)

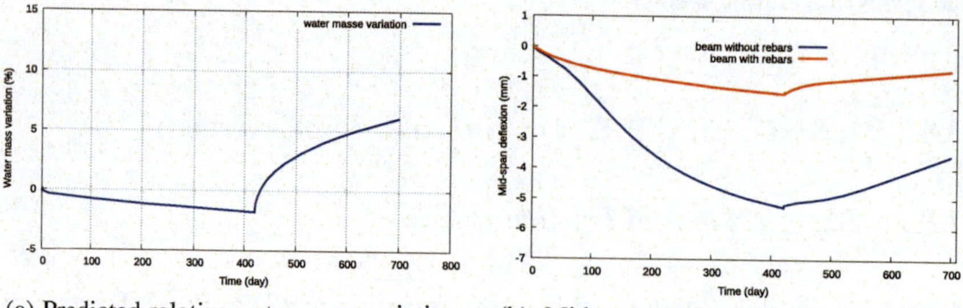

(a) Predicted relative water mass variation

(b) Mid-span deflections of the unreinforced (blue) and reinforced (red) beams

Fig. 23.9 P8: Structures; Reinforced concrete beams

3. Nedjar, B., Le Roy, R.: An approach to the modeling of viscoelastic damage. Application to the long-term creep of gypsum rock material. Int. J. Numer. Anal. Methods Geomech. **37**, 1066-1078 (2013)
4. Malbois, M., Nedjar, B., Lavaud, S., Rospars, C., Divet, L., Torrenti, J.-M.: On DEF expansion modelling in concrete structures under variable hydric conditions. Construct. Build. Mater. **207**, 396–402 (2019)

Chapter 24
Benchmark Study Results: Hydro-Québec

Simon-Nicolas Roth

24.1 P0: Finite Element Model Description

The model used by Hydro-Québec Production was developed targeting large hydraulic structures such as dams. It is used by engineers to determine if the hydraulic structures are safe despite the presence of the alkali-aggregate reaction and to predict the long-term behavior as well as its performance for different loading scenarios including seismic loads. The model may use simplifying assumptions to reduce the number of parameters while ensuring that these assumptions are on the conservative side. The code was developed to be embedded inside the finite element software ANSYS using *User Programmable Features* (UPF). The approach used to implement all the physics required to model AAR in hydraulic structures and to ensure the greatest flexibility while remaining within the framework provided by the commercial software is to program a new element type (commonly named *UserElement*).

To promote the convergence of the solution that contains several non-linearities and to reduce the computation time, several physical phenomena have been decoupled. The swelling caused by AAR and the creep effects are therefore considered as deformations similar to initial strains.

The thermal effects are considered as independent of the problem non-linearities. Therefore, the thermal problem can be computed independently from the structural problem solution. The saved results that contain the yearly temperature distribution can then be applied as a function of the time step used during solution of the structural problem. Generally four to six time steps per year are used for dams located in nordic climates, where the temperature vary from −30 °C in the winter to 30 °C in the summer.

S.-N. Roth (✉)
Hydro-Québec, Direction Expertise Barrages et Infrastructures, Vice-présidence Planification, Stratégies et Expertises, Montréal (Québec), Canada
e-mail: Roth.Simon-Nicolas@hydro.qc.ca

© RILEM 2021

V. E. Saouma (ed.), *Diagnosis & Prognosis of AAR Affected Structures*, RILEM State-of-the-Art Reports 31,
https://doi.org/10.1007/978-3-030-44014-5_24

Finally, the hygral effects including poroelastic effects are decoupled. The coupling is weak, hence, the use of segregated solvers reduce the complexity of the equation system. For each load case, the water saturation distribution in the concrete is updated, considering that a damaged material has a permeability different from a sound material. This also makes possible to update the uplift pressures in the structure, these often being responsible for instabilities in hydraulic structures.

The next sections will give information about the model, including AAR constitutive model, damage-plasticity model, creep model, reinforcement consideration and hygral diffusion.

24.1.1 Constitutive Model for AAR

The constitutive model for AAR is largely inspired by [1]. Stemming from the different weights assigned to each of the three principal stress directions, the AAR model is anisotropic. The initial strain tensor is given by expression:

$$\left[\dot{\boldsymbol{\varepsilon}}^{AAR}\right] = \mathbf{R} \begin{bmatrix} W_1 & 0 & 0 \\ 0 & W_2 & 0 \\ 0 & 0 & W_3 \end{bmatrix} \dot{\varepsilon}^{AAR} \mathbf{R}^T \tag{24.1}$$

where $\mathbf{R}$ is a tensor consisting of the eigenvectors associated with the stress tensor eigenvalues, $W_{1\ldots3}$ are the weights to consider AAR expansion constrained by compression and redirected in other less constrained principal directions and $\dot{\varepsilon}^{AAR}$ is the incremental free volumetric AAR strain given by:

$$\dot{\varepsilon}^{AAR} = \Gamma_t(w_{cr})\Gamma_c(I_\sigma) f_{S_w}(S_w)\dot{\xi}(\xi, T)\varepsilon_{max}^{AAR} \tag{24.2}$$

where $\Gamma_t(w_{cr})$ is the attenuation due to micro-cracks in tension, $\Gamma_c(I_\sigma)$ is the attenuation due to micro-cracks in compression, $f_{S_w}(S_w)$ is the attenuation due to concrete saturation, $\dot{\xi}(\xi, T)$ is the progress rate of the reaction and ε_{max}^{AAR} is the maximum free swelling. The kinetic of the reaction is taken from the work of Larive [2] and expressed in rate form as:

$$\dot{\xi}(\xi, T) = \frac{e^{\frac{t}{\tau_c(T)}}\left(e^{\frac{\tau_l(T)}{\tau_c(T)}} + 1\right)}{\tau_c\left(e^{\frac{t}{\tau_c(T)}} + e^{\frac{\tau_l(T)}{\tau_c(T)}}\right)^2} \tag{24.3}$$

where $\tau_l(T)$ and $\tau_c(T)$ are respectively the latency and characteristic times that are dependent on the temperature T (defined in Kelvin). These were defined in [3]. The retardation effect of the hydrostatic compressive stress modifies the latency time $\tau_l(T)$ such that:

$$\tau_l(T) = f(I_\sigma, f'_c)\tau_l(T_0)\exp\left[U_l\left(\frac{1}{T} - \frac{1}{T_0}\right)\right] \tag{24.4}$$

where T_0 is the reference temperature, U_l is the activation energy required to trigger the reaction for latency time, f'_c is the material compressive strength and $f(I_\sigma, f'_c)$ is defined as:

$$f(I_\sigma, f'_c) = \begin{cases} 1, & \text{if } I_\sigma \geq 0 \\ 1 + \dfrac{4I_\sigma}{9f'_c}, & \text{if } I_\sigma < 0 \end{cases} \tag{24.5}$$

where I_σ is the first invariant of the stress tensor ($I_\sigma = \sigma_1 + \sigma_2 + \sigma_3$)

Since most dams have a linear swelling rate and the calibration of Larive's kinetics is a complex task, linear kinetics is often used. The maximum free swelling parameter, which is difficult to characterize, can be replaced by a fixed maximum swelling rate ($\dot{\varepsilon}^{AAR}_{max}$) calibrated on the instrumentation data of the structure. In this case, $\dot{\xi}(\xi, T)\varepsilon^{AAR}_{max}$ in Eq. 24.2 is replaced by $\theta(T)\dot{\varepsilon}^{AAR}_{max}$, defined as:

$$\theta(T) = e^{\dfrac{Ea}{R}\left(\dfrac{1}{T_{\text{ref}} + 273} - \dfrac{1}{T + 273}\right)} \tag{24.6}$$

where Ea is the activation energy (approximately 50000 J/(mol K)) and R is the ideal gaz constant (8.314 J/(mol K)).

To account for gel absorption in macrocracks in tension, $\Gamma_t(w_{cr})$ is defined as:

$$\Gamma_t(w_{cr}) = \begin{cases} 1, & \text{if } w_{cr} \leq \gamma_t w_c \\ \Gamma_r + (1 - \Gamma_r)\gamma_t \dfrac{w_c}{w_{cr}}, & \text{if } \gamma_t w_c < w_{cr} \end{cases} \tag{24.7}$$

where γ_t is the fraction of the crack opening displacement beyond which gel is absorbed by the crack, Γ_r is a residual AAR retention factor, w_{cr} is the crack opening computed (see Sect. 24.1.6) and w_c is the maximum crack opening displacement on the tensile softening curve given by $w_c = 4.444 G_F / f_t$, with G_F the fracture energy and f_t the tensile strength of concrete.

To predict the observed ASR volumetric expansion rate under compressive stresses, $\Gamma_c(I_\sigma)$ is defined [4] as:

$$\Gamma_c(I_\sigma) = \begin{cases} 1, & \text{if } I_\sigma \geq 0 \\ 1 - (I_\sigma / 3\bar{\sigma}_v)^2, & \text{if } 0 > I_\sigma \geq 3\bar{\sigma}_v \\ 0, & \text{if } I_\sigma < 3\bar{\sigma}_v \end{cases} \tag{24.8}$$

where $\bar{\sigma}_v$ is a parameter that stands for the volumetric stress under which ASR expansion would be totally suppressed.

Gravity dams usually have only a small layer of exposed concrete that is unsaturated. This thickness is often considered negligible and fully saturated conditions are usually assumed. However, the cracking pattern computed on this layer may be different from that computed imposing unsaturated conditions on the exposed faces. Since model calibration is often performed on the basis of the measured displacements and the observed crack pattern located primarily on the exposed concrete, it is important to apply the correct boundary conditions for calibration. This remark is even more valid for thin arch dams where the unsaturated layer may be large compared to the thickness of the dam. For AAR cases in power plants, the interior of the building is often dry and hot because of the generators. The dry surface is similar to a cap that retains the AAR reaction, but its surface often contains many cracks caused by the swelling of the most hydrated body below the surface. Therefore, transient saturation conditions should be applied on the exposed surfaces and a constitutive model for diffusion of concrete saturation should be considered in the analyses. This constitutive model will be presented in Sect. 24.1.5. The effect of concrete saturation on AAR is based on the reference [5] and is defined by the relation:

$$f_{S_w}(S_w) = \frac{1 + b_1 e^{-b_2}}{1 + b_1 e^{-b_2 Sw}} \tag{24.9}$$

where b_1 and b_2 define the maximum free axial expansion achievable in different moisture conditions. Due to the presence of $S_w \leq 1$ in the evolution law, the limit value $\xi = 1$ corresponding to the complete development of the reaction, can only be obtained for fully saturated concrete. It should be noted that the latency and the characteristic times should also be affected by the concrete saturation, however to reduce the number of parameters, these were supposed to be independent of the concrete saturation.

The last element required to define the AAR volumetric expansion distribution are the three weights components (W_1, W_2, W_3) from Eq. 24.1. Defining $W_{n\,[3\times16]}$ as a table storing the three columns and the sixteen lines defined on Fig. 24.1b. Given the principal stress vector defined by σ_1, σ_2 and σ_3, the weights are assigned in function of the three principal stresses.

Starting with $S_k = \sigma_1$, $S_l = \sigma_2$ and $S_m = \sigma_3$, the four quadrant nodes must be defined and stored in a vector $\text{Node}[i]_{i=1...4}$. The interpolation functions are then defined as:

$$\begin{aligned} N_1 &= (S_l - a_e)(S_m - b_e)/(a \cdot b) \\ N_2 &= (a_b - S_l)(S_m - b_e)/(a \cdot b) \\ N_3 &= (a_b - S_l)(b_b - S_m)/(a \cdot b) \\ N_4 &= (S_l - a_e)(b_b - S_m)/(a \cdot b) \end{aligned} \tag{24.10}$$

with $\{a_b, a_e\}$ respectively the begining and end limits of the abscissa axis for the quadrant (see Fig. 24.1a) and analogously $\{b_b, b_e\}$ for the ordinate axis. The variable $a = a_e - a_b$ corresponds to the dimension of the quadrant in the abscissa direction and $b = b_e - b_b$ is the dimension of the quadrant in the ordinate direction. To define the column in table W_n in which the weight is taken and to interpolate its value, the vector W_k is defined such that:

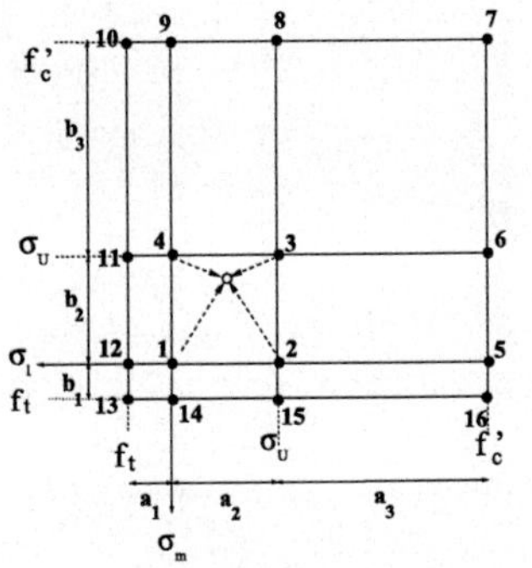

(a) Interpolation domain

Node			Weights		
No.	σ_l	σ_m	$\sigma_k \geqslant 0$	$\sigma_k = \sigma_u$	$\sigma_k = f'_c$
1	0.	0.	1/3	0.	0.
2	σ_u	0.	1/2	0.	0.
3	σ_u	σ_u	1.	1/3	0.
4	0.	σ_u	1/2	0.	0.
5	f'_c	0.	1/2	0.	0.
6	f'_c	σ_u	1.	1/2	0.
7	f'_c	f'_c	1.	1.	1/3
8	σ_u	f'_c	1.	1/2	0.
9	0.	f'_c	1/2	0.	0.
10	f'_t	f'_c	1/2	0.	0.
11	f'_t	σ_u	1/2	0.	0.
12	f'_t	0.	1/3	0.	0.
13	f'_t	f'_t	1/3	0.	0.
14	0.	f'_t	1/3	0.	0.
15	σ_u	f'_t	1/2	0.	0.
16	f'_c	f'_t	1/2	0.	0.

(b) Weigths

Fig. 24.1 Interpolation domain and weights

$$W_k = \begin{cases} [1, 0, 0], & \text{if } S_k \geq 0 \\ [(S_k - \bar{\sigma}_v)/(0 - \bar{\sigma}_v), 1 - W_1, 0], & \text{if } S_k \geq \bar{\sigma}_v \\ \left[0, (S_k - f'_c)/(\bar{\sigma}_v - f'_c), 1 - W_2\right], & \text{else} \end{cases} \tag{24.11}$$

Then for each quadrant node i, the weight is defined as:

$$\tilde{W}_i = \sum_{j=1}^{3} W_k[j] \cdot W_n[\text{Node}[i], j] \tag{24.12}$$

Finally, the weight is interpolated using the shape functions N defined earlier and the weights $\tilde{W}_i$ such that:

$$W_1 = \sum_{i=1}^{4} N_i \tilde{W}_i \tag{24.13}$$

This procedure is repeated with $S_k = \sigma_2$, $S_l = \sigma_3$ and $S_m = \sigma_1$ and $S_k = \sigma_3$, $S_l = \sigma_1$ and $S_m = \sigma_2$ to obtain respectively W_2 and W_3.

24.1.2 *Constitutive Model for Damage-Plasticity*

The modeling of damage caused by AAR is achieved with a damage-plasticity model [6]. Other models were tested including anisotropic damage models, however stability of the solution for complex problems was found to be an important issue. The concrete damage-plasticity model (CDPM2) was found to be stable and very few Newton-Raphson iterations are required to achieve sufficient convergence on large scale problems.

The model is introduced briefly in this section. The stress for the anisotropic damage plasticity model is defined as

$$\boldsymbol{\sigma} = (1 - \omega_{\mathrm{t}})\,\bar{\boldsymbol{\sigma}}_{\mathrm{t}} + (1 - \omega_{\mathrm{c}})\,\bar{\boldsymbol{\sigma}}_{\mathrm{c}} \tag{24.14}$$

where $\bar{\boldsymbol{\sigma}}_{\mathrm{t}}$ and $\bar{\boldsymbol{\sigma}}_{\mathrm{c}}$ are the positive and negative parts of the effective stress tensor $\bar{\boldsymbol{\sigma}}$, respectively, and ω_{t} and ω_{c} are two scalar damage variables, ranging from 0 (undamaged) to 1 (fully damaged).

The effective stress $\bar{\boldsymbol{\sigma}}$ is defined according to the damage mechanics convention as:

$$\bar{\boldsymbol{\sigma}} = \mathbf{D}_{\mathrm{e}} : \left(\boldsymbol{\varepsilon} - \boldsymbol{\varepsilon}_{\mathrm{p}}\right) \tag{24.15}$$

24.1.2.1 Plasticity

The yield surface is described by the Haigh-Westergaard coordinates, the volumetric effective stress $\bar{\sigma}_{\mathrm{V}}$, the norm of the deviatoric effective stress $\bar{\rho}$, the Lode angle $\bar{\theta}$ and the hardening variable κ_{p}. The yield surface is given by:

$$\begin{aligned} f_{\mathrm{p}}(\bar{\sigma}_{\mathrm{V}}, \bar{\rho}, \bar{\theta}, \kappa_{\mathrm{p}}) = {} & \left\{ [1 - q_{\mathrm{h1}}(\kappa_{\mathrm{p}})] \left(\frac{\bar{\rho}}{\sqrt{6} f_c'} + \frac{\bar{\sigma}_{\mathrm{V}}}{f_c'} \right)^2 + \sqrt{\frac{3}{2}} \frac{\bar{\rho}}{f_c'} \right\}^2 \\ & + m_0 q_{\mathrm{h1}}^2(\kappa_{\mathrm{p}}) q_{\mathrm{h2}}(\kappa_{\mathrm{p}}) \left[\frac{\bar{\rho}}{\sqrt{6} f_c'} r(\cos\bar{\theta}) + \frac{\bar{\sigma}_{\mathrm{V}}}{f_c'} \right] - q_{\mathrm{h1}}^2(\kappa_{\mathrm{p}}) q_{\mathrm{h2}}^2(\kappa_{\mathrm{p}}) \end{aligned} \tag{24.16}$$

where m_0 is the friction parameter, q_{h1} and q_{h2} are hardening laws and κ_{p} is a hardening variable. The shape of the deviatoric section is controlled by the Willam-Warnke function:

$$r(\cos\bar{\theta}) = \frac{4(1 - e^2)\cos^2\bar{\theta} + (2e - 1)^2}{2(1 - e^2)\cos\bar{\theta} + (2e - 1)\sqrt{4(1 - e^2)\cos^2\bar{\theta} + 5e^2 - 4e}} \tag{24.17}$$

where, e is the eccentricity parameter. The friction parameter m_0 is given by:

$$m_0 = \frac{3\left(f_c'^2 - f_t^2\right)}{f_c' f_t} \frac{e}{e + 1} \tag{24.18}$$

The flow rule is non-associative and is defined as:

$$\boldsymbol{m} = \frac{\partial m_{\mathrm{g}}}{\partial \bar{\sigma}_{\mathrm{V}}} \frac{\boldsymbol{\delta}}{3 f_c'} + \left(\frac{3}{f_c'} + \frac{m_0}{\sqrt{6}\bar{\rho}} \right) \frac{\bar{\boldsymbol{s}}}{f_c'} \tag{24.19}$$

The variables q_{h1} and q_{h2} are functions of the hardening variable κ_{p}. The first hardening law q_{h1} is defined as:

$$q_{\mathrm{h1}}(\kappa_{\mathrm{p}}) = \begin{cases} q_{\mathrm{h0}} + (1 - q_{\mathrm{h0}})\left(\kappa_{\mathrm{p}}^3 - 3\kappa_{\mathrm{p}}^2 + 3\kappa_{\mathrm{p}}\right) - H_{\mathrm{p}}\left(\kappa_{\mathrm{p}}^3 - 3\kappa_{\mathrm{p}}^2 + 2\kappa_{\mathrm{p}}\right) & \text{if } \kappa_{\mathrm{p}} < 1 \\ 1 & \text{if } \kappa_{\mathrm{p}} \geq 1 \end{cases} \tag{24.20}$$

The second hardening law q_{h2} is given by:

$$q_{\mathrm{h2}}(\kappa_{\mathrm{p}}) = \begin{cases} 1 & \text{if } \kappa_{\mathrm{p}} < 1 \\ 1 + H_{\mathrm{p}}(\kappa_{\mathrm{p}} - 1) & \text{if } \kappa_{\mathrm{p}} \geq 1 \end{cases} \tag{24.21}$$

The evolution rate for the hardening variable is given by:

$$\dot{\kappa}_{\mathrm{p}} = \frac{\|\dot{\boldsymbol{\varepsilon}}_{\mathrm{p}}\|}{x_{\mathrm{h}}(\bar{\sigma}_{\mathrm{V}})}\left(2\cos\bar{\theta}\right)^2 = \frac{\dot{\lambda}\|\boldsymbol{m}\|}{x_{\mathrm{h}}(\bar{\sigma}_{\mathrm{V}})}\left(2\cos\bar{\theta}\right)^2 \tag{24.22}$$

where $x_{\mathrm{h}}(\bar{\sigma}_{\mathrm{V}})$ is a hardening ductility measure.

24.1.2.2 Damage

Damage is initiated when the maximum equivalent strain in the history of the material reaches the threshold $\varepsilon_0 = f_{\mathrm{t}}/E$. The equivalent strain expression for general triaxial stress states is defined as:

$$\tilde{\varepsilon} = \frac{\varepsilon_0 m_0}{2}\left(\frac{\bar{\rho}}{\sqrt{6}f_c'} r(\cos\theta) + \frac{\bar{\sigma}_{\mathrm{V}}}{f_c}\right) + \sqrt{\frac{\varepsilon_0^2 m_0^2}{4}\left(\frac{\bar{\rho}}{\sqrt{6}f_c'} r(\cos\theta) + \frac{\bar{\sigma}_{\mathrm{V}}}{f_c'}\right)^2 + \frac{3\varepsilon_0^2\bar{\rho}^2}{2\,f_c'^2}} \tag{24.23}$$

Tensile damage is described by a stress-inelastic displacement exponential softening law. For the compressive damage variable, an evolution based on an exponential stress-inelastic strain law is used and is given by:

$$\sigma = f_{\mathrm{t}}\exp\left(-\frac{\varepsilon_{\mathrm{i}}}{\varepsilon_{\mathrm{fc}}}\right) \text{ if } 0 < \varepsilon_{\mathrm{i}} \tag{24.24}$$

where $\varepsilon_{\mathrm{fc}}$ is an inelastic strain threshold which controls the initial inclination of the softening curve.

The history variables κ_{dt1}, κ_{dt2}, κ_{dc1} and κ_{dc2} depend on a ductility measure x_{s}, which takes into account the influence of multiaxial stress states on the damage evolution. This ductility measure is given by

$$x_{\mathrm{s}} = 1 + (A_{\mathrm{s}} - 1)\,R_{\mathrm{s}} \tag{24.25}$$

where R_s is

$$R_s = \begin{cases} -\frac{\sqrt{6}\bar{\sigma}_V}{\bar{\rho}} & \text{if } \bar{\sigma}_V \leq 0 \\ 0 & \text{if } \bar{\sigma}_V > 0 \end{cases} \tag{24.26}$$

and A_s is a model parameter.

24.1.3 Constitutive Model for Reinforcement

The addition of rebars as discrete bars coupled to the solid finite element mesh is an option often used for the numerical analysis of reinforced concrete. However, this adds many degrees of freedom to the problem and has the effect of increasing the computation time. Therefore, a good alternative is to model the reinforcements using an *embedded rebar* approach. In this approach, the stiffness matrix of the bars is superimposed on that of the element:

$$K^e = \int_{\Omega_p} \mathbf{B}_p^{e\ T} \mathbf{C}_p \mathbf{B}_p^e \, d\Omega_p + \sum_{i=1}^{n_{\text{rebar}}} \int_{\Omega_{r_i}} \mathbf{B}_p^{e\ T} \mathbf{T}_{r_i}^T \mathbf{C}_{r_i} \mathbf{T}_{r_i} \mathbf{B}_p^e \, d\Omega_{r_i} \tag{24.27}$$

with $\mathbf{B}_p^e$ the displacement-strain matrix for the parent element, $\mathbf{C}_p$ the stress-stain constitutive relation for the parent element, Ω_p the volume of the parent element, $\mathbf{T}_{r_i}$ the transformation matrix of the local-global configuration elasticity matrix for the reinforcement and Ω_{r_i} the volume of the reinforcement inside the parent element. To simplify the constitutive law of the reinforcements, bi-linear steel model with rupture at maximum strain is considered. It is known that tension-stiffening can have a significant effect on the concrete near the rebar. However, given the deterioration of concrete caused by AAR, this effect is neglected and further research is needed to define the effect of rebar in AAR-affected concrete, particularly in relation to the tension stiffening effect.

24.1.4 Constitutive Model for Creep

The Kelvin–Voigt model is used to consider the effects of long-term reversible creep. The model consists of a Newtonian damper and an elastic spring connected in parallel and its constitutive relation is expressed as a linear first-order differential equation [7]:

$$\epsilon^{\text{ve}}(t + \Delta t) = e^{-\frac{E_{\text{ve}}}{\eta_{\text{ve}}}\Delta t} \epsilon^{\text{ve}}(t) + \frac{\mathbf{A}\boldsymbol{\sigma}(t)}{E_{\text{ve}}} \left(1 - e^{-\frac{E_{\text{ve}}}{\eta_{\text{ve}}}\Delta t}\right) \tag{24.28}$$

where $\boldsymbol{\sigma}$ is the effective stress, E_{ve} is the modulus of the rheological model and η_{ve} is the viscosity. The Poisson's ratio coefficient matrix is given by:

$$\mathbf{A} = \begin{bmatrix} 1 & -\mu & -\mu & 0 & 0 & 0 \\ -\mu & 1 & -\mu & 0 & 0 & 0 \\ -\mu & -\mu & 1 & 0 & 0 & 0 \\ 0 & 0 & 0 & 2(1+\mu) & 0 & 0 \\ 0 & 0 & 0 & 0 & 2(1+\mu) & 0 \\ 0 & 0 & 0 & 0 & 0 & 2(1+\mu) \end{bmatrix} \tag{24.29}$$

where μ is the Poisson's ratio.

24.1.5 Constitutive Model for Hygral Diffusion

To describe the fluid flow in an interstitial medium, two hypotheses can be considered; (i) a fully saturated medium, and (ii) an unsaturated medium. Under the hypothesis of fully saturated medium, the fluid movement is governed by the law relating the velocity vector $\mathbf{v}$ to the driving force $-\nabla p_l$. Linearly relating $\mathbf{v}$ to $-\nabla p_l$ and neglecting body loads is the simplest form that this law can take which leads to the expression of Darcy's law. For an unsaturated medium, the relation between the velocity vector and the driving force is nonlinear and depends on the saturation level. An unsaturated porous medium is composed of three constituents: a solid skeleton, a liquid-water phase (index l), and a gas phase (index g). The gas phase is composed of water vapor (index v) and dry air (index a). To define the conservation equation in an unsaturated medium, the flow velocity $\mathbf{w}$ for each constituent must be defined [8]:

$$\mathbf{w}_l = -\rho_l \frac{K}{\mu_l} k_{rl}(S_l) \nabla p_l \tag{24.30a}$$

$$\mathbf{w}_v = -\rho_v \frac{K}{\mu_g} k_{rg}(S_l) \nabla p_g - \left(\frac{M_v}{RT}\right) D_{va}(T, p_g) f(S_l, \phi) \nabla C_v \tag{24.30b}$$

$$\mathbf{w}_a = -\rho_a \frac{K}{\mu_g} k_{rg}(S_l) \nabla p_g - \left(\frac{M_a}{RT}\right) D_{va}(T, p_a) f(S_l, \phi) \nabla C_a \tag{24.30c}$$

where R is the ideal gas constant, T the absolute temperature, ρ_α the density of constituent α and p_α is the pressure of constituent α. The permeability of constituent α is reduced by a relative permeability factor $k_{r\alpha}(S_l)$, which is a function of the degree of saturation S_l. The diffusion coefficient of water vapor in wet air D_{va} (cm/s) is given by:

$$D_{va}(T, p_g) = 0.217 p_{atm} \left(\frac{T}{T_0}\right)^{1.88} \tag{24.31}$$

with T_0 the reference temperature (273 K) and $p_{atm} = 101325$ Pa. The function $f(S_l, \phi)$ represents the resistance factor to gaseous diffusion in a porous medium compared to the case of free diffusion and takes the form [9]:

$$f(S_l, \phi) = \phi^{4/3}\,(1 - S_l)^{10/3} \tag{24.32}$$

The saturation degree S_α of constituent α is defined by:

$$S_\alpha = \frac{\phi_\alpha}{\phi}; \qquad \sum_\alpha S_\alpha = 1 \tag{24.33}$$

where ϕ denotes the total porosity and ϕ_α the fraction occupied by α ={liquid, vapor, air}. The mass conservation equations of each constituent together with the state equations form the modelling basis of water transfer in unsaturated porous media. Solving this system of highly nonlinear equations makes the development of numerical models complex. Therefore, depending on the solid material involved in the water transfer, some assumptions can be made to simplify the model. Assuming the gas pressure remains constant and equal to the atmospheric pressure during the transport process greatly simplifies the equations. This assumption is based on the fact that due to the high conductivity of the vapor-air phase, any change in its pressure, disappears almost immediately as air transport occurs almost instantaneously compared to the time scale of all other phenomena [10]. Moreover, for low permeability materials, drying is ultimately achieved by transporting moisture in its liquid form. The diffusion of the vapor quickly becomes inactive for the transport of moisture [10]. Hence only the conservation equation formulated as the moisture transport in liquid form is required:

$$\frac{\partial \phi \rho_l S_l(p_l)}{\partial t} = -\nabla \cdot \mathbf{w}_l \tag{24.34}$$

This yields the final non linear diffusion equation for application on low permeability materials at constant gas pressure:

$$\phi \frac{\partial S_l(p_l)}{\partial t} = \nabla \cdot \left(k_{rl}(S_l) \frac{K}{\mu_l} \nabla p_l \right) \tag{24.35}$$

where K is the permeability and μ_l the fluid viscosity. An expression of the often-admitted relative permeability factor $k_{rl}(S_l)$ is given by the model of Mualem [11]:

$$k_{rl}(S_l) = S_l^{1/2} \left(1 - \left(1 - S_l^{1/m}\right)^m\right)^2 \tag{24.36}$$

The imposition of boundary conditions requires a relation between the relative humidity h_r and the degree of saturation achieved using Kelvin's law (24.37):

$$\frac{\rho_l RT}{M_v} \ln h_r = -p_c \qquad h_r = \frac{p_v}{p_{vs}} \tag{24.37}$$

with p_c the capillary pressure, p_v the vapor pressure and p_{vs} the saturation vapor pressure. At the macroscopic scale, the capillary pressure between two fluids is introduced as the pressure difference between the non-wetting fluid and the wetting fluid due to the phenomenon of surface tension. In our case, the wetting fluid is water and the non-wetting fluid is the gas. Therefore, the capillary pressure is defined as:

$$p_c = p_g - p_l \tag{24.38}$$

The expression linking the liquid saturation to the capillary pressure depends on the material, the saturating fluids and the temperature. The relationship is not the same when the material is adsorbing or desorbing water. This phenomenon results in a sorption curve that differs from the desorption curve. A model often used in the literature for desorption can be found by using the Mualem model [11]:

$$p_c(S_l) = a\left(S_l^{-1/m} - 1\right)^{1-m} \tag{24.39}$$

The values of the parameters are $a = 18.6$ MPa and $m = 0.44$ for a standard concrete with a compressive strength f_c'=50 MPa [12]. From this equation, for a 90% degree of saturation, the capillary pressure is 9 MPa. In comparison, the pressure of the gas phase is approximately 0.1 MPa. It is therefore possible to make the assumption that:

$$p_c \approx -p_l \tag{24.40}$$

Solving Eq. 24.35 with the relations 24.36, 24.39, 24.40, and imposing the liquid pressure at the boundaries using 24.37 and 24.40, allows to compute the water transfer in unsaturated porous media inside a structure as a function of time.

The equation system of the structural displacement and the fluid transport can be solved using either a segregated or a coupled algorithm. The coupled problem leads to the solution of an unsymmetrical and often ill-conditioned matrix system that may demand considerable computational efforts, especially when the solved system is nonlinear. Using a segregated algorithm it is possible to solve the nonlinear flow problem with less computational effort. This latter approach was preferred over the strong coupling algorithm. Hence, two equation system are solved independently. First the nonlinear diffusion problem is solved to get the pressure distribution $p = p_c = -p_l$ in the structure:

$$\int_\Omega \mathbf{H}^T \phi \frac{\partial S_l(p_l)}{\partial t} d\Omega + \int_\Omega \nabla \mathbf{H}^T \mathbf{K} \nabla p d\Omega = 0 \tag{24.41}$$

Following the solution of this equation, the structural problem is solved by taking care to add the initial strain caused by AAR and the interstitial pressure inside the structure. The negative liquid pressures caused by concrete drying (which is associated to shrinkage) are neglected in the solution. This results in the second equation system that must be solved:

$$\int_{\Omega} \mathbf{B}^{\mathrm{T}} \boldsymbol{\sigma} \mathrm{d}\Omega = \int_{\Gamma} \mathbf{N}^{\mathrm{T}} \bar{\mathbf{t}} + \int_{\Omega} \mathbf{N}^{\mathrm{T}} \mathbf{b} \mathrm{d}\Omega + \int_{\Omega} \mathbf{B}^{\mathrm{T}} \boldsymbol{\sigma}_{p} \mathrm{d}\Omega + \int_{\Omega} \mathbf{B}^{\mathrm{T}} \mathbf{D} \boldsymbol{\varepsilon}^{\mathrm{init}} \mathrm{d}\Omega \tag{24.42}$$

where $\boldsymbol{\varepsilon}^{\mathrm{init}}$ is the strain caused by AAR and creep, $\boldsymbol{\sigma}_p^T = b(p, p, p, 0, 0, 0)$ and b is the Biot coefficient. Depending on the size of the problem, the pressure is not necessarily computed every time step. Moreover subiterations between the structural and the diffusion problems are not performed during the solution of a time step, therefore the problem is coupled in a weak sense. The effect of structural damage on the diffusion is considered by modifying the permeability K in Eq. 24.35.

According to [13] changes in concrete permeability can be split in two parts. The first part is where micro cracking linked to low-level damage (0 to 15%) occurs, and the second is where macro-cracks occurs beyond the peak load. Based on experimental results, this reference frame gives a damaged-permeability relationship in the prepeak phase of concrete (valid for damage values between 0 and 0.18) as an exponential function:

$$k_D = k_0 \exp\left((\alpha \omega_{\mathrm{t}})^{\beta}\right) \tag{24.43}$$

with $\alpha = 11.3, \beta = 1.64$ and ω_{t} is the first principal tensile damage scalar. For serious damage, the permeability of a crack is given by the Hele-Shaw flow (Eq. 24.44) (also called the plane-Poiseuille flow):

$$k_{cr} = \frac{g w_{cr}^2}{12\nu} \tag{24.44}$$

where g is the acceleration due to gravity, w_{cr} is the crack opening and ν is the fluid kinematic viscosity. The crack opening is computed using the approach given in Sect. 24.1.6.

A single mathematical law, based on the law of mixtures which makes it possible to describe the evolution of permeability from the initiation of micro cracks until the opening of the macro crack [14, 15] can be given by:

$$\log(k_m) = (1 - \omega_{\mathrm{t}}) \log\left(k_D^F\right) + \omega_{\mathrm{t}} \log\left(k_{cr}\right) \tag{24.45}$$

where k_D^F is a limited Taylor expansion of the exponential relation. Assuming that the permeability change in the damaged material is isotropic, the permeability matrix from Eq. 24.41 is given by $\mathbf{K} = \delta_{ij} k_m / \mu$. Similarly, the Biot coefficient increases such that:

$$b = b_0 + (1 - b_0)\omega_{\mathrm{t}} \tag{24.46}$$

with b_0 the initial Biot coefficient.

24.1.6 Crack Opening Computation in the Continuous Damage Model

Computing crack opening in a damaged material is not trivial when continuous damage models are used. The strain must somehow be scaled by a reference length. This length can be determined analogously to the characteristic length or the localization limiter required for the so-called mesh-adjusted softening modulus technique or crack band approach. Often, this length is computed with the volume of the representative element ($l_{rve} = \sqrt[3]{V}$). However, more elaborated reference lengths can be computed taking into account the crack direction using the approaches found in [16–19].

The crack opening of the damaged element is computed using a local XFEM solution as proposed in [20]. The displacement solution of the standard FEM solution is used to constrain the displacements of the element (Fig. 24.2) while the formulation of the local stiffness matrix is computed using XFEM. Hence, the additional unknowns stemming from the discontinuous functions at the element nodes can be computed. Using these unknowns, it is possible to compute the crack opening without introducing a reference length. The system of equations to solve at the element level is given by:

$$\begin{bmatrix} \int_{\Omega} \mathbf{B}^{\mathrm{T}}\mathbf{C}\mathbf{B}\,\mathrm{d}\Omega & \int_{\Omega^{\star}} \mathcal{H}\mathbf{B}^{\mathrm{T}}\mathbf{C}\mathbf{B}\,\mathrm{d}\Omega^{\star} \\ \int_{\Omega^{\star}} \mathcal{H}\mathbf{B}^{\mathrm{T}}\mathbf{C}\mathbf{B}\,\mathrm{d}\Omega^{\star} & \int_{\Omega^{\star}} \mathcal{H}\mathbf{B}^{\mathrm{T}}\mathbf{C}\mathbf{B}\,\mathrm{d}\Omega^{\star} + \int_{\Gamma_d} \mathbf{M}^{\mathrm{T}}\mathbf{T}\mathbf{M}\,\mathrm{d}\Gamma_d \end{bmatrix} \begin{Bmatrix} \bar{u} \\ \hat{u} \end{Bmatrix} = \begin{Bmatrix} f_{\bar{u}}^{int} \\ 0 \end{Bmatrix} \tag{24.47}$$

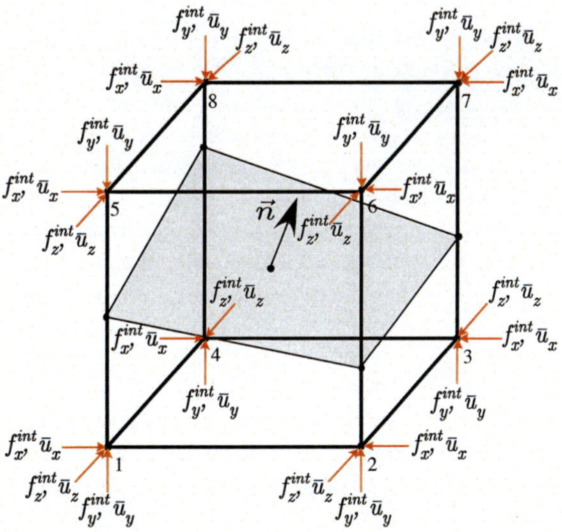

Fig. 24.2 Constrained element for crack opening computation

in which **M** comprises the shape function **N** evaluated on the discontinuity surface, **C** corresponds to the stress-strain constitutive relation computed using the undamaged material and $\mathcal{H}$ is the Heaviside function defined as:

$$\mathcal{H}(\phi) = \begin{cases} 1, \ \phi > 0 \ (\text{in } \Omega^+ \subset \Omega) \\ 0, \ \text{else} \quad (\text{in } \Omega^- \subset \Omega) \end{cases} \quad \Omega = \Omega^+ \cup \Omega^- \tag{24.48}$$

where ϕ corresponds to the level-set function defined as positive for the nodes located in the direction positive from the discontinuity plane (in the direction above the plane oriented with vector $\mathbf{n}$, corresponding to the Ω^+ portion). For the computation of these level-sets function, the discontinuity is placed at the element centroid and oriented with the eigenvector $\mathbf{n} = \left\{\varepsilon_{1\,x}\,,\,\varepsilon_{1\,y}\,,\,\varepsilon_{1\,z}\right\}^T$ of the largest principal strain $\max(\lambda(\boldsymbol{\varepsilon}))$ and $\phi = \mathbf{n} \cdot (c_n - \bar{c})/||\mathbf{n}||$, where c_n and $\bar{c}$ are respectively the coordinates of the node n and the element centroid. The matrix **T** contains the cohesive force evaluated on the discontinuity.

$$\mathbf{T} = \Big((1 - \omega_{\mathrm{t}})\, f_t \Big) (\mathbf{n} \otimes \mathbf{n}) \tag{24.49}$$

The damage parameter from the continuous damage model ω_{t} is directly used. To simplify crack opening computation, the cohesive law used only considers mode I cracks. The internal loads $f_{\tilde{u}}^{int}$ (the right hand side of 24.47) are evaluated using the relation $f_{\tilde{u}}^{int} = \mathbf{B}^{\mathrm{T}}\boldsymbol{\sigma}$. Finally, the crack opening is computed as $w_{cr} = ||\sum_{j=1}^{N_{nodes}} N_j(x)\hat{u}_j||, \ x \in \Gamma_d$

24.1.7 *Alteration of Concrete Properties with AAR Reaction*

It is known that concrete properties change with the evolution of AAR [21–24]. However, following experimental tests [25], it was noted that the modifications of mechanical properties of a concrete affected by the AAR can not be directly associated with a particular expansion level because the behavior of concrete depends on the materials involved and the components involved in the reaction. Depending on the nature of the aggregates, micro-cracking can be produced at the paste/aggregate interface or inside aggregates according to the location of gel formation. In an existing structure, the state of confinement will influence the orientation of the micro-cracking and the intensity of the degradation of mechanical properties. Therefore defining the parameters to consider the alteration of concrete properties as a function of AAR expansion in a numerical model is not a trivial task. It is generally accepted that the tensile strength and the elastic modulus reduce in a concrete affected by AAR. On the other hand, most tests on compressive strength agree that it is not negatively affected by the reaction. Simple relations are generally used to consider properties change such as those proposed in [1]:

$$E(t) = E_{t_0}\Big(1 - (1 - \beta_E)\,\xi(t)\Big)$$
$$f_t(t) = f_{t_0}\Big(1 - \big(1 - \beta_f\big)\,\xi(t)\Big) \tag{24.50}$$

with E_{t_0} and f_{t_0} respectively the original elastic modulus and tensile strength, β_E and β_f are the corresponding residual factors. These must be considered with care and more research is needed on the effect of these equations, especially their use when concrete damage is modeled using a nonlinear constitutive relation in the numerical model.

24.2 P1: Constitutive Model

P1 seeks to capture the nonlinear response of concrete when subjected to a load history that covers the nonlinear response in tension and compression. Simulation is conducted on a 16x32 cm cylinder (the same mesh will be used for all test problems and is shown on Fig. 24.3). Two simulations are conducted, the first without AAR, and for a strain history given by:

$$0 \Rightarrow 1.5\frac{f_t}{E} \Rightarrow 0 \Rightarrow 3\frac{f_t}{E} \Rightarrow 1.5\epsilon_c \Rightarrow 0 \Rightarrow 3\epsilon_c \tag{24.51}$$

and the second for an identical strain history which is however preceded by a 0.5% AAR expansion. The parameters are $\epsilon_c = -0.002$, $f_t = 3.5$ MPa, $f_c' = -37.3$ GPa

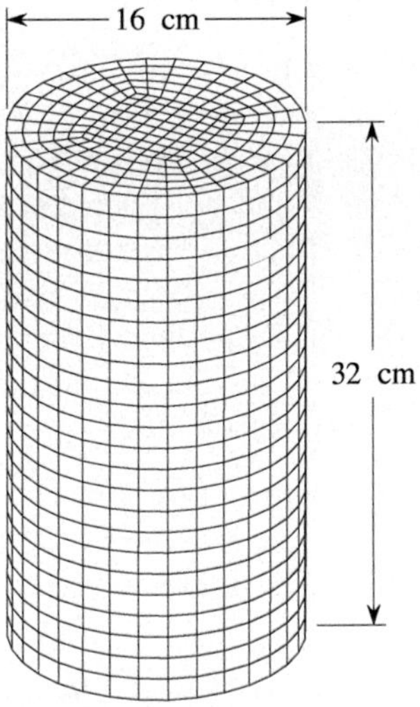

Fig. 24.3 Cylinder mesh used for the analyses

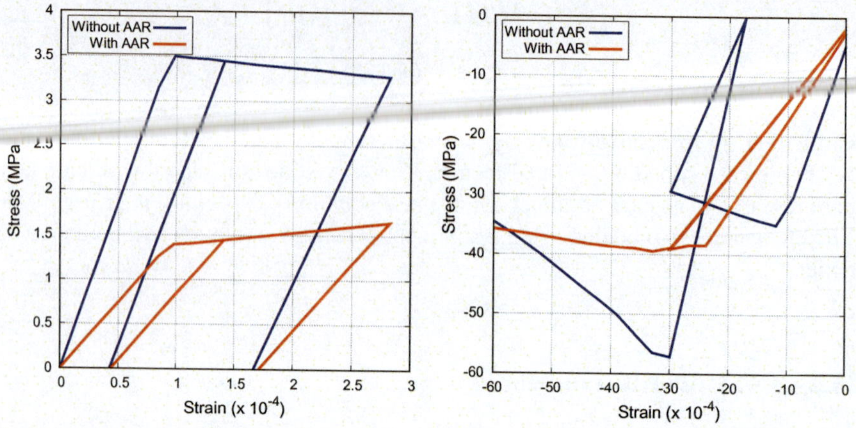

Fig. 24.4 Results of P1: constitutive model

and $G_F = 100$ N/m. With the properties degradation provided by the formulators, β_E and β_f are taken equal to 0.4.

Figure 24.4 plots the stress-strain curve for tension (24.4a) and compression (24.4b). The effect of plasticity is clearly visible as the curves dont come back to the origin point. Moreover the effect of properties degration as a function of AAR can also be observed on these two figures.

24.3 P2: Drying and Shrinkage

The shrinkage is not considered in the constitutive model as it's effect is not included in the state of practice for dam stability re-evaluation. However hygral diffusion problem can be solved. The creep model was calibrated to fit the data provided by the formulators with $k = 6 \times 10^{-10}$ m/s, total porosity of $\phi = 0.165$ and initial saturation of 0.85% (converted to liquid pressure using Eq. 24.39 and relation 24.40, with $a = 18 \times 10^6$ and $b = 0.51282$). Hence the results shown on Fig. 24.5 give the loss of mass of the cylinder as a function of time for a case where the cylinder is immersed in water and for a case where the cylinder is subjected to a 30% humidity boundary conditions.

24.4 P3: Basic Creep

The longitudinal and radial displacements for a 13 by 24 cm cylinder (the mesh shown on Fig. 24.3 is scaled) subjected to 10 and 20 MPa axial compression are plotted on Fig. 24.6a. The creep model was calibrated to fit the data provided by

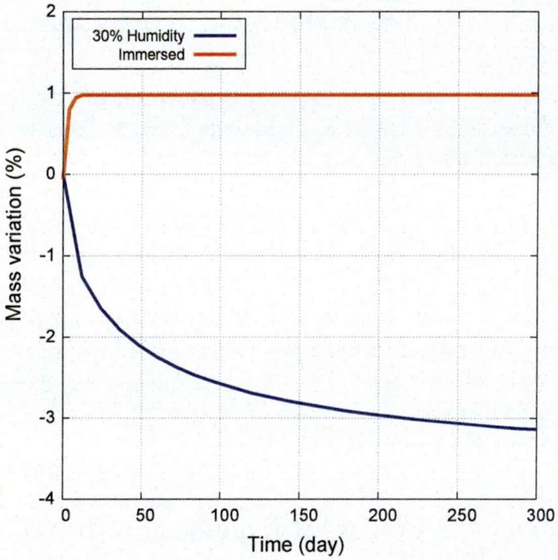

Fig. 24.5 Results of P2: mass variation

the formulators with $E_{\text{ve}} = 0.6E_0$ and $\eta_{\text{ve}} = 100$ days. Thereafter, a load history is applied. During the first 16 first weeks, a 5 MPa compression is applied in the axial direction. Afterwards, the load is increased 10 MPa at the 16th week. This compression load is maintained until the 40th week. Finally, the load in the axial direction is reduced to 5 MPa in compression. The results for this load history are given on Fig. 24.6b for the axial and the radial strain histories.

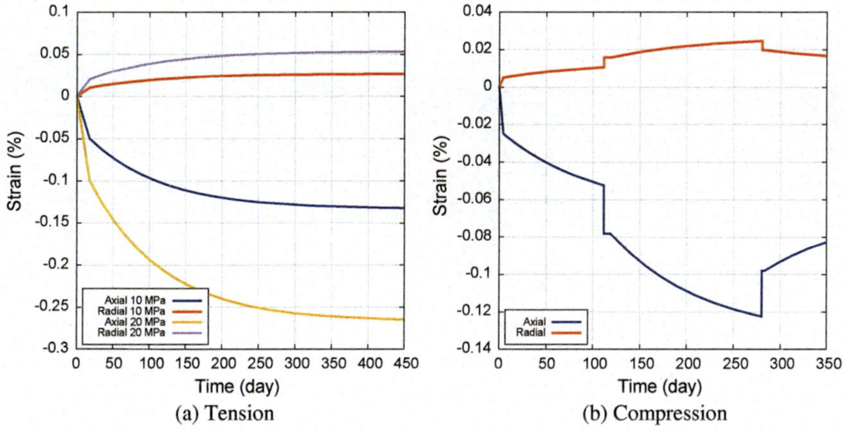

Fig. 24.6 Results of P3: basic Creep

24.5 P4: AAR Expansion; Temperature Effect

The fully saturated $13x24$ cm cylinder, under free expansion is analysed for three temperatures histories: at constant $T = 23$ °C, at constant $T = 38$ °C, and with a temperature variation given by:

$$T(\text{day}) = \frac{T_{max} - T_{min}}{2} \sin\left(2\pi \frac{t/7 - 16}{52}\right) + \frac{T_{max} + T_{min}}{2} \tag{24.52}$$

with $T_{min} = 0$ °C and $T_{max} = 25$ °C. The model parameters were fitted on the curves provided by the formulators with $\tau_l = 40$ days, $\tau_c = 20$ days and $\varepsilon_{max}^{AAR} = 0.00545$. The results for the different temperature histories are given on Fig. 24.7 for the axial strain histories.

24.6 P5: Free AAR Expansion; Effect of RH

The $16x32$ cm cylinder is used to show the effect of concrete saturation on the AAR expansion. Two saturation conditions are used: one fully immersed and one at 30% RH. The initial conditions and parameters for the diffusion problem are taken from Sect. 24.3. As explained earlier, the diffusion problem is weakly coupled with the AAR problem. Similarly to what was done in Sect. 24.3, no shrinkage is considered. The axial strains are given on Fig. 24.8a for these two saturation conditions. It must be noted that because the initial saturation is at 0.85% RH, AAR expansion occurs

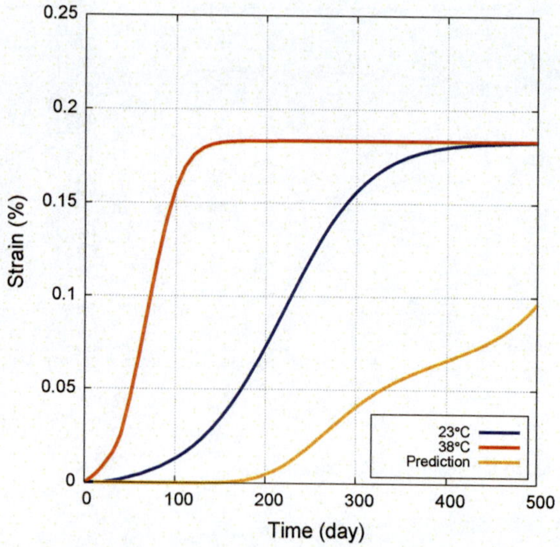

Fig. 24.7 Results of P4: temperature Effect

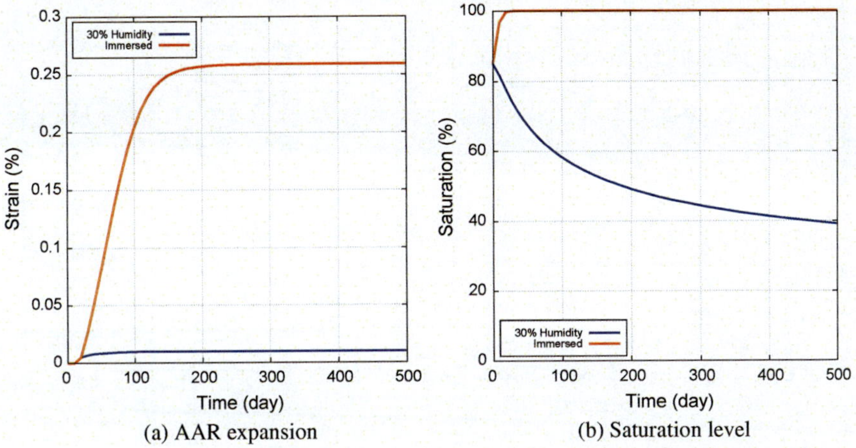

Fig. 24.8 Results of P5: effect of RH

in the first days before the sample dries to 30% RH. Therefore, it is normal to see a small expansion of the cylinder, even if it is expected to be inhibited. The saturation level at the center of the specimen as a function of time is given on Fig. 24.8b.

24.7 P6: AAR Expansion; Effect of Confinement

The effect of confinement on a $13x24$ cm cylinder affected by AAR expansion is achieved assuming a temperature of 38 °C and fully saturated conditions. The following configurations are computed:

- No vertical stress, no confinement (free swelling);
- Vertical stress of 10 MPa, no confinement;
- No vertical stress, concrete cast in a 5 mm thick steel container;
- Vertical stress of 10 MPa and concrete cast in a 5 mm thick steel container.

Creep was neglected to isolate the effect of AAR expansion. The contact between the steel jacket and the concrete cylinder has perfect slip conditions modeled using contact elements. The model parameters were fitted on the curves provided by the formulators with $\tau_l = 140$ days, $\tau_c = 40$ days, $\varepsilon_{max}^{AAR} = 0.00288$ and $\bar{\sigma}_v = -9.7$ MPa. Figures 24.9a and 24.9b give respectively the radial and axial strains as a function of time for the different confinement configurations.

References

1. Saouma, V., Perotti, L.: Constitutive model for alkali aggregate reactions. ACI Mater. J. **103**(3), 194–202 (2006)

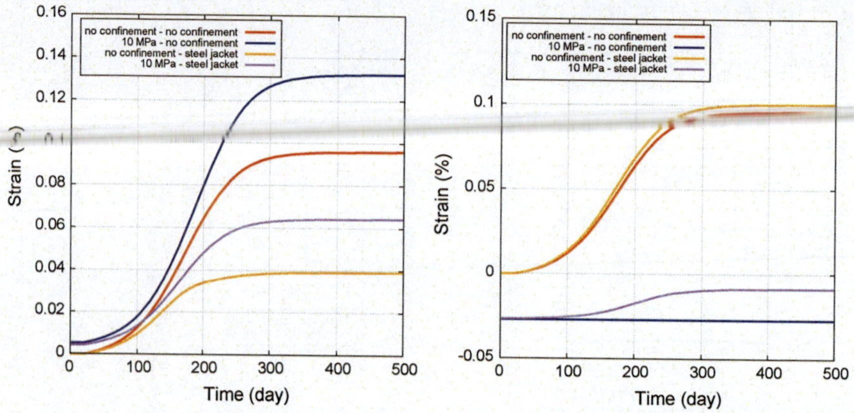

(a) Radial (b) Axial

Fig. 24.9 Results of P6: Effect of Confinement

2. Larive, C.: Apports combinés de l'experimentation et de la modélisation à la comprehension del'Alcali-Réaction et de ses Effets Mécaniques. PhD thesis (1998). Paris: Laboratoire Central des Ponts et Chaussées. https://hal.inria.fr/docs/00/52/06/76/PDF/1997TH_LARIVE_C_NS20683.pdf
3. Ulm, F., Coussy, O., Li, K., Larive, C.: Thermo-chemo-mechanics of ASR expansion in concrete structures. ASCE J. Eng. Mech. **126**(3), 233–242 (2000)
4. Liaudat, J., Carol, I., López, C.M., Saouma, V.: ASR expansions in concrete under triaxial confinement. Cem. Concr. Compos. **86**, 160–170 (2018)
5. Comi, C., Pignatelli, R.: A three-phase model for damage induced by asr in con-crete structures. In: IV International Conference on Computational Methods for Coupled Problems in Science and Engineering (2011)
6. Grassl, P., Xenos, D., Nyström, D., Rempling, R., Gylltoft, K.: CDPM2: A damage-plasticity approach to modelling the failure of concrete. Int. J. Solids. Struct. **50**(24), 3805–3816 (2013). issn: 0020–7683
7. Pan, J., Feng, Y., Jin, F., Zhang, C.: Numerical prediction of swelling in concrete arch dams affected by alkali-aggregate reaction. Eur. J. Environ. Civil Eng. **17**(4), 231–247 (2013). https://doi.org/10.1080/19648189.2013.771112
8. Mainguy, M., Coussy, O., Eymard, R.: Modélisation des transferts hydriques isothermes en milieu poreux : application au séchage des matériaux à base de ciment. Tech. rep. 32. Paris, FRANCE: Laboratoire central des ponts et chaussées, 1999
9. Millington, R.J.: Gas diffusion in porous media. Science **130**(3367) (1959). https://doi.org/10.1126/science.130.3367.100-a, pp. 100-102
10. Mainguy, M., Coussy, O., Baroghel-Bouny, V.: Role of air pressure in drying of weakly permeable materials. J. Eng. Mech. **127**(6), 582–592 (2001). https://doi.org/10.1061/(ASCE)0733-9399
11. van Genuchten, M.T.: A closed form equation for predicting the hydraulic conductivity of unsaturated soils. Soil Sci. Soc. Am. J. **44**(5), 892–898 (1980)
12. Baroghel-Bouny, V., Mainguy, M., Lassabatere, T., Coussy, O.: Characterizationand identification of equilibrium and transfer moisture properties for ordinary and high-performance cementitious materials. Cem. Concr. Res. **29**(8), 1225–1238 (1999)
13. Picandet, V., Khelidj, A., Bastian, G.: Effect of axial compressive damage on gas permeability of ordinary and high-performance concrete. Cem. Concr. Res. **31**(11), 1525–1532 (2001)

14. Pijaudier-Cabot, G., Dufour, F., Choinska, M.: Permeability due to the increase of damage in concrete: from diffuse to localized damage distributions. J. Eng. Mecha. **135**(9), 1022–1028 (2009). https://doi.org/10.1061/(ASCE)EM.1943-7889.0000016
15. Bouhjiti, D.E.-M., Ezzedine El Dandachy, M., Dufour, F., Dal Pont, S., Briffaut, M., Baroth, J., Masson, B.: New continuous strain-based description of concrete's damage-permeability coupling. In. J. Numer. Anal. Methods Geomech. **42**(14), 1671–1697 (2018). https://doi.org/10.1002/nag.2808
16. Oliver, J.: A consistent characteristic length for smeared cracking models. Int. J. Numer. Methods Eng. **28**(2), 461–474 (1989). https://doi.org/10.1002/nme.1620280214
17. Govindjee, S., Gregory, J., Simo, J.: Anisotropic modelling and numerical simulation of brittle damage in concrete. Int. J. Numer. Methods Eng. **38**(21), 3611–3633 (1995). https://doi.org/10.1002/nme.1620382105
18. Slobbe, A., Hendriks, M., Rots, J.: Systematic assessment of directional mesh bias with periodic boundary conditions: Applied to the crack band model. Eng. Fract. Mech. **109**, 186–208 (2013). issn: 0013-7944
19. Morenon, P.: Modélisation des réactions de gonflement interne des bétonsavec prise en compte des couplages poro-mécaniques et chimique". PhD thesis (2017). LMDC—Laboratoire Matériaux et Durabilité des constructions
20. Roth, S.-N., Léger, P., Soulaïmani, A.: Strongly coupled XFEM formulation for non-planar three-dimensional simulation of hydraulic fracturing with emphasis on concrete dams. Comput. Methods Appl. Mech. Eng. **363**, 112899 (2020). https://doi.org/10.1016/j.cma.2020.112899
21. ISE.: Structural Effects of Alkali-Silica Reaction—Technical Guidance Appraisal of Existing Structures. 11 Upper Belgrave Street, London SW1X 8BH: Institution of Structural Engineers (ISE), 1992
22. Esposito, R., Anaç, C., Hendriks, M., Çopuroğlu, O.: Influence of the alkali-silica reaction on the mechanical degradation of concrete. J. Mater. Civil Eng. **28**(6), 04016007 (2016). https://doi.org/10.1061/(ASCE)MT.1943-5533.0001486
23. Dolen, T.P.: Materials properties model for aging concrete. Tech. rep. Dam Safety Program Report No. DSO-05-05. USBR, Colorado.: Bureau of Reclamation, 2005
24. Dolen, T.P.: Selecting strength input poarameters for structural analysis of aging concrete dams. In: Proceedings of the 31st Annual USSD Conference. San Diego, California, 2011
25. Giaccio, G., Zerbino, R., Ponce, J., Batic, O.: Mechanical behavior of concretes damaged by alkali-silica reaction. Cem. Concr. Res. **38**(7), 993–1004 (2008)

Chapter 25
Benchmark Study Results: Merlin/Colorado

Victor Saouma and M. Amin Hariri-Ardebili

25.1 Finite Element Model Description

The AAR model of the author is an uncoupled one, that is the constitutive model is in no way affected by the AAR which itself is considered to be an initial strain (akin of temperature), which grafts itself on the mechanical one. It is implemented in [1], and a complete "validation" of the code with the RILEM benchmark is separately published [2]. This section will describe first the AAR model yielding to the expression of the AAR strain tensor which is accounted for.

25.1.1 AAR Model

25.1.1.1 Premises

Two different aspects of mathematical modeling of AAR in concrete may be distinguished: (1) The kinetics of the chemical reactions and diffusion processes involved, and (2) The mechanics of fracture that affects volume expansion and causes loss of strength, with possible disintegration of the material [3].

The proposed model [4, 5] is driven by the following considerations:

1. AAR is a volumetric expansion, and as such can not be addressed individually along a principal direction without due regard to what may occur along the other two orthogonal ones.

V. Saouma (✉) · M. A. Hariri-Ardebili
Department of Civil Engineering, University of Colorado, Boulder, CO, USA
e-mail: saouma@colorado.edu

M. A. Hariri-Ardebili
e-mail: Mohammad.HaririArdebili@Colorado.EDU

© RILEM 2021

V. E. Saouma (ed.), *Diagnosis & Prognosis of AAR Affected Structures*,
RILEM State-of-the-Art Reports 31,
https://doi.org/10.1007/978-3-030-44014-5_25

2. Kinetics component is taken from the work of [6, 7].
3. AAR is sufficiently influenced by temperature to account its temporal variation in an analysis.
4. AAR expansion is constrained by compression, and is redirected in other less constrained principal directions. This will be accomplished by assigning "weights" to each of the three principal directions.
5. Relatively high compressive or tensile stresses inhibit AAR expansion due to the formation of micro or macro cracks which absorb the expanding gel.
6. High compressive hydrostatic stresses slow down the reaction.
7. Triaxial compressive state of stress reduces but does not eliminate expansion.
8. Accompanying AAR expansion is a reduction in tensile strength and elastic modulus.

25.1.1.2 Expansion Curve

One of the most extensive and rigorous investigation of AAR has been conducted by [6] who tested more than 600 specimens with various mixes, ambiental and mechanical conditions. Not only did the author conduct this extensive experimental investigation, but a numerical model has also been proposed for the time expansion of the concrete. In particular, a thermodynamical based model for the expansion evolution is developed, and then calibrated with the experimental data, Fig. E.1.

$$\xi(t,\theta) = \frac{1 - e^{-\frac{t}{\tau_C(\theta)}}}{1 + e^{-\frac{(t-\tau_l(\theta))}{\tau_C(\theta)}}} \tag{25.1}$$

where τ_l and τ_c are the latency and characteristic times respectively. The first corresponds to the inflexion point, and the second is defined in terms of the inter-subsection of the tangent at τ_L with the asymptotic unit value of ξ. In a subsequent work, [7] have shown the thermal dependency of those two coefficients:

$$\begin{aligned}\tau_l(\theta) &= \tau_l(\theta_0)\exp\left[U_l\left(\frac{1}{\theta} - \frac{1}{\theta_0}\right)\right]; \qquad U_l = 9{,}400 \pm 500K \\ \tau_c(\theta) &= \tau_c(\theta_0)\exp\left[U_c\left(\frac{1}{\theta} - \frac{1}{\theta_0}\right)\right]; \qquad U_c = 5{,}400 \pm 500K\end{aligned} \tag{25.2}$$

expressed in terms of the absolute temperature ($\theta°\text{K} = 273 + T°\text{C}$) and the corresponding activation energies. U_l and U_c are the activation energies minimum energy required to trigger the reaction for the latency and characteristic times respectively. To the best of the authors knowledge, the only other tests for these values were performed by [8] who obtained values within 20% of Larive's, and dependency on types of aggregates and alkali content of the cement has not been investigated. Hence, in the absence of other tests, those values can be reasonably considered as representative of dam concrete also.

25.1.1.3 Volumetric Expansion

Hence, the general (uncoupled) equation for the incremental free volumetric AAR strain is given by

$$\dot{\varepsilon}_V^{AAR}(t) = \Gamma_t(f_t'|w_c, \sigma_I|COD_{\max})\Gamma_c(\overline{\sigma}, f_c')g(h)\dot{\xi}(t,\theta)\, \varepsilon^{\infty}\big|_{\theta=\theta_0} \tag{25.3}$$

where COD is the crack opening displacement, $\xi(t,\theta)$ is a sigmoid curve expressing the volumetric expansion in time as a function of temperature and is given by Eq. 25.1, ε^{∞} is the laboratory determined (or predicted) maximum free volumetric expansion at the reference temperature θ_0, Fig. E.1.

The retardation effect of the hydrostatic compressive stress manifests itself through τ_l. Hence, Eq. 25.2 is expanded as follows

$$\tau_l(\theta, \theta_0, I_\sigma, f_c') = f(I_\sigma, f_c')\tau_l(\theta_0)\exp\left[U_l\left(\frac{1}{\theta} - \frac{1}{\theta_0}\right)\right] \tag{25.4}$$

where

$$f(I_\sigma, f_c') = \begin{cases} 1 & \text{if } I_\sigma \geq 0. \\ 1 + \alpha\frac{I_\sigma}{3f_c'} & \text{if } I_\sigma < 0. \end{cases} \tag{25.5}$$

and I_σ is the first invariant of the stress tensor, and f_c' the compressive strength. Based on a careful analysis of [9], it was determined that $\alpha = 4/3$. It should be noted, that the stress dependency (through I_σ) of the kinetic parameter τ_l makes the model a truly coupled one between the chemical and mechanical phases. Coupling with the thermal component, is a loose one (hence a thermal analysis can be separately run), $0 < g(h) \leq 1$ is a reduction function to account for humidity given by

$$g(h) = h^m \tag{25.6}$$

where h is the relative humidity [10]. However, one can reasonably assume that (contrarily to bridges) inside a dam $g(h) = 1$ for all temperatures.

$\Gamma_t(f_t'|w_c, \sigma_I|COD_{\max})$ accounts for AAR reduction due to tensile cracking (in which case gel is absorbed by macro-cracks).

$$\text{Smeared Crack}\begin{cases} \text{No } \Gamma_t = \begin{cases} 1 & \text{if } \sigma_I \leq \gamma_t f_t' \\ \Gamma_r + (1-\Gamma_r)\gamma_t\frac{f_t'}{\sigma_I} & \text{if } \gamma_t f_t' < \sigma_I \end{cases} \\ \text{Yes } \Gamma_t = \begin{cases} 1 & \text{if } \text{COD}_{\max} \leq \gamma_t w_c \\ \Gamma_r + (1-\Gamma_r)\gamma_t\frac{w_c}{\text{COD}_{\max}} & \text{if } \gamma_t w_c < \text{COD}_{\max} \end{cases} \end{cases} \tag{25.7}$$

where γ_t is the fraction of the tensile strength beyond which gel is absorbed by the crack, Γ_r is a residual AAR retention factor for AAR under tension. If an elastic model is used, then f_t' is the the tensile strength, σ_I the maximum principal tensile stress. On the other hand, if a smeared crack model is adopted, then $\text{COD}_{\max}$ is the maximum crack opening displacement at the current Gauss point, and w_c the maximum crack opening displacement in the tensile softening curve [11].

Concrete pores being seldom interconnected, and the gel viscosity relatively high, gel absorption by the pores is not explicitly accounted for. Furthermore, gel absorption by the pores is accounted for by the kinetic equation through the latency time which depends on concrete porosity. The higher the porosity, the larger the latency time.

Γ_c in turns accounts for the reduction in AAR volumetric expansion under compressive stresses (in which case gel is absorbed by diffused micro-cracks) [9]:

$$\Gamma_c = \begin{cases} 1 & \text{if } \overline{\sigma} \leq 0. \text{ Tension} \\ 1 - \frac{e^{\beta}\overline{\sigma}}{1+(e^{\beta}-1.)\overline{\sigma}} & \text{if } \overline{\sigma} > 0. \text{ Compression} \end{cases} \tag{25.8}$$

$$\overline{\sigma} = \frac{\sigma_I + \sigma_{II} + \sigma_{III}}{3 f_c'} \tag{25.9}$$

Whereas this expression will also reduce expansion under uniaxial or biaxial confinement, these conditions are more directly accounted for below through the assignment of weights.

25.1.1.4 AAR Strain Redistribution

The third major premise of the model, is that the volumetric AAR strain must be redistributed to the three principal directions according to their relative propensity for expansion on the basis of a weight which is a function of the respective stresses. Whereas the determination of the weight is relatively straightforward for triaxial AAR expansion under uniaxial confinement (for which some experimental data is available), it is more problematic for biaxially or triaxially confined concrete.

Given principal stress vector defined by $\sigma_k, \sigma_l, \sigma_m$, weights are assigned in function of the three principal stresses, Fig. 25.1. These weights will control AAR volumetric expansion distribution.

It should be noted that the proposed model will indeed result in an anisotropic AAR expansion. While not explicitly expressed in tensorial form, the anisotropy stems from the different weights assigned to each of the three principal directions.

25.1.1.5 Degradation

This deterioration being time dependent, a time dependent model that mirrors the expansion is adopted.

$$E(t, \theta) = E_0 [1 - (1 - \beta_E) \xi(t, \theta)] \tag{25.10}$$

$$f_t'(t, \theta) = f_{t,0}' \left[1 - \left(1 - \beta_f\right) \xi(t, \theta)\right] \tag{25.11}$$

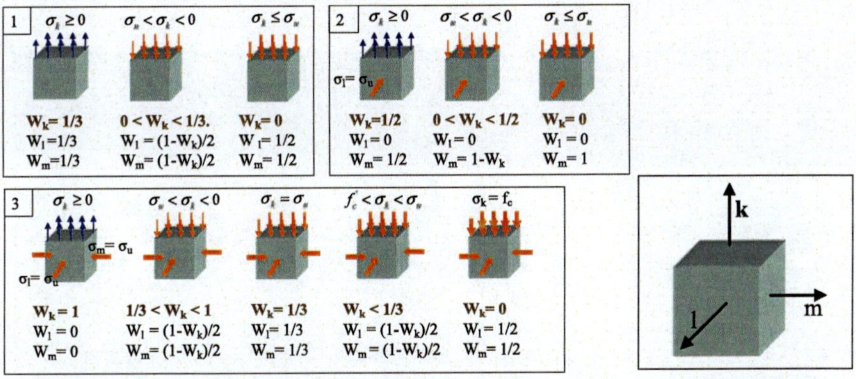

Fig. 25.1 Weight of volumetric AAR redistribution in selected cases

where E_0 and $f'_{t,0}$ are the original elastic modulus and tensile strength, β_E and β_f are the corresponding residual fractional values when ε_{AAR} tends to $\varepsilon^{\infty}_{AAR}$.

25.1.2 Concrete Constitutive Models

Whereas our AAR model could be coupled with any (including linear elastic) constitutive model, the last one in Merlin is based on a fracture-plastic one for concrete continuum (smeared crack model) and on a fracture mechanics based one for discrete cracks.

The structural model, has two constitutive models: (a) one for distributed failures (smeared crack model) implemented in the spirit of plasticity Červenka and Červenka [12]; and (b) one for discrete cracks implemented in the spirit of "Fracture Mechanics" Multon [13].

25.2 P1: Constitutive Model

25.2.1 Problem Description

As previously mentioned, Merlin's constitutive model is completely disassociated from AAR's, and is first tested in this section. Hence, P1 seeks to capture the nonlinear response of concrete when subjected to a load history covering both tension and compression. Simulation is conducted for a 16 × 32 cm cylinder shown in Fig. 25.2 (same mesh will be used for all test problems).

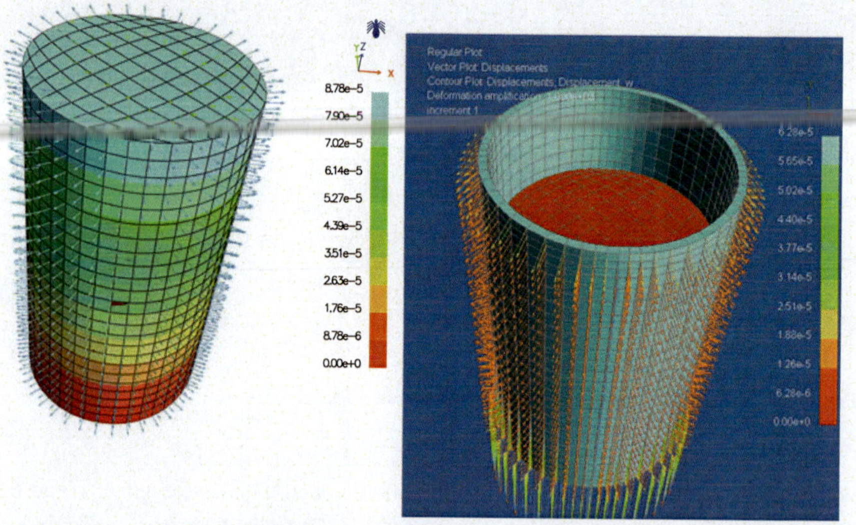

(a) Mesh of concrete cylinder **(b) Mesh of concrete cylinder and steel jacket**

Fig. 25.2 Finite element model for concrete cylinders

25.2.2 Simulations

Two simulations are conducted, the first without AAR, and for a strain history given by

$$0 => 1.5\frac{f_t'}{E} => 0 => 3\frac{f_t'}{E} => 1.5\epsilon_c => 0 => 1.5\epsilon_c \tag{25.12}$$

and the second for an identical strain history which is however preceded by a AAR expansion.

Figure 25.3 plots the load-displacement curve at the top of the cylinder. In both cases, the curve load-displacement at the top of the cylinder surface is plotted. The AAR expansion vs time is also plotted.

First, we observe the model nonlinear response with a peak compressive strength of about −38 MPa, and an onset of nonlinearity of about −13 MPa. The tensile strength of 3.5 MPa is also reduced by the specified $\beta_t = 0.4$ to about 1.4 MPa, and finally the elastic modulus degradation of β_E is also clearly captured.

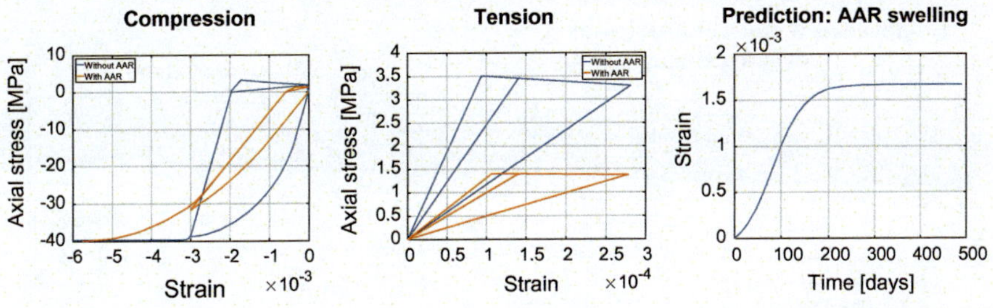

(a) Compression (b) Tension (c) AAR Swelling

Fig. 25.3 Results of constitutive model

25.3 P2: Drying and Shrinkage

Problem was not addressed as our finite element code does not have a hygral model. This can be partially alleviated by having layers of concrete with reduced expansion on the surface to account for drying shrinkage.

25.4 P3: Creep

In the absence of an explicit creep model in Merlin, creep is indirectly accounted for through a time varying creep coefficient ($\Phi(t)$) as follows:

$$\sigma(t) = \frac{E_0}{1+\phi}\varepsilon(t) \Rightarrow \phi(t) = \frac{E_0 \epsilon(t)}{\sigma(t)} - 1 \tag{25.13}$$

and at each time step the young modulus is modified according to

$$E(t) = \frac{E_0}{1+\phi(t)} \tag{25.14}$$

25.4.1 Simulations

Using a time varying creep coefficient calibrated from Fig. 21.5, a 13 × 24 cm cylinder concrete cylinder is investigated for the stress variation shown in Fig. 21.6.

Traction was applied on the frictionless top of the cylinder.

In Fig. 25.4a we examine numerical and experimental axial strain:

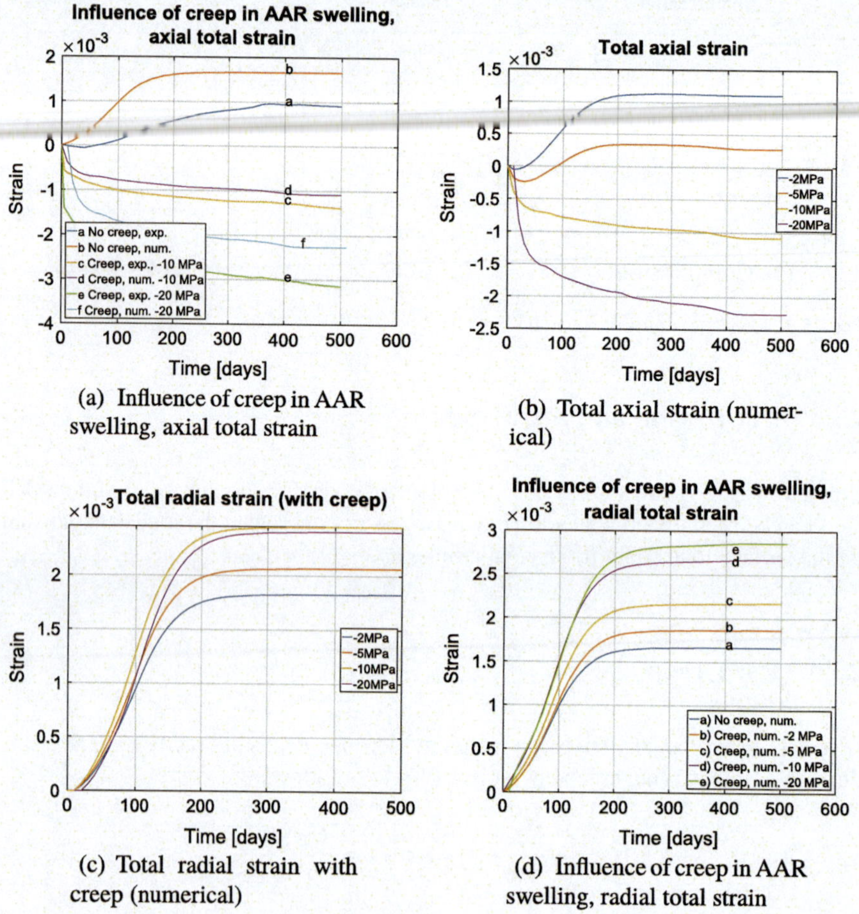

(a) Influence of creep in AAR swelling, axial total strain

(b) Total axial strain (numerical)

(c) Total radial strain with creep (numerical)

(d) Influence of creep in AAR swelling, radial total strain

Fig. 25.4 Numerical results of calibration for Creep; Part 1

- In the absence of creep, the experimental (a) and numerical (b) results without creep are reasonably close and in the absence of an axial compressive stress they are highest.
- Amongst experimental results, largest swelling is (a) (no compressive stress), followed by (c) and (e) (with axial stresses of -10 and -20 MPa respectively).
- Amongst numerical predictions, in descending order of expansion: (b) with no axial stresses followed by (c), (d), and (f), where the corresponding axial stresses are 0, -10 and -20 MPa respectively.

In Fig. 25.4b the -2 MPa stress is still too low to overcome the AAR expansion, and thus it is the only case where a positive strain takes place. For stresses higher than -10 MPa, the AAR is zero and the combined elastic and AAR strain are thus

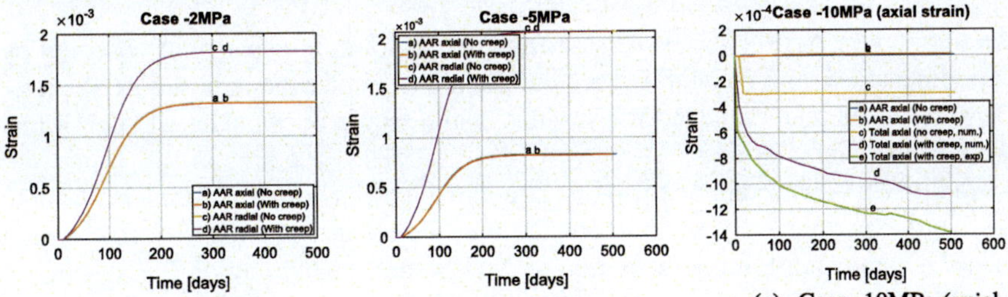

(a) Case -2MPa (b) Case -5MPa (c) Case -10MPa (axial strain)

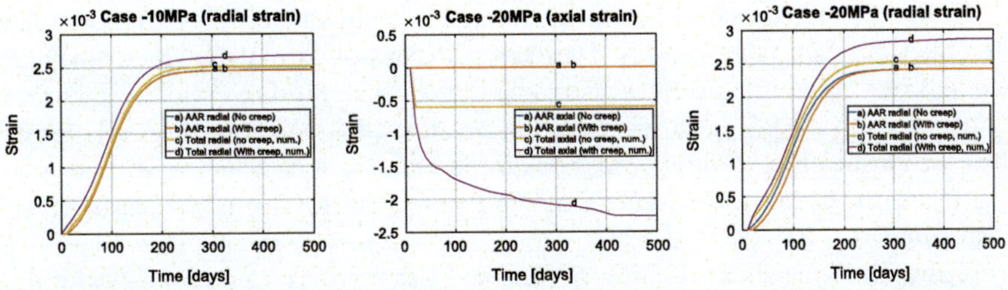

(d) Case -10MPa (radial strain) (e) Case -20MPa (axial strain) (f) Case -20MPa (radial strain)

Fig. 25.5 Numerical results of calibration for Creep, Part 2

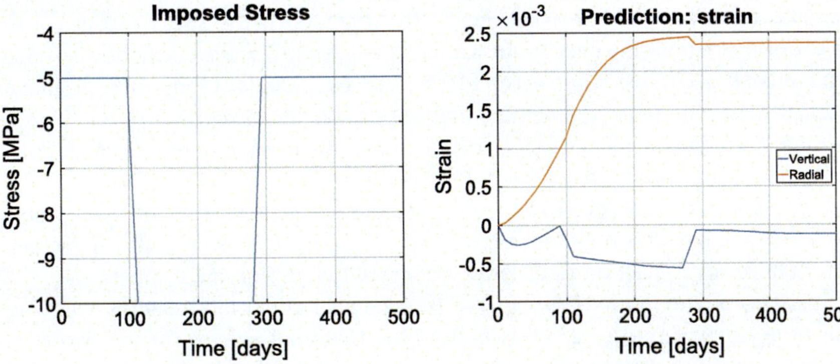

(a) Imposed Stress (b) Axial and radial strains

Fig. 25.6 Numerical results of prediction

well into the negative range, while for -5 and -10 MP the net axial strain is almost nil.

In Fig. 25.4c we examine the radial strain. In this axi-symmetric problem, we note that, with an imposed axial stress of -10 and -20 MPa, both experimental and numerical strains are about equal to $2.5\ \times 10^{-3}$ which is half ε^{∞} thus reinforcing the notion that AAR's strain redistribution (or anisotropic expansion) observed by experimentalists and the author's model. Then the smaller the imposed axial stress, the smaller the final radial AAR's strain is, and it would be about equal to ε^{∞} when there is no creep.

In Fig. 25.4d we examine the radial strain in this axi-symmetric problem, we note that both experimentally and numerically they are about equal to 2.5×10^{-3} which is half ε^{∞}, thus reinforcing the notion of AAR's strain redistribution (or anisotropic expansion) observed by researchers and embedded in the author's model. It should be noted that the radial strain is also mildly affected by the Poisson's radial expansion.

We then examine each stress value separately. In Fig. 25.5a, −2 MPa shows that the effect of creep is almost nil. The largest expansion is radially (between third and half of ε^{∞} and the lowest is also radially (less than third of ε^{∞}). Creep has no influence on the axial strain which is still positive (that is the expansion is larger than the elastic/creep contraction), nor on the radial strain, which is still positive too (that is the expansion is larger than the Poisson's effect due to the elastic/creep contraction).

Figure 25.5b the stress is now −5 MPa, and observations are the same as for the preceding case of −2 MPa; however, the larger imposed stress accentuate them.

Figure 25.5c, d corresponds to the axial strain under −10 MPa, (c) with axial strains and (d) with radial strains. Curves (a) and (b) in Fig. 25.5c shows that the creep doesn't have any impact on the vertical strain, which is correct: there is not supposed to be any AAR's expansion in the direction if the stress exceeds −10 MPa. Curve (c) gives the total strain without creep, so this is the elastic strain (once again, there is no AAR's strain here). Finally, curves (d) and (e) show numerical and experimental strain in axial direction, they are reasonably close.

For Fig. 25.5d, we note that creep doesn't have any impact, except through Poisson's effect. Note that we reach half of ε^{∞} i.e. 2.5×10^{-5}.

Figure 25.5e, f correspond to the axial strain under −20 MPa, (e) with axial strains and (f) with radial strains. Same conclusions can be made as with −10 MPa: creep doesn't have any impact on AAR strains, except through Poisson's effect.

Finally, a prediction for the response of a cylinder subjected to an time varying axial stress shown in Fig. 25.6a is performed. Using an average of the two $\phi(t)$ (corresponding to −10 and −20 MPa), response is shown in Fig. 25.6.

First, the vertical elastic strain, compounded by creep decreases down to a minimum of about -2.5×10^{-4} at about 20 days. At that point, AAR's expansion rate is almost nil, smaller than the contraction due to creep, and thus the strain decreases. Then the AAR's expansion starts, and the strain is increasing again, AAR's expansion is starting to overcome the elastic strain.

Then the axial stress is increased from −5 to −10 MPa, and the elastic strain compounded with creep causes a further contraction. At −10 MPa, the AAR axial expansion is completely inhibited (and redirected in the radial direction), and all strain increase is solely due to creep. When the stress is again dropped from −10 to −5 MPa there is a rebound, and from that point onward both creep contraction and (reduced) AAR's expansion are at work. However, at that point, the propensity for AAR has been exhausted, and mst of it occurred along the radial direction (which is close to 2.5×10^{-3} at the end), so it cannot compensate for the elastic strain. In the radial direction, we observe an opposite behavior. First, due to the basic AAR's expansion in this direction, plus the redirected AAR's expansion between 100 and 300 day, and finally because of Poisson's effect.

25.5 P4: AAR Expansion; Temperature Effect

25.5.1 Simulations

Three simulations are performed:

Validation By simulating the free expansion at 23 and 38 °C for which Fig. 21.7 shows the experimental data (the large variability should be noticed).
Prediction for a harmonic temperature variation given by

$$T \text{ (days)} = \frac{T_{\max} - T_{\min}}{2} \sin(2\pi \frac{t/7 - 16}{52}) + \frac{T_{\max} + T_{\min}}{2} \tag{25.15}$$

where $T_{\max}$ and $T_{\min}$ are 25 °C and 0 °C respectively, Fig. 21.8.

Figure 25.7a shows the predicted expansion versus time for the two temperature. As anticipated, expansion is much faster at the higher temperature. Furthermore, the

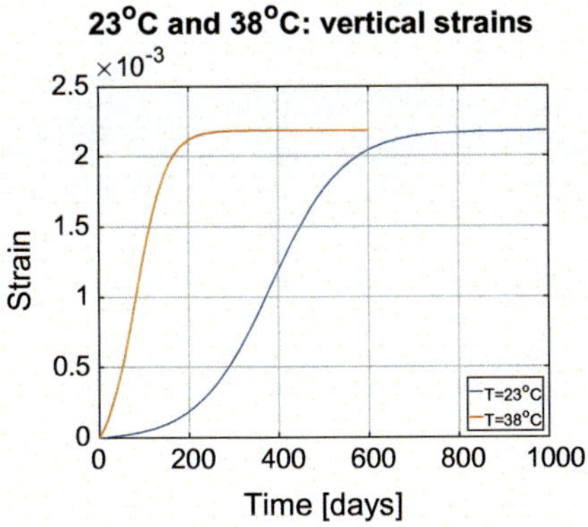

(a) Vertical strains at 23 °C and 38 °C

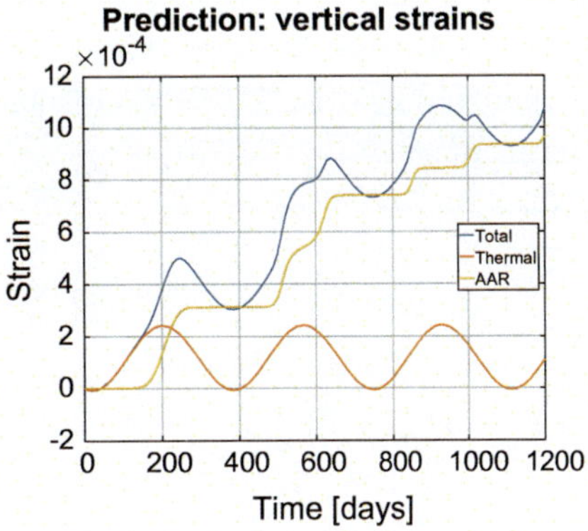

(b) Predicted response to an harmonic temperature variation

Fig. 25.7 Numerical results for calibration and prediction for the effects of temperature

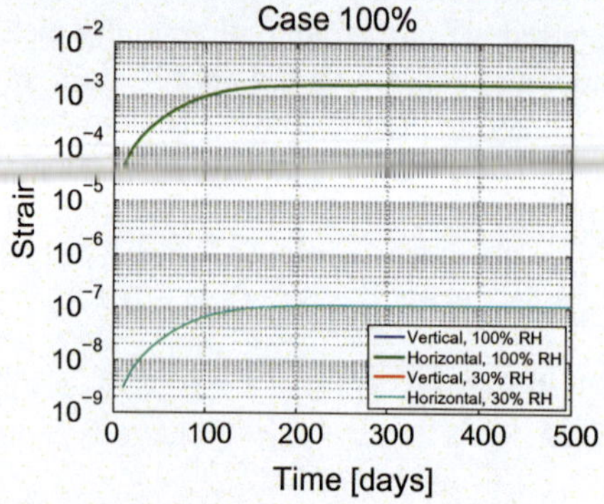

(a) Expansion in terms of RH

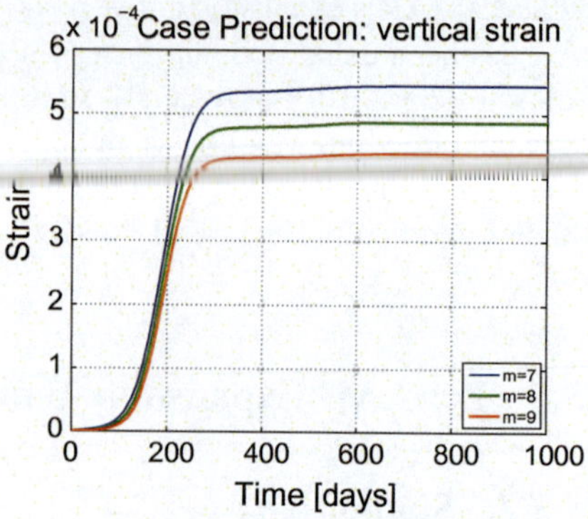

(b) Effect of the variation in m

Fig. 25.8 Calibration: effect of relative humidity

numerically predicted AAR strain are close to the experimentally obtained one from which critical data was calibrated (ε^{∞}, τ_{lat} and $\tau char$), Fig. 21.7.

Results for the prediction are shown in Fig. 25.7b. The harmonic thermal strain ($\epsilon_{th} = \alpha \cdot \Delta T$) is first given, and the one caused by AAR is simply given by $\epsilon_{AAR} = \epsilon_{total} - \epsilon_{th}$.

We note that the AAR strain is flat for low temperature (and thus the plateau), and the total strain increases with time due to the combined effects of AAR expansion and temperature. The decreases are driven by the decrease in the temperature.

25.6 P5: Free AAR Expansion; Effect of RH

Adopting the model of [10], the variation of RH is acounted as follows:

$$\varepsilon^{RH} = RH^{m}\varepsilon^{100\%} \tag{25.16}$$

where m is determined to equal 8 through a regression analysis of experimental data.

25.6.1 Simulations

Again, three simulations have been performed:

Calibration Two analyses with an external relative humidity of 100 and 30% using the experimental dataset of [9], as shown in Fig. 21.9a and b, respectively.

The impact of RH on AAR swelling has been modeled as a modification of the final volumetric AAR strain according to the equation: $\epsilon^{AAR}(t) = \epsilon^{AAR}_{\infty} \cdot RH^{8}(t)$. A

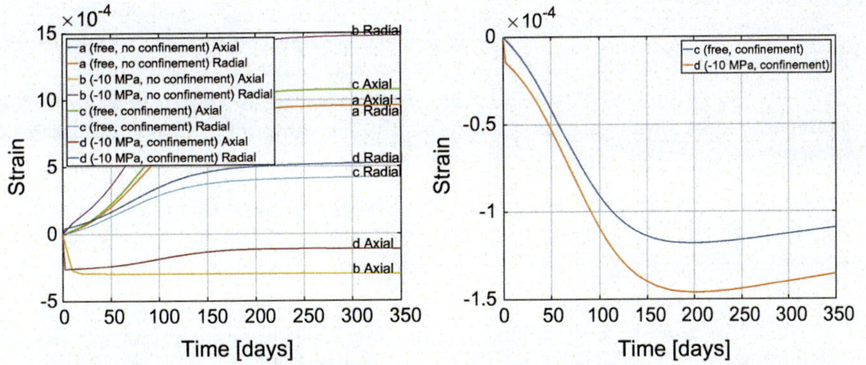

(a) Calibration, concrete axial and radial strains

(b) Calibration, steel radial strains

Fig. 25.9 Calibration; effect of confinement

constant relative humidity of 80% will thus lead to multiplying the final volumetric AAR strain by 0.17.

The factor 8 however, has not yet been well established, which led to simulating the Prediction case using three different values of this factor: 7, 8, and 9.

The results of the first simulation are shown in Fig. 25.8a for expansion in terms of RH; as expected at 30% RH, expansion is practically nil. Figure 25.8b shows the effect of factor m on this expansion: the higher the exponent value, the lower the level of expansion. Let's point out that in this case, no attempt was made to calibrate input data with experimental results, and the final AAR-induced strain was set to 0.5% just like in most of the previous simulations.

25.7 P6: AAR Expansion; Effect of Confinement

25.7.1 Simulations

Five simulations are performed:
Calibration Based on Multon's thesis in which four cases are considered:

- (a) Free expansion, no confinement.
- (b) −10 MPa vertical stress, no confinement.
- (c) Free expansion, confinement.
- (d) −10 MPa vertical stress, confinement.

Prediction for the variable stress history with confinement.

Confinement is provided by a 5 mm steel cylindrical jacket inside which the concrete is cast.

Numerically, concrete and steel are separated by an interface element which allows for axial deformation, and which allows for the steel to act as a confinement.

For the calibration, concrete axial and radial strains are shown in Fig. 25.9a, whereas the steel strains are shown in Fig. 25.9b. Examining the concrete strains, and in descending strain order, we observe that:

1. (b) Radial, -10 MPa, no confinement. Since AAR is inhibited in the axial direction by the -10 MPa axial stress, it is entirely redirected in the radial direction. The AAR's strain is about 15×10^{-4} corresponding to a total volumetric AAR strain of 30×10^{-4} which is approximately equal to the specified $\varepsilon^{\infty} = 28.8 \times 10^{-4}$.
2. (c) Axial, free with confinement. In this case it is the radial expansion which is inhibiting the expansion, and hence it is redirected in the free axial one. As expected the magnitude is about twice.
3. (a) Axial and radial, free no confinement; they are both equal to approximately 9×10^{-4} at 350 days. This is indeed one third of the specified $\varepsilon^{\infty} = 0.288\%$ since we have an unconstrained isotropic expansion.
4. (d) -10 MPa confinement radial; Following an initial increase due to Poisson effect, some swelling occurs but is partially inhibited.
5. (c) Radial free confinement; unlike previous case, there is no initial strain, and a gradual increase in swelling. Swelling is reduced as most of it occurs in the axial direction.
6. (d) -10 MPa axial confinement the initial compressive strain corresponds approximately to the elastic one (σ/E or 2.7×10^{-4}, then as a result of AAR swelling it rebounds (specially that due to confinement, it can only expand axially).
7. (b) -10 MPa axial no confinement; -10 axial confinement the initial compressive strain corresponds approximately to the elastic one (σ/E or 2.7×10^{-4} however since there is no confinement all the AAR expansion is redistributed in the radial direction (contrarily to the preceding case).

As to the steel radial strains they reflect the gradual AAR's induced (swelling) radial strains in the confining jacket.

As to the prediction, Fig. 25.10a shows the applied stress history, and Fig. 25.10b the corresponding strains. Concrete undergoes an initial elastic axial deformation due to the -5 MPa traction, then it expands due to the AAR. When the -10 MPa traction is applied, there is an elastic strain, and at that point the AAR is practically nil as the concrete is axially subjected to a stress equal to the threshold limiting value. When the -10 MPa is dropped to -5 MPa, there is again an elastic "rebounding" and the AAR is nil as it has been exhausted by that time.

The concrete radial strain is primarily driven by AAR (Poisson effect is shown but almost negligible). Though partially constrained by the steel jacket, expansion is mostly in the radial direction in this case since axial expansion is constrained. Finally, the steel radial strain reflects the concrete time dependent expansion.

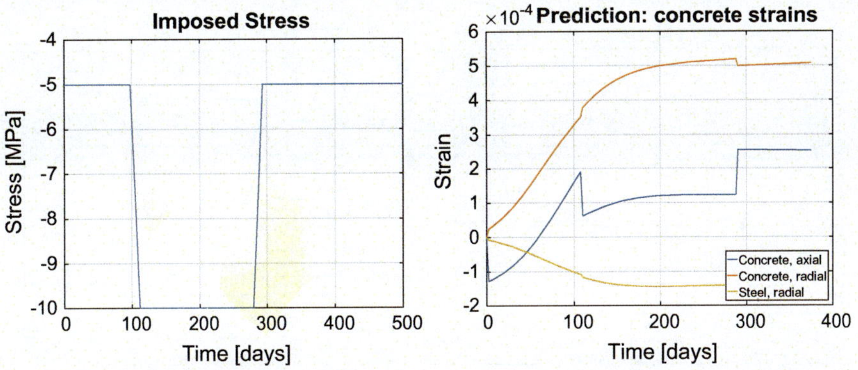

(a) Imposed Stress (b) Axial and radial strains

Fig. 25.10 Prediction; effect of confinement

25.8 P7: Effect of Internal Reinforcement

Internal reinforcement inhibits expansion and AAR induced cracks would then align themselves with the direction of reinforcement as opposed to the traditional "map cracking" [14].

Concrete is modeled by its nonlinear constitutive model, and a linear elasto-plastic model is used for the steel.

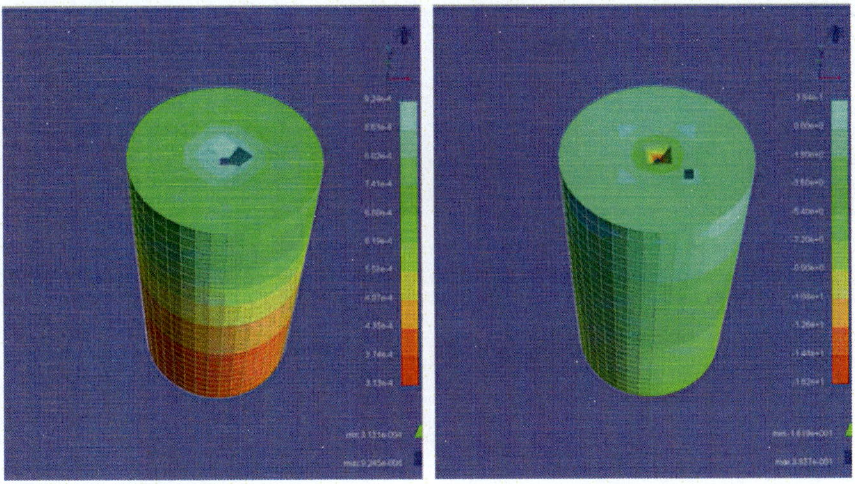

(a) Axial strain (b) Axial stress

Fig. 25.11 Effects of reinforcement on AAR

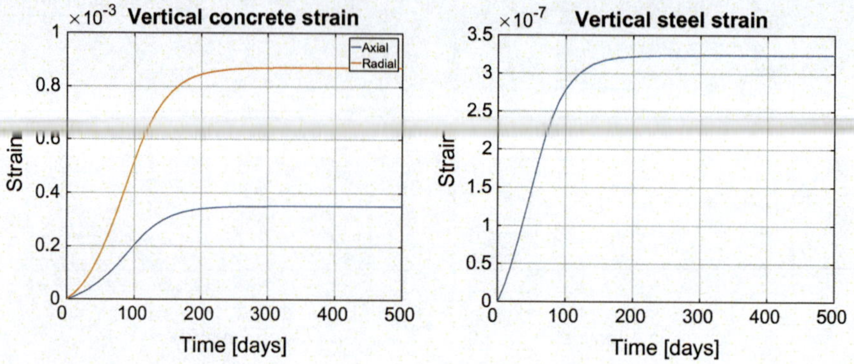

(a) Concrete (b) Steel

Fig. 25.12 Effect of internal reinforcement

Figure 25.11 provides some snapshots of the mesh, and note that at about 5 mm away of the rebar, its effect on the AAR is almost nil. The steel axial stress is quite small, $\simeq$0.063 MPa.

Concrete strains, Fig. 25.11 are indeed restrained in the axial direction, and most of the expansion is in the radial. $\varepsilon_{axial} + 2\varepsilon_{radial} = (0.35 + 2(0.85)) \times 10^{-3} = 20.5 \times 10^{-4}$ which is approximately equal to the specified ε^{∞} (Fig. 25.12).

25.9 P8: Reinforced Concrete Beam

Not modeled, as Merlin does not have a hygral model.

25.10 P9: AAR Expansion; Idealized Dam

Using the fitting data of P6, and an friction angle of 50° for concrete against concrete, and zero cohesion, we consider two cases:

Slot Cut Simulation Performed on a 2D mesh.
Slot cut closure performed on a 3D mesh.

25.10.1 2D Slot Cut

A "proof of concept" approach was followed to capture the impact of a slot cut in a 2D nonlinear analysis (both concrete and joint) with dimensions corresponding to those of the Benchmark.

The model is shown in Fig. 25.13a and consists of four parts: (a) a non reactive concrete on the left; (b) reactive concrete on the right; (c) a volume defining the slot inside the reactive concrete; and (d) a zero thickness joint element between the first and third group.

Results are shown in Fig. 25.13b and clearly show that the following essential features were captured following the slot cut: (a) there is a drastic decrease in the original 10 mm slot as the concrete is allowed to more freely expand; and (b) there is a decrease in the resulting lateral force exercised by the concrete expansion. On the other hand, should there have been no slot cut, then the gap would remain essentially the same, however the lateral confining force would keep on increasing.

For both analysis, the specified temperature and relative humidity is the one of the concrete surface. Zero flux condition between dam and foundation. Reference base temperature of the dam is 20 °C.

1. x, y, z displacements of point A.
2. F_x, F_y and F_z resultant forces on the fixed lateral face versus time (25 years). Assume the typical yearly variations of external air temperature and pool elevation shown in Figs. 21.8 and 21.15, respectively.

This model seeks to capture: (a) general finite element program capabilities in modeling the joint response; (b) ease (or difficulty in preparing the input data file for a realistic problem; and (c) coupling of the various parameters.

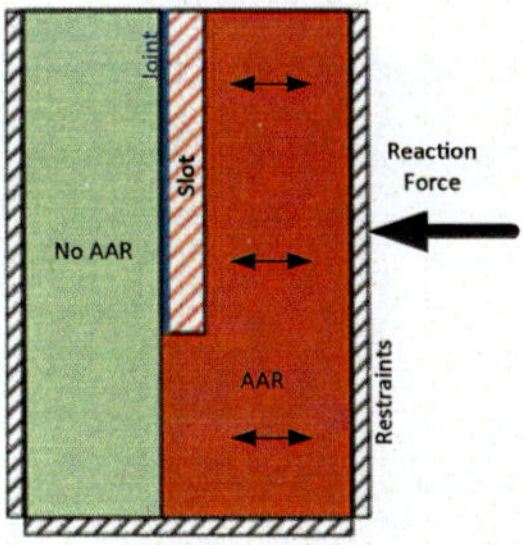

(a) Schematic

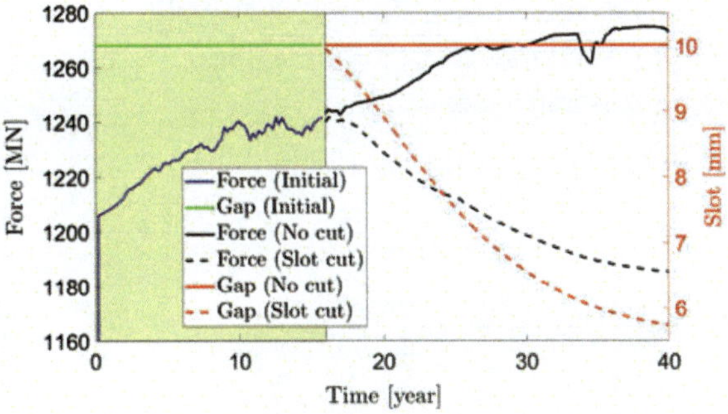

(b) Results

Fig. 25.13 Simulation of slot cut in a dam

25.10.2 3D Simulation of Post-cut Gap Closure

The 3D mesh of the model of Fig. 21.14 is shown in Fig. 25.14.

In this first analysis, the dam is subjected only to AAR, and for the 10 mm slot, with AAR, we seek to determine the slot thickness in terms of time (as it is reduced by AAR), Fig. 25.15a and the corresponding contact stresses, Fig. 25.15b.

From these plots, we conclude that

1. The slot is initially completely open (COD = 0), and gradually the COD reaches -10 mm which is precisely the thickness of the slot. At that point, the interface element is activated, and there can be no more expansion.

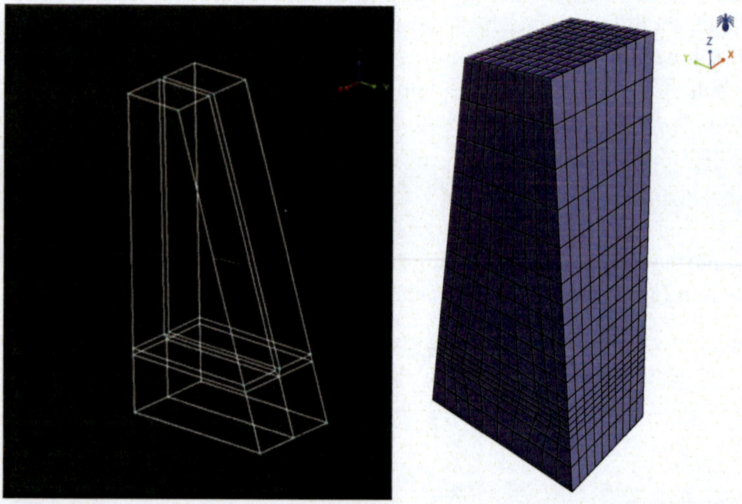

(a) 3D finite element mesh outline (exagerated slot width)

(b) 3D finite element mesh

Fig. 25.14 3D finite element mesh

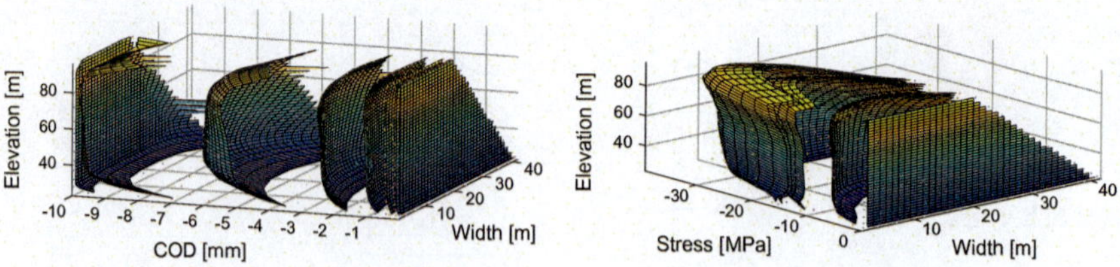

(a) Slot thickness for time: 0.167, 1.67, 3.33, 5, 6.67, 8.33, 10 and 11.67 years

(b) Slot normal stresses for time: 8.33, 11.67, 15, 18.33 and 121.67 years

Fig. 25.15 3D response of a dam subjected to AAR

2. The concavity of the surfaces is to be noted. There is more expansion in the center than on the edges. Eventually the expansion is entirely uniform and we have a nearly full contact at −10 mm.
3. Similarly, there is less expansion on the top than in the bottom.
4. The bottom has also reduced expansion due to the problem formulation as it is constrained to the bottom concrete.
5. Similarly, the stresses are zero in the beginning as there is no contact.
6. Gradually the stress increase, and we observe the same concavity as the one noted for the COD. In other words, stresses are much higher in the center than on the edges.
7. The concavity remains present even after many years, stresses will be higher in the central part than on the edges.

From these observations, we conclude that we were we able to capture the "true" response of the slot, to anticipate the time of closure and to map the corresponding stresses. A salient observation is that the COD/stress state can be quite complex, and thus great care should be exercised in planning for and after slot cutting.

25.11 P10: Reinforced Concrete Panel

Full disclosure: this analysis was undertaken as part of a sponsored project, and a probabilistic based analysis was performed [15]. Only results of the deterministic one are reported.

Six different analyses were performed, Table 25.1 with a finite element mesh composed of 1,001 nodes and 720 quadrilateral elements, Fig. 25.16a. Two sets of reinforcements are modeled at the top and the bottom of the specimen. In the Merlin finite element software, the rebars can be modeled easily by defining the the start and end nodes. Then, Merlin automatically applies the mesh on the rebars. The finite element mesh of the rebars for the panel is shown in Fig. 25.16b.

In all models, the panel is experienced only the ASR expansion and no external load (directly or indirectly) is applied to the panel. The models are expected to expand

Table 25.1 Different types of the models for the panel

ID	Reinf.	Loading	BC on xz plane	BC on yz plane	Material model
P1	No	ASR	x =y = z = 0	x = y = z = 0	Linear elastic
P2	No	ASR	x = y = 0	x = y = 0	Linear elastic
P3	Yes	ASR	x = y = z = 0	x = y = z = 0	Linear elastic
P4	Yes	ASR	x = y = 0	x = y = 0	Linear elastic
P5	Yes	ASR	x = y = z = 0	x = y = z = 0	Non-linear
P6	Yes	ASR	x = y = 0	x = y = 0	Non-linear

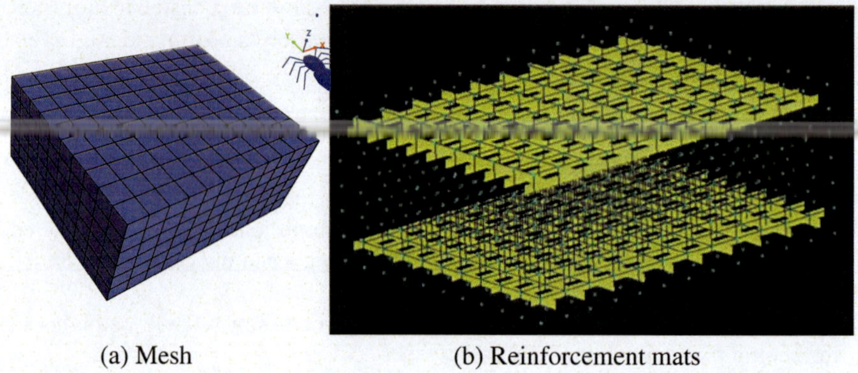

Fig. 25.16 Details of the finite element mesh

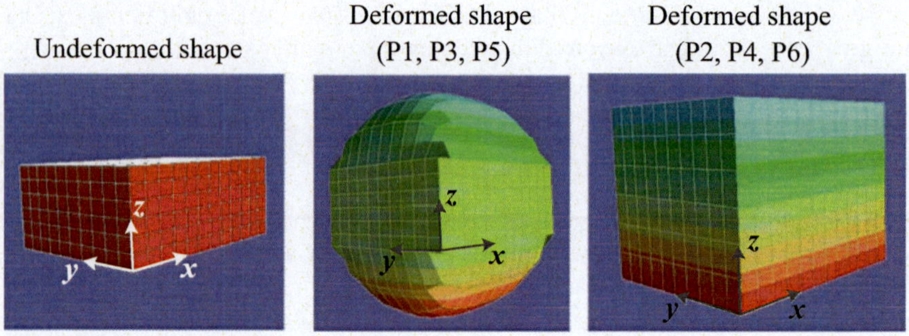

Fig. 25.17 Comparison of the deformed and un-deformed shapes of the panel

only in z direction. Figure 25.17 compares the un-deformed and deformed shape of the panel under two boundary conditions explained in Table 25.1. Based on Fig. 25.17, restricting the side-walls in all three directions leads to expansion of the panel z direction. In this condition, the maximum deformation belongs to the middle point in the upper and lower faces. On the other hand, this figure shows the condition in which the side-walls are only restricted in x and y directions (not z). In this condition the model has a uniform expansion in both +z and −z direction. Considering the symmetry of the panel, in both boundary models the panel shows the same responses along the positive and negative z axis.

In order to investigate the structural responses of the panel, two index points in the center of the panel. The first is on the top surface and the other in the middle of the thickness.

Figure 25.18 shows the progressive failure of the panel model P5 under ASR expansion. As seen, the cracking first starts at the Inc = 109 at the corners of the panel (where is higher stresses is expected). Also, the starting increment (109) corresponds to the previously discontinuity in the stress time histories of the concrete and reinforcement. The cracking is first propagate along the two opposite corner of

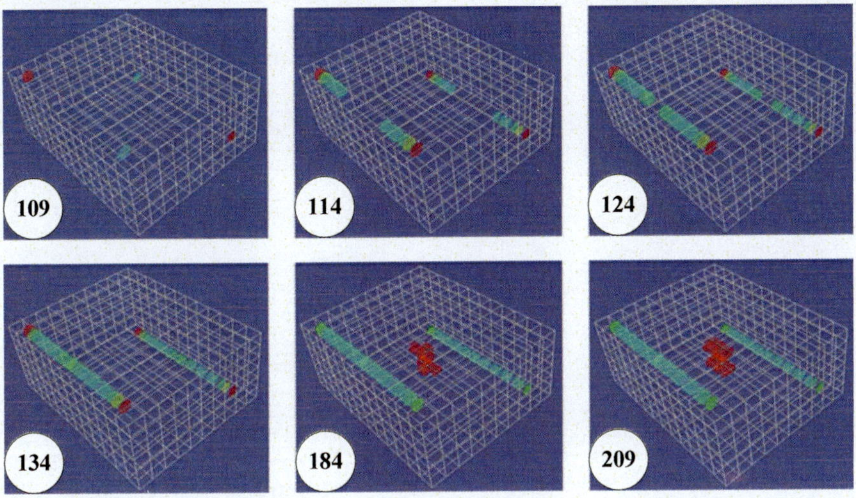

Fig. 25.18 Progressive failure in panel model p5 (x^- : (0, 0, 0); x^+ : (0, 0, 0); y^- : (0, 0, 0); y^+ : (0, 0, 0)) under ASR expansion

the panel (it is symmetry). Another set of cracks appear in the center of the panel (around the Index-2) at Inc = 184. This corresponds to the stress reduction in Index-2 at practically the same increment.

25.11.1 ASR + Shear Load

The first six models are only based on ASR expansion. Those are followed by P7 to P17 based on different combination of linear/nonlinear models and applied incremental displacement load (to impose a shear load). Models P7 to P12 all are based on linear elastic concrete, with and without reinforcement effect (not studied here). P13, P14 ad P15 are based on nonlinear concrete model (smeared crack model), in the presence of reinforcement, and different boundary conditions. P13 is, in fact, the reference model for this group (no ASR expansion is applied). P14 and P15 are different in the boundary conditions applied under the ASR expansion. P14's BC is identical to P5 and the one in P15 is similar to P6.

Figure 25.19 shows the load-displacement curve of three cases. As seen, the load-displacement curve is nonlinear in all cases (nonlinear material assumption). Based on this figure, the initial slope of the P13 is more than P14 and P15, showing that the initial ASR expansion leads to softening of the panel (due to cracking). There is a discontinuity in the capacity curve of the panel P13 at $\Delta = 2.14$ mm. This point corresponds to the initiation of the first set of the cracks in the diagonal form. There is not such a sudden jump or reduction in two other curves (P14 and P15) because they already experienced some cracking before applying the incremental displacement.

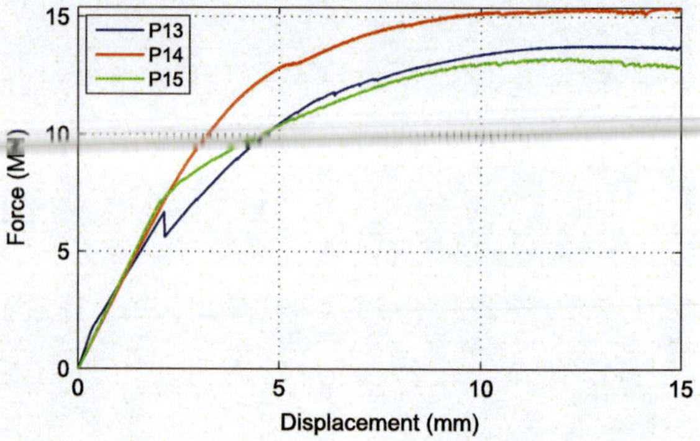

Fig. 25.19 Load-displacement curve for the panel P13, P14 and P15

The capacity curve of the panel P14 is more than the P13, while the capacity curve of the P15 is less than P13. Both the P14 and P15 experience ASR expansion and some cracking before applying the incremental displacement. The differences of the capacity curves can be attributed to the boundary conditions applied in the ASR part. In the panel P14, full confinement of the model under ASR expansion make the panel strength. Based on Fig. 25.18, the panel does not have critical cracking under ASR expansion. Also, most of the cracking is concentrated at the center which experience lower tensile stresses under incremental divorcement. On the other hand, panel P15 was free to expand along its thickness and thus had more cracked elements initially (before applying the incremental displacement). Therefore, this model has a lowest capacity curve among the three cases.

Figure 25.20 shows the progressive failure mode of the panel model P13 under incremental displacement. As seen, the cracking starts at the upper right corner of the panel and then lower left corner. These two points correspond to the high tension area. The cracking proceeds in vertical direction along the walls. Then, a sudden diagonal cracked area appears in the panel which is believed to be the main failure mode. At this time, the crack pattern looks like "N" letter. Further cracking develops around this main path.

Figure 25.21 shows the progressive failure mode of the panel model P14 under incremental displacement. The cracking of the panel at the last increment of the ASR expansion is taken as an initial condition for the incremental displacement. As seen, the failure mode in this model is completely different from P13 (the reference model). Considering initial cracking at the center of panel, the rest of the cracks propagate in the diagonal form in both directions. It makes the crack pattern looks like "X" letter. Further cracking develops around this main path.

Figure 25.22 shows the progressive failure mode of the panel model P15 under incremental displacement. The cracking of the panel at the last increment of the ASR expansion is taken as an initial condition for the incremental displacement. Based on this figure (and having in mind the boundary conditions of the P15 under ASR expansion), it can be seen that the panel experience almost a uniform cracking along its thickness under ASR expansion. This failure mode is completely different from

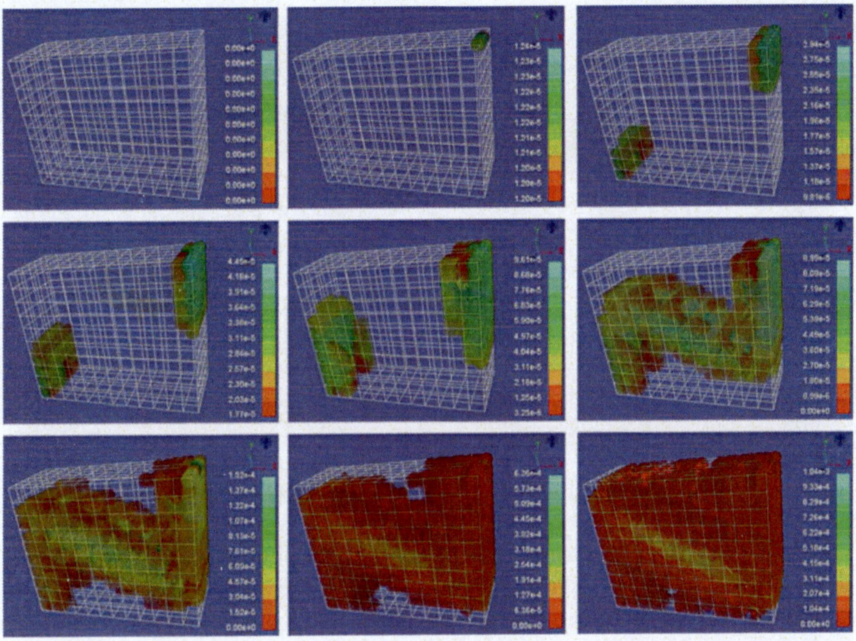

Fig. 25.20 Progressive failure of the panel P13 under incremental displacement

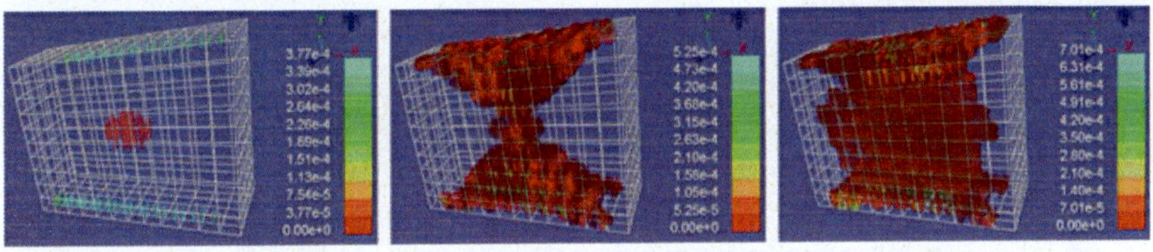

Fig. 25.21 Progressive failure of the panel P14 under incremental displacement

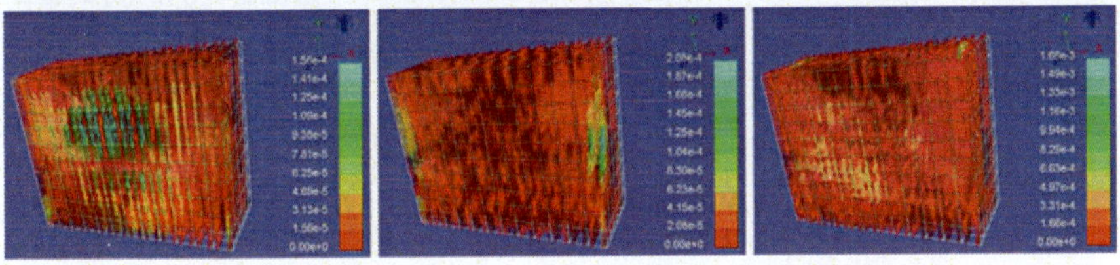

Fig. 25.22 Progressive failure of the panel P15 under incremental displacement

P13 (the reference model) and P14 (the other ASR-affected model). It is not possible to define a specific failure mode in this model, because even under the incremental displacement the model has more or less uniform cracking (or opening of already cracked elements).

25.12 P-11: Nuclear Containment Structure

This last analysis addresses the last benchmark problem, however there are some minor suble differences; (1) we used dynamic intensifying acceleration function (so-called ETAF), while the benchmark is based on static intensifying load; and (2) we used the results of six ETAFs to reduce the uncertainty and dependency to external load while the benchmark is deterministic.

Again, full disclosure: this analysis was also performed within the context of an NRC sponsored research and has been separately published in [16] and [17].

The structure will be first subjected to 40 years of ASR expansion followed by multiple seismic excitation (with or without ASR induced damage), and results will be compared with the response of the NCVS subjected to the same seismic excitations but without prior ASR expansion (Fig. 25.23B, C, and A respectively).

The selected and partially buried NCVS is schematically shown in Fig. 25.24a. Note that only the concrete underneath the soil level will be subjected to ASR (as a result of the high relative humidity likely to be present in the surrounding foundation).

The potential secondary stresses induced by the uplift forces (caused by the eccentricity of the resulting inertial force), Fig. 25.24b, will be mitigated by the insertion of cohesive based joint elements [13] where necessary. A 3D continuum model, Fig. 25.24c is prepared. Reinforcement is modeled as "smeared" by altering the stiffness matrices of those continuum elements they cross. A 0.5% reinforcement was assumed in both directions.

It is assumed that the NCVS operates for 40 years during which it undergoes a relatively mild total expansion of 0.3% uniformly distributed over the "contaminated" zone as an additional internal strain.

Accompanying this expansion are two levels of concrete degradation zero and 30% reduction of E and f_t' after 40 years. The 40 years expansion is simulated in two weeks increments assuming a constant temperature and RH. The external average temperature at the site is estimated to be 11 °C (external face of NCVS), the internal temperature is in turn estimated to be 25 °C. Hence, an average mean yearly

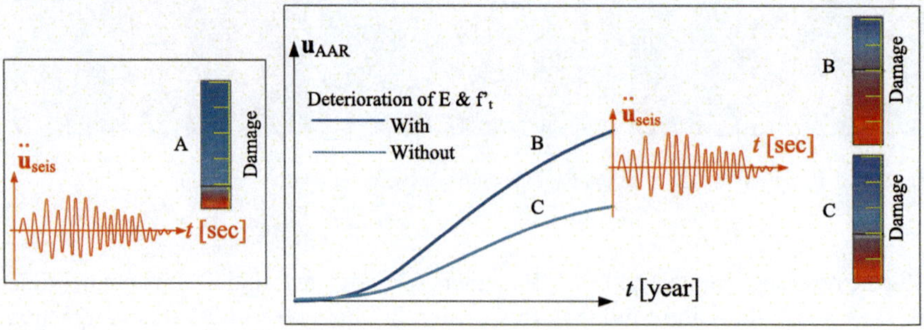

Fig. 25.23 Three scenarios of investigation: A: No ASR; B: ASR with 40% damage; and C: ASR without damage

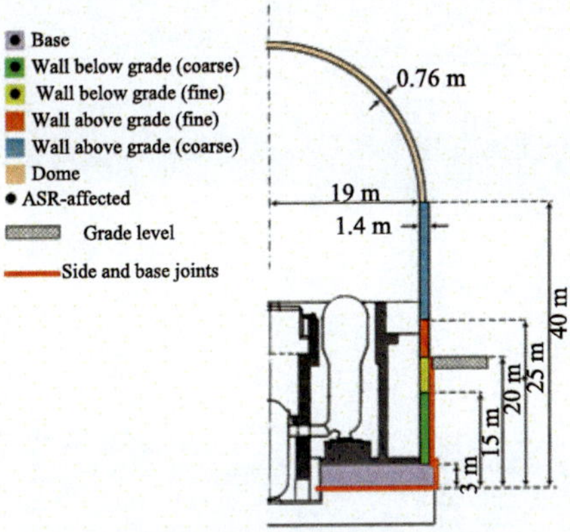

(a) Geometry and dimensions

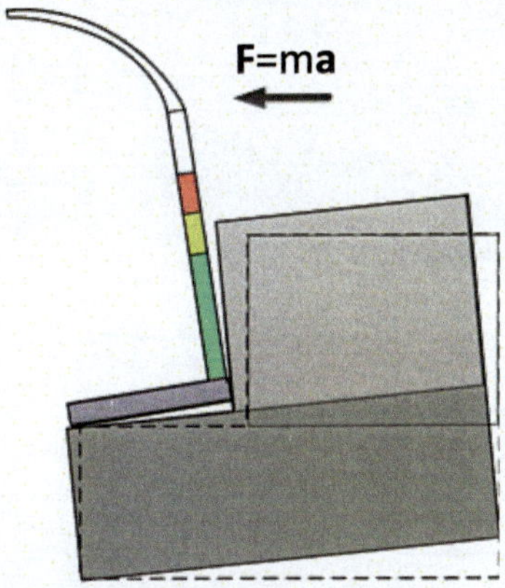

(b) Potential uplift modeled by joint elements

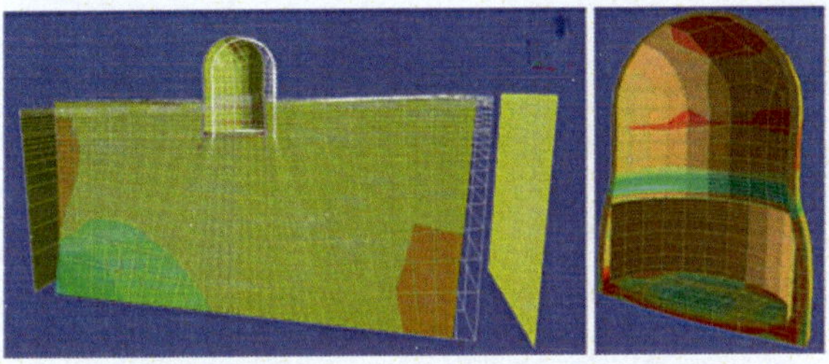

(c) Half 3D model with lateral viscous dampers and container mesh (with ASR expansion)

Fig. 25.24 Geometry, material groups, role of joint elements, and finite element mesh

temperature of $(25 + 11)/2 = 18$ °C is assumed. Note that in a more refined analysis, the temperature distribution across the wall should be considered, and monthly average temperatures should also be refined.

Rock (both the foundation and lateral) is assumed to be linear elastic. Interface joints are placed around and below the NCVS to capture potential uplift of lateral separation of the container from the adjacent rock.

Six different randomly generated ETAFs were used. Schematically, all the ETAFs are analogous as they are all based on a random white noise (Fig. 25.25).

Three sets of analyses were performed: (1) Static + ASR, (2) Static + Dynamic, and (3) Static + ASR + Dynamic (100 for the stochastic ground motions, and three for each of the six ETA).

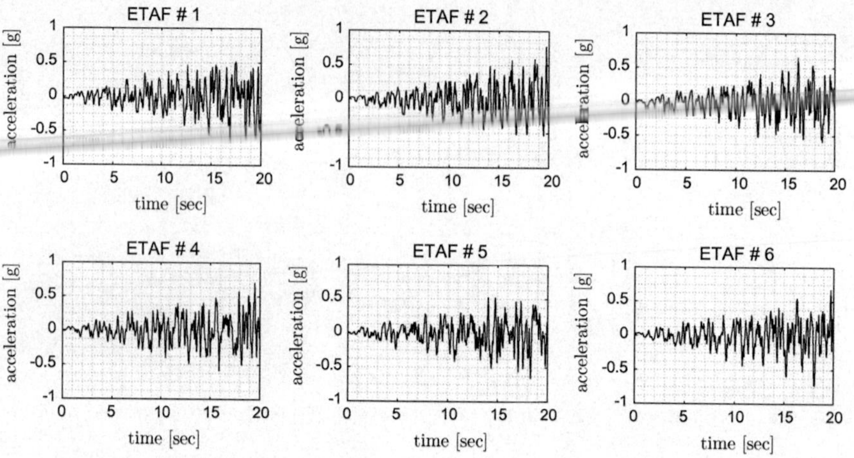

Fig. 25.25 Six adopted ETAF

25.12.1 *Static + ASR Analyses*

In this first analysis, 40 years of ASR in the NCVS is simulated. Figure 25.26a shows swelling of the container along with a closeup on the concrete-rock separation. Clearly ASR expansion interacts with the structure in what may be a priori counter-intuitive: (a) the mat expands in a concave shape due to the structural constraints of the cylindrical vessel, Fig. 25.26b-1; (b) the wall pushes against the adjacent rock, but is constrained by both the base mat and the upper portion of the enclosure not subjected to expansion, Fig. 25.26b-2, causing strong curvature, joint opening, and ensuing stress discontinuities; and (c) sliding of the NCVS itself, Fig. 25.26b-3. Furthermore, the evolution of concrete cracks is shown in Fig. 25.26c. It should be noted that cracking starts at the central region of the mat base and along a ring on the wall next to soil level.

25.12.2 *Impact of ASR on Capacity Curves*

Impact of ASR on the structural response of the NCVS can now be ascertained by comparing "Static + Seismic" with "Static + ASR + Seismic" for displacement and stresses for six different ETA functions. Three sets of simulations are compared: (a) Static + dynamic analysis (Referred to Dyn. in the plots); (b) Static + ASR with degradation of f_t' and E over time + dynamic analysis; and (c) Static + ASR (without material degradation) + dynamic analysis.

Displacements: The absolute value of the (horizontal) displacements corresponding to peaks in (the six) ETAFs is first extracted. The mean of those six ETAFs for

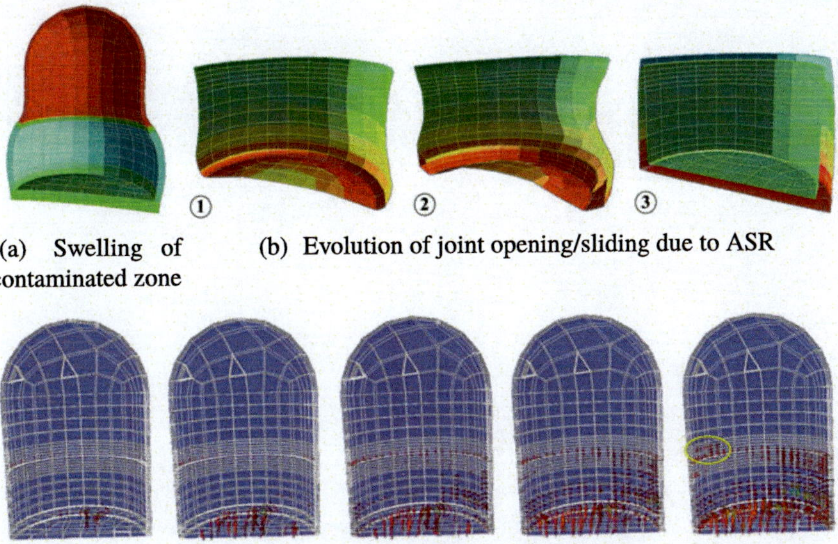

(a) Swelling of contaminated zone

(b) Evolution of joint opening/sliding due to ASR

(c) Internal crack propagation (left to right)

Fig. 25.26 Response of NCVS under static + ASR analysis after 40 years

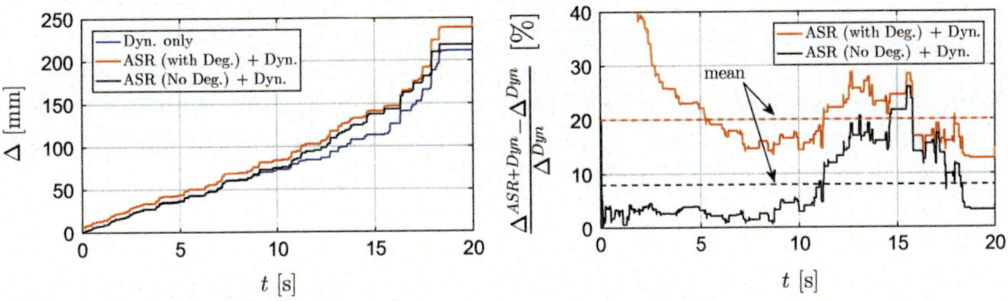

(a) Displacement profile at the top

(b) Displacements normalized and mean differences

Fig. 25.27 ETA-based displacements and the mean differences

each of three assumptions are computed, Fig. 25.27a. These are ramping curves as the dynamic acceleration is indeed defined as a linearly increasing one. To better grasp the impact of ASR, results are normalized with respect to the one without ASR (dynamic only), Fig. 25.27b. The deviations are time-dependent and, as expected, model with ASR degradation is much more impacted than the one without. On average, and for this case study, ASR with degradation results in ~20% change, whereas the one without has ~8% variation with respect to the "Dyn. only" model. If material degradation is ignored (which is an erroneous abstraction) displacements are still lower than those cases without ASR, but greater than ASR with degradation. Note that discrepancy with respect to the case without

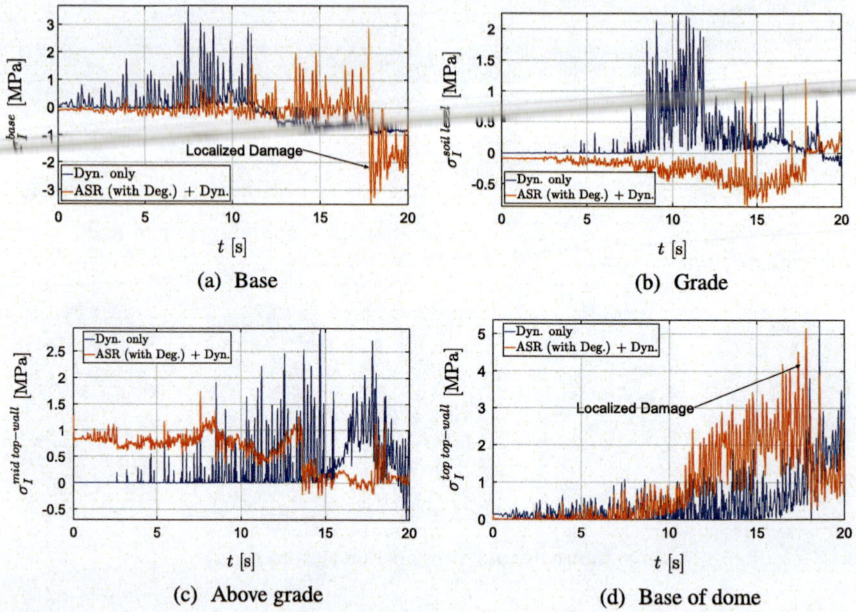

Fig. 25.28 Principal stresses capacity curves

ASR starts at around 9 s (i.e., until this point the ASR had little impact on deformation). The impact of ASR (with and without degradation) is time-dependent due to the complexities of the internal stress states induced by it or resulting from the seismic excitation, Fig. 25.27b.

Stresses: Time history of maximum principal stresses are recorded. The ASR affected responses result in higher stresses than those without degradation, and additional substantial damages will be induced by the ASR (with internal damage accounted for). At the base, Fig. 25.28a, maximum principal stresses are positive (cracking) and attenuate with time. Stresses are at first low when ASR dominates, but then suddenly increase with a localized damage at time $\simeq$17 s. At the grade elevation, Fig. 25.28b, stresses are much higher without ASR, and then gradually decrease with no indication of failure. Note that the tensile strength is 3.1 MPa. On the other hand, in the presence of prior ASR expansion, the stresses are negative, and a sudden localized failure appears at $t = 14$ s. For a point above grade, Fig. 25.28c, stresses are higher in the absence of ASR and there is indication of a localized failure at $t = 15$ s. In the presence of ASR, the failure is delayed to about 17 s. Finally, at the base of the dome, Fig. 25.28d, the ASR stresses are substantially higher than without and localized failure occurs around 17 s. For this case, ASR has reduced the stresses at the base, but substantially increased them at the base of the dome. Indeed, stress attenuation with time is the direct result of a nonlinear analysis where upon cracking there is a substantial stress redistribution resulting in localized stress reduction.

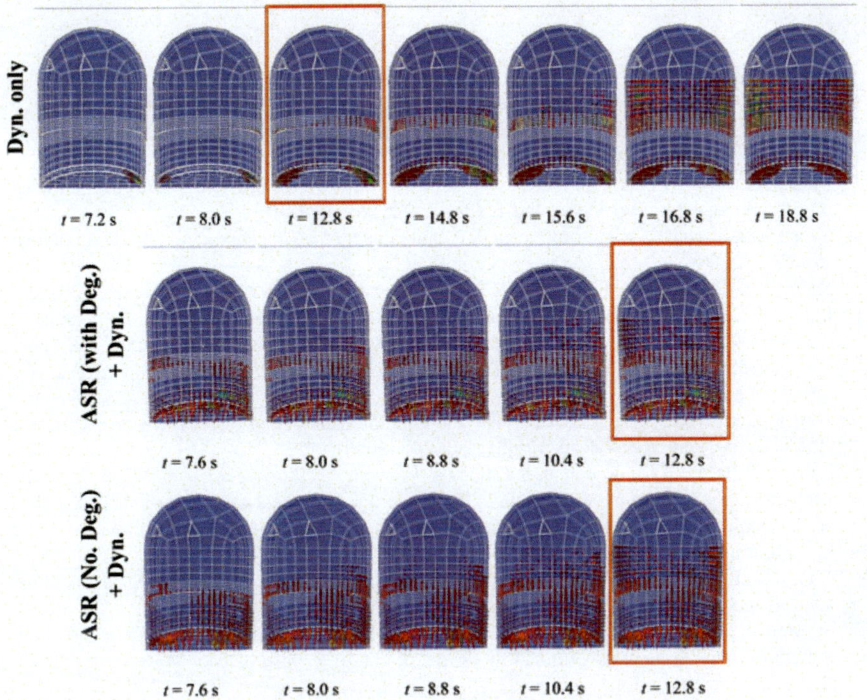

Fig. 25.29 Crack profile from a sample ETA simulation at identical time steps

Cracking: of the structure is shown Fig. 25.29 at different times. In general, the crack pattern of ASR affected models are different and the previous observations are qualitatively confirmed by the crack profiles. Indeed, the damage index (DI), ratio of the cracked sections to the total area, is highest when ASR (with damage) preceded the seismic excitation. The ASR has a much higher impact of that portion of the NCVS below grade than above (where no ASR is modeled).

The sound NCVS experiences the major cracks at the soil level at about 12.8 s, while at the corresponding time the ASR-affect NCVS had already some major cracks though all the top-wall. Results of ETA analysis prove that "endurance" of the NCVS is reduced when it is subjected to initial ASR.

Acknowledgements The assistance of Mr. Antoine Tixier in the analyses of P1-P8 is gratefully acknowledged.

References

1. Saouma, V., Cervenka, J., Reich, R.: Merlin Finite Element User's Manual (2010)
2. Saouma, V., Hariri-Ardebili, M.: Merlin Validation & Application of AAR Problems. Technical report. University of Colorado, Boulder (2019)
3. Bažant, Z., Zi, G., Meyer, C.: Fracture mechanics of AAR in concretes with waste glass particles of different sizes. J. Eng. Mech. ASCE **126**(3), 226–232 (2000)
4. Saouma, V., Perotti, L.: Constitutive model for alkali aggregate reactions. ACI Mater. J. **103**(3), 194–202 (2006)
5. Saouma, V.: Numerical Modeling of Alkali Aggregate Reaction, 320 pp. CRC Press (2013)
6. Larive, C.: Apports Combinés de l'Experimentation et de la Modélisation á la Comprehension del'Alcali-Réaction et de ses Effets Mécaniques. Ph.D. thesis. Laboratoire Central des Ponts et Chaussées, Paris (1998). https://hal.inria.fr/docs/00/52/06/76/PDF/1997TH_LARIVE_C_NS20683.pdf
7. Ulm, F., Coussy, O., Li, K., Larive, C.: Thermo-chemo-mechanics of ASR expansion in concrete structures. ASCE J. Eng. Mech. **126**(3), 233–242 (2000)
8. Scrivener: Personal Communication (2005)
9. Multon, S.: Evaluation expérimentale et théorique des effets mécaniques de l'alcali-réaction sur des structures modèles. Technical report. Etudes et recherches des Laboratoires des ponts et chaussées, Série Ouvrages d'art OA46 (2004)
10. Capra, B., Bournazel, J.: Modeling of induced mechanical effects of alkali-aggregate reactions. Cem. Concr. Res. **28**(2), 251–260 (1998)
11. Wittmann, F., Rokugo, K., Brühwiler, E., Mihashi, H., Simonin, P.: Fracture energy and strain softening of concrete as determined by means of compact tension specimens. Mater. Struct. **21**, 21–32 (1988)
12. Cervenka, J., Cervenka, V.: Three dimensional combined fracture-plastic material model for concrete. In: 5th U.S. National Congress on Computational Mechanics, Boulder, CO (1999)
13. Cervenka, J., Chandra, J., Saouma, V.: Mixed mode fracture of cementitious bimaterial interfaces; Part II: Numerical simulation. Eng. Fract. Mech. **60**(1), 95–107 (1998). https://doi.org/10.1016/S0013-7944(97)00094-5
14. Mohammed, T., Hamada, H., Yamaji, T.: Alkali-silica reaction-induced strains over concrete surface and steel bars in concrete. ACI Mater. **100**(2), 133–142 (2003)
15. Hariri-Ardebili, M.A., Saouma, V., LePape, Y.: Independent Modeling of the Alkali-Silica Reaction: Mock-up Test Block. Technical report. ORNL/TM-2016/537. Oak Ridge National Laboratory, Oak Ridge, TN 37831 (2016)
16. Saouma, V.: Probabilistic Based Nonlinear Seismic Analysis of Nuclear Containment Vessel Structures with AAR. Technical report. Final Report to NRC, Grant No. NRC-HQ-60-14-G-0010, Task 3-B. University of Colorado, Boulder (2017)
17. Saouma, V.E., Hariri-Ardebili, M.: Seismic capacity and fragility analysis of an ASR-affected nuclear containment vessel structure. Nucl. Eng. Des. **346**, 140–156 (2019)

Chapter 26
Benchmark Study Results: University of Tokyo

Yuya Takahashi

26.1 P0: Basic Principles of the Model and Its Implementation

A multi-scale chemo-hygral computational system (DuCOM-COM3, [1, 2]) has been developed and used in the simulations. Figure 26.1 shows the summary of this system and it conducts a three-dimensional multi-scale analysis of structural concrete and also can consider recently the multi-ionic equilibrium [3]. A model of ASR and the mechanistic actions that accompany multi-directional cracking [4, 5] were developed based on the poro-mechanical scheme of the solid-liquid two-phase interaction model and nonorthogonal crack-to-crack interaction modeling [6] Temperature, relative humidity (RH), water content, potassium ions and sodium ions in pore solution can be linked with the reaction modeling of silica gels.

Multi-scale poro-mechanics has been developed by Takahashi et al. [7]. Here, the ASR gel is treated as a medium filling cracks and micro-voids and migrating over the volume through voids and cracks. The point of formulation is to combine the kinematics of pore media with the solid skeleton deformation and fracture.

Figure 26.2 shows the computational constitutive material's modeling of ASR gel generation and its contribution to expansion. Based on the chemical equations for ASR, the rate of ASR is formulated as a function of the alkali concentration, updated free water in the unit referential volume, and the reactive aggregate's content, as expressed by Eq. (a).

The control coefficient of the alkali silica reaction rate denoted by k in Eq. (a) is arranged to identify the characteristics of each aggregate with different minerals through inverse sensitivity analyses, because the reactivity can change greatly with the aggregate type as well as the aggregate size (or specific surface area). Thus,

Y. Takahashi (✉)
Department of Civil Engineering, School of Engineering, The University of Tokyo, 7-3-1 Hongo, Bunkyo-ku, Tokyo 113-8656, Japan
e-mail: takahashi@concrete.t.u-tokyo.ac.jp

© RILEM 2021

V. E. Saouma (ed.), *Diagnosis & Prognosis of AAR Affected Structures*, RILEM State-of-the-Art Reports 31,
https://doi.org/10.1007/978-3-030-44014-5_26

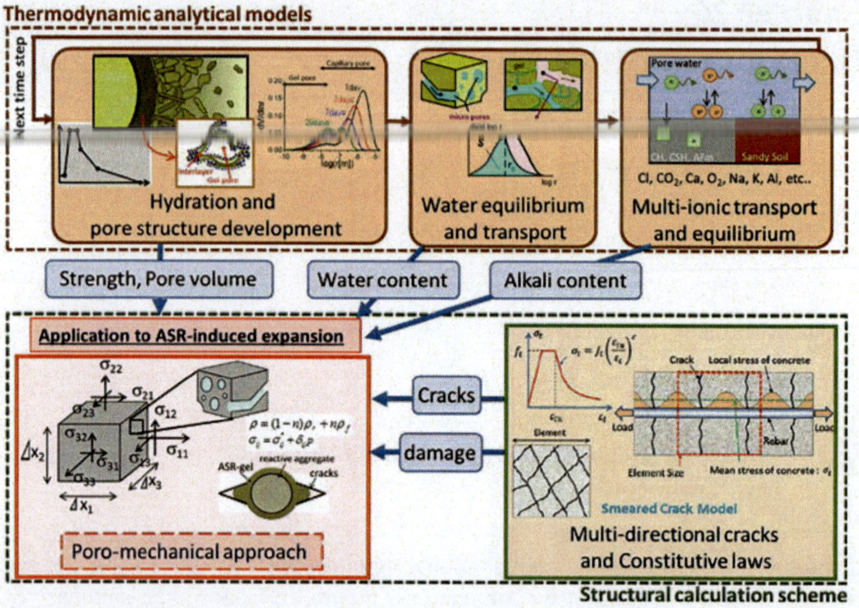

Fig. 26.1 Multi-scale chemo-hygral computational scheme

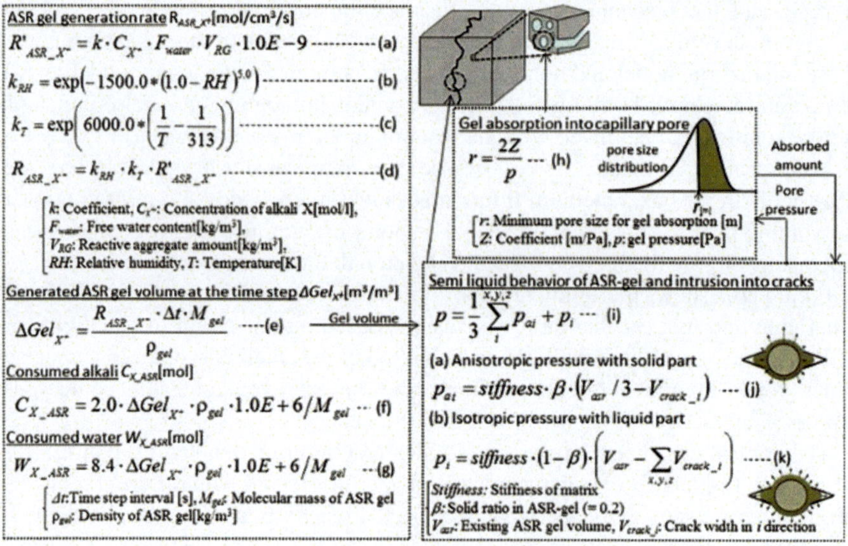

Fig. 26.2 Calculation scheme for ASR gel generation and stress formations

this parameter represents the reactivity of aggregate phase. The effects of the RH is formulated in Eq. (b) based upon the experimental facts that the ASR almost terminates below 80% of the relative humidity. The temperature dependency is also taken into account in Eq. (c) based upon the conventional Alenius principle. This formula indicates that the reaction is descending according to drying specified by RH and is accelerated under elevated temperature.

The generated ASR gel volume is calculated by using the volume-compatibility Eq. (e) with reference to $X_2Si_2O_5(H_2O)_{8.4}$ as the ASR gel molecular formula for each alkali, X (=Na or K) and 1.6 kg/m^3 as the averaged density of the created gels denoted by ρ_{gel}. This linear formulae linked with Eq. (a) are obvious in consideration of the mass balance and the definition of time interval denoted by Δt. The consumed alkali and water are also calculated in terms of the mass conservation. The mass change rate expressed by Eq. (f) and Eq. (g) is included in the sink term of the discretized mass conservation formulae and computationally solved step by step.

Central to the theorem is that the water and alkali contents of the pore solution, which are control parameters for computing the reaction rate by Eq. (d), are set as global variables for both the thermodynamic analytical system (DuCOM; [2]) and the 3D meso-scale structural analytical system (COM3; [6]). These global schemes are to search nonlinear finite element solutions to satisfy the equilibrium of mass and the momentum conservation. On the basis of generated gel volume, the stress formation can be automatically calculated by simultaneously solving the stress equilibrium and the deformational compatibility. Here, some parts of the created gels can contribute to stress formation, but the other parts do not. The silica gel is partially absorbed into the capillary pores and the amount of absorbed gel can be calculated by using Kelvin's Eq. (h) in Fig. 26.2, which is a function of the gel pressure and the surface tension of the pore solution. The resultant remaining gel may substantially contribute to the stress formation. Regarding the gel-oriented internal pressure, the solid-liquid characteristics of ASR gels are considered.

To express the state in which a solid and liquid coexist, parameter β is introduced, to indicate the ratio of the solidified phase volume to those of the total ASR gel. Under certain stress states, the solidified part of the produced ASR gel can expand around aggregate particles uniformly even under an anisotropic pressure distribution and creates the partial stresses given by Eq. (j), in which $V_{asr}/3$ indicates the term of free solid expansion strain in all directions and $V_{crack,i}$ is the smeared crack strain in i-th direction of the Cartesian coordinate. Then, the effective stress component is associated with the deviation of the two strain components.

The liquefied part of gels expands without shear rigidity under the isotropic pressure expressed by Eq. (k), in which $(V_{asr} - \Sigma V_{crack,i})$ means the deviation of the ASR volume expansion and the increased volume of concrete solid skeleton. Then, this deviation may create the hydro-static pressure. Parameter β is tentatively assumed to be 0.2 for the first assumption with a sensitivity analysis and this parameter governs the anisotropic expansions under 3D confined conditions. The volumetric stiffness of the ASR gel is tentatively supposed to be the same as that of the condensed water.

Other details of the modeling can be referred to the previous paper by Takahashi et al. [7]. Based on these developed models, we conducted P1, P2, P5 and P6 in the benchmark problems.

26.2 P1: Constitutive Model Calibration

Figure 26.3a and b shows the stress-strain relationships of concrete cylinder without and with 0.5% of AAR expansion, respectively. 38.4 MPa of compression strength, 3.5 MPa of tension strength, and 37.3 GPa of elastic modulus are inputted. Imposed strain histograms designated in benchmark analyses are inputted in each case. 0.5% of AAR expansion is quite large and cause much cracks, which leads to the obvious decrement in both the stiffness and the peak stresses. Here, the tension transfer of ASR-gel inside cracks is not considered and tension stress is much lower for tension strain.

26.3 P2: Drying and Shrinkage for Non-reactive Concrete

Figure 26.4a shows the longitudinal strains of hardened concrete under different water supply conditions. Figure 26.4b shows the referential experimental results. In the simulation, coupled material-mechanical simulation is conducted with DuCOM-COM3. 50% of W/C is set and 28-days sealed curing before the exposure is calculated first because the pore structures and the water contents at 28 days (before drying or other exposures) are important to simulate the succeeding strain progresses. The strains under different water supply conditions were simulated based on the moisture equilibrium and transport model [8]. The length at 28 days are considered as zero and the following strains are plotted in the figures.

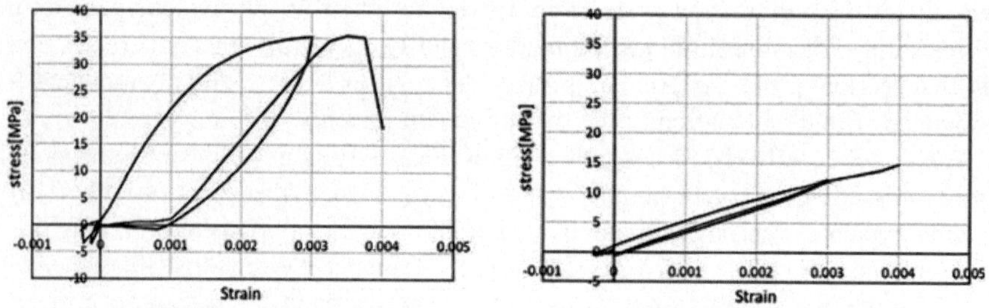

(a) Without AAR expansion

(b) With 0.5% AAR expansion

Fig. 26.3 The stress-strain relationships

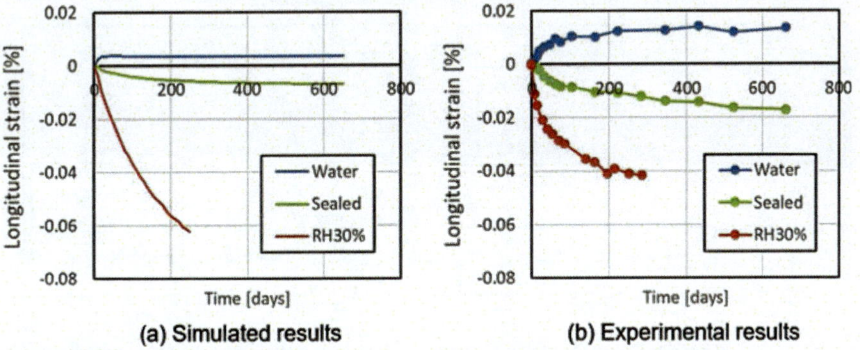

Fig. 26.4 The simulated and experimented strain under various water supply conditions

Fig. 26.5 The strain prediction under the variation of external RH

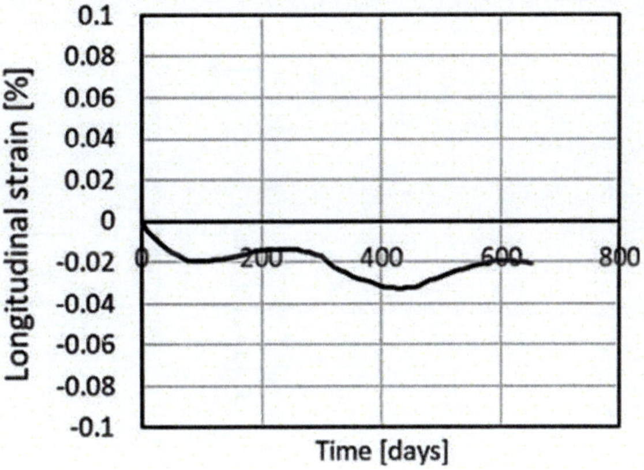

Figure 26.5 shows the predicted results of longitudinal strains from 28 days under the variation of external RH shown in the following equation.

$$RH\ (Week) = \frac{RH_{\max} - RH_{\min}}{2} \sin\left(2\pi \frac{t-16}{52}\right) + \frac{RH_{\max} - RH_{\min}}{2} \quad (26.1)$$

where $RH_{\max}$ and $RH_{\min}$ are equal to 95% and 60% respectively.

26.4 P5: Free AAR Expansion: Effect of RH

Figure 26.6a shows the longitudinal strains of AAR concrete under different water supply conditions. Figure 26.6b shows the referential experimental results. 50% of W/C is set and 28-days sealed curing before the exposure is calculated then different water supply conditions were applied. Sensitivity analyses for fully wet conditions

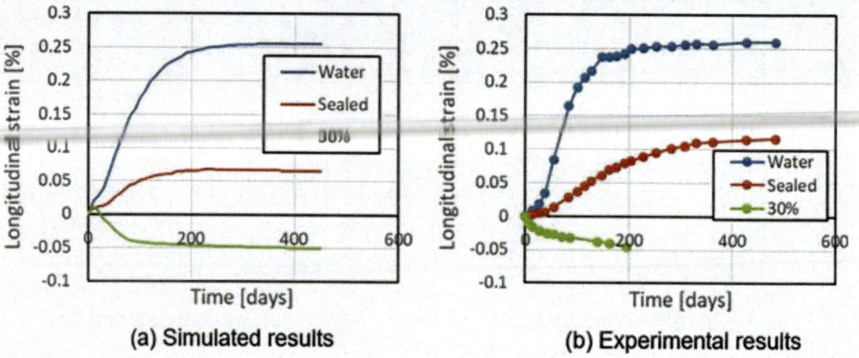

Fig. 26.6 The simulated and experimented strain pf AAR concrete under various water supply conditions

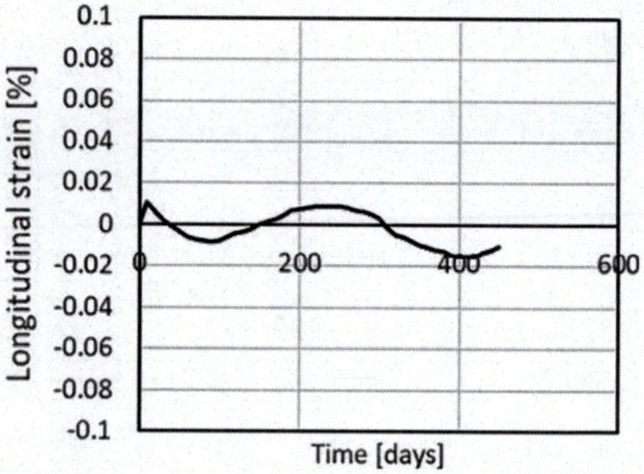

Fig. 26.7 The strain prediction of AAR concrete under the variation of external RH

are conducted to fix the parameter *k*, which was used for different water supply conditions. With the model to consider RH effects on ASR progress (Eq. (b) in Fig. 26.3), different strain progress in the experiment can be captured.

Figure 26.7 shows the predicted results of longitudinal strains of AAR concrete from 28 days under variation of external RH shown in Eq. 26.1.

26.5 P6: AAR Expansion; Effect of Confinement

Figure 26.8 shows the simulated and referential experimental results of expansive strains of cylindrical concrete for no vertical stress and no confinement case. Figure 26.8a is the simulation results and Fig. 26.8b is the experimental results. And Fig. 26.9 shows the strains for 10 MPa of vertical stress and no confinement case. The anisotropy in expansions is calculated considering the semi-liquid behavior of ASR-

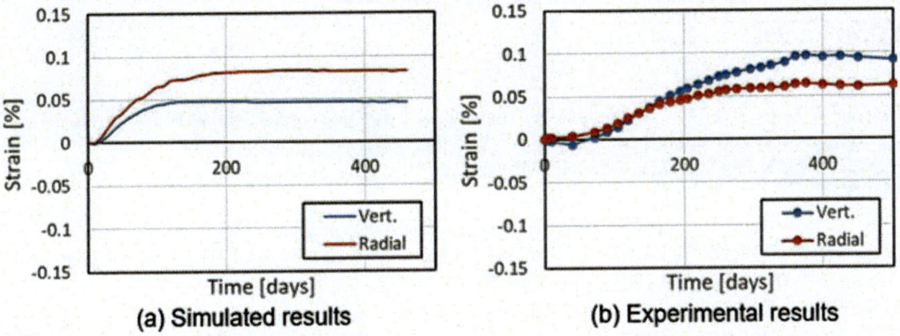

Fig. 26.8 Vertical and radial strains for no vertical stress and no confinement case

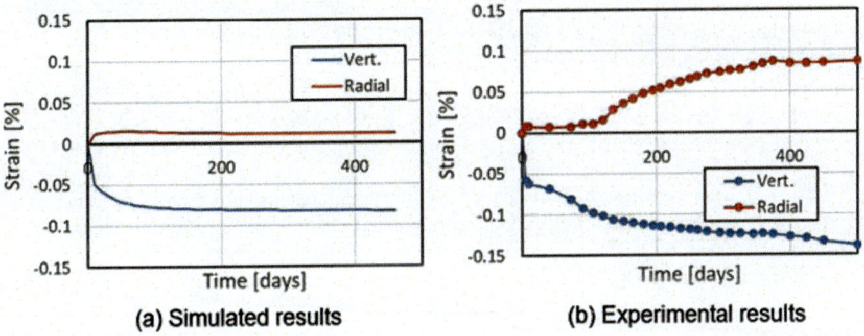

Fig. 26.9 Vertical and radial strains for 10 MPa vertical stress and no confinement case

Fig. 26.10 Vertical and radial strains under designated stress histories

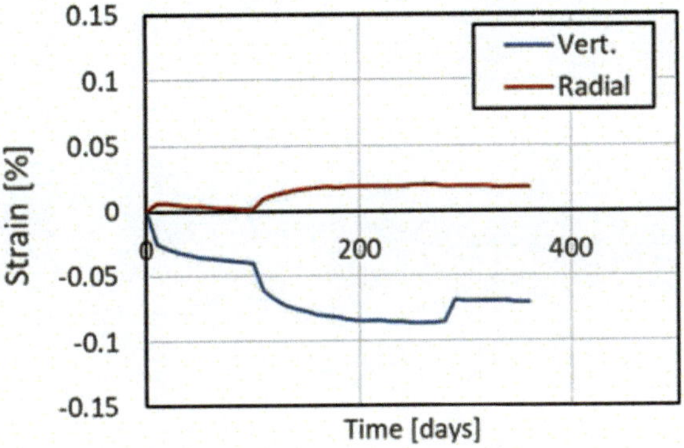

gel (Eq. (i) in Fig. 26.3) and due to the value 0.2 of parameter β, similar anisotropic expansion can be reproduced in confined case.

Figure 26.10 shows the predicted simulation results of vertical and radian expansion strain under designated stress histories in benchmark analyses.

References

1. Maekawa, K., Ishida, T.: Modeling of structural performances under coupled environmental and weather actions. Mater. Struct. **35**(10), 591–602 (2002)
2. Maekawa, K., Ishida, T., Kishi, T.: Multi-scale Modeling of Structural Concrete. CRC Press (2008)
3. Elakneswaran, Y., Ishida, T.: Development and verification of an integrated physicochemical and geochemical modelling framework for performance assessment of cement-based materials. J. Adv. Concr. Technol. **12**(4), 111–126 (2014)
4. Maekawa, K., Fujiyama, C.: Rate-dependent model of structural concrete incorporating kinematics of ambient water subjected to high-cycle loads. In: Engineering Computations. Emerald Group Publishing Limited (2013)
5. Takahashi, Y., Shibata, K., Maekawa, K.: Chemo-hygral modeling and structural behaviors of reinforced concrete damaged by alkali silica reaction. In: Proceedings of Asian Concrete Federation-2014, pp. 1274–1281 (2014)
6. Maekawa, K., Okamura, H., Pimanmas, A.: Non-linear Mechanics of Reinforced Concrete. CRC Press (2003)
7. Takahashi, Y., Ogawa, S., Tanaka, Y., Maekawa, K.: Scale-dependent ASR expansion of concrete and its prediction coupled with silica gel generation and migration. J. Adv. Concr. Technol. **14**(8), 444–463 (2016)
8. Ishida, T., Maekawa, K., Kishi, T.: Enhanced modeling of moisture equilibrium and transport in cementitious materials under arbitrary temperature and relative humidity history. Cem. Concr. Res. **37**(4), 565–578 (2007)

Appendix A
From Laboratory to Field

Barbara Lothenbach and Andreas Leemann

A.1 Effect of Temperature on Hydrated Cements and Pore Solution

The increased temperatures used in ASR testing accelerate not only the formation of ASR products but affect also the composition of the hydrates and of the pore solution. While in a Portland cement hydrated at ambient temperature the presence of C-S-H, portlandite, ettringite and monocarbonate can be expected, at temperatures above 50 °C, ettringite and monocarbonate destabilise to monosulfate and calcite as shown in Fig. A.1 resulting in an increased porosity, as well as in an increase of sulfate and a decrease of aluminium concentrations in the pore solutions [1–3]. In Portland cement pastes, the alkali and hydroxide concentrations are not significantly affected by temperature (although pH values measured at high temperature will be lower—even at the same hydroxide concentrations—due to the strong increase of the ionic product of water, K_W, with temperature).

At 60 and 80 °C not only the destabilisation of ettringite and monocarbonate can be expected, but in addition an enhanced formation of siliceous hydrogarnet leading to a destabilisation of monosulfate. The formation of aluminium containing siliceous hydrogarnet is limited at ambient temperatures due to kinetic hindrance [4, 5]. At increased temperatures, however, hydrogarnet formation and the destabilisation of monosulfate was observed in Portland cements [1, 5] as well as in fly ash-blended cements [6] as shown in Fig. A.2.

B. Lothenbach · A. Leemann (✉)
Empa, Swiss Federal Laboratories for Materials Testing and Research,
Laboratory for Concrete and Construction Chemistry, Überlandstrasse 129,
8600 Dübendorf, Switzerland
e-mail: andreas.leemann@empa.ch

B. Lothenbach
e-mail: barbara.lothenbach@empa.ch

© RILEM 2021

V. E. Saouma (ed.), *Diagnosis & Prognosis of AAR Affected Structures*,
RILEM State-of-the-Art Reports 31,
https://doi.org/10.1007/978-3-030-44014-5

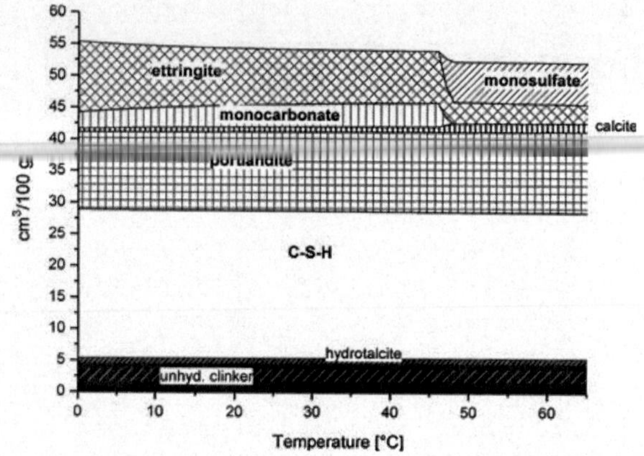

Fig. A.1 Calculated volume of solids in a hydrated cement paste as a function of temperature. Adapted from [2]

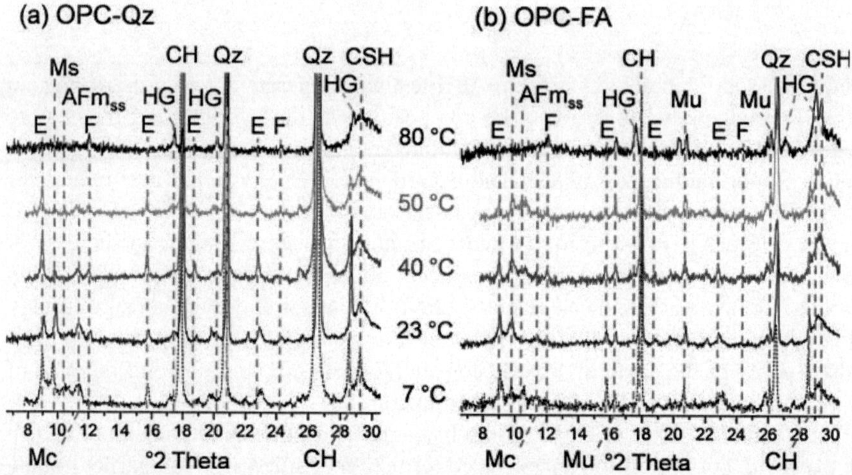

Fig. A.2 Comparison of XRD data of **a** OPC-Qz and **b** OPC-FA after 180 days of hydration at 7, 23, 40, 50 and 80 °C. *E* ettringite, *MS* monosulphate, *AFmss* solid solution of AFm phases, *CH* portlandite, *HG* siliceous hydrogarnet, *F* ferrite, *Mu* mullite, *Qz* quartz. From [6]

In blended cements, higher temperatures lead to a higher degree of fly ash or slag reaction [6, 7], resulting in a decrease of Ca/Si in C-S-H and in an increased uptake of aluminium in the C-S-H. Alkali concentrations at early times can be higher due to the increased reaction of fly ash at higher temperature. In the long term, however, alkali concentrations tend to be lower [6] as more C-S-H is present and as low Ca/Si C-S-H is able to bind more alkali ions [8, 9]. Temperature was found to increase sulfate and silicate concentrations and lower calcium concentration in Portland cement-ly ash blends as shown in Fig. A.3.

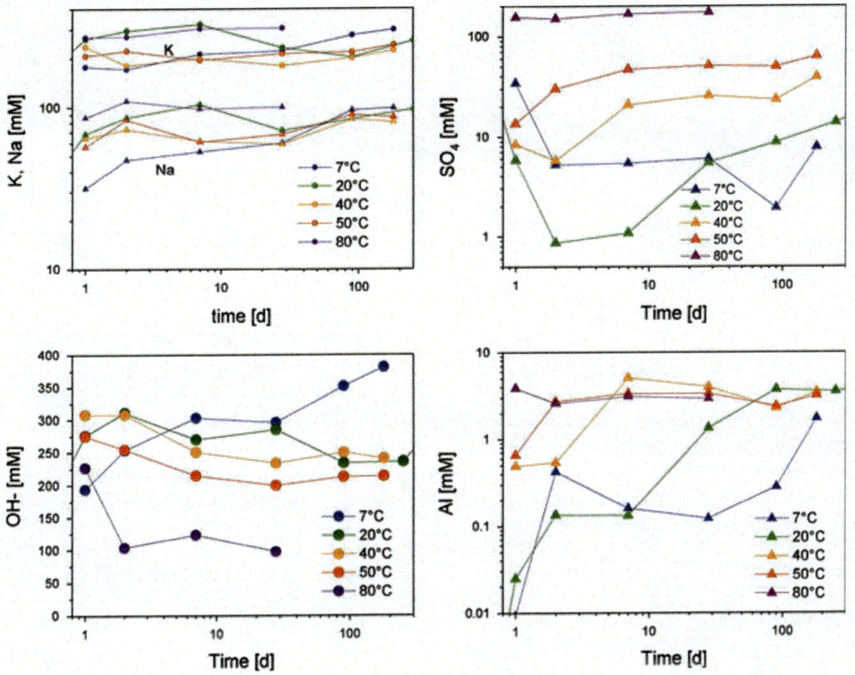

Fig. A.3 Effect of temperature on measured concentrations in and OPC-FA cement. Data from [6]

For concrete exposed to 38 and 60 °C during ASR tests similar observations have been made as for blended cements: after longer reaction times a decrease of alkali and hydroxide concentrations at higher temperature has been observed due to the increased reaction of the aggregate at 60 °C [10, 11].

A.2 Effect of Temperature on Silica, Quartz and Feldspar Solubility

Elevated temperatures will affect not only the composition of the hydrate assemblage and of the pore solution in hydrated cements, increase the reactivity of the aggregate, but will influence also the solubility of silica-based-aggregates. The solubility of quartz and amorphous silica [12] increases by a factor 3 to 4 from 20 to 60 °C as also shown in Fig. A.4. Also the solubility of alkali-feldspars increases strongly with temperature; 4 times higher silicon concentrations are expected at 80 °C than at 20 °C in equilibrium with K-feldspar or albite. However, as increased temperature affects both the solubility of silicates in the aggregates and of siliceous supplementary cementitious materials (SCM), the relative effect on ASR-induced at ambient and elevated temperatures cannot be assessed. Similar drastic changes in solubility could

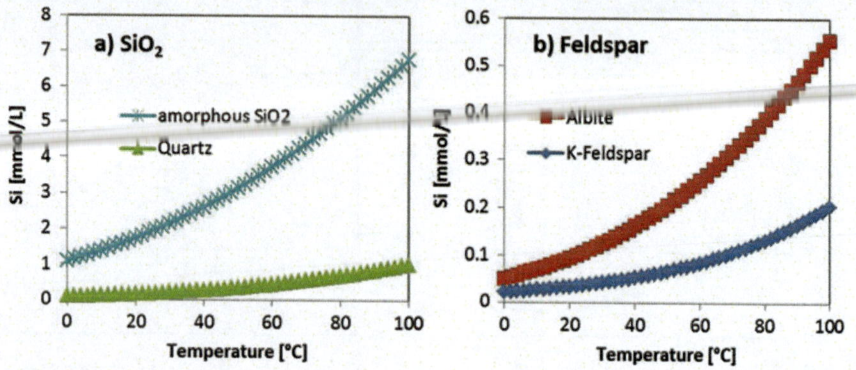

Fig. A.4 Effect of temperature on the solubility of **a** quartz and amorphous silica, and **b** K-feldspar and albite (for clarity only silicon concentrations are shown). Calculated with the thermodynamic calculations were carried out using the Gibbs free energy minimization software GEMS 3.5 [13]. The general thermodynamic database [14, 15] was expanded with thermodynamic data from the SUPCRT database for albite and K-feldspar [16]

also occur for the ASR products. Based on our limited knowledge on ASR products it is presently not possible to assess whether the ASR products formed at increased temperature will have the same properties as those formed at 20 °C or even lower temperatures.

A.3 Effect of Temperature on Expansion

The previous section provides information about the effect of temperature both on hydrated cement including pore solution and on SiO_2 solubility. In order to establish a link to experimentally determined expansion of mortar and concrete with reactive aggregates, a short overview of data on accelerated tests and concrete exposed to natural conditions published in literature is given. Details on the specific tests and mix designs can be found in the references. In has to be pointed out that expansion limits of specific tests only provide in binary answer with "passed" or "not passed" for a given mortar or concrete. Therefore, the determined expansion values should be taken into account as well in the following comparisons, independently of their classification based on the limit values.

A comparison between the microbar test with an autoclave treatment at 150 °C and the concrete prism test (CPT) at 60 °C shows a negative correlation, Fig. A.5. It has been demonstrated in [17] that the aggregates dissolving in the microbar test and the relative extent of mineral dissolution differ both from the CPT and from structures produced with identical aggregates as used in both tests. These data are further confirmed by an analysis of a large data set spanning several years of microbar and concrete prism testing [18], which clearly indicates that the reliability of the

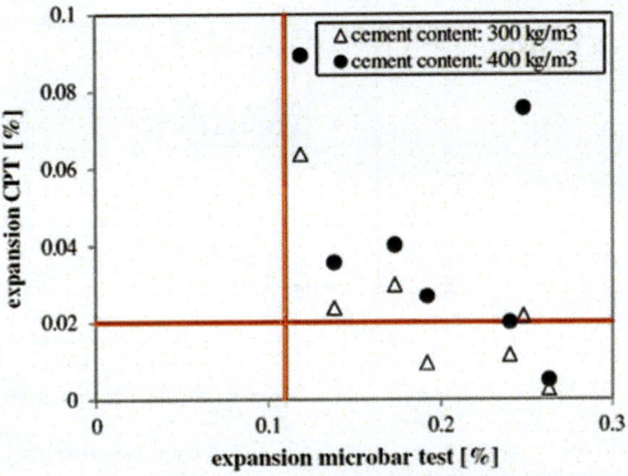

Fig. A.5 Expansion in the CPT test (60 °C) as a function of expansion in the microbar test using seven different aggregates (autoclave treatment at 150 °C) [19]. The red lines indicate the limit values of expansion

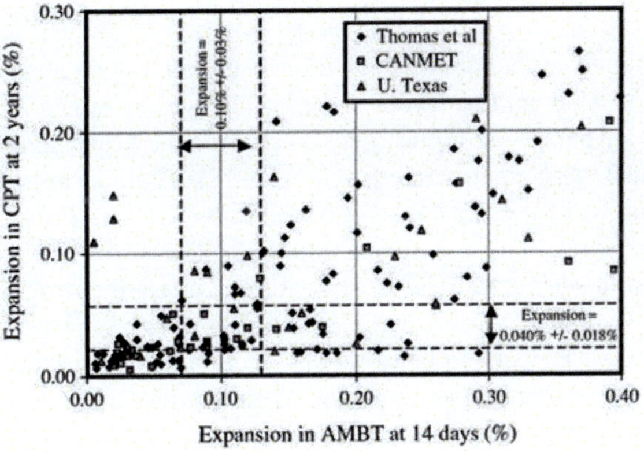

Fig. A.6 Comparison between expansion at 2 years in CPT (38 °C) and expansion at 14 days in the AMBT [20]. The dotted lines indicate the limit values of expansion including the range of values defined as "inconclusive"

microbar test to identify reactive aggregates and to assess the degree of reactivity of aggregates is poor.

The accelerated mortar bar test (AMBT) is conducted at 80 °C incorporating an immersion of the samples in 1 M NaOH. The expansion of the mortar bars does not correlate well with the expansion determined in the CPT at 38 °C using identical aggregates, Figs. A.6 and A.9. In particular, a fair number of aggregates above the limit value of expansion in the AMBT cause no expansion above the limit value in the CPT. The opposite case can occur as well in a few cases. Microstructural analysis leads to similar findings as in the case of the microbar test: the types of aggregates dissolving and the amount of dissolved minerals differ from the CPT and concrete structures (Leemann, unpublished data).

A compilation of data by Thomas, Fournier, Folliard, Ideker, and Shehata [20] indicates a reasonably good correlation between the expansion in the CPT at 38 and 60 °C, Fig. A.7. Still, it has to be noted that some types of aggregates show a differing behaviour in the CPT at 38 and 60 °C. As an example, the CPT at 60 °C is better

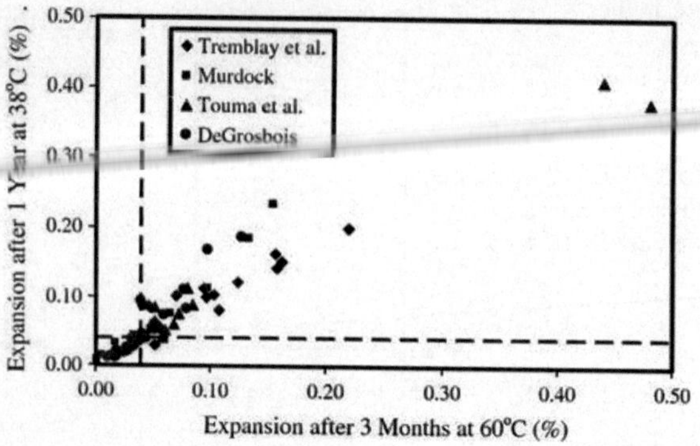

Fig. A.7 Comparison of expansions in the concrete prism test at 38 and 60° for concrete without preventive measures [20]. The dotted lines indicate the limit values of expansion

suited to match the behaviour of granite aggregates in the field, while the CPT at 38 °C should be favoured for basalt aggregates [21].

There are several studies comparing the expansion obtained with accelerated CPT and exposed concrete blocks [22–28]. There seems to be a reasonably good correlation of the expansion obtained in the CPT at 38 °C and the exposed blocks using concrete with measures against ASR, Fig. A.8. In the case of concrete produced with low or high alkali cement, the correlation differs depending on the CPT used, Fig. A.9.

A comparison between the results in CPT and the degree of ASR developing in structures is challenging due to the scarcity of data. ASR test data on concrete used for a given structure later affected by ASR often do not exist or are not available anymore. In order to bypass this problem the concrete mix design and aggregate source of several ASR-affected structures were identified by analysing concrete cores [19]. Afterwards, identical aggregates and cements with comparable composition were used to produce concrete and conduct the CPT at 60 °C. The expansion shows a reasonable correlation with the crack-index determined in components of the structures with the most pronounced damage, Fig. A.10.

A.4 Summary

An increase of temperature has a significant impact on the products of cement hydration, the degree of reaction of mineral additions like fly ash or slag, the composition of pore solution and the solubility of the minerals present in aggregates.

The data presented show that accelerated test conducted at ≥80 °C may at best be used to identify potentially reactive aggregates but are unsuitable to assess the magnitude of expansion reached in the CPT or in concrete structures. The comparison between the expansion in the concrete prim tests at 38 and 60 °C, exposed blocks and

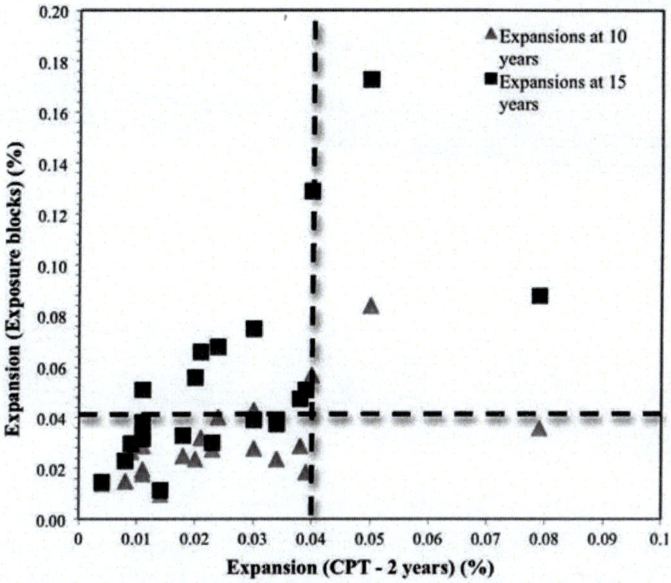

Fig. A.8 Comparison of expansion in the CPT at 2 years (38 °C) to expansion in outdoor exposure blocks at 10 and 15 years for mixtures containing mitigation measures for ASR [24]. The dotted lines indicate the limit values of expansion

Fig. A.9 Expansion of concrete prisms (differing storing temperatures and conditions) and mortar bars plotted against the 15-year expansions of exposure blocks produced with high or low-alkali cement [26]. The lines indicate the limit values of expansion

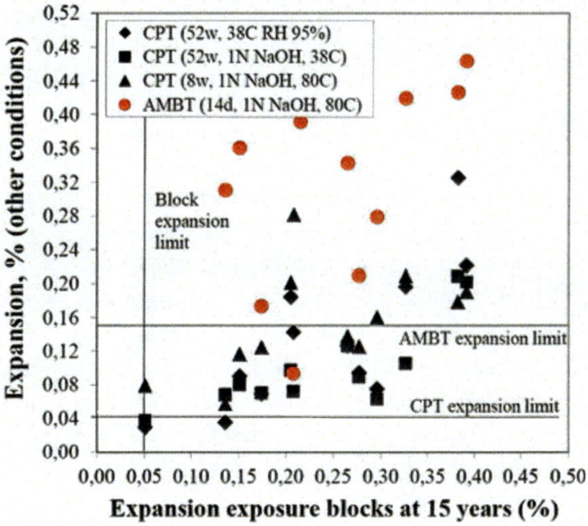

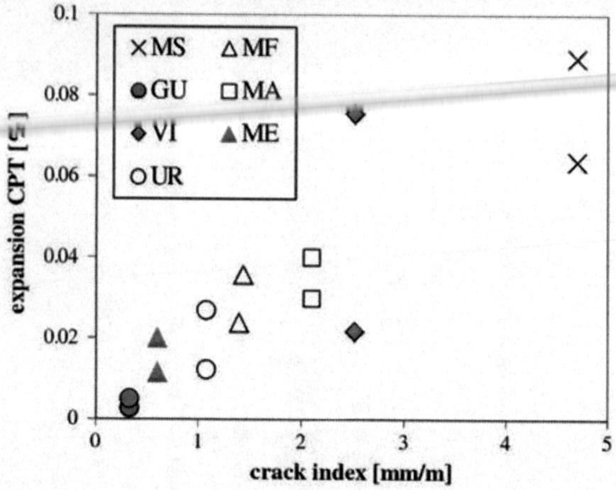

Fig. A.10 Expansion in the CPT at 60 °C (cement content of 300 and 400 kg/m^3) as a function of the crack-index determined on structures containing the same aggregates as used in the CPT [19]

degree of damage in structures indicates that these accelerated tests give a reasonable assessment of the expansion developing in exposed blocks and concrete structures. However, it has to be emphasized that the differences in the mineralogy and texture of reactive aggregates are significant leading to aggregate-specific behaviour at varying temperature that often impedes to draw general conclusions. The same applies to supplementary cementitious materials, whose effectiveness to suppress ASR depends on their specific properties and may vary with test temperature, test duration and alkali level of the concrete.

References

1. Lothenbach, B., Winnefeld, F., Alder, C., Wieland, E., Lunk, P.: Effect of temperature on the pore solution, microstructure and hydration products of Portland cement pastes. Cement Concrete Res. **37**(4), 483–491 (2007). https://doi.org/10.1016/j.cemconres.2006.11.016. http://www.sciencedirect.com/science/article/pii/S0008884606003103. ISSN: 0008-8846
2. Lothenbach, B., Winnefeld, F., Alder, C., Wieland, E., Lunk, P.: Effect of temperature on the pore solution, microstructure and hydration products of Portland cement pastes. Cement Concrete Res. **37**(4), 483–491 (2007). https://doi.org/10.1016/j.cemconres.2006.11.016. http://www.sciencedirect.com/science/article/pii/S0008884606003103. ISSN: 0008-8846
3. Lothenbach, B., Matschei, T., Moschner, G., Glasser, F.: Thermodynamic modelling of the effect of temperature on the hydration and porosity of Portland cement. Cement Concrete Res. **38**(1), 1–18 (2008). https://doi.org/10.1016/j.cemconres.2007.08.017. http://www.sciencedirect.com/science/article/pii/S0008884607001998. ISSN: 0008-8846
4. Bach, T., Cau-Dit-Coumes, C., Pochard, I., Mercier, C., Revel, B., Nonat, A.: Influence of temperature on the hydration products of low pH cements. Cement

Concrete Res. **42**(6), 805–817 (2012). https://doi.org/10.1016/j.cemconres.2012.03.009. http://www.sciencedirect.com/science/article/pii/S0008884612000622. ISSN: 0008-8846

5. Dilnesa, B., Lothenbach, B., Renaudin, G., Wichser, A., Kulik, D.: Synthesis and characterization of hydrogarnet Ca3(AlxFe1-x)2(SiO4)y(OH)4(3-y). Cement Concrete Res. **59**, 96–111 (2014). https://doi.org/10.1016/j.cemconres.2014.02.001. http://www.sciencedirect.com/science/article/pii/S000888461400043X. ISSN: 0008-8846
6. Dilnesa, B.Z., Wieland, E., Lothenbach, B., Dahn, R., Scrivener, K.L.: Fe-containing phases in hydrated cements. Cement Concrete Res. **58**, 45–55 (2014). https://doi.org/10.1016/j.cemconres.2013.12.012. http://www.sciencedirect.com/science/article/pii/S0008884614000118. ISSN: 0008-8846
7. Deschner, F., Lothenbach, B., Winnefeld, F., Neubauer, J.: Effect of temperature on the hydration of Portland cement blended with siliceous fly ash. Cement Concrete Res. **52**, 169–181 (2013). https://doi.org/10.1016/j.cemconres.2013.07.006. http://www.sciencedirect.com/science/article/pii/S0008884613001671. ISSN: 0008-8846
8. Escalante, J.I., Gómez, L.Y., Johal, K.K., Mendoza, G., Mancha, H., Ménendez, J.: Reactivity of blast-furnace slag in Portland cement blends hydrated under different conditions. Cement Concrete Res. **31**(10), 1403–1409 (2001). https://doi.org/10.1016/S0008-8846(01)00587-7. http://www.sciencedirect.com/science/article/. ISSN: 0008-8846
9. Hong, S., Glasser, F.: Alkali binding in cement pastes: Part I. The CS-H phase. Cement Concrete Res. **29**(12), 1893–1903 (1999). https://doi.org/10.1016/S0008-8846(99)00187-8. http://www.sciencedirect.com/science/article/pii/S0008884699001878. ISSN: 0008-8846
10. L'Hôpital, E., Lothenbach, B., Scrivener, K., Kulik, D.A.: Alkali uptake in calcium alumina silicate hydrate (C-A-S-H). Cement Concrete Res. **85**, 122–136 (2016). https://doi.org/10.1016/j.cemconres.2016.03.009. http://www.sciencedirect.com/science/article/pii/S0008884616303088. ISSN: 0008-8846
11. Bérubé, M., Tremblay, C., Fournier, B., Thomas, M., Stokes, D.: Influence of lithium-based products proposed for counteracting ASR on the chemistry of pore solution and cement hydrates. Cement Concrete Res. **34**(9) (2004). H. F. W. Taylor Commemorative Issue, pp. 1645–1660. https://doi.org/10.1016/j.cemconres.2004.03.025. http://www.sciencedirect.com/science/article/pii/S0008884604001528. ISSN: 0008-8846
12. Tremblay, C., Bérubé, M., Fournier, B., Thomas, M., Folliard, K.: Effectiveness of lithium-based products in concrete made with Canadian natural aggregates susceptible to Alkali-Silica reactivity. ACI Mater. J. (2007)
13. Gunnarsson, I., Arnórsson, S.: Amorphous silica solubility and the thermodynamic properties of H4SiO_4 in the range of 0_ to 350_C at Psat. Geochimica et Cosmochimica Acta **64**(13), 2295–2307 (2000). https://doi.org/10.1016/S0016-7037(99)00426-3. http://www.sciencedirect.com/science/article/pii/S0016703799004263. ISSN: 0016-7037

14. Kulik, D., Wagner, T., Dmytrieva, S., Kosakowski, G., Hingerl, F., Chudnenko, K., Berner, U.: GEM-Selektor geochemical modeling package: revised algorithm and GEMS3K numerical kernel for coupled simulation codes. Comput. Geosci. **17**(1), 1–24 (2013). https://doi.org/10.1007/s10596-012-9310-6. ISSN: 1573-1499
15. Thoenen, T., Kulik, D.: Nagra/PSI Chemical Thermodynamic Data Base 01/01 for the GEM-Selektor (V.2- PSI) Geochemical Modeling Code: Release 28-02-03. Tech. rep. Internal Report TM-44-03-04 (2003). http://gems.web.psi.ch/GEMS3/doc/pdf
16. Thoenen, T., Hummel, W., Berner, U., Curti, E.: The PSI/Nagra Chemical Thermodynamic Data Base 12/07. Tech. rep. Report 14-04. Paul Scherer Institute, Villigen, Switzerland (2014). https://www.psi.ch/les/DatabaseEN/PSI-Bericht
17. Helgeson, H.: Summary and critique of the thermodynamic properties of rock-forming minerals. Am. J. Sci. A **287**, 1–229 (1979). https://ci.nii.ac.jp/naid/10026604989/en/
18. Leemann, A., Merz, C.: An attempt to validate the ultra-accelerated microbar and the concrete performance test with the degree of AAR-induced damage observed in concrete structures. Cement Concrete Res. **49**, (2013)
19. Merz, C., Leemann, A.: Validierung der AAR-Prüfungen für Neubau und Instandsetzung. Tech. rep. AGB 2005/023 and AGB 2006/003. ASTRA-Bericht, Bern, Switzerland (2012)
20. Merz, C., Leemann, A.: Assessment of the residual expansion potential of concrete from structures damaged by AAR. Cement Concrete Res. **52**, 182–189 (2013)
21. Thomas, M., Fournier, B., Folliard, K., Ideker, J., Shehata, M.: Test methods for evaluating preventive measures for controlling expansion due to alkali-silica reaction in concrete. Cement Concrete Res. **36**(10), 1842–1856 (2006)
22. Santos Silva, A., Fernandes, I., Soares, D., Custódio, J., Ribeiro, A.C., Ramos, V., Medeiros, S.: Portuguese experience in ASR aggregate assessment. In: Proceedings of the 15th International Conference on Alkali-Aggregate Reaction (ICAAR). Sao Paolo, Brasil (2016)
23. Fournier, B., Ideker, J.H., Folliard, K., Thomas, M., Nkinamubanzi, P., Chevrier, R.: Effect of environmental conditions on expansion in concrete due to alkali-silica reaction (ASR). Mater. Char. **60**(7), 669–679 (2009)
24. Borchers, I., Müller, C.: Seven years of field site tests to assess the reliability of different laboratory test methods for evaluating the alkali-reactivity potential of aggregates. In: Proceedings of the 14th International Conference on Alkali-Aggregate Reaction (ICAAR). Austin, Texas, USA (2012)
25. Ideker, J., Drimalas, T., Bentivegna, A., Folliard, K., Fournier, B., Thomas, M., Hooton, R., Rogers, C.: The importance of outdoor exposure site testing. In: Proceedings of the 14th International Conference on Alkali-Aggregate Reaction (ICAAR). Austin, Texas, USA (2012)
26. MacDonald, C., Rogers, C., Hootan, R.D.: The relationship between laboratory and filed expansion: observations at the Kingston outdoor exposure site for ASR

after 20 years. In: Proceedings of the 14th International Conference on Alkali-Aggregate Reaction (ICAAR). Austin, Texas, USA (2012)
27. Fournier, B., Chevrier, R., Bilodeau, A., Nkinamubanzi, P., Bouzoubaa, N.: Comparative field and laboratory investigations on the use of supplementary cementing materials (SCMs) to control alkali-silica reaction (ASR) in concrete. In: Proceedings of the 15th International Conference on Alkali-Aggregate Reaction (ICAAR). Sao Paolo, Brasil (2016)
28. Kawabata, Y., Yamada, K., Ogawa, S., Martin, R., Sagawa, Y., Seignol, J.F., Toutlemonde, F.: Correlation between laboratory expansion and field expansion of concrete: prediction based on modified concrete expansion test. In: Proceedings of the 15th International Conference on Alkali-Aggregate Reaction (ICAAR). Sao Paolo, Brasil (2016)

Appendix B
Estimation of Past Expansion Based on Crack Index

Tetsuya Katayama

B.1 Scope

This test method concerns how to directly estimate the past expansion of ASR-affected concrete structures based on field survey and laboratory examinations.

B.2 Principles and Methodology

The method of estimating the amount of past expansion of ASR-affected concrete is based on the measurement of crack index, i.e. total crack width per unit length of concrete. Two approaches are possible, Fig. B.1, i.e. field inspection of concrete surface for macroscopic cracks, and laboratory petrographic examination of interior concrete for microscopic cracks. Results are fitted with Larive's equiation [1] to characterize S-shaped curves.

B.3 Field Inspection

B.3.1 Crack Index On-Site

The extent of past expansion of the concrete can be roughly estimated by measuring the crack index [2] of the concrete structure by visual inspection in the field. Crack index is defined as total width of cracks per unit length of concrete (mm/m). To

T. Katayama
Taiheiyo Consultant Co., Ltd., Sakura, Japan
e-mail: katayama@catv296.ne.jp

© RILEM 2021

V. E. Saouma (ed.), *Diagnosis & Prognosis of AAR Affected Structures*,
RILEM State-of-the-Art Reports 31,
https://doi.org/10.1007/978-3-030-44014-5

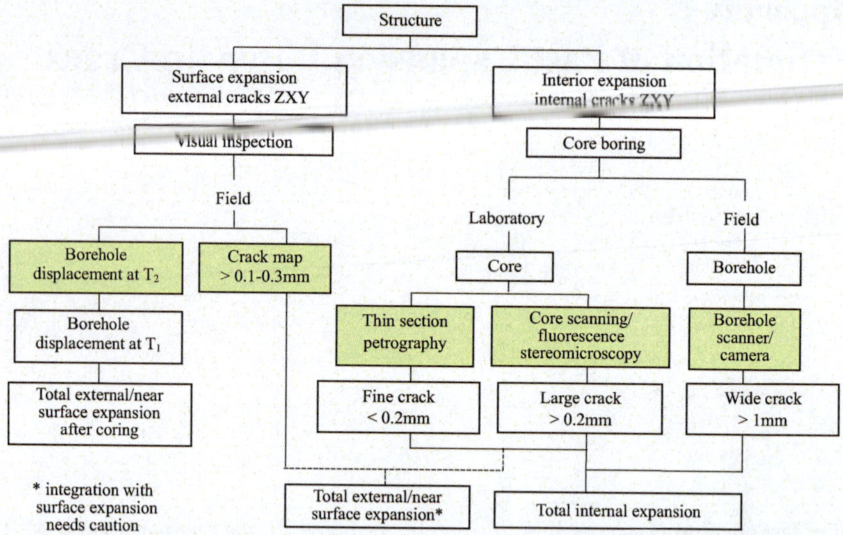

Fig. B.1 Crack indices in various scales obtained from field and laboratory examinations, Katayama (in preparation)

directly compare with monitored expansion (%) of structures, crack index should better be expressed in %. The following notes may be useful [3]. Because fine cracks (e.g. <0.2 mm) are not recorded in the routine visual inspection, they could be supplemented by petrographic examination of concrete core specimens (vertical or horizontal direction, if necessary).

Beside this crack index method, Shrimer [4] measured DRI on a polished core and presented a good correlation between the visually observed deterioration ratings of concrete structures (e.g. 0–7) and DRI values (0–1000).

B.3.1.1 Apparatus

For visual inspection of concrete surface (Fig. B.2), tape measure, crack scale and recording camera are necessary.

B.3.1.2 Procedures

Area for measurement: Crack width on concrete surface is recorded by on-site visual inspection with a recording unit of 1 × 1 m. Crack distribution can be mapped by marking on a vinyl sheet coved on the concrete surface, or by chalking on the concrete surface, or by reading high-resolution photographic images.

(a) Excessive expansion leading to the rupture of rebar in a RC foundation (30 years old) [33]

(b) Rapture of rebar in a RC foundation (30 years old), [33]

(c) Random map-cracking, footing of a viaduct (Quebec)

(d) Parallel horizontal ctacking, T-shape beam of a bridge pier (Japan)

(e) Grid-like cracking, colum of a building (Ontario)

Fig. B.2 Field inspection of external cracks, Katayama (unpublished results)

Measuring direction: Crack index should be measured in both vertical (Z) and horizontal (X, Y) directions in RC structures and PC members. This is because cracks in these structures often exhibit parallel (Fig. B.2d or grid-like (Fig. B.2e) distribution due to constraint by reinforcing steel bars, Fig. B.5b or prestressing tendons. Except for random map-cracking Fig. B.2c, results for the horizontal lines and vertical lines are averaged separately, and should be distinguished.

Interval of transects: Measurement should be made on each traversing line, drawn an interval of 25–30 cm. In RC structures, cracks appear in a cycle of 25–30 cm controlled by the spacing of reinforcing steel bars, Fig. B.2e, B.6a. This means that on-site visual inspection of 1 m square on the concrete surface covers 3–4 cycles of cracking which is a minimal size to get an averaged crack index.

Crack width: Minimum crack width identifiable in the field by the visual inspection depends on the condition of concrete. In the routine inspection of road bridges, minimal requirement for recording is 0.2 mm for PC structures and 0.3 mm for RC structures. However, where conditions permit, crack width down to a class <0.2 mm (actually >0.05 mm) are actually recorded. In structures where fine cracks are filled by exudations, a larger lower limit (e.g. 0.3 mm) is required. Maximum crack width within each recording unit is a useful indication for the crack index (e.g. width 5–10 mm) (Fig. B.15).

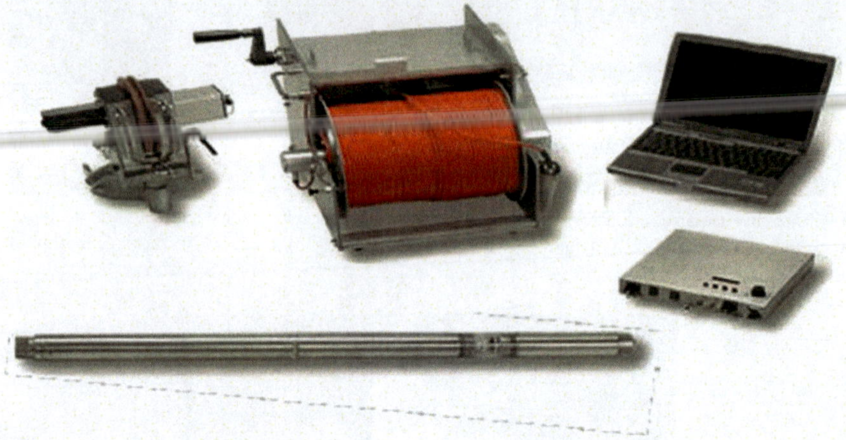

Fig. B.3 Borehole camera (scanner), image courtesy of K. Ishii [5]

B.3.1.3 Supplementary Method

For cracks in the interior of large structures, including dam, pile, footing, pier, etc., borehole scanner or borehole camera (Fig. B.3) is useful in recording wide cracks that appear on the wall of borehole (Fig. B.4a–c). This is because, along wide cracks, core samples are often fractured by the rotation of a core bit inducing tapered end faces (Fig. B.4d), and a long intact core with wide cracks suitable for measurement of macroscopic crack index is hardly recovered from the bore hole. This type of apparatus can obtain continuous panoramic image (360°) of borehole walls with a wide range of diameter (e.g. 4–12 cm) and depth (more than 100 m), sufficient for inspections of concrete structures [5]. To obtain a clear image of the borehole, cement slimes attached on the borehole wall should be cleansed by circulating clean water immediately after the boring before the slimes start to solidify.

B.3.2 Displacement After Coring

If expansion of concrete continues, displacement of concrete may occur, as an opening between the repair mortar of borehole and host concrete (Fig. B.6b). By combining the crack index and displacement around bore holes, expansion curve can be drawn.

(a) Parallel horizontal cracks in concrete (dia. 90mm oblique hole). Image courtesy of K. Ishii

(b) Bedding mortar layers on the horizontal joint after green cutting (dia. 66mm vertical hole). >55 years old. Image courtesy of K. Ishii

(c) Horizontal cold joint without green cutting of the laitance after concrete casting (dia. 66mm vertical hole). >75 years old. Image courtesy of K. Ishii

(d) Fracture along cracks and induced damage of core ends during extraction (dia. 66mm, vertical). ASR-affected footing of pier. Katayama

Fig. B.4 Internal cracks and joint planes within dam concretes detected by borehole camera (scanner) [5] and secondary damage by coring

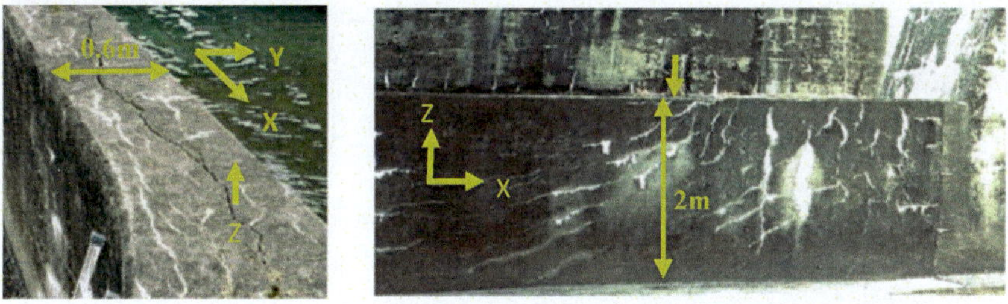

(a) axial cracks in a hydraulic structure (Japan)

(b) Cracks grading from parallel (thin wall) to random map-like (massive part), with position of coring in a hydraulic structure (35 years old)

Fig. B.5 Crack orientation. Adapted from [3]

(a) Grid-like cracking that follows rebar configuration (pillar head)

(b) Continuing expansion of concrete after coring (slab, water intake tower)

Fig. B.6 Late-expansive ASR in a hydropowerplant (gneiss) [3]

Table B.1 Calculation of past expansion based on macroscopic displacement and crack growth around boreholes, hydraulic structures by late-expansive ASR [3]

structure construction 1968		direc-tion	0-27 years	27-35 years	35 years		35-44 years
			expansion (1) = (3)-(2)	expansion (2) from gap after first coring	expansion (3) from crack index	max. crack width	expansion (4) from gap after second coring
O1	training dyke (spillway)	Y	1.03%	4.8mm/0.6m =0.80%	11mm/0.6m =1.83%	10mm	repaired
O2	pillar, mid-slope terrace	Y	0.05%	2.6mm/2m =0.13%	3.5mm/2m =0.18%	3mm	repaired
F1	pillar, mid-slope terrace	Y	0.30%	5.0mm/2m =0.25%	11mm/2m =0.55%	10mm	repaired
F2	training dyke (fish ladder)	Y	0.22%	8.8mm/m =0.88%	11mm/m =1.10%	10mm	repaired 2.5mm/1m=0.25 %

B.3.3 Data Processing

To define the expansion curve after Larive's equation with three parameters (τ_L, τ_C, ε^{∞}, at least three independent data are necessary. Brunetaud et al. [6] equation is good for depicting a steeper slope at longer age [7]. However, as it has 5 unknown parameters requiring more than 5 independent expansion data, it is not used here. In fitting expansion data with the Larive's sigmoidal curve, following cases are possible [3].

1. Two past coring (Fig. B.7a: F2): expansion data for three different times can be obtained, which enables to define a sigmoidal expansion curve by fitting (katayama17).
2. Only one past coring (Fig. B.7a: O1, O2, F1): expansion data for two different times can be obtained. Therefore, to obtain the third data, the following assumptions are possible.
   a. Use the same expansion ratio. Only one concrete (F2) has a perfect data set of three expansion, Fig. B.7a. Hence, the same expansion development ratio for two time intervals, (age 44–35)/(age 35–27), is applied to other concretes undergoing ASR of the same rock type aggregate.

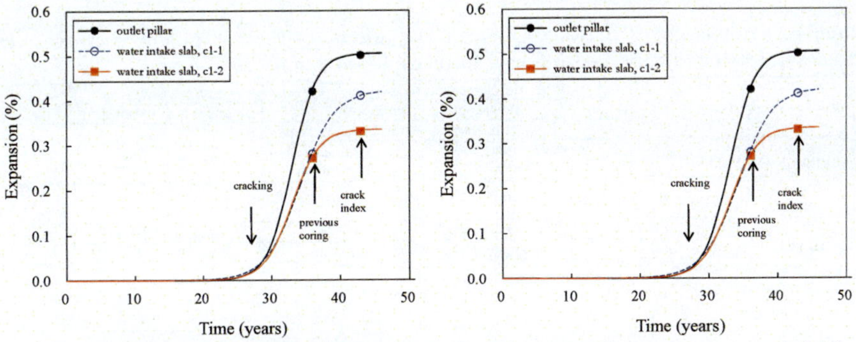

(a) assumption: same expansion development ratio (between age 44-35 years and age 35-27) as the site with a perfect data set (F2). hydraulic structures (Y direction) [31]

(b) Assumption: expansion was 0.05% when cracking was first identified. hydraulic structures [31]

Fig. B.7 Expansion curves of late-expansive ASR, based on displacement and crack index around repaired bore holes

b. Use the first occurrence of cracking. If there is a record when the first cracking occurred, then the expansion of concrete at that time can be postulated to be 0.04–0.05% (Fig. B.7b).
c. Use the data of accelerated concrete core expansion test. Final expansion can be estimated by the fitting with the Larive equation. This is detailed in Table 6.7. To estimate potential residual expansion of structure, past expansion of field structures was combined, Fig. 6.6b.

B.3.4 Cracking After Strengthening—Effect of Double Cylinder

Cracking occurs on the concrete lining due to expansion of inner concrete. In this case, total crack width appearing on the outer lining layer is equal to the expansion of the peripheral length of the inner concrete cylinder [8], Fig. B.7. When the lining concrete is a reinforced one, total crack width is dispersed into a number of small cracks. This simple relationship can be used for assessing expansion rate of a concrete pier that has occurred after reinforcement or repair work was done, that is:

$$\begin{aligned}\text{Total crack width X (mm)} &= \text{peripheral expansion X (mm)}\\ &= 3.14 \times (D \times 1,000) \times \varepsilon/100\\ &= 31.4 \times D \times \varepsilon(\%) \qquad \text{(B.1)}\end{aligned}$$

Table B.2 Estimation of the past expansion of ASR-affected concrete after lining a new concrete member for strengthening or repair Katayama (unpublished results)

cracks			outer concrete (strengthened or repaired)		inner concrete	
			non-reactive but cracked		reactive and expanding	
expansion (%) =100x(Σ1+Σ2)			lining concrete crack index (1) Σ1 (mm/m)	patching concrete crack index (2) Σ2 (mm/m)	unexposed to the concrete surface	total
time after repair	expansion (%)					
10 years	horizontal	X	2.0mm/m=0.20%*	absent	equal to 0.20% *	0.20
	horizontal	X	1.4mm/m=0.14%*	absent	equal to 0.14% *	0.14
	vertical	Z				

where $3.14 \times D$ corresponds to the peripheral length and ε (%) is the expansion, and D the pier diameter in m. When the entire periphery of the pier is measured:

$$\varepsilon(\%) = X(\text{mm})/(31.4 \times D) \quad \text{(B.2)}$$

When a partial length of the periphery (1 m) is measured:
$\varepsilon(\%)$ = [average crack width X (mm)] × [crack number n (1/m)]/10 = crack index (mm/m) × 100.

For example, when $X = 0.35$ mm and $n = 4$/m, then crack index = $0.35 \times 4/1000$ = 0.0014 (mm/m), and $\varepsilon = 0.14$ % (Table B.2).

Then, the peripheral expansion of the pier of diameter 5 m is 22 mm (= 31.4 × 5 m × 0.14%), and the above frequency of cracking corresponds to 63 cracks (= 22/0.35 mm) on the entire circumference of the pier.

B.4 Laboratory Work

B.4.1 Crack Index Under the Microscope

Fine cracks can be identified either on the cut surface of core specimens impregnated with fluorescent epoxy resin under UV light, or on polished thin sections using transmitted/reflected light, Table B.3. The range of the crack widths should not be overlapped to avoid duplication. Usually crack width is indicative of the expansion of concrete, but drying shrinkage cracks may occur within mortar around aggregate rich in clay minerals. The contribution of such cracks should be subtracted from the measured value of the crack width. Cracks filled with ASR gel and calcite veins are not impregnated by the fluorescent dye, hence caution is necessary not to underestimate the crack density of concrete.

Table B.3 General approach to estimate past expansion of a hydraulic structure (petrographic stages iv, v, vi, Appendix D). (F1 in Table B.4) Adapted from [3, 9]

cracks			expansion crack			shrinkage crack	
			field	laboratory			
			visual inspection	core section	polished thin section		
			tracing or linear traverse	fluorescence, UV light**	transmitted/reflected light		
crack width			macro cracks (>0.2mm)	fine cracks (<0.2mm)	micro cracks (<0.2mm)		
crack index (%) =100x[Σ1+(Σ2 or Σ3)-Σ4]			crack index (1) Σ1 (mm/m)	crack index (2) Σ2 (μm/mm)	crack index (3) Σ3 (μm/mm)	crack index (4) Σ4 (μm/mm)	total
age (y)	expansion (%)						
35	horizontal	X			1.3μm/mm =0.13% max 0.66%		0.68%*** max 1.21%
		Y	5.5mm/m=0.55%	not applied			
	vertical	Z					
	drying shrinkage					negligible	
27 previous coring	horizontal	X			1.3μm/mm =0.13%		0.43%***
		Y	3.0mm/m=0.30%*	not applied			
	vertical	Z					
	drying shrinkage					negligible	

*from the aperture between the infilled mortar and previous borehole, *** used for Table 8.4, Figure 8.6
**alternative to microscopy if fluorescent resin was impregnated well
base line for measurement: L=1m for visual inspection, L=100mm for fluorescence and microscopy

(a) Section of double cylinder, composed of a center concrete with reactive Spratt's limestone and enveloping mortar layer of non-reactive sand (38°C) [38]

(b) Cracking of non-reactive outer layer induced by the expansion of the reactive center cylinder [38]

(c) crack propagation through mortar from the reacted center cylinder [38]

(d) Cracked RC lining constructed for anti-earthquake strengthening of a concrete pier undergoing ASR [39]

Fig. B.8 Double cylinder effect

B.4.1.1 Apparatus

A wide range of apparatus can be used to measure the crack index from macroscopic to microscopic level. They are common to those used for petrographic examinations, e.g. stereobinocular microscope, recording camera, etc. If necessary, UV light is used to highlight cracks within concrete, in which epoxy resin with fluorescent dye is vacuum impregnated (Fig. B.8).

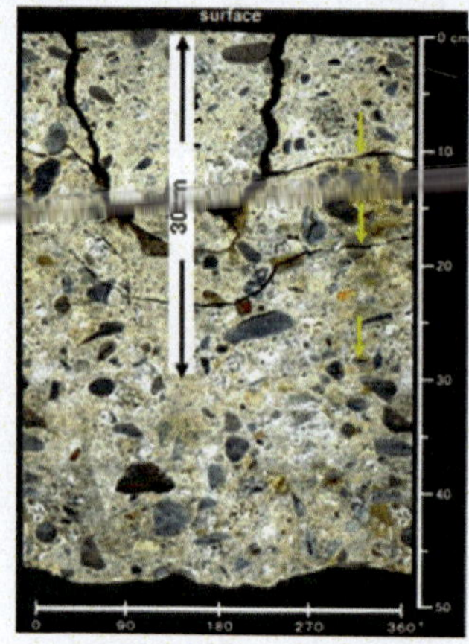

(a) Scanned immage of core, hydraulic structure (late-expansive ASR, greywacke, gneiss)

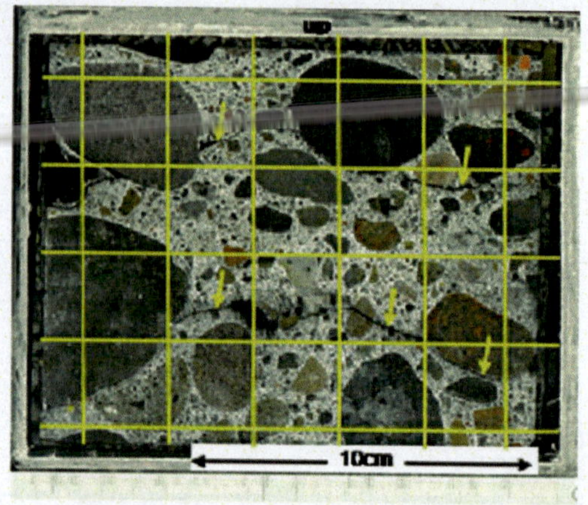

(b) Large-area thin section of core, upper level of arch dam (early-expansive ASR, rhyolite), parallel horizontal cracks mostly debonding pebbles, in dark-field illumination. Adapted from [31]

Fig. B.9 Oriented cracks formation in concrete, vertical cores

B.4.1.2 Procedures

Sectioning cores: It is recommended to cut core samples closely correlated with the original directions (e.g. X, Y, Z) in the structure (Fig. B.10). When a set of small thin sections are used instead of large area thin section, crack indices obtained for each direction should be averaged. For the purpose of measuring crack index, thin sections should not be concentrated around reacted aggregate particles, but should be random along the same direction.

Grid line transects: Where main crack system occurs in parallel direction, measure the crack width on two traversing lines (e.g. horizontal and vertical), at least 5 lines set in each direction (Fig. B.9b).

Area of measurement: It is desirable to measure a wide area (e.g.10 × 10 cm) when aggregate is very coarse. Preparation of a large-area thin section of this size (Fig. B.9b) and of large amount of small thin sections with equivalent total area is expensive and time-consuming. Hence, stereomicroscopy on polished concrete section with a large area (e.g. longitudinal section of core) can be an alternative when the resolution of fine cracks is compromised.

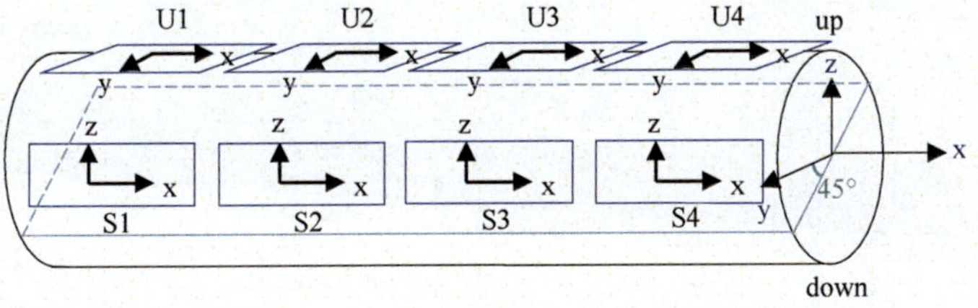

Fig. B.10 Thin sectioning for measurement of three dimensional crack indices (ZXY), horizontal core, Katayama (in preparation)

Table B.4 Detail of total crack index of hydraulic structures: on-site and microscopy (35 years old) [3]

structure	crack index (%) = (mm/m)x100 on-site	microscopy		total	direction	structure	crack index (%) = (mm/m)x100 on-site	microscopy		total	direction
	>0.2 mm a	>0.2 mm	<0.2 mm b	a+b			>0.2 mm a	>0.2 mm	<0.2 mm b	a+b	
O1	1.83	0.00	0.27	2.10	Y	F1	0.55	0.77	0.66	1.21	Y
	0.47*	0.43	0.42	0.89	Z			0.46	0.76		Z
O2	0.18	0.27	0.37	0.55	Y	F2	1.10	0.28	1.11	2.21	Y
		0.00	0.47		Z			2.64	0.86		Z

*core scanning, approximate direction

B.4.2 *Integration with Petrographic Examination*

Even if monitoring data is absent, crack indices of concrete (%) from on-site visual inspection for macrocracks may give a measure of past expansion of the concrete surface. Thin section microscopy for microcracks are thought to compensate the visual inspection and approximate the extent of past expansion.

In thin-wall structures (e.g. Fig. B.2, Tables B.1, B.3 and B.4), crack index of the inner concrete by thin section microscopy varies considerably (Table B.3) and could exceed 1%, whereas total crack index of the concrete surface by visual inspection (>0.2 mm) and inner concrete by microscopy (<0.2 mm) could exceed 2% (Fig. B.11) [3]. This is subject to several factors (Table B.3), including orientation, depth, stress, relative position to rebars, etc.). However, there is argument that the crack system on the concrete surface (generally widely open) is different from the interior of the R/C structure, e.g. depth >20 cm [10]. Hence crack indices obtained from different depths should be handled with care.

B.5 Expansion Before Monitoring Started

Where monitoring data of expansion or displacement of concrete structures are available, it is recommended to estimate the total expansion that had already been generated before the monitoring or coring for examination was made. From the monitoring data, it is evident that most of the expansion curves do not produce a straight line.

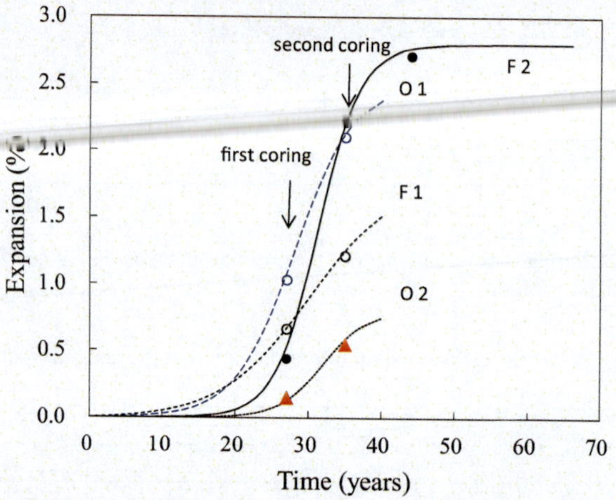

Fig. B.11 Total expansion curves of late-expansive ASR estimated from displacement around repaired bore holes and crack index from onsite and microscopy, assuming the same expansion ratios with F2. hydraulic structure [3]

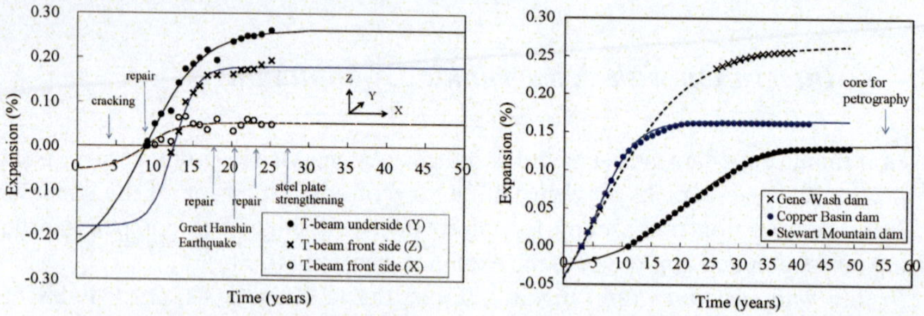

(a) Anisotropic expansion of RC T-beam, viaduct pier (bronzite andesite), adapted from [42]

(b) Vertical expansion at arch midcrest, US arch dams (rhyolite), adapted from [43]

Fig. B.12 Monitored expansion of early-expansive ASR fitted to Larive's sigmoidal curve [3]

By fitting with the Larive's equation, S-shaped expansion curves can be obtained. From Fig. B.12 it can be seen that ASR-affected structures had already expanded considerably when monitoring started [3].

Early-expansive ASR With the early-expansive aggregate (andesite, Fig. B.12a; rhyolite, Fig. B.12b), expansion of 0.03–0.05% was generated within 5 years when cracking of concrete was noticed. T-beam of a bridge pier presents highly anisotropic expansion due to oriented configuration of reinforcement [11]. It is noteworthy that expansion of US dams has ceased within 40 years after construction [12].

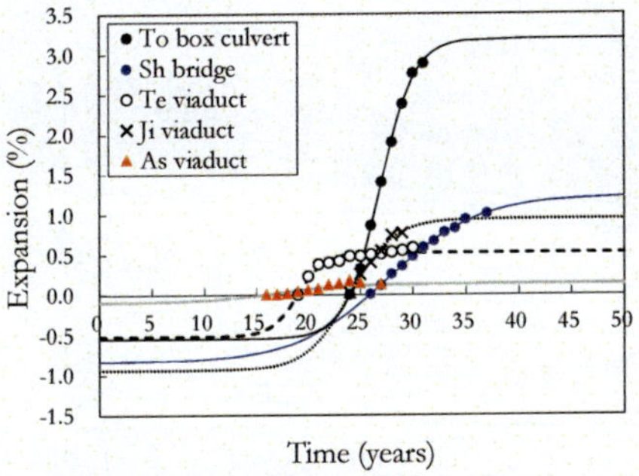

(a) Large expansion before monitoring started, bridge abutments (andesite and granitic rocks), drawn using data from Nomura, Komatsubara, Kuroyanagi, and Torii [44]

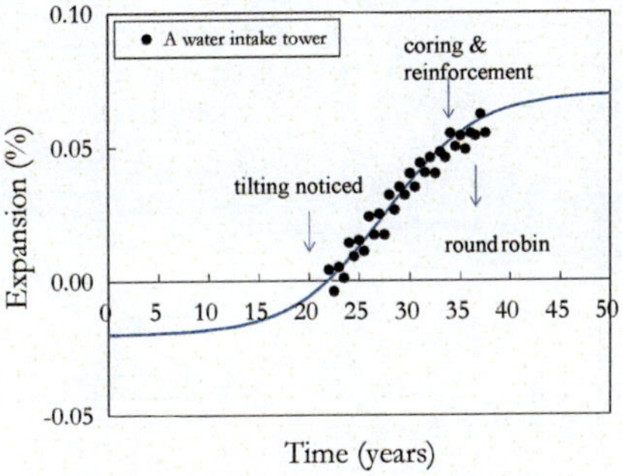

(b) Slow vertical expansion of water-intake tower (andesite and nearly low-alkali cement), drawn using data from Kubo, Shibata, Sannoh, and Torii [45]

Fig. B.13 Monitored expansion of early- and late-expansive ASR, fitted to Larive's sigmoidal curve [3]

Early- and late-expansive ASR At the time when monitoring and petrographic examination started [9], expansion of concrete had already reached 0.5–1% in some structures (andesite, granitic rocks, Fig. B.13a) and continued additional expansion of 0.5–1% [13]. This is important information because it enables to correlate and interpret actual expansion of field structures with the progress (stage) of ASR in concrete, based on the microtextures observed in the petrographic examination [3]. In some structure containing reactive andesite, expansion started slowly, more than 20 years after construction [14]. This may be due partly to a lower level of cement alkali used in concrete.

B.6 Expansion Versus Petrographic Stage of ASR

This aspect has been clarified and is shown in Fig. B.14 [3].

B.7 Precision and Bias

Crack index: This may vary considerably according to the coverage length, maximum crack width, and method of examination (visual inspection, stereomicroscopy, polarizing microscopy, etc.). The larger the coverage length, the smaller will be the crack index (Fig. B.15). In general, beams and slabs placed in horizontal direction produce the largest expansion or crack index in the vertical direction.

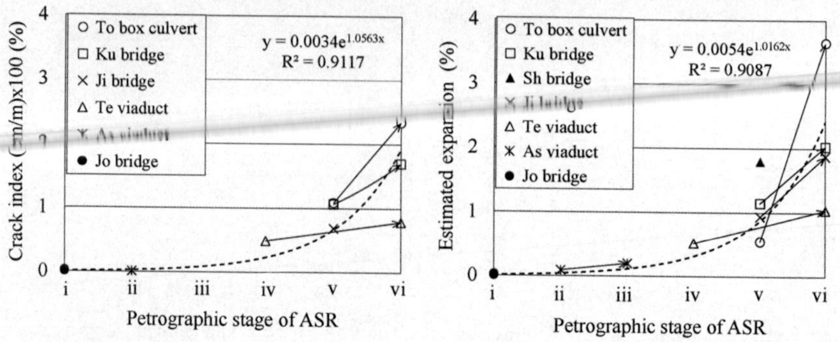

(a) Crack index from visual inspection plus thin section petrography

(b) past expansion from fitting the monitoring expansion data by Nomura, Komatsubara, Kuroyanagi, and Torii [44]

Fig. B.14 progress of ASR in 10 years, as past expansion and petrographic stage, bridges from 1999 to 2010 [3]

Possible discrepancy: This may occur when the values of expansion or crack indices obtained with different levels of concrete length are directly compared, e.g. entire structure (20–60 m), concrete member (2–6 m), concrete core (20–60 cm), thin section (3–10 cm).

Inhomogeneity of cracking: Possibly, smaller part of concrete is much more influenced by reacted aggregate particles, local expansion, cracking and laminations influenced by rebars and structural framework.

Fitting to S-shaped curve: Larive's equation is convenient, with only three parameters to determine, but may not fit well and may underestimate the final expansion of some curves that have an increasing tendency at later age. If sufficient field survey data is available, other equation (Brunetaud equation) could be used.

B.8 Limitations and Applicability to Analysis

B.8.1 Limitations

1. It is important to confirm that the cracking of concrete was due to ASR based on petrographic examination.
2. CI measured on the concrete surface does not represent internal expansion of concrete, particularly below the level of reinforcing steel bars. CI should be distinguished by depth and orientation ZXY.
3. CI may underestimate past expansion up to 0.04%, because cracks in concrete generally become evident when expansion exceeds this limit.

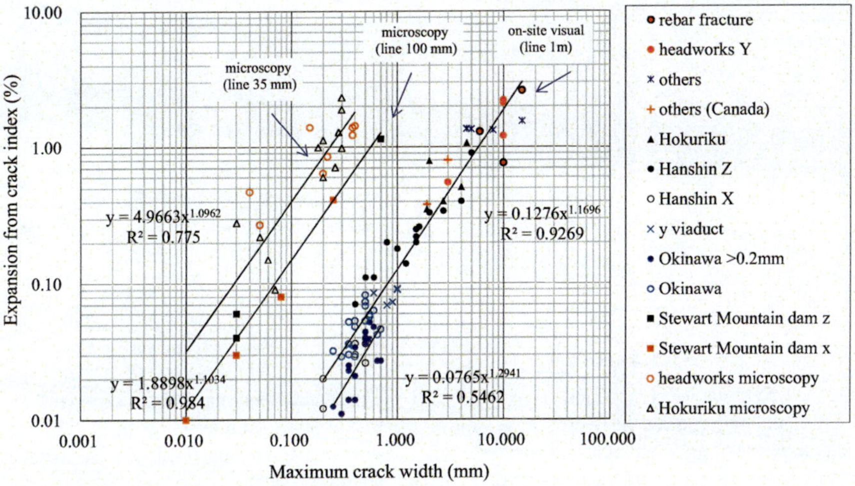

Fig. B.15 Crack indices vs maximum crack width different unit coverage length and methods [3], drawn using data partly from [16]

4. In highly distressed structure, measurement of CI of core samples may underestimate expansion of structure, because extraction of intact core sample containing wide cracks (e.g. >2 mm) is difficult.
5. In a structure where large horizontal cracks develop, the extent of missing wide cracks in the vertical core should be confirmed by a borehole scanner/camera.
6. With hydraulic structures subject to drying/wetting, measurement of crack width is difficult because carbonate exudations cover the crack openings.

B.8.2 Applicability

1. Crack index (CI) can be used to quantify both drying shrinkage and expansion of concrete (ASR, freeze/thaw, sulfate attack, DEF, etc.).
2. CI can be used to directly estimate past expansion of concrete in three directions ZXY, even monitoring data is unavailable.
3. CI on the concrete surface can be used for monitoring the progress of ASR expansion of concrete structure.
4. CI correlates with other expansion parameters used in field inspection, such as crack density (CD: total crack length per unit area: L m/m^2) (Fig. B.16a) and maximum crack width (Fig. B.16b).
5. CI measurement seems to be suitable for RC structure, because expansion and peripheral cracks of the concrete cover increase abruptly at a certain degree of CD (35 m/m^2) (Fig. B.16c) [16].

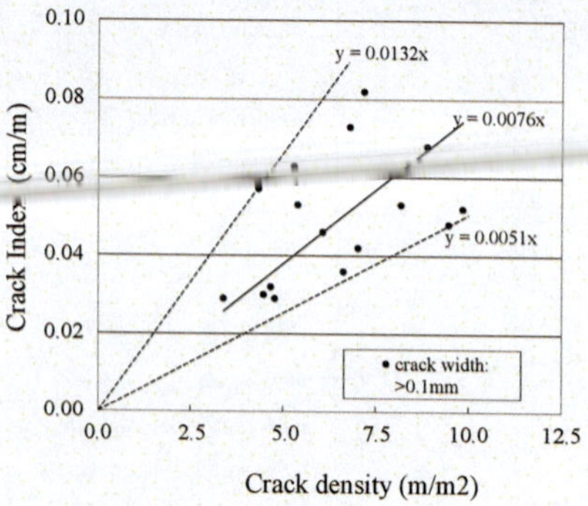

(a) Crack index vs. crack density on RC abutment undergoing mild ASR of late-expansive sand aggregate (visual surface expansion >0.03%)

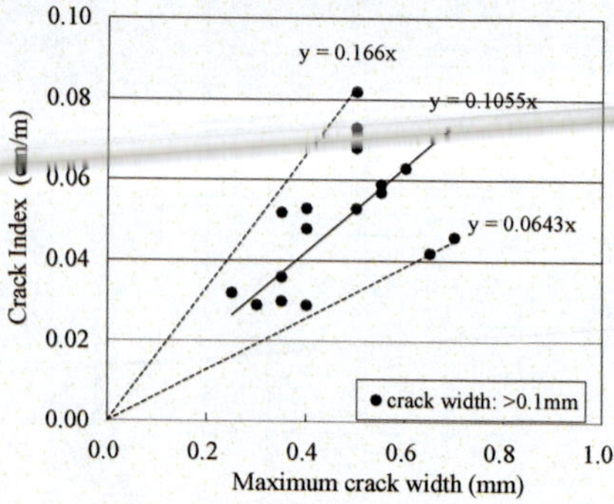

(b) Crack index vs. maximum crack width on RC abutment undergoing mild ASR of late-expansive sand aggregate (visual surface expansion >0.03%)

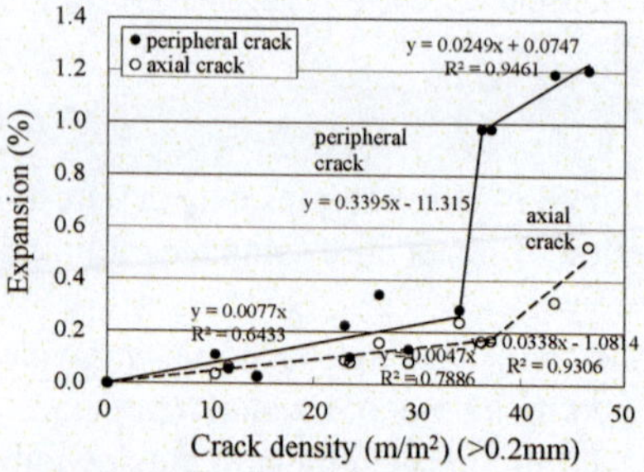

(c) Abrupt increase of peripheral cracks on RC blocks at a crack density around 35m/m2, drawn using data from Yamamura, Kojima, Kuzume, and Okamoto [47]

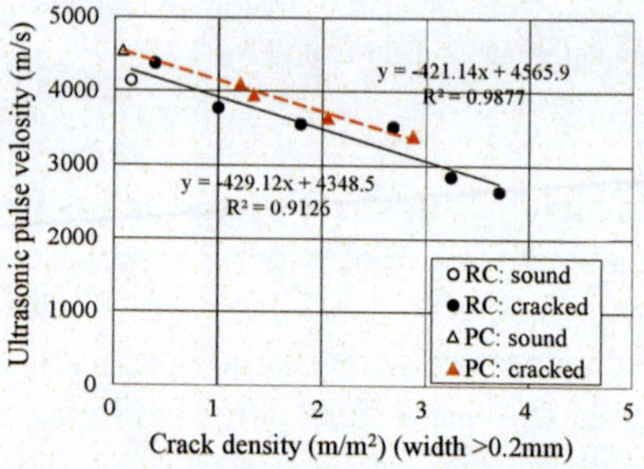

(d) Crack density vs. in-situ ultrasonic pulse velocity of T-beams of viaduct piers undergoing early-expansive ASR due to bronzite andesite in Japan. Redrawn averaging data from [48]

Fig. B.16 Crack index and related parameters related to expansion (Katayama, in preparation)

6. CI possibly correlates with ultrasonic pulse velocity (V) through the correlation with CD. In ASR-affected T-beams of viaduct piers, a pronounced relationship has been known between CD and V, although the latter fluctuates up to ± 600 m/s (Fig. B.16a) [17].

References

1. Larive, C.: Apports Combinés de l'Experimentation et de la Modélisation à la Comprehension del'Alcali-Réaction et de ses Effets Mécaniques. Ph.D. thesis.

Laboratoire Central des Ponts et Chaussées, Paris (1998). https://hal.inria.fr/docs/00/52/06/76/PDF/1997TH_LARIVE_C_NS20683.pdf

2. Fournier, B., Bérubé, M., Folliard, K., Thomas, M.: Report on the Diagnosis, Prognosis, and Mitigation of Alkali-Silica Reaction (ASR) in Transportation Structures. Tech. rep.FHWA-HIF-09-004, p. 154. Federal Highway Administration (2010)
3. Katayama, M.: An attempt to estimate past expansion of concrete based on petrographic stage of alkali-silica reaction. In: Proceedings of the 39th International Conference on Cement Microscopy, pp. 217–236. Toronto, Canada (2017)
4. Shrimer, F.: Development of the damage rating index method as a tool in the assessment of alkali-aggregate reaction in concrete: A critical review. In: Fournier, B., Bérubé, M. (eds.) Symposium on Alkali-Aggregate Reaction in Concrete, pp. 391–401. Canada, Montreal (2006)
5. Ishii, K.: Image Data of Borehole Camera (Scanner) from Concrete Dams. Unpublished data, RaaX (2019)
6. Brunetaud, X., Divet, L., Damidot, D.: Delayed ettringite formation: suggestion of a global mechanism in order to link previous hypotheses. In: Proceedings of the 7th CANMET/ACI International Conference on recent Advances in Concrete technology, pp. 63–76 (2004)
7. Kawabata, Y., Yamada, K., Ogawa, S., Sagawa, Y.: Prediction of ASR expansion of field-exposed concrete based on accelerated concrete prism test with alkali-wrapping. Cement Sci. Concrete Technol. **69**, 496–503 (2015). (in Japanese)
8. Kobayashi, H., Kawano, H., Mirama, A., Tsujiko, M.: Study on the simple testing of ASR—Double cylinder method. Tech. rep. Report No.2587. Public Works Research Institute Ministry of Construction (1988) (in Japanese)
9. Katayama, T., Tagami, M., Sarai, Y., Izumi, S., Hira, T.: Alkali-aggregate reaction under the influence of deicing salts in the Hokuriku district. Japan. Mater. Char. **53**, 105–122 (2004)
10. Katayama, T., Sarai, Y., Higashi, Y., Honma, A.: Late-expansive alkalisilica reaction in the Ohnyu and Furikusa headwork structures, central Japan. In: Proceedings of the 12th International Conference on Alkali-Aggregate Reaction (ICAAR), pp. 1086–1094. Beijing, China (2004)
11. Horie, K., Matsumoto, K., Sasaki, K., Niina, T.: An approach to maintenance of ASR-affected piers of Hanshin Expressway. Concrete J. (Japan Concrete Institute, JCI) **48**(1), 113–116 (2010) (in Japanese)
12. Tuthill, L.H.: Alkali-silica reaction-40 years later. What Happened? Where are we? What now? Concrete Int. **4**(04), 32–36 (1982)
13. Nomura, M., Komatsubara, A., Kuroyanagi, M., Torii, K.: Evaluation of petrographic features of reactive aggregates in the Hokuriku district and residual expansion of core. Concrete J. (Japan Concrete Inst. JCI) **33**(1), 953–958 (2011) (in Japanese)
14. Kubo, T., Shibata, T., Sannoh, C., Torii, K.: The reinforcement of an ASR affected intake tower using post-tensioned tendons. In: Proceedings of the 15th International Conference on Alkali-Aggregate Reaction (ICAAR). Sao Paulo, Brazil (2016)

15. H. E. P. Corporation.: Committee report on the investigation of alkali aggregate reaction. Tech. rep. See p. 17. Hanshin Expressway Public Corporation & Hanshin Expressway Management Technology Center (1986) (in Japanese)
16. Yamamura, S., Kojima, T., Kuzume, K., Okamoto, Y.: Experimental study on the relationships between structural members and AAR cracking. Proc. Japan Concrete Inst. **30**(1), 1011–1016 (2008) (in Japanese)
17. H. E. T. Center: Maintenance Manual of ASR-affected Structures. Appendix 2. Criteria for judging ASR-affected bridges (2007) (in Japanese)
18. Katayama, T.: Late-expansive ASR in a 30-year old PC structure in eastern Japan. In: Proceedings of the 14th International Conference on Alkali-Aggregate Reaction (ICAAR). paper 030411-KATA-05. Austin, Texas, USA (2012)
19. Grattan-Bellew, P.: Samples sent to T. Katayama (1992)
20. Katayama, T.: Alkali-aggregate reaction. In: Maruyama, H. (ed.) Repair and Reinforcement of Concrete Structures. Tech Information S.C. Ltd., Tokyo pp. 253–262 (2016). (in Japanese)

Appendix C
Assessment of the Alkali-Budgets in Concrete

Tetsuya Katayama

C.1 Scope

This method covers assessment of the mass balance of alkalis in hardened concrete, including contributions of each concrete-making constituent called "alkali-budgets" [1]. In the context of assessing future maximum potential expansion, it can be used to correct final expansion obtained from the accelerated core expansion test. Part of the method has been used to estimate the alkali amount of Portland cement in old concrete structures without records even more than 100 years ago [2]. Examination of the alkali-budgets has been applied to nuclear power plants in Japan since it was proposed in a JNES guideline (inherited by present NRA) to diagnose ASR in NPP concretes [3], as well as to highway structures since the early 2,000s [1, 4].

C.2 Principles and Methodology

Alkali-budgets in concrete are estimated to distinguish contribution of alkalis from cement, coarse aggregate, and other sources based on SEM/EDS analysis of unhydrated clinker phases and wet-chemical analyses of water-soluble alkalis of the bulk concrete and coarse aggregate. The likelihood of future expansion due to ASR, based on accelerated concrete core expansion test, was described in Chap. 6.

In the past, water-soluble alkali in concrete was attributed solely to cement (Fig. C.1a). Nowadays, additional water-soluble alkali is known to originate from many sources including aggregates, supplementary cementing materials, chemical admixtures and external deicing salt and sea water (Fig. C.1b). As a result, aged

T. Katayama
Taiheiyo Consultant Co., Ltd., Sakura, Japan
e-mail: katayama@catv296.ne.jp

© RILEM 2021

V. E. Saouma (ed.), *Diagnosis & Prognosis of AAR Affected Structures*,
RILEM State-of-the-Art Reports 31,
https://doi.org/10.1007/978-3-030-44014-5

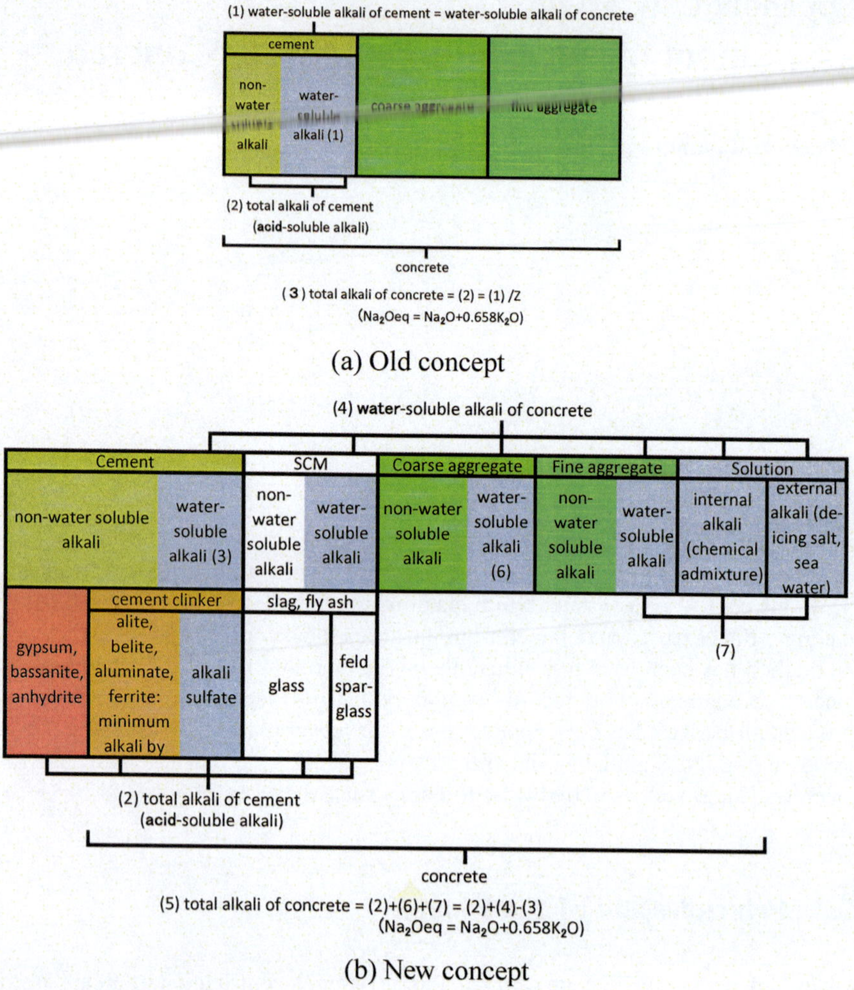

Fig. C.1 Concepts on the alkali-budgets in concrete (adapted from [5]

concretes may contain total alkalis higher than those expected from the original cement, sometimes exceeding Na_2O_{eq} 1 kg/m^3 (Table C.4). Alkali-budgets in concrete are assessed in two steps, combining petrography-based analyses of cement alkalis (phase i, Table C.4) and wet-chemical analyses of water-soluble alkalis of the bulk concrete and separated aggregate (phase ii, Table C.4).

Several assumptions are necessary, because in most cases construction records about mix proportion and cement compositions are either lost or unavailable. For the original content of each clinker phase not exactly known, rounded values of potential content after Bogue calculation for normal Portland cement (e.g. type I) may be used if special anomaly or deviation (e.g. low C_3S, low C_3A: moderate-heat

Table C.1 Minimum alkali content of clinker (unhydrated cement) in the concrete pavement, 22 years old (dolomitic limestone), [6]

	SiO_2	TiO_2	Al_2O_3	Fe_2O_3	MnO	MgO	CaO	Na_2O	K_2O	SO_3	Total	Na_2Oeq
alite	24.84	0.04	0.59	0.36	0.27	1.14	68.79	0.43	0.00	0.06	96.51	0.30
	25.01	0.24	0.72	0.47	0.00	1.11	69.59	0.14	0.05	0.26	97.60	
belite	31.74	0.15	1.22	2.06	0.00	0.48	59.65	0.62	0.87	0.00	96.79	0.89
	32.83	0.15	0.58	0.37	0.24	0.00	61.07	0.35	0.37	0.46	96.42	
aluminate	6.73	0.45	23.15	10.54	0.00	1.23	49.67	3.22	1.70	0.00	96.70	4.87
	7.75	0.28	23.00	7.15	0.00	0.65	51.26	3.91	2.27	0.08	96.36	
ferrite	4.51	0.18	19.63	25.03	0.00	3.55	45.29	0.58	0.48	0.25	99.50	0.56
	4.58	1.09	15.61	24.95	0.72	2.75	45.98	0.12	0.16	0.15	96.11	
alkali content								0.66	0.37			0.90

Assumption: alite 60%, belite 20%, aluminate 10% and ferrite 10%

alite belite aluminate ferrite

$Na_2O = (0.29x0.60) + (0.49x0.20) + (3.57x0.10) + (0.35x0.10) = 0.66\%$

$K_2O = (0.03x0.60) + (0.62x0.20) + (1.99x0.10) + (0.32x0.10) = 0.37\%$

$Na_2Oeq = 0.66+0.658x0.37 = 0.90\%$ contribution from water-soluble alkali sulfate is ignored in this table

Table C.2 Water-soluble alkali of major rock types of the coarse aggregate extracted from aged concrete structures in Japan, Katayama (unpublished results)

		aggregate separated from old structures (> 31 years)															
rock type	major	andesite**		greywacke								granite	gneiss	green rock	rhyolitic welded tuff	mudstone	hornfels
	minor	dacite		mudstone				gneiss		green rock	conglomerate	-	granite	greywacke	-	chert	gneiss
water-soluble alkali Na_2Oeq kg/m³ *		1.4-1.0	0.9-0.6	1.3	1.1	1.0	1.0	0.8	0.3	0.7	0.7-0.6	1.0-0.4	0.7	0.7	0.6	0.6-0.5	0.3
age (years)		31	31	ca 40	54	40	39	54	44	ca 40	45	43	42	41	38	36	42
structure		Ko	Ko	X	Ur	T	M	Ur	Fu	Y	Sk	Ta	Sr	T	Mu	Ku	Sr

* General Project Method (<0.3mm, 40°C, water/solid=10, shaking 30 minutes) with shaking (160/min, amplitude 3cm)

[shaded] Hydraulic structure **contains ASR gel veins within aggregate

Portland cement) is undetected by the polarizing microscopy or SEM observation. Water-soluble alkali sulfates are readily dissolved into water and are not preserved in the hardened concrete. Hence a missing contribution of alkalis from the alkali sulfates has to be estimated, applying an empirical conversion factor. Results of analyzing cement (Table C.1), separated aggregate (Table C.2) and total alkali (Table C.4) are shown later.

C.3 Minimum Alkali Content of Clinker

C.3.1 Sampling and Sample Preparation

Sampling and cutting: Concrete core sample (typically, diameter 10 cm, length 20 cm) is longitudinally cut into two half cylinders. One half is used to determine water-soluble alkali content of the bulk concrete, and the other half is to separate coarse aggregate for determining its water-soluble alkali content. Weathered surface portion of concrete structure (e.g. carbonated area) should not be used for analysis.

Thin sectioning: Polished thin sections are prepared cutting and grinding with abrasives. In this process, leaching of alkalia should be avoided in terms of using non-polar solutions (lubricant oil, cleansing liquid, ultrasonic cleaner, draft chamber).

C.3.2 Apparatus

- Cutting and grinding machine (concrete cutter, grinding machine, thin section machine, polishing machine).
- Polarizing microscope.
- Well-calibrated energy-dispersive spectrometer (EDS) attached to a scanning electron microscope (SEM) or electron-probe microanalyzer (EPMA) should be used for quantitative analysis (Fig. C.2).

C.3.3 Experimental Procedures

Selection of unhydrated cement particles: Unhydrated cement particles in hardened concrete should be identified in thin sections under the crossed polar in polarizing microscopy, then subjected to SEM observation, (Fig. C.3, to ensure that unhydrated portions (looking bright) for EDS analysis are exposed on the polished thin section (surrounded by the dark hydrated portions). To ensure good average analysis, cement particles from various parts of original clinker nodules (annealed interior and quenched outer shell) should be covered by increasing the number of particles for analysis, Table C.1.

EDS analysis: Unhydrated cement clinker phases (alite, belite, aluminate, ferrite, glass, etc.) in the cement particle are analyzed quantitatively using a well-adjusted EDS or an EPMA calibrated with proper standard materials and a proper correction software. Polished thin section is used for quantitative EDS analysis, Fig. C.3. Results are presented without normalizing to analytical total of 100%. Analytical data from various cement particles in concrete should be averaged.

C.3.4 Calculation Process

Numbers (1) to (5) in the following calculations refer to the numbers (1) to (5) in Table C.4.

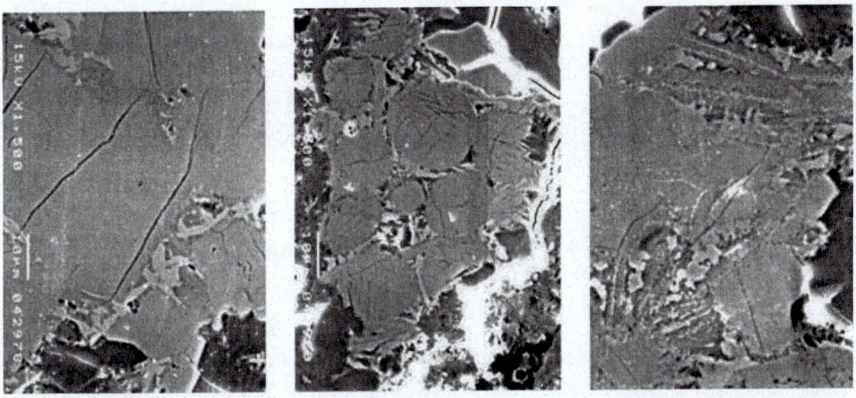

(a) Alite (b) Belite (c) Elongated interstitial aluminate and ferrite

Fig. C.2 Unhydrated cement particles in concrete used to estimate the minimum content of alkali in cement clinker and cement. SEM photographs on polished thin section

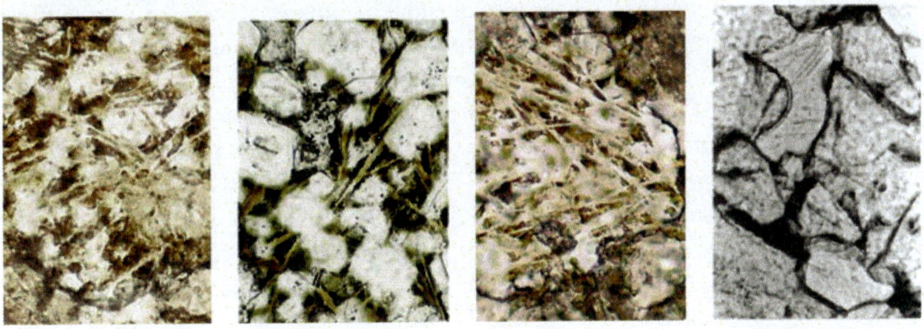

(a) Elongated alkali aluminate in concrete, Stewart Mountain dam, Arizona (b) Elongated alkali aluminate in concrete, RH Saunders dam, Ontario (c) Elongated alkali aluminate in concrete, Beauharnois dam, Quebec (d) Alkali sulfates in old cement clinker, 4) filling voids and forming exsolution cross-lamellae of arcanite K_2SO_4 and glaserite $3K_2SO_4$ Na_2SO_4. Polarizing microscopy

Fig. C.3 Major sources of water-soluble alkalis in high-alkali cement responsible for ASR

1. Minimum alkali content of clinker (1) (%) = (alkali content of alite) × (content of alite) + (alkali content of belite) × (content of belite) + (alkali content of aluminate) × (content of aluminate) + (alkali content of ferrite) × (content of ferrite) + (alkali content of interstitial glass) × (content of interstitial glass)
    - Roughly, alite (60%), belite (20%), aluminate (10%) and ferrite (10%). Alkali-aluminate may replace cubic aluminate. The amount of interstitial glass should be determined by SEM observation on polished thin section.
    - For a moderate-heat cement or classical Portland cement produced by a bottle kiln, appropriate content should be assumed.
2. Total alkali of clinker (2) (%) = k1 × minimum alkali of clinker (1) (%) Empirically, k1 = 1.3–1.4, a compensation factor of water-soluble alkalis from alkali-sulfates which dissolved upon mixing concrete
    a. This ratio may vary locally and by time depending on the change in the production method and raw mix of cement plant. Calibration is necessary when both cement and concrete are available.
3. Total alkali of cement (3) (%) = k2 × total alkali of clinker (2) (%) = k3 × minimum alkali of clinker (1) (%), k2 is a dilution factor of alkali due to the addition of gypsum in cement. k3 is an overall conversion factor, being 1.2–1.3, to convert minimum alkali of clinker (1) to total alkali of cement.
4. Total alkali of cement (4) (kg/m^3) = k3 × [minimum alkali of clinker (1) (%)/100] × cement mix (kg/m^3)
5. Water-soluble alkali of cement (kg/m^3) = k4 × [total alkali of cement (%)/100] × cement mix (kg/m^3), k4 is a ratio of water-soluble to total (acid soluble) alkali, empirically 0.5–0.6 according to the condition of the General Project Method (size <0.3 mm, water/solid = 10, 40 °C, 30 min)
    - When a blended cement with fly ash or blastfurnace slag is identified in concrete with a substitution ratio (z%), the amount of water-soluble alkali of the cement should be multiplied by (100-z)/100. assuming a simple dilution (e.g. 30%) by these supplementary cementing materials (SCM). If no information is available on the extent of alkali release from the SCM in question (Fig. C.1b), just ignore a contribution of water-soluble alkali from them for convenience.
    - In other extraction methods [7] (size <100 mesh = 0.15 mm, water/solid = 10, room temperature, shaking 5 min) and modified versions which apply boiling water, extraction efficiency should be different.
    - In estimating the alkali budgets in concrete, it is important to consistently use the same analytical conditions (temperature, grain size, extraction time, etc.) and avoid acid not to overestimate the amount of soluble alkalis from the aggregate and SCM in concrete.

C.3.5 Precision and Bias

In aged concretes, unhydrated cement particles, even being optically sound, are more or less affected by partial hydration and leaching of alkalis along cracks and by carbonation on concrete surfaces.

It is therefore desirable to adopt analytical data of each cement clinker phase with a total exceeding 96%. By doing so, fluctuation of compositions (cation numbers per unit cell of each phase), calculated fixing oxygen numbers inherent to each phase (alite O = 5, belite O = 4, aluminate O = 6, ferrite O = 5), generally falls within an acceptable range of 1–2%.

Conversion factor of frequentlly used 1.3 for total alkali, inclusive of water-soluble alkali sulfates, has a relative error of 15%, according to comparison with about 20-yeas-old concrete with known cement composition.

C.4 Total Alkali Content of Concrete

C.4.1 Sampling and Sample Preparation

Water-soluble alkali by warm-water extraction. Concrete sample or separated coarse aggregate is crushed into the grain size of max 300 μm. Crushed sample reduced to 10 g is mixed with 100 ml of distilled warm water (40 °C) and subjected to shaking for 30 minutes in a thermostatically controlled warm bath (40 °C) [8]. It is important not to rotate the sample to avoid secondary crushing and peeling off of the aggregate surface during shaking. Then the solution is filtrated to determine concentrations of liberated Na and K ions by atomic absorption spectrometry.

C.4.2 Apparatus

Crusher: jaw crusher, vibration mill, agate mortar, sieve (50 mesh = aperture 0.3 mm).

Extractor: thermostatically controlled warm bath (with agitation table), Erlenmeyer flask (200–250 ml, polyethylene), Buchner funnel, filter paper (grade B), aspirator.

Analytical equipment: atomic absorption spectrometer, chemical balance (sensitivity 10 mg), glass wares (whole pipet, measuring cylinders, measuring flask, etc.).

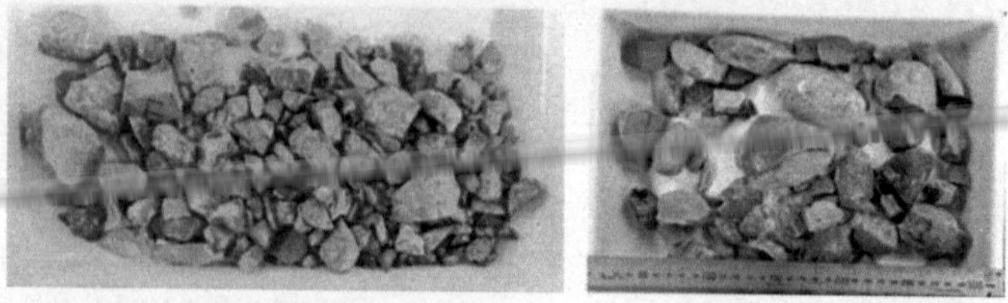

(a) crushed stone from a concrete building (dia.5cm by L20cm)

(b) gravel in a concrete dam (dia. 10cm by L20cm)

Fig. C.4 Coarse aggregate separated from the concrete core specimens

C.4.3 Experimental Procedures

Separation of coarse aggregate and crushing: To avoid leaching of alkalis during sample preparation, particles of coarse aggregate should be separated physically, e.g. by hand chipping from the concrete (Fig. C.4), and not by the selective dissolution by acids (hydrochloric, nitric, salicylic, formic, oxalic, gluconic, etc.). Crush the concrete sample or the extracted coarse aggregate into the maximum size of # 50 (0.3 mm), homogenize repeatedly, then reduce it to a representative 10 g.

Agitation, extraction and filtration: Put 10 g of the reduced sample (concrete or extracted aggregate) into 100 ml of water (40 °C) in the Erlenmeyer flask (200–250 ml), and agitate for 30 min on the warm bath (40 °C). One of shaking conditions can be 160 times/min, amplitude 3 cm. Filtrate the sample with a filter paper (grade B) on the Buchner funnel using an aspirator into the Erlenmeyer flask (200 ml).

Chemical analysis: Dilute the filtrated sample solution to 1 L and measure the concentration of Na and K (mg/L) using an atomic absorption spectrometer with calibrated standards of Na and K concentrations. Calculate the concentration of Na and K (mg/g) in the 1 g of concrete or aggregate: Na (mg/g) = 0.01× Na (mg/L), K (mg/g) = 0.01× K (mg/L). Following conversions should be made: Na (%) = (1/10) × Na (mg/g), K (%) = (1/10) × K (mg/g). Na_2O (%) = 1.348 × Na (%), K_2O (%) = 1.205 × K (%), Na_2O_{eq} (%) = Na_2O (%) + 0.658 × K_2O (%).

C.4.4 Calculation Process

Numbers (6) to (9) in the following calculations refer to numbers (6) to (9) in Table C.4.

6. Water-soluble alkali of concrete [(6)] (kg/m^3) = [water-soluble alkali of concrete (%)/100] × unit weight of concrete (kg/m^3).

- Concentration of water-soluble alkalis of the bulk concrete (%) is converted to the concentration in the unit volume of concrete (kg/m^3) by multiplying the unit mass of bulk concrete.
- If construction record is available, concrete mix (coarse aggregate, fine aggregate, cement, water, etc.) can be used to calculate alkali-budgets in concrete. If not, concrete mix should be estimated by the combination of particular wet chemical analyses, or assumed properly on the empirical basis (e.g. 2,250–2,350 kg/m^3).

7. Water-soluble alkali of separated coarse aggregate $^{(7)}$ (kg/m^3) = [water-soluble alkali of separated coarse aggregate (%)/100] × mix proportion of coarse aggregate in concrete (kg/m^3).

   - Similar calculation process is applied to the separated coarse aggregate by using either the unit mass of coarse aggregate in concrete, or the measured mass of separated coarse aggregate and concrete core specimen.
   - In aged concrete, part of alkali in the aggregate converts to water-soluble by an interaction with Ca ions in cement paste. For instance (Table C.2), in aged dams over 30 years old, Na_2O_{eq} 1.3–0.5 kg/m^3 increased in greywacke rich in illite mica and feldspars; Na_2O_{eq} 1.4–0.6 kg/m^3 in volcanic rocks containing feldspars and glass.
   - Releasable alkali from aggregate before use. For comparison, Table C.3 shows water-soluble alkali of unused aggregates, including greywackes from the bedrock at Mactaquac dam (sampled 1992) and in the pit gravel aggregate used for bridges (sampled 1986) near Saint John, New Brunswick, Canada. It can be seen that, in the course of time, part of alkalis in these aggregates, even not detectable in this method, becomes water-soluble, about 1 kg/m^3 of Na_2O_{eq} (Table C.2) being about 4–5 times that of the original state (Table C.3). As a result, in a concrete dam constructed with about 2 kg/m^3 of cement alkali, a total of 1 kg/m^3 of water-soluble alkalis could be supplied from greywacke aggregate thus exceeding the 3 kg/m^3 threshold for accelerating ASR.
     Ideally, releasable alkali of aggregate should be extracted in alkaline solution with a composition resembling pore water in concrete (e.g. for extraction of K^+, Na^+ saturated $Ca(OH)_2$) [9], because warm water extracts only negligible amount of water-soluble alkalis from unused aggregate.

8. Other water-soluble alkali $^{(8)}$ (kg/m^3) = water-soluble alkali of concrete $^{(6)}$ – water-soluble alkali of cement $^{(5)}$ – water-soluble alkali of separated coarse aggregate $^{(7)}$.

   - The other water-soluble alkali in concrete comes from fine aggregate, deicing salt, sea water, and chemical admixture used, or from NaOH added in concrete prism or from NaOH solution absorbed by the concrete core in the alkali immersion test.
   - If the rock type of fine aggregate is common to the coarse aggregate, water-soluble alkali of the fine aggregate could be estimated from that of the coarse aggregate with proper assumption.

Table C.3 Water-soluble alkali of major rock types of unused coarse aggregate, Katayama (unpublished results)

		unused aggregate										
rock type	major	andesite	slate	greywacke		siltstone	mudstone		siliceous shale	lime stone	[illegible] welded tuff	chert
	minor	-	-	-	argillite	-	chert hornfels	siliceous shale	chert hornfels	-	-	-
water-soluble alkali Na_2Oeq kg/m³ *		0.3	0.3	0.3	0.2	0.2	0.2	0.2	0.1	0.1	0.1	0.05
locality		Na	SJ, NB	SJ, NB	Mactaquac	SJ, NB	Ot	Ka	Ri	Ts	Ak	Se

* General Project Method (<0.3mm, 40°C, water/solid=10, shaking 30 minutes) with shaking (160/min, amplitude 3cm) as 1150kg/m³

Table C.4 Alkali-budgets in ASR-affected concretes recalculated from [1, 4]

	clinker (unhydrated)			cement				coarse aggregate (separated)			others*	concrete			
	minimum alkali			total alkali				water-soluble alkali							alkali
	measured (EDS)			estimated				measured			estm.	measured			estm.
	Na_2O	K_2O	Na_2O_{eq}	Na_2O_{eq}				Na_2O	K_2O	Na_2O_{eq}		Na_2O	K_2O	Na_2O_{eq}	Na_2O_{eq}
	%		%	%		kg/m³		%		kg/m³		%		kg/m³	kg/m³
1	0.62	0.36	0.84	1.09	1.01	2.67	1.34	Na_2O_{eq} 0.040		0.41	1.15**	Na_2O_{eq} 0.123		2.90	4.24
2	0.50	0.22	0.64#	0.83	0.77	2.62	1.57	0.011	0.011	0.14	0.68***	0.066	0.058	2.39	3.44
3	0.43	0.16	0.53#	0.69	0.64	2.16	1.30	0.015	0.013	0.20	0.00	0.042	0.035	1.50	2.36
			(1)	(2)	(3)	(4)	(5)			(7)	(8)			(6)	(9)
	petrography-based (phase 1)							wet chemical analysis-based (phase2)							

\# interstitial glass is present (8)=(6)-(5)-(7), (9)=(4)+(7)+(8)

*contribution of water-soluble alkali from fine aggregate, deicing salt, sea water and chemical admixture, depending on the environment of the structure

**alkali from deicing salt is major source, ***alkali from fine aggregate (slate) is major source

9. Total alkali of concrete [(9)] (kg/m³) = total alkali of cement [(4)] + water-soluble alkali of separated coarse aggregate [(7)] + other water-soluble alkali [(8)].
   - Finally, total alkali content of concrete is calculated by summing up total cement alkali, water-soluble alkalis from the coarse aggregate and other sources. As a reporting format of alkali-budgets in concrete, an example is shown in Table C.4. This total alkali of concrete always exceeds the total cement alkali.
   - "Other water-soluble alkali" contains contribution from fine aggregate, deicing salt, sea water and chemical admixture, depending on the environment of the structure. The amount of added alkali (as NaOH) in concrete prism and absorbed alkali from NaOH solution in the concrete core are also included, and can be estimated from this test.
   - In the cold region, 1 kg/m³ of Na_2O_{eq} could be supplied to concrete by deicing salt.

Table C.4 presents alkali-budgets of ASR-affected concrete viaducts at ages of 19–25 years with total cement alkali in concrete Na_2O eq < 3.0 kg/m³. Viaduct 1 has total alkali Na_2O eq > 4 kg/m³, contributed from water-soluble alkali of deicing salt and aggregate. Structures 2 and 3 contained limestone coarse aggregate and were free of de-icing salt. Viaduct 2 represents late-expansive ASR of the sand aggregate in which reactive slate particles liberated water-soluble alkali, whereas viaduct 3 contained highly reactive cristobalite in the andesite present in a pessimum proportion, which reacted with total alkali of concrete Na_2O eq over 2 kg/m^3.

C.4.5 Implication of Alkali-Budgets

Estimation of alkali-budgets provides evidence of long-term supply of soluble alkalis from the aggregate in the hardened concrete.

1. Early-expansive aggregate with a small pessimum proportion (opal, cristobalite, tridymite) expands deleteriously, even if the total concrete alkali including the contribution of water-soluble alkali from the aggregate is lower than 2.5 kg/m^3.
2. As much as 1 kg/m^3 of water-soluble alkali can be supplied from particular rock types of aggregate (greywacke, andesite, granite: Table C.2). This means that the alkali content of a mass of concrete originally with less than 2.0 kg/m^3 can come to exceed Na_2O eq 3 kg/m^3.
3. As a result, expansion of concrete will continue for a long time, particularly with late-expansive aggregate such as greywacke and granite.

C.4.6 Precision and Bias

Analytical error of pulverized sample is small. However, increased extraction temperature, extraction time, water/solid ratio [10] and fineness of pulverized sample result in higher extraction ratio of alkalis. Main error originates from inhomogeneous distribution of coarse aggregate in concrete, small size of core sample, and application of inappropriate unit mass of concrete and coarse aggregate. Usually, the amount of the recovered coarse aggregate by hand-picking is lower than the designed mix.

C.4.7 Long-Term Alkali Release of Alkali from Aggregate in Concrete

Unlike the quick extraction method of alkalis in warm water [8] release of alkali ions from alkali-bearing minerals into pore solution is enhanced by the presence of calcium ion and OH^- ions, increased temperature and time for reaction, as well as by reduced particle size of the aggregate.

In basic solutions, for instance, alumino-silicate tetrahedral framework decomposes more at higher pH, as shown in Eq. C.1 for albite [11] and in Eq. C.2 for muscovite [12]. This means that OH^- concentration will decrease when alkali ion (Na^+) is liberated, just like ASR of reactive silica with alkali ions. In concrete, Ca^{2+} ions are present, liberated aluminate ion and silicate ion may form C_4A hydrates and CSH gel respectively, while a hydrogarnet katoite $Ca_3Al_2(SiO_4)_{3-x}$-$x(OH)_{4x}$ (x = 1.5–3) will form at higher temperature.

$$NaAlSi_3O_8 + OH^- \rightarrow Na^+ + Al(OH)_4^- + SiO_4^{4-} \quad \text{Albite} \tag{C.1}$$

$$KAl_2(Si_3AlO_{10})(OH)_2 + 7H_2O + 5OH^- \rightarrow K^+ + 3Al(OH)_4^- + 3SiO(OH)_3^- \quad \text{Muscovite} \tag{C.2}$$

Even non-reactive aggregate, e.g. nepheline syenite, may release alkali and enhance expansion of reactive aggregate in concrete prism at 38 °C [13] and mortar microbar at 80 °C [14]. With the late-expansive aggregate, like granites, coarse fraction (12.5–20 mm) containing feldspars and micas kept in alkaline solution (38 °C, 1M NaOH + saturated $Ca(OH)_2$ or 1M KOH + saturated $Ca(OH)_2$) releases alkalis of Na_2O_{eq} 0.12% in 65 weeks, corresponding to 2.0% of alkalis of the aggregates [15].

Fine fraction of a granite aggregate (<4 mm) in contact with alkali solutions for 13 weeks (0.7M NaOH + saturated $Ca(OH)_2$ or 0.7M KOH + saturated $Ca(OH)_2$) releases 4.9% (38 °C) to 5.9% (60 °C) of alkalis of the aggregate, which corresponds to Na_2O_{eq} 5.1–6.1 kg/m^3 in concrete prism, higher than cement alkali [9]. Fine aggregate of alkali feldspar microcline (<4.75 mm), constituting 25% of the aggregate in the ASTM C227 mortar bar (low alkali cement) kept at 38 °C for 10 months, releases alkalis of 0.14 mol/L (2–3% of alkalis in microcline), whereas 30–50% of the released alkalis is removed by ASR products in the reacted calcined flint [16]. Powders of orthoclase and albite, when kept in a slurry with $Ca(OH)_2$ at 80 °C for 26 days, release as much as 35% of alkalis, and these feldspars are decomposed into C_4A hydrates (<50 °C) and/or hydrogarnet (>50 °C) [17]. In the autoclaving condition (150 °C, 1 day), vigorous decomposition of alkali-rich rocks in alkaline solution (KOH + saturated $Ca(OH)_2$) occurs, leaving calcium-alkali-bearing products [14] with approximate compositions of Ca_4K_2 (Si,Al)$_4$ Ox.nH_2O from nepheline and Ca_3K_3 (Si,Al)$_4$ Ox.nH_2O from K-feldspar. This is not a result of ion exchange between Ca ion and alkali ion, but a decomposition of these minerals at hydrothermal condition.

C.5 Limitations and Applicability to Analysis

C.5.1 Assumptions Used in Calculations

Water-soluble alkali ratio of cement: wherever unhydrated cement particles are present in concrete, EDS analysis can be used to estimate the minimum amount of cement alkali. However, in converting the minimum cement alkali to total cement alkali (the ratio of water-soluble alkali sulfate to other cement minerals in clinker), and the total cement alkali to water-soluble alkali of cement, fixed conversion factors were postulated for convenience. This can be the source of error in estimating cement-originated alkalis in concrete, since these ratios may vary considerably according to the production methods of cement (e.g. raw mix, kiln type, processing of kiln dust and gas, etc.). Hence, the above conversion factors for old concretes should be chosen to conform with the local situations of cement production in the past.

Recovery ratio of aggregate: the recovery of coarse aggregate for analysis by chipping hardened concrete does not always reach 100%, and could be over 10% less than the designed mix. Water-soluble alkalis originated from the coarse aggregate may have been fixed within this aggregate and/or exuded into the surrounding mortar portion in several forms (ASR gel, CSH gel). Because of this, relative proportions of estimated alkalis (coarse aggregate, fine aggregate and others) in concrete could vary depending on the recovery method of the coarse aggregate. However, the total amount of alkalis of concrete (cement, aggregates and other sources) in the alkali budgets in concrete does not change, and this total alkali should be used for assessing the potential for future expansion of ASR in concrete.

Extraction of water-soluble alkali: the condition for extracting water-soluble alkali (40 °C, 30 min, $< 300 \mu m$: General Project Method 3.3 [8] is even milder than the long-term extraction methods based on combined alkali-$Ca(OH)_2$ solutions that simulate the chemistry of pore solution in concrete to see ultimate release of alkalis. Hence, the examples of concrete structures (less than 55 years old, Table C.2) from which water-soluble alkalis of separated coarse aggregate were determined may not be old enough. Then, the total water-soluble alkalis exclusive of the cement origin (coarse aggregate, fine aggregate, deicing salt, sea water, etc.), namely, the additional alkalis (up to Na_2O_{eq} 1.5 kg/m^3), could be underestimated.

C.5.2 *Applicability*

Explanation of ASR for originally low-alkali concrete: when concrete contains early-expansive aggregate with highly reactive silica minerals, additional alkalis can accelerate ASR in field concretes, even if the total cement alkali or the total concrete alkali is below the threshold of expansion (Na_2O_{eq} 3 kg/m^3). Thus, the calculation of the alkali budgets clarifies the origin of additional alkalis, and the reason why concrete made with relatively low-alkali cement produced deleterious expansion.

Correction of final expansion with alkali content: alkali ions supplied from the aggregate and NaCl solutions (sea water and deicing salt) do not increase OH- ion concentration, but raise alkali concentrations of pore solutions in concrete. This accelerates the expansion of concrete prisms/cylinders and field concretes exposed to humid atmosphere. It is therefore desirable to estimate the amount of equivalent alkali (Na_2O_{eq}) that was gradually supplied to concrete structures after construction, then correct the final expansion of concrete. By contrast, in the alkali immersion test, alkali content of concrete reportedly does not significantly affect the expansion. However, difference in the alkali level of concrete before and after the immersion of core specimens into alkali solution should better be corrected for expansion.

References

1. Katayama, T., Tagami, M., Sarai, Y., Izumi, S., Hira, T.: Alkali-aggregate reaction under the influence of deicing salts in the Hokuriku district. Japan. Mater. Char. **53**, 105–122 (2004)
2. Ando, Y., Katayama, T., Hirono, S., Sato, T., Noguchi, T.: Oldest example of alkali-silica reaction in Japan-Tokyo bay sear forts No. 2 and No. 3 constructed 120–130 years ago. In: Proceedings of the 15th International Conference on Alkali-Aggregate Reaction (ICAAR). Sao Paulo, Brazil (2016)
3. Nakano, M.: Proposal of investigation methods of alkali aggregate reaction on concrete for nuclear power plant. Tech. rep. JNES-RE-2013-2050, p. 29. Japan Nuclear Energy Safety Organization (JNES) (2014) (in Japanese)
4. Katayama, T., Oshiro, T., Sarai, Y., Zaha, K., Yamato, T.: Late-expansive ASR due to imported sand and local aggregates in Okinawa Island, southwestern Japan. In: Proceedings of the 13th International Conference on Alkali-Aggregate Reaction (ICAAR), pp. 862–873. Trondheim, Norway (2008)
5. Katayama, T.: Petrographic study of alkali-aggregate reactions in Concrete. Ph.D. thesis. University of Tokyo (2012)
6. Katayama, T., Grattan-Bellew, P.E.: Petrography of the Kingston experimental sidewalk at age 22 years - ASR as the cause of deleteriously expansive, so-called alkali-carbonate reaction. In: Proceedings of the 14th International Conference on Alkali-Aggregate Reaction (ICAAR). Austin, Texas, USA (2012)
7. ASTM C114-18: Standard test methods for chemical analysis of hydraulic cement. Tech. rep. West Conshohocken, PA (2018)
8. General Project Method 3.3: Method of analyzing water-soluble alkalis in concrete. In: Development of Technology to Improve the Durability of Concrete Structures, vol. 2, pp. 141–142. Ministry of Construction, and National Land Developing Technology Research Center (1988)
9. Menendez, E., Garcia-Roves, R., Aldea, B., Ruiz, S.: Alkali release of aggregates of different solutions and conditions of test. In: Proceedings of the 15th International Conference on Alkali-Aggregate Reaction (ICAAR). Paper No. 138. Sao Paolo, Brazil (2016)
10. Nishizaki, I., Moriya, S.: Some experiments on the method of analyzing alkalis in concrete. Concrete J. (Japan Concrete Inst. JCI) **8**, 73–76 (1986) (in Japanese)
11. Blum, L., Stillings, A.E.: Feldspar dissolution kinetics. Chemical weathering rates of silicate minerals. Rev. Mineral. **31**, 291–346 (1995)
12. Knaus, K., Wolery, T.: Muscovite dissolution as a function of pH and time at 70_C. Geochimica et Cosmochimica Acta **53**, 1493–1501 (1989)

13. Durand, B.: More results about the use of lithium salts and mineral admixtures to inhibit ASR in concrete. In: Proceedings of the 11th International Conference on Alkali-Aggregate Reaction (ICAAR), pp. 623–632. Quebec, Canada (2000)
14. Wang, Y., Deng, M., Tang, M.: Alkali release from aggregate and the effect on ASR expansion. Mater. Struct. **41**(1), 159–171 (2008)
15. Soares, D., Silva, A., Mirao, J., Fernandes, I., Menendez, E.: Study on the factors affecting alkalis release from aggregates into ASR. In: Proceedings of the 15th International Conference on Alkali-Aggregate Reaction (ICAAR). Paper No 135. Sao Paolo, Brazil (2016)
16. Constantiner, D., Diamond, S.: Alkali release from feldspars into pore solutions. Cement Concrete Res. **33**, 549–554 (2003)
17. van Aadt, J., Visser, S.: Reaction of Ca(OH)2 and of Ca(OH)2 + CaSO4.2H2O at various temperatures with feldspars in aggregates used for concrete making. Cement Concrete Res. **8**, 677–682 (1978)

Appendix D
Petrographic Based Assessment

Tetsuya Katayama

Reviewed by D. Rothstein

D.1 Scope

This method assesses the advancement of alkali-silica reactions (ASR) in concrete structures and susceptibility to ASR, based on petrography (identification of reactive minerals and reaction type of the aggregate, stage of reaction and grade of deterioration). The method can also be used to evaluate laboratory concrete specimens as to what manner reactions took place in the acceleration process applied to determine their expansion potential.

D.2 Principles and Methodology

The method described here has been used to various extents to diagnose ASR and the extent of past expansion in several concrete structures in Japan. Some of the results are published by [1–3]. The stage of ASR in concrete is determined based on the microscopic textures observed in thin sections of concrete, because the history of development of ASR is recorded as crack propagation from the reacted aggregate with reaction products in concrete, as well as by the crack index Table D.1.

Advantages:

1. Can identify type of aggregate and reactive minerals.
2. Can determine past expansion of concrete by measuring crack index, in conjunction with the field inspection.

T. Katayama
Taiheiyo Consultant Co., Ltd., Sakura, Japan
e-mail: katayama@catv296.ne.jp

© RILEM 2021

V. E. Saouma (ed.), *Diagnosis & Prognosis of AAR Affected Structures*,
RILEM State-of-the-Art Reports 31,
https://doi.org/10.1007/978-3-030-44014-5

Table D.1 Petrographic examination for the study of ASR in concrete, (compiled by Kayama);

item	subject	early-expansive ASR	late-expansive ASR	methodology
type of ASR	early-, or late-expansive	identification of reactive minerals		polarizing microscope
pessimum proportion (vol%)	reactive rock type			
	rock types of the aggregate	the ratio of reactive rock types		core-scanning, point counting
past expansion (%)	crack index	total crack width/unit length of concrete		visual inspection with crack scale, microscopy
progress of ASR field concrete test specimen	reaction rims, cracks and ASR gel	identification of the stages i),ii),iii),iv), v) and vi)		polarizing microscopy
	composition of ASR gel and rosette-like products	plot on the [Ca/Si]-[Ca]/[Na+K] diagram		quantitative EDS analysis
		distribution of alkalis		elementary mapping
reaction ratio (vol%)	aggregate (opal, glass)	thickness of reaction rim/original radius		reflecting microscope, SEM
	reactive mineral (microcrystalline quartz)		reacted area/ original area of pseudomorph	SEM (BSE image), elemental mapping on polished thin section
amount of reaction products (vol%)	ASR gel and rosette-like products	area of gel/unit area of concrete polished thin section		
alkali budgets in concrete (kg/m^3) Appendix B	cement alkali	unhydrated clinker minerals		petrography, quantitative EDS analysis
	alkali from coarse aggregate	water-soluble alkali of separated aggregate		water-soluble alkali (General Progect method)
	alkali from fine aggregate and deicing salt	water-soluble alkali of bulk concrete. Aggregate, and alkali of cement		
	leaching of alkali			
content of mineral admixtures (vol%)	fly ash, blast furnace slag	area of admixtures/area of hydration products		elemental mapping by EDS or EPMA
correlation with sigmoidal curves Chapter 8	classification of rock types for reactivity	ε_∞, τ_L, τ_C of CPT for each rock type (38°C, 60°C, Na_2Oeq 5-5.5kg/m^3)		comparison with fitting data

3. Cement type can be evaluated, even if concrete is 100 years old and lacking construction records.
4. ASR can be detected before external cracks appear in concrete.
5. Can tell what reactions actually took place in the laboratory specimens.
6. Provides background data for qualitative assessment of future potential expansion.

Disadvantages:

1. Need to secure representativeness of core samples and thin sections in concrete.
2. Analysis of ASR gel needs expensive facilities and skill.

D.3 Sampling and Sample Preparation

Polished thin sections (usually 25 mm by 35 mm, 15–20 μm thick) Fig. D.1b, or large area thin section, Fig. D.1c are made from concrete core samples taken from portions showing representative deterioration, as classified in the field inspection. Core samples, wrapped with a cling film to avoid drying and secondary carbonation, are used for petrographic examination. They are cut into pieces to contain reacted aggregate particles which generally show a wet appearance due to exudation of sol/gel during storage for 1–2 days at ambient temperature. Polished thin sections are prepared using a non-polar solvent to avoid leaching of alkalis during sample preparation.

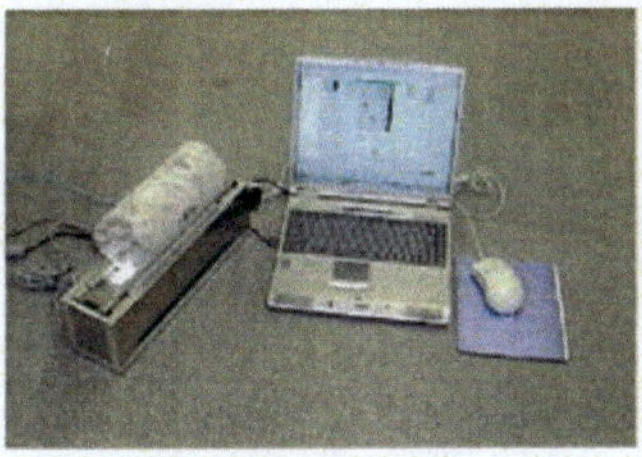

(a)Core scanner

(b) uncovered polished thin sections

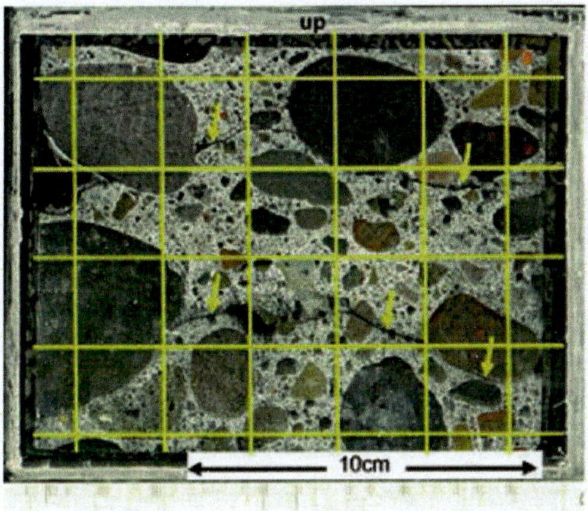

(c) Large-area thin section of core from arch dam (early-expansive ASR, rhyolite), with mesh for measuring crack index in vertical (Z) and horizontal (X-Y) directions. Parallel horizontal cracks mostly debonding pebbles, in dark-field illumination. Adapted from [31]

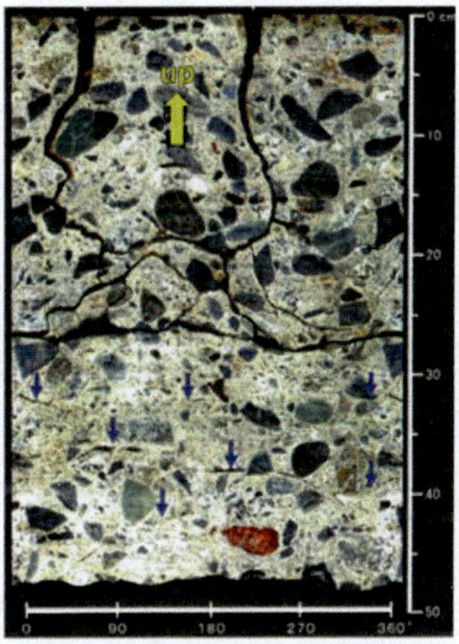

(d) Scanned image (360 °) of core from intensely deteriorated hydraulic structure, with vertical tension crack growing to the surface and numerous horizontal compressive cracks (arrows) inside [40]

Fig. D.1 Laboratory tests

D.4 Apparatus

Stereo-binocular microscope, Fig. D.2a, polarizing/reflecting/fluorescence microscope, Fig. D.2b and scanning electron microscope (SEM) equipped with quantitative energy-dispersive spectrometer (EDS), Fig. D.2c are the most popular equipment used for petrography of ASR-affected concrete.

Core scanner can be used for recording and modal analysis of coarse aggregate, Fig. D.1a, d.

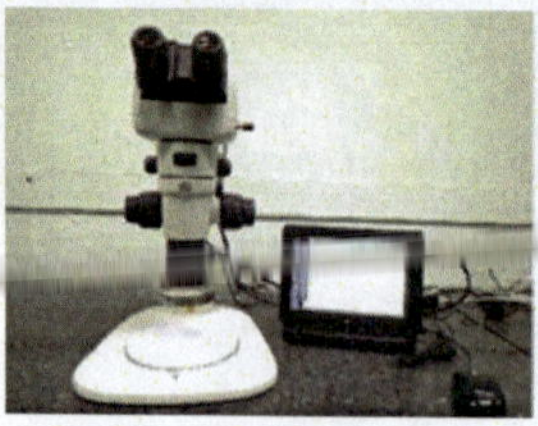

(a) Stereo-binocular microscope

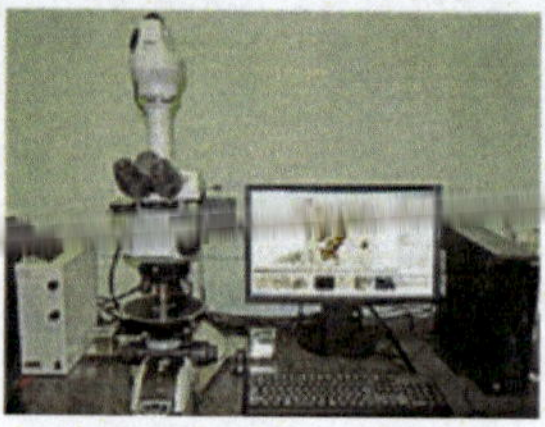

(b) Polarizing /reflecting /fluorescence microscopy

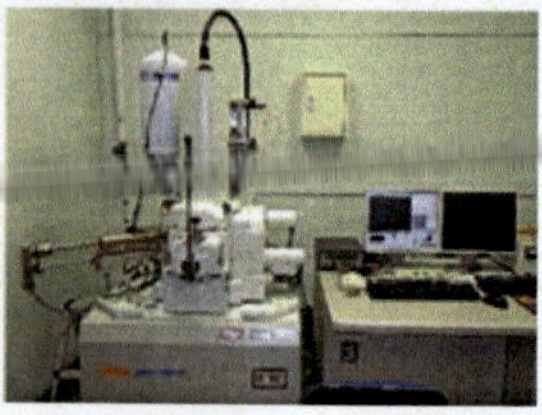

(c) SEM/EDS for qantitative analysis

Fig. D.2 Laboratory tests Equipments

D.5 Experimental Procedures

D.5.1 Petrographic Examination

D.5.1.1 Phase 1: General Background

Expansion speed of rock types: ASR can be roughly classified into two types: early-expansive ASR and late-expansive ASR. Early-expansive ASR is "classical ASR." Late-expansive ASR was formerly called "slow/late-expanding alkali-silicate/silica-reaction" in the CSA standard in the 1980s and 1990s [4], whose deleterious reactivity cannot be judged by the conventional mortar bar test [5]. For convenience, this type of ASR was renamed as "late-expansive ASR" by Katayama [6].

Figure D.3 characterizes the potential reactivity of representative rock types as evaluated by the concrete prism test (CPT). Expansion data were compiled from worldwide literature, confined to tests performed mainly at 38°C and alkali content between Na_2Oeq 5.1–5.5 kg/m^3, based on the compiled data of this test by the author (Table 6.9).

Although CPT is subject to drying up and leaching of alkali in the laboratory, expansion is often underestimated, it can be used to compare expansion speed of each rock type (to reach deleterious expansion in concrete, 0.04%). However, expansion speed has a wide variation within certain rock types (e.g. quartzite, gneiss, granite) due to the variety of texture and reactive minerals they contain.

1. Early-expansive: < 50 days
2. Moderately-expansive: 50–90 days
3. Late-expansive: 90–180 days
4. Extremely late-expansive: > 180 days

As shown in Table D.2, the most problematic rock types that cause ASR in large dams and power plants in the world are extremely late-expansive rocks (quartzite, gneiss, granite) and late-expansive rocks (greywacke, siltstone, quartz sandstone). It is generally difficult to obtain reliable expansion data for these rock types in the concrete prism test, because their expansion does not come to an end at 38°C within

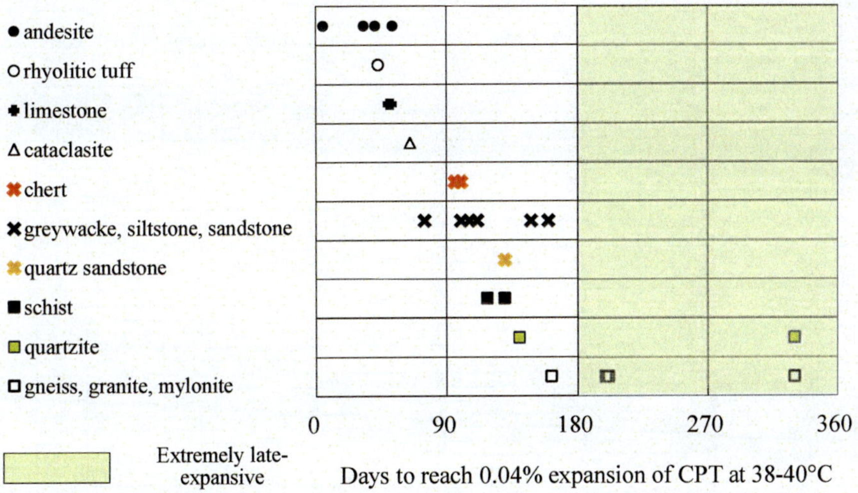

Fig. D.3 Time to reach 0.04% expansion of major rock types from 8 countries in the concrete prism test (mostly 38 °C, Na_2O_{eq} 5.1–5.5 kg/m^3, one andesite 5.8 kg/m^3), see Tables 6.9 and D.2, Katayama (unpublished results). Data sources: andesite: JCI (1991), Yamada, Karasuda, Ogawa, Sagawa, Osako, Hamada and Isneini (2014), Tanaka, Yamada, Sagawa and Ogawa (2015); tuff: Berube, Dupont, Pigeon and Stoian (1992); limestone: Fournier, Chevrier, de Grosboir, Lisella, Folliard, Ideker, Shehata, Thomas and Baxter (2004), Nixon and Lane (2006); cataclasite: Nixon and Lane (2006); chert: Tanaka, Yamada, Sagawa and Ogawa (2015); greywacke, siltstone, greywacke/argillite: Berube, Dupont, Pigeon and Stoian (1992), Nixon and Lane (2006), Fournier, Chevrier, de Grosboir, Lisella, Folliard, Ideker, Shehata, Thomas and Baxter (2004); sandstone: Jensen (1993); quartz sandstone: Durand (2000); schist: Durand (2000), Haha, Gallucci, Guidoum and Scrivener (2007); metagreywacke: Berube, Dupont, Pigeon and Stoian (1992); limestone: Fournier, Chevrier, de Grosboir, Lisella, Folliard, Ideker, Shehata, Thomas and Baxter (1992); quartzite: Jensen (1993), Sanchez, Kuperman, Helene and Kihara (2008); gneiss: Sanchez, Kuperman, Helene and Kihara (2008): granite: Sibbick and Page (1992), Sanchez, Kuperman, Helene and Kihara (2008); gneiss: Sanchez, Kuperman, Helene and Kihara (2008); mylonite: Jensen (1993)

one year as specified in the test, and there is much evidence of underestimation of the potential expansion due to leaching of alkali and drying during the storage of the concrete prism specimens in the air with >95% relative humidity.

D.5.1.2 Phase 2: Identification of the Type of ASR and Reactive Rocks

Thin section petrography using a polarizing microscope is performed to identify the type of rocks and alkali-reactive minerals in concrete, and whether the aggregate in concrete is producing early-expansive ASR, late-expansive ASR or intermediate of these (Table D.3).

In older strata, highly reactive minerals (opal, cristobalite, tridymite and glass) have been recrystallized into stable quartz through diagenesis [7] or, further, metamorphism. In active volcanic belts where the geothermal gradient is high, e.g. circum-

Table D.2 The types of reactive aggregates in various dams and nuclear power plants, compiled by Katayama

	rock type	structure		country	core expansion test at 50*-60°C			outdoor condition of structure			ref
					τ_l [illegible]	τ_c [illegible]	ε^{∞} %	τ_l years	τ_c years	ε^{∞} %	
[illegible] expansive	andesite	Friant	dam	California, USA							[1]
		[illegible]	NPP	Japan				6.0	2.3	0.09	[2][32]
	rhyolite/andesite	Copper Basin	dam	California, USA				5.5	3.3	0.20	[1][32]
		Gene Wash	dam	California, USA				5.0	8.0	0.31	
		Parker	dam	Calif./Arizona, USA							[3]
		Stewart Mountain	dam	Arizona, USA				20.3	6.3	0.16	[4][32]
Moderate expansive	limestone	RH Saunders	dam	Ontario, Canada							[5]
		Riviere des Prairies	dam	Quebec, Canada							[6]
		Hull 2	dam								
		Gentiilly 2	NPP								[8]
	?	Poglia	dam	Italy				6.7	2.5	0.35	[9]
Late-expansive	greywacke	Fontana	dam	North Carolina, USA							[10]
		Petty Harbor	dam	Newfoundland, Canada							[11]
		Malay Falls	dam	Nova Scotia, Canada							[12]
		Hemmings	dam	Quebec, Canada							[6]
		Dinas	dam	UK							[7]
		Maentwrog	dam								[13]
	greywacke/argillite	Mactaquac	dam	New Brunswick, Canada							[14]
	greywacke/gneiss	Ohnyu	headwork	Japan							[15]
		Furikusa	headwork								
	argillite	Lady Evelyn Lake	dam	Ontario, Canada							[16]
	slate	Drum Afterbay	dam	California, USA							[17]
	phyllite	Buck	dam	Virginia, USA							[18]
		Lac Mitis	dam	Quebec, Canada							[6]
		Temiscouata	dam								
	quartz sandstone	Beauharnois	dam								[19]
	quartzite	Santa Luzia	dam*	Portugal	112.5	56.5	0.085-0.18				[20]
		Pracana	dam								[21]
		Paul Sauer		South Africa							[17]
		Keerom									
		Poortjieskloof									
		Roode elsburg									
		Stompdrift									
		Furnas	dam, spillway, power house	Brazil	30	50	0.15	100	30	0.20	[22]
		Pedras	dam								[23]
		Jaguara	water intake/ power house/spillway								
	metaquartzite	Paugan	dam	Quebec, Canada							[6]
		Sept Chutes	dam								
	quartzitic diorite	Rapides Blanc	dam								
	?	Isola	dam	Italy	62	22	0.355				[24]
	mylonite	Traicao	water intake/power house	Brazil							[23]
	granite/mylonite	Sperillen	dam	Norway							[25]
		Sarpfossen	dam								
	gneiss/schist	Castelnau	dam	France							[26]
	gneiss	Chambon	dam								
		St.Narcisse	dam	Quebec, Canada							[6]
		Gabelle	dam								
		Pålsby	dam	Norway							[25]
		Grotte	dam								
		Kamburu	spillway	Kenya							[27]
		Kariba	dam	Zimbabwe-Zambia				25	25	0.28	[28]
		Jurupara	dam	Brazil							[23]
		Joanes II									
		Ilha dos Pombo	dam/spillway								
		Santa Branca									
		Rio das Pedras									
		Peti	water intake /spillway								
		Sa Carvalho	power house								
	granite-gneiss	Pedro Bight	dam/spillway	Brazil							[23]
		Paulo Afonso 1-4	water intake/power house								
		Moxoto	spillway/power house								
	gneissic granite	Dartmouth	dam	Victoria, Australia							[33]
	granite	Maury	dam	France							[26]
		Salas	dam	Spain							[29]
		Bilings Pedra	water intake	Brazil							[23]
		Rihand	dam	India							[17]
		Seabrook	NPP	New Hampshire, USA							[4]
	granitoid	Chelsea	dam	Quebec, Canada							[6]
		Daniel Johnson	dam								
		Farmers	dam								
		Grand-Mere	dam								
		La Tuque	dam								
		Rapides des Quinze	dam								
		Shawinigan 2	dam								
	diorite/granite	Hirakud	dam	India							[30]
	siliceous rock	Temple-sur-Lot	dam	France					84-522		[31]

(shaded) extremely late-expansive aggregate

Data sources: [1] Blanks and Meissner (1945), [2] Takakura et al. (2005), [3] Blanks (1941), [4] Meissner (1948), [5] Danay et al. (1993), [6] Berard and Roux (1986), [7] Thomas et al. (1992), [8] Pietruszczak (2011), [9] Saouma et al. (2007), [10] Abraham and Sloan (1979), [11] Katayama and Bragg (1996), [12] Gillot et al. (1973), [13] Blackwell et al. (1996), [14] Hayward et al. (1988), [15]Katayama et al. (2004), [16] Dolar-Mantuani (1969), [17] Charlwood et al. (1992), [18] Kammer and Carlson (1941), [19] Gocevski and Pietruszczak (1999), [20] Gomes et al. (2004), [21] Silva and Fernandes (2011), [22] Fairbairn et al. (2004), [23] Andriolo (2000), [24] Saouma (2014), [25] Jensen (1993), [26] Millet et al. (1985), [27] Watson (1985), [28] Pan et al. (2013), [29] Del Hoyo and Guerreiro (1985), [30] Murthy et al. (1985), [31] Sellier et al. (2009), [32] Katayama (2017), [33] Victorian Field Trip Notes (1996)

Table D.3 The type of ASR and reactive minerals which should be identified in the petrographic examination, (compiled by Katayama)

type	reactive mineral		pessimum proportion*	reactive rock type	geologic age		alteration zone**
early-expansive	opal		<5%	diatomite, silicified rock from active hot spring	Cenozoic	Pliocene	slightly altered zone (opal-A zone)
				opaline mudstone/shale/tuff/sandstone/flint/limestone		Pliocene, Miocene	smectite zone (opal CT zone)
	cristobalite, tridymite		<10%	andesite, dacite, rhyolite (primary cristobalite and tridymite)	Cenozoic	Quaternary	unaltered zone
						Pleistocene, Pliocene	slightly altered zone
				altered andesite, altered rhyolite		Miocene	smectite zone
moderately-expansive	volcanic glass	rhyolitic glass, dacitic glass, andesitic glass	100%	andesite, dacite, rhyolite, obsidian, rhyolitic welded tuff, interstitial glass in basalt, in andesite and in dacite	Cenozoic	Quaternary	unaltered zone
		hydrated rhyolitic glass		perlite		Pleistocene	slightly altered zone
				pitchstone		Miocene	smectite zone
	chalcedony		20%	chert, flint, siliceous limestone	Cenozoic to Paleozoic	Miocene & earlier	mixed-layer clay zone (chalcedonic/cryptocrystalline quartz zone)
late-expansive	cryptocrystalline quartz		50%	chert, flint, siliceous limestone,			
				devitrified rhyolite, rhyolitic welded tuff, mudstone	Mesozoic, Paleozoic	Cretaceous & earlier	chlorite zone
	microcrystalline quartz		>80%	sandstone, greywacke, siltstone, argillite, cataclasite	Mesozoic, Paleozoic, Precambrian		chlorite zone (microcrystalline quartz zone)
				quartz sandstone with quartz overgrowth	Paleozoic, Precambrian		
				metachert, metarhyolite, metagreywacke, slate, phyllite			
				schist, hornfels, cataclasite			(granular quartz zone)
				quartzite, gneiss	Precambrian		(granular quartz zone)
				granite, charnockite, mylonite			

*Katayama (1997), **Katayama & Kaneshige (1987), Katayama & Futagawa (1989), Katayama, Helgason & Olafsson (1996), Katayama (2016)

(shaded) extremely late-expansive

Pacific region [8], and mid-Atlantic ridges [9], the smectite zone is within the Cenozoic strata (Miocene), whereas in old platforms around shields in the continental regions, this zone goes down to the lower Paleozoic strata (e.g. Ordovician bentonite beds). Thus, chalcedony (which characterizes an even lower horizon than the smectite zone) may occur in older strata than those in the active regions in the world.

D.5.1.3 Phase 3: The Stage of ASR in Concrete

The progress of ASR in concrete can be classified into five or six petrographic stages, based on microscopic textures of reaction products and propagation of cracks [1]. They are common in both early-expansive ASR and late-expansive ASR in field concretes [2, 3, 11]. The following stages can be seen Table D.4.

Figure D.4 shows the typical sequence of ASR mainly in the early-expansive andesite aggregates in Japan. Reaction rims are not evident on the cut or polished surface of concrete prism, particularly with particles of the late-expansive aggregates (e.g. quartzite, schist, gneiss, granite), Fig. D.4a. They can be confirmed on the fracture surface. Exudation of ASR sol/gel into the surrounding cement paste is characterized by darkening of the contacting paste (rimming, halo) under transmitted light, Fig. D.4b. At this stage, capillary pores within aggregate and surrounding paste are filled with sol, but this exudation is less evident in the late-expansive rock types. In the concrete prism test, late-expansive aggregates may not develop apparent reac-

Table D.4 Petrographic stage of ASR, compiled from [1, 10]

stage	site	development of microscopic textures
i	aggregate	• formation of reaction rim. • no cracks
ii	cement paste	• halo of ASR sol/gel in cement paste around the reacted aggregate. • no cracks • air voids adjacent to the reacted aggregate may be filled with ASR gel through permeation
iii	aggregate	• formation of gel-filled cracks within reacted aggregate • no cracks in cement paste
iv	cement paste	• propagation of gel-filled cracks from reacted aggregate into cement paste
v		• widening of gel-filled cracks. • air voids distant from the reacted aggregate are lined with ASR gel migrated along cracks in cement paste
vi		• formation of network of gel-filled cracks connecting the reacted aggregate

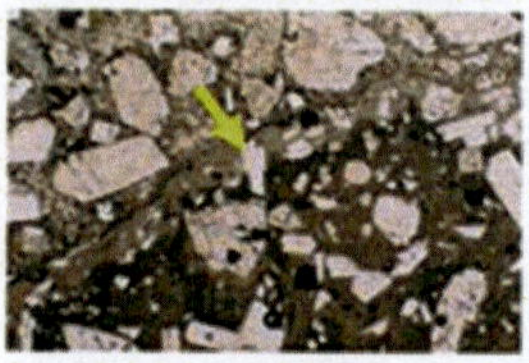

(a)Stage i) Reaction rim within aggregate

(b) Stage ii) Halo of ASR gel in cement paste around the aggregate

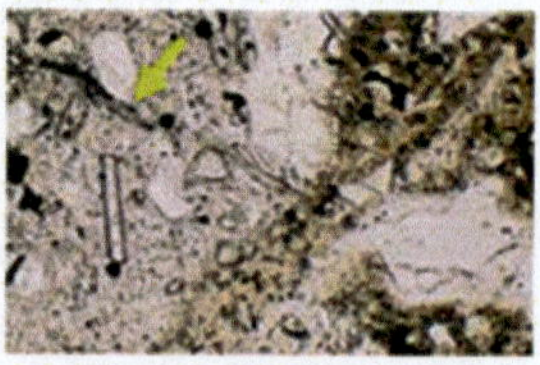

(c) Stage iii) Crack within reacted aggregate

(d) Stage iv) Gel-filled crack extending from the aggregate into cement paste

(e) Stage v) Void-filling gel along the crack distant from the aggregate

(f) Stage vi) Network of gel-filled cracks between aggregate

Fig. D.4 Sequence of textural development in ASR of mostly the early-expansive andesite aggregates in field concretes, [1, 10]

tion rims, gel-filled cracks or gel-filled voids in the cement paste. This is because, as laboratory expansion tests are designed to make a minimal engineering judgement whether the aggregate is potentially deleterious or not, they terminate within an acceptable time limit (e.g. 150 days to 1 year) before ASR develops into a mature state. For this reason, ASR in the laboratory specimens is not pronounced compared with field deteriorated concrete structures with age over 40 years.

After the appearance of cracks within cement paste, i.e. stages (iv) (Fig. D.4d), (v) (Fig. D.4e), and (vi) (Fig. D.4f), crack width or the crack index (total crack width (mm/m) per unit length of concrete) will serve as an indicator of expansion or the distress caused by ASR.

In about 10 years, ASR proceeded from stage (v) to the most advanced stage (vi) and late-expansive gneiss and granite took part in ASR (Table D.5). Table D.6 is an

Table D.5 Progress of petrographic stage of ASR in 10 years in thin section. Simplified from Katayama [1]

structure	aggregate	rock type	1999 progress of ASR →						2008 (2010) progress of ASR →							
			i	ii	iii	iv	v	all	i	ii	iii	iv	v	vi	all	*
Ku 1975	gravel	andesite			x	x	x	v	x	x	x	x	x		v	3
		rhyolite							x							
		rhyolitic tuff	xx													
		gneiss									x	x				
	sand	andesite			xx	x		iv	x	x	x	x	x	x	vi	
		glassy rhyolite			x											
		rhyolitic tuff							x							
Te 1980	gravel	andesite	x		x	x	(x)	iv	x	x	xx	x	x	x	vi	3
		granite									x	x				
	sand	andesite			xx	x		iv	x		x	x	x		v	
		rhy.welded tuff							x							
Jo 1980	sand	rhyolitic tuff						i	x						i	0
		rhy.welded tuff	x						x							

xx abundant; x common; (x) rare; * Field severity of ASR: 3 conspicuous; 2 moderate; 1 weak; 0 none.

Table D.6 Petrographic assessment of ASR in field-deteriorated concrete, adapted from Katayama [11]

Aggregate			Progress of ASR →					Grade of ASR	
			Aggregate	Paste	Aggregate	Cement paste			
			i	ii	iii	iv	v		
Type	Reaction	Rock type	Reaction rim	Sol/gel halo	Crack-filling gel		Void-filling gel	Microscopy	Field
Gravel	Early-expansive	Dacite	x	x	xx	xx	xx	3	3**
	Late-expansive	Meta-rhyolite	x					1	
		Rhyolitic tuff	xx					1	
		Silicified tuff	x	x	x	x		2	
		Sandstone		x	x	x		2	
		Siltstone			x	x		2	
		Mudstone		x	x	x		2	
		Chert	x	x	x	x	x	2	
Sand	Early-expansive	Andesite	x		xx	xx		3	
		Dacite	x	x	x	x		2	
		Glassy rhyolite*	x	x	xx	xx		3	
		Pumiceous tuff		x	xx	xx		3	
		Andesitic tuff	x	x	xx	xx		3	
	Late-expansive	Meta-andesite	x					1	
		Meta-rhyolite	x					1	
		Rhyolitic welded tuff	x					1	
		Silicified tuff	x		xx	x		2	
		Sandstone	x	x	x	x		2	
		Siltstone	x					1	
		Mudstone	x	x	x			1	
		Tuffaceous mudstone	x	x	x	x		2	
		Siliceous mudstone	x		x	x	x	2	
		Chert	x		x	x		2	
Overall evaluation								3	

* pop-out after coring, ** rupture of rebar, Occurrence of ASR: xx conspicuous, x common, (x) rare Grade of deterioration of ASR: Microscopy: 1 (weak), 2 (moderate), 3 (intense); Field: 0 (cryptic stage), 1 weak (developing stage), 2 moderate (acceleration stage), 3 intense (acceleration stage, deterioration stage)

example of petrographic assessment of ASR in a deteriorated concrete containing ruptured rebar. For the severity of the damage of ASR in concrete, see Table D.9.

D.5.1.4 Phase 4: Identification of Alkali-Reactive Minerals

After completing observation under the polarizing microscope, SEM observation on the same polished thin section is done to confirm the very point of reaction of the reactive minerals within reacted aggregate, Fig. D.5. It is important to identify the species of the reacting mineral in concrete, as well as to assess whether the type of the reaction is early-expansive or late-expansive. This can be done by identifying early-expansive silica minerals, because they are highly alkali-reactive having small pessimum proportions, they can produce ASR even if they are present in a small proportion at a lower alkali level of concrete, i.e. $Na_2Oeq < 3.0\,kg/m^3$.

D.5.1.5 Phase 5: Compositional Trends of ASR Gel and Evolutional Stage of ASR

Analytical points necessary for quantitative EDS analysis of ASR gel and CSH gel are shown in Table D.7.

Progress of ASR in Field Concrete

Compositions of ASR gel and CSH gel are determined by EDS analysis, distinguishing alite, belite, their hydrates, ASR gels (crack-filling within aggregate, crack-filling within cement paste, air voids) and their crystalline products (rosette-like crystals), Tables D.7, D.8). Results are plotted on the [Ca/Si]-[Ca]/[Na+K] diagram, distinguishing three compositional lines (ASR gel, CSH gel after alite, CSH gel after belite). Reaction products and the trend of evolution are common, irrespective of the rock types, whether ASR is early-expansive or late-expansive, or field concrete or laboratory specimen under accelerated test.

Type I evolution: During migration of ASR gel along cracks from the reacted aggregate into cement paste, this gel absorbs Ca ions from cement paste and loses alkalis, approaching a "convergent point" at which apparent chemical equilibrium is reached with CSH gel in the cement paste [3, 12], whereby losing expansion potential. At the initial stage of ASR external cracks are not evident on concrete, compositional range of ASR gel is only halfway to the convergent point (Fig. D.6a).

Progress of ASR in the accelerated CPT (80 °C, 1N NaOH)

Laboratory accelerated expansion tests (high temperature, high alkali content, alkali immersion) is able to reproduce all the processes of ASR in aged affected concrete structures, although reaction is not completed. Both ASR gels (Fig. D.7f) and rosette-like crystals (Fig. D.7d, e), similar in texture and compositions to those in the aged structures, are formed in the laboratory tests (Fig. D.7a–c). According to quantitative EDS analysis on the polished thin section, compositions and crystallinity of reaction

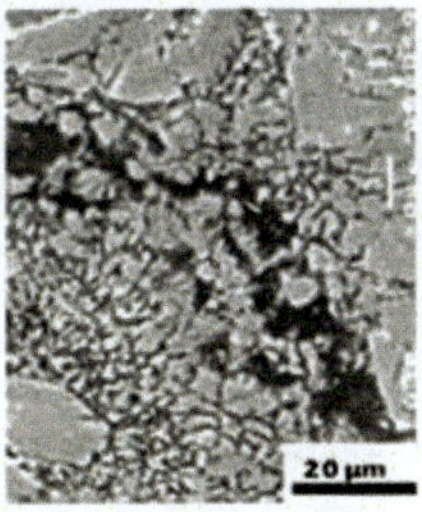

(a) Early-expansive cristobalite in andesite

(b) Reacting platy tridymite in andesite

(c) Late-expansive microcrystalline quartz in gneiss

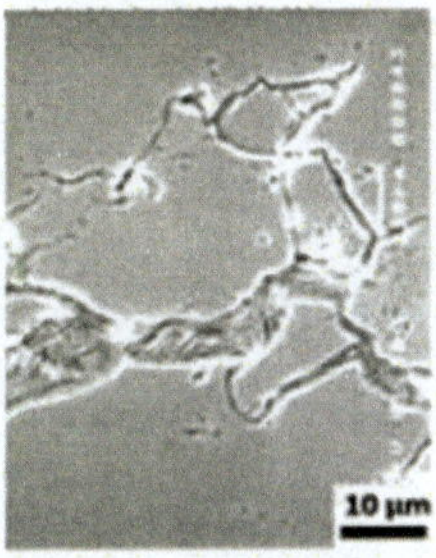

(d) Reacting microcrystalline quartz in gneiss

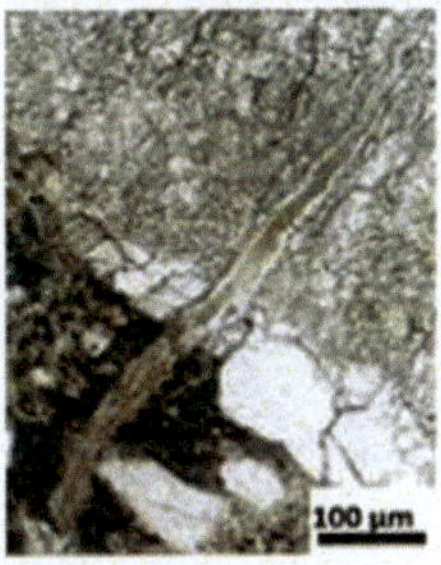

(e) Late-expansive cryptocrystalline quartz in limestone

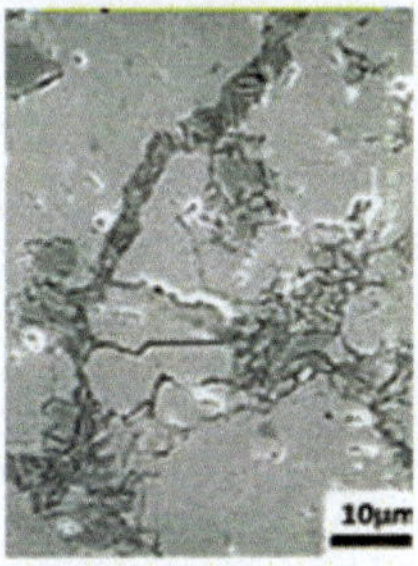

(f) Reacting cryptocrystalline quartz in limestone

(g) Late-expansive cryptocrystalline quartz in chert

(h) Reacting cryptocrystalline quartz in chert

Fig. D.5 Reacting silica minerals in the ASR-affected field concretes from [10, 12]

Table D.7 Analytical points necessary for quantitative EDS analysis of ASR gel and CSH gel in concrete

	analysis on polished thin section (thickness 15-20 μm)							
	ASR gel/rosette crystals*			cement hydrates				
	crack-filling		void-filling	alite series		belite series		
point for analysis	in aggregate	in cement paste	in air void	unhydrated alite	CSH gel after alite	unhydrated belite	CSH gel after belite	deduction from EDS analysis
	x	x	x	x	x	x	x	
compositional line	ASR gel*			alite hydrates		belite hydrates		convergent point
	x			x		x		x
	plot on the [Ca/Si]-[Ca]/[Na+K] diagram							

* rosette-like crystals should be distinguished when present

Table D.8 Compositions of ASR gel and CSH gel in concrete as analyzed by EDS on polished thin section to plot on the [Ca/Si]-[Ca]/[Na+K] diagram. Type I evolution, box culvert, 22 years old (andesite) [12]

	position of analysis		SiO_2	TiO_2	Al_2O_3	Fe_2O_3	MnO	MgO	CaO	Na_2O	K_2O	SO_3	Total	[Ca/Si]	[Ca]/[Na+K]
ASR gel	crack in aggregate	rosette-like crystal	44.46	0.20	0.26	1.12	0.32	0.00	10.09	1.68	7.82	0.00	66.64	0.26	0.87
			48.13	0.21	0.00	0.00	0.20	0.00	10.80	1.67	6.14	0.00	67.14	0.24	1.05
		gel	46.83	0.00	0.00	0.23	0.19	0.00	15.04	2.89	4.87	0.17	70.21	0.34	1.36
			44.61	0.00	0.00	0.00	0.00	0.00	19.60	2.22	4.57	0.02	71.03	0.47	2.07
	crack in paste	gel	46.23	0.33	0.00	0.16	0.01	0.00	33.85	1.17	1.35	0.53	83.64	0.78	9.13
			34.52	0.00	0.00	0.06	0.14	0.00	32.67	0.26	0.29	0.00	67.95	1.01	39.7
	in air void	gel	32.19	0.00	0.00	0.00	0.00	0.00	36.20	0.07	0.25	0.00	68.71	1.20	88.2
			31.08	0.00	0.68	0.56	0.00	0.00	35.72	0.08	0.00	0.07	68.20	1.23	243
CSH gel	hydration of alite	alite	24.50	0.57	0.98	1.33	0.32	0.84	68.70	0.34	0.36	0.15	98.11	3.01	64.6
			24.80	0.25	0.47	0.75	0.04	1.02	67.89	0.35	0.04	0.00	95.62	2.93	99.9
		gel after alite	26.02	0.39	1.45	0.00	0.00	0.04	45.58	0.16	0.00	1.51	75.15	1.88	155
			26.73	0.35	2.03	0.35	0.20	1.17	40.85	0.00	0.30	0.83	72.80	1.64	114
	hydration of belite	belite	32.27	0.11	1.59	0.91	0.00	0.00	61.42	0.57	0.25	1.14	98.25	2.04	46.4
			31.60	0.32	1.66	0.52	0.00	0.00	57.87	0.93	0.59	1.19	94.67	1.96	24.4
		gel after belite	22.86	0.48	2.04	0.97	0.24	0.00	36.79	0.11	0.06	0.32	63.87	1.72	138
			24.48	0.01	1.91	1.04	0.04	1.06	34.49	0.18	0.00	0.57	63.78	1.51	104

products change regularly depending on their position in the aggregate and cement paste in concrete with the following tendency:

- Alkali-rich gel: abounds in cracks of reacted aggregate, often crystallizes into rosette-like materials.
- Calcium-rich gel: generally amorphous and occurs in cracks within cement paste and in air void.

Early literature data of ASR gel were based on semi-quantitative EDS analysis on the fracture surface of concrete, and analytical total was normalized to 100%. Because such analysis was performed independently of thin section petrography under the polarizing microscope, without distinguishing the positions of analysis, obtained data was not linked with information of the microtexture and minerals in the aggregate. Hence such analytical results and preparations were not useful and unreliable.

It should be noted, however, that in the laboratory concrete prism tests performed under fixed environment (generally at 38–60 °C), the stage of ASR does not proceed to the end point of reaction (i.e. the convergent point) (Fig. D.6c). This is because engineering expansion tests terminate within acceptable short terms (e.g. <1 year),

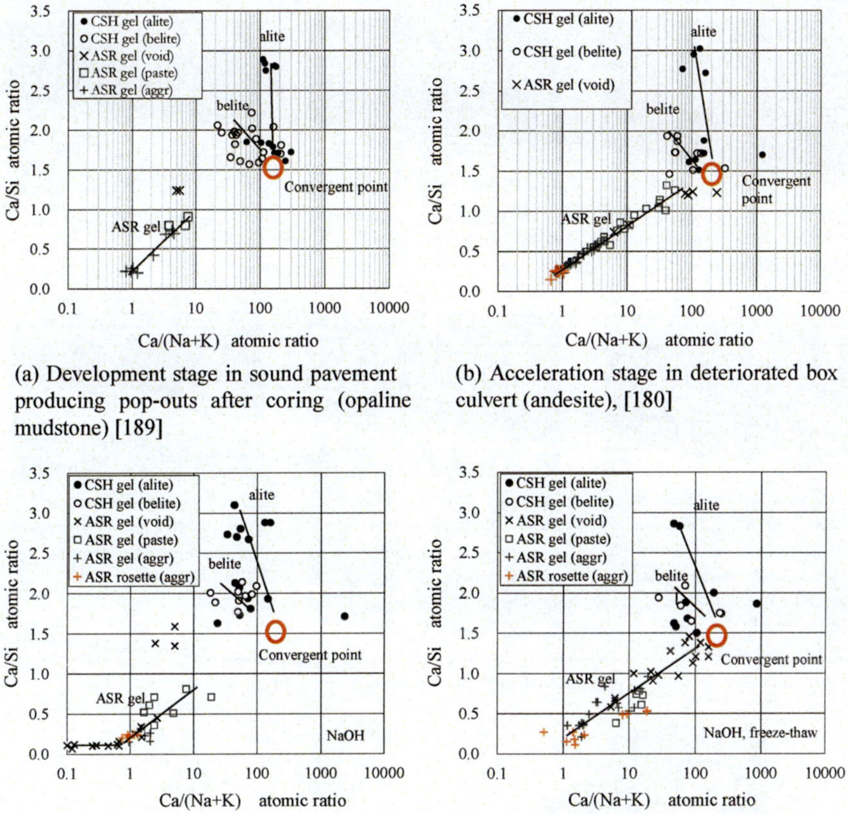

(a) Development stage in sound pavement producing pop-outs after coring (opaline mudstone) [189]

(b) Acceleration stage in deteriorated box culvert (andesite), [180]

(c) concrete prism + 80 °C 1N NaOH immersion (sandstone). Specimen prepared by Langdon [208] supplied by Bragg. Adapted from [178]

(d) Concrete prism + 80 °C 1N NaOH immersion + 300 freeze-thaw cycles (siltstone). Specimen prepared by Langdon [208], supplied by Bragg. Adapted from [178]

Fig. D.6 Type I evolution of ASR gel Similar compositional trends in field concretes and accelerated concrete prism tests

and do not always reproduce advanced reaction which is often met in aged field concrete structures.

Effect of weathering of field concrete

Aged field concrete present compositional trend line of ASR gel of the type I evolution containing alkali-rich rosette-crystals (Fig. D.8). Weathered concretes present either the type II (Fig. D.9) or the type III evolution (Fig. D.10), with Ca-rich ASR gel and alkali-leached rosette crystals [3, 12]. Typical compositional range of ASR gel is shown in Chap. 1, Table 1.3.

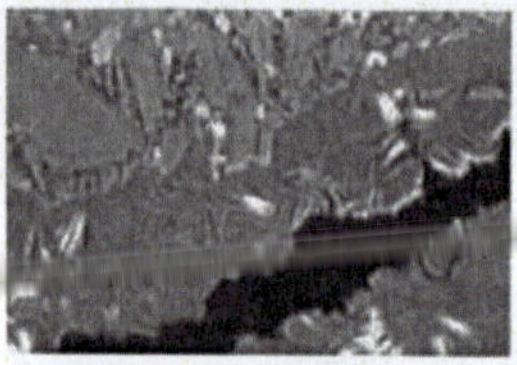

(a)ASR gel vein converted to rosette-like crystals in andesite aggregate [180]

(b)ASR gel vein in cement paste (andesite)

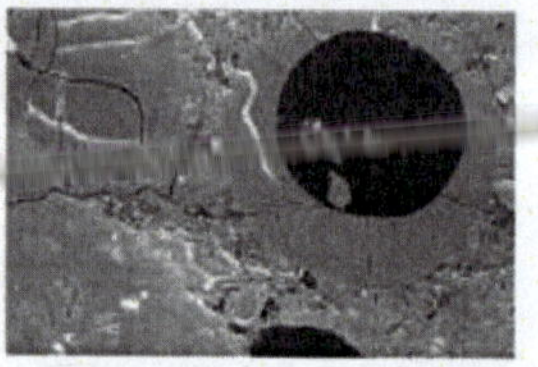

(c)ASR gel filling air voids (andesite)

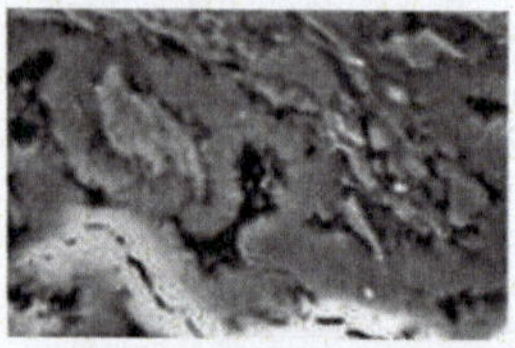

(d) CSA concrete prism immersed in 80°C 1N NaOH solution: vein of rosette-like crystals in the siltstone aggregate [178]

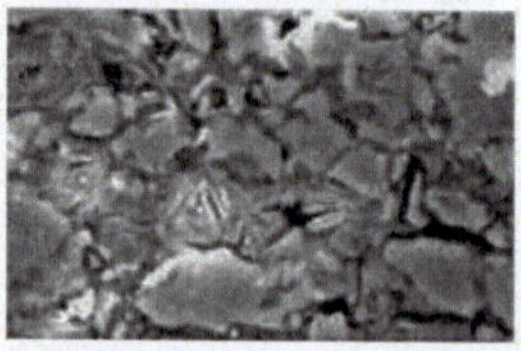

(e) CSA Concrete prism + 80°C 1N NaOH + 300 freeze-thaw cycles: rosettes around microcrystalline quartz in sandstone (greywacke), [178]

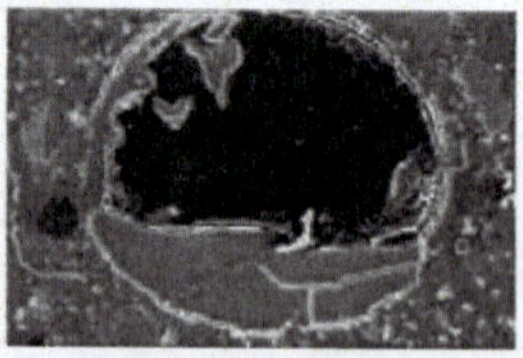

(f) CSA Concrete prism + 80°C 1N NaOH + 300 freeze-thaw cycles: amorphous gel filling air void

Fig. D.7 Similarity of ASR gel in field concretes and accelerated concrete prism tests. SEM images on polished thin section. Specimens (d),(e) and (f) prepared by Langdon [13]

(a) Late-expansive ASR of old red sandstone (71 years old)

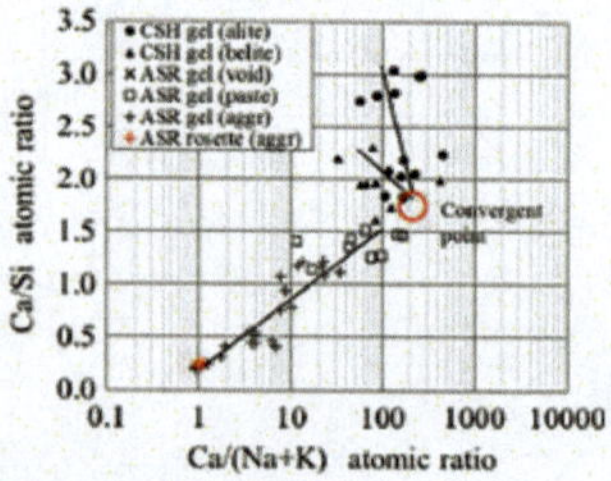

(b) Single compositional line leading to the convergent point to CSH gel

Fig. D.8 Type I evolution of ASR gel. Petty Harbor dam, Newfoundland [12, 14]

D.6 Report

Table D.6 is a reporting format for the petrographic assessment of ASR in field concrete, which contains information of coarse aggregate, fine aggregate, rock types, distinction of early-expansive or late-expansive, 5-stage propagation of ASR, and 3-grade deterioration of field structures.

Table D.8 shows an example of compositions of ASR gel (crack-filling of aggregate, cement paste, and void-filling) and CSH gel (alite, belite). Figures D.7, D.8

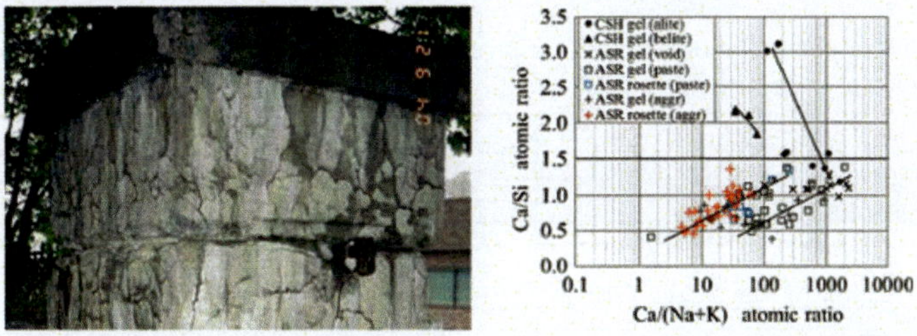

(a) Late-expansive intense ASR of sandstone and rhyolitic welded tuff, plus weak freeze/thaw (71 years old)

(b) Two parallel compositional lines representing both fresh part far from cracks and weathered part along cracks

Fig. D.9 Type II evolution of ASR gel. Gate of a residence, St. John's, Newfoundland [12, 14]

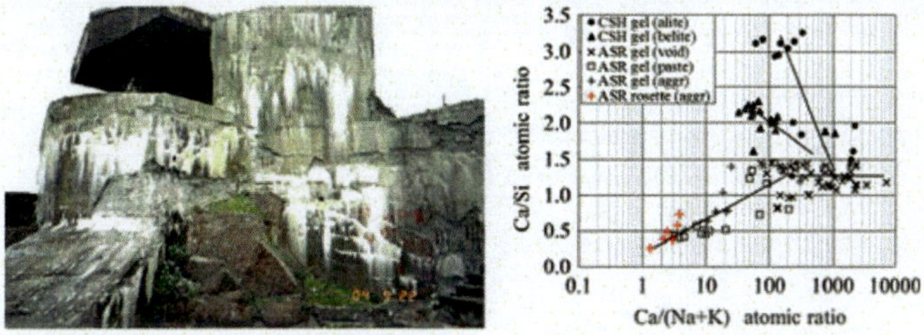

(a) Late-expansive ASR of sandstone and siltstone, plus intense freeze/thaw (51 years old)

(b) Deflected line represents severe weathering, i.e. carbonation of CSH gel and leaching of alkalis from ASR gel

Fig. D.10 Type III evolution of ASR gel. Fort Amherst, Newfoundland [12, 14]

and D.9 present some examples of plotting the EDS data on the [Ca/Si]–Ca/[Na+K] diagram.

D.7 Interpretation

D.7.1 Petrographic Examination

D.7.1.1 Phase 1: Stage of ASR

Table D.9 correlates the relationship between the conceptual evolutionary stages of ASR (cryptic, development, acceleration, and deterioration), and the six petrographic stages of ASR.

1. Cryptic stage: this is a dormant period before expansion starts, and the concrete surface looks sound. However, ASR is taking place internally, forming reaction

Table D.9 General correlation between the progress of ASR and petrographic stages in field concretes. Katayama (unpublished results)

item	process	evolution of ASR time →						
		induction stage (no cracks)		development stage		acceleration stage		deterioration stage
reaction	silica dissolution by diffusion	x	x	x	x	x	x	x
	formation of ASR gel	x	x	x	x	x	x	x
	filling of interstices by ASR gel		x	x	x	x	x	x
	exudation of ASR sol/gel into paste		x	x	x	x	x	x
	expansion of ASR gel by water-uptake			x	x	x	x	x
	crystallization of ASR gel into rosette				x	x	x	x
	alteration of ASR gel into CSH gel						x	x
expansion	expansion rate		(x)	small	increas-ing	before max	after max	decreasing
	expansion pressure		negligible	small	moderate	large	large	small
	total displacement by expansion			negligible	small	increasing	increasing	max
petrographic stages	i) reaction rims of aggregate	x	x	x	x	x	x	x
	ii) sol/gel rimming of paste around the reacted aggregate		x	x	x	x	x	x
	gel pool within reacted aggregate		x	x	x	x	x	x
	iii) cracking of reacted aggregate			x	x	x	x	x
	iv) crack propagation into cement paste				x	x	x	x
	increase of gel-filled crack width					x	x	conspicuous
	v) precipitation of gel into air voids along cracks distant from aggregate					x	x	conspicuous
	vi) network formation of gel-filled cracks in concrete						x	conspicuous
	delamination and fracturing						x	conspicuous
rebar	debonding						x	x
	fracture						(x)	x

rims and sol/gel exudation by permeation through micro-porosity within aggregate. If the alkali level of concrete is sufficient for a particular rock type to proceed ASR, and alkali contribution from the aggregate is expected in the humid environment, ASR will continue.

2. Development stage: this ranges from crack formation of the aggregate due to accumulation of the expansion force within aggregate, to the initial stage of cracking of cement paste that extends from the cracked aggregate. Compositional range of ASR gel is narrow, far from the "convergent point" with CSH gel.
3. Acceleration stage: Expansion and cracking of concrete proceed steadily at a high rate, increasing the width of gel-filled cracks that extend from the reacted aggregate. Precipitation of ASR gel along cracks distant from the reacted aggregate becomes pronounced. Lamination and fracture of concrete are characteristic of the later stage.
4. Deterioration stage: Expansion rate decreases due to equilibrium of ASR in concrete, but displacement and crack width of concrete reach a maximum, resulting in possible fracture of rebar. ASR gel within reacted aggregate is replaced by Ca ion in the cement paste, converting to CSH gel.

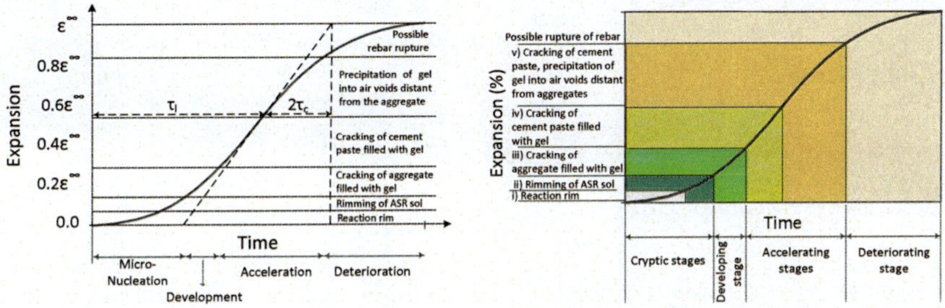

(a) Saouma, Martin, Hariri-Ardebili, and Katayama [15]

(b) Petrographic stages of ASR and expansion curve of concrete (late-expansive ASR) adapted from Katayama [10]

Fig. D.11 Petrographic and evolutionary interpretation of ASR

D.7.1.2 Phase 2: Interpretation of ASR Relevant to the Expansion Curve

Figure D.11a [15] presents a general relationship between the evolutionary stage and petrographic stage of ASR, with superimposed sigmoidal expansion curve after Larive (1998) [24] with the following three parameters for expansion (ε^{∞}) and ASR time constants (τ_L, τ_C).

$$\varepsilon(t) = \varepsilon^{\infty} \frac{1 - \exp(-t/\tau_C)}{1 + \exp(-(t - \tau_L)/\tau_C)} \tag{D.1}$$

where ε^{∞}: ultimate asymptotic expansion (%), τ_L: latency time (years), τ_C: characteristic time (years).

The above relationship with the thresholds of concrete failure (cracking, rebar fracture) could vary considerably according to the rate of expansion (early- or late-expansive ASR), type of aggregate (coarse or fine), mix design (pessimum proportion, alkali content), concrete (reinforced or not), etc., because mechanical properties of concrete are independent from chemical reaction.

Figure D.11b correlates the five petrographic stages (i, ii, iii, iv, v) of the late-expansive ASR in field concrete (Table D.5) with the evolutionary stage of ASR on the sigmoid [10]. In this figure, the highest expansion field with possible rebar fracture corresponds to the stage vi in Table D.4. By contrast, concrete prism in the accelerated condition (e.g. 60 °C, alkali-wrapping) may not present a sigmoidal curve but a simple convex curve without a marked inflection point, i.e. latency time $\tau_L = 0$.

The terminology used in the kinetics and in concrete durability does not necessarily conform. Although it has generally been accepted that expansion of concrete starts after reaching the latency time (τ_L), monitoring data of structures indicates that substantial expansion starts well before τ_L (Fig. B.13a) [1]. Because the period around τ_L is characterized by the largest expansion rate, this can be called the acceleration stage. In general, expansion rate decreases at time after $\tau_L + 2\tau_C$, i.e. the deterioration stage.

D.8 Precision and Bias

Representativeness: Manifestation of ASR is highly subject to the inhomogeneity of concrete, accessibility to water, constraint by structural elements, etc. Expansion of RC and PC structures differs by orientation (Z,X,Y) which is often largest in the vertical direction (Z), forming parallel horizontal cracks originally related to the casting direction of fresh concrete, Fig. B.12a. Hence, intensity of cracking (crack index, crack density) in thin sections may also differ considerably, depending on the orientation and position examined (surface or interior, near to rebar or not). To minimize the scattering of data, orientation for observation should be unified and the area of thin section should be increased. When average expansion is necessary, data of three directions may be averaged.

Recognition of ASR: This could differ greatly by petrographers when concrete is not severely affected by ASR. In a round robin test of core (dia. 15 cm × L30 cm) from a deteriorated structure with small estimated past expansion (e.g. 0.08%) [1], two laboratories reported ASR in both gravel and sand of andesite, while one laboratory identified ASR only in gravel, and another laboratory denied the occurrence of ASR even detecting calcium silicate gel within cracks and air-voids by EDS analysis [15]. Hence, a minimal guideline of petrography of ASR is necessary for diagnosis (Table D.5), selection of analysis points (Table D.7), preparation of EDS data (Figs. D.6, D.8, D.9, and D.10), and assessment of the progress of ASR (Table D.9).

Petrographic stage of ASR: There has been agreement that similar petrographic stages exist, and that they correlate well with expansion of extracted cores and by monitoring [1–3, 16–18]. However, with the late-expansive aggregates, early stage of ASR is likely to be missed, but a full sequence can be confirmed by SEM observation on polished thin section. Air void-filling ASR gel forms at several stages, ranging from early stage penetration into cement paste around reacted aggregate to late stage migration along crack.

Reactivity of rock types: General tendency of reactivity of each rock type recognized by concrete prism test (Fig. D.3) would apply to mass concrete including dam (Table D.3). However, reactivity of quartzite and andesite (Fig. D.3, Table D.2) may vary considerably depending on the crystallization in the geologic time.

Pessimum-causing minerals: Conventional thin sections (thickness 30 μm) used by geologists are too thick to identify fine grained (mainly $< 10\,\mu$m) pessimum-causing minerals opal, cristobalite, and tridymite, and cryptocrystalline quartz. To identify them, very thin polished thin sections (e.g. 15 μm thick) are necessary for combined polarizing microscopy with SEM-EDS analysis [19]. The most reactive opal may no longer remain in old structures (e.g. >20 years.

Quality of EDS analysis: Accuracy of EDS data (relative error) is thought to be 1% for oxides in 50 wt.% level and 10% for 0.5 wt.% (20% for 0.1 wt%), whereas precision of data (repeatability) rarely deviates more than 1-sigma, which are comparable to conventional wet chemical analysis down to the range of 0.1–0.2 wt.% [20]. With ASR gel, compositional trend line of gel can be safely drawn at the center of scattered data points in the range of [Ca]/[Na^+K]<400 (e.g. at CaO = 35 wt.%, Na_2O = 0.05 wt.% or K_2O = 0.03 wt.%). In the weathered concrete where alkali-leaching is intense, a plateau of Ca/Si can be seen with [Ca]/[Na+K]>400, which can be detect well by WDS.

Convergent point: To depict a "convergent point" between ASR gel and CSH gel on the [Ca/Si]–[Ca]/[Na+K] diagram, EDS data with analytical total >65 wt.% of ASR products and >70 wt.% of CSH gel and cement minerals (alite and belite) can be used. Counting loss is due to water contained, porosity and penetration of electron beam into very thin rosette-like crystals ($< 0.5\mu$m thick) and CSH gel.

Relationship between time, expansion and petrographic stage: The boundaries in Fig. D.11 could vary considerably affected by a number of factors, including the type of ASR (early-expansive or late-expansive, at a pessimum proportion or not), ASR accelerator (accessibility to water, cement alkali, alkali from aggregate), ASR inhibitor (air content, stress of structure, reinforced or not), etc. (Table D.10).

D.9 Limitations and Applicability to Analysis

D.9.1 Progress of ASR

Highly cracked or laminated areas around rebars in concrete tend to escape sampling core specimens for a petrographic study of ASR. Polarizing microscopy cannot be replaced by oversimplified petrographic methods (digital microscope on sawn concrete surface, application of film scanner to thin sections), because the latter miss fine cracks within cement paste thereby underestimating the progress of ASR. The [Ca/Si]–Ca/[Na+K] diagram is not applicable to products of pozzolanic reaction, typically CASH gel that forms from clay minerals and silica minerals in argillaceous aggregate.

D.9.2 Degree of Expansion

Past expansion versus petrographic stage: once petrography confirms ASR in concrete, past expansion of concrete can be estimated by measuring crack index (mm × 100/m (%)). According to the monitoring of concrete structures over 10 years, expansion may present a sigmoidal curve with an abrupt rise (Fig. D.12a). Petrographic examinations at a corresponding interval (10 years) enable to see the

Table D.10 Comparisons of evolutionary interpretations of ASR, compiled by Katayama

<table>
<tr><td rowspan="7">see Table D.9</td><td colspan="8">evolutionary stage of ASR</td><td>reference curve</td></tr>
<tr><td colspan="2">cryptic stage</td><td colspan="3">development stage</td><td colspan="2">acceleration stage</td><td>deterioration stage</td><td rowspan="6">sigmoid
boundaries depend on the rock type
(early, or late expansive ASR)</td></tr>
<tr><td>early</td><td>late</td><td colspan="2">early</td><td>late</td><td>early</td><td>late</td><td></td></tr>
<tr><td colspan="8">petrographic stage of ASR</td></tr>
<tr><td>i</td><td>ii</td><td colspan="2">iii</td><td>iv</td><td colspan="2">v</td><td>(vi)</td></tr>
<tr><td>reaction rim</td><td>exudation of ASR sol gel pool</td><td colspan="2">cracking of aggregate</td><td>cracking of cement paste</td><td>widening of crack filled with gel</td><td>widening of crack, void filling distant from aggregate</td><td>delamination, possible rebar fracture</td></tr>
<tr><td rowspan="5">Saouma et al. (2015)
modified</td><td colspan="2">micro-nucleation</td><td colspan="2">develop-ment</td><td colspan="3">acceleration</td><td>deterioration</td><td rowspan="5">sigmoid
(late-expansive ASR)
(chert)</td></tr>
<tr><td colspan="8">petrographic stage of ASR</td></tr>
<tr><td>i</td><td>ii</td><td colspan="2">iii</td><td colspan="2">iv</td><td>v</td><td>(vi)</td></tr>
<tr><td><1/15 E_∞</td><td>1/15-2/15E_∞</td><td colspan="2">2/15-4/15E_∞</td><td colspan="2">4/15-8/15E_∞</td><td>8/15-13/15E_∞</td><td>13/15E_∞ <</td></tr>
<tr><td>reaction rim</td><td>rimming of ASR sol in paste around reacted aggregate</td><td colspan="2">cracking of aggregate filled with gel</td><td colspan="2">cracking of cement paste filled with gel</td><td>widening of gel-filled cracks in cement paste; precipitation of gel into air voids distant from aggregate</td><td>possible rebar fracture</td></tr>
<tr><td rowspan="4">Lemar-chand et al. (2014)</td><td colspan="2">initiation</td><td colspan="5">development phase</td><td>rest</td><td rowspan="4">sigmoid
Larive (1998) concrete cylinder
French limestone</td></tr>
<tr><td colspan="2">< 0.1E_∞</td><td colspan="5">0.1-0.9E_∞</td><td>0.9E_∞ <</td></tr>
<tr><td colspan="2">< (1/2)T_L</td><td colspan="5">(1/2)T_L- (3/2)T_L, Tc=0.5T_L</td><td>(3/2)T_L <</td></tr>
<tr><td colspan="2">“topochemical reaction” at the aggregate-cement paste interface; dissolution of reactive silica from the aggregates; initial filling process</td><td colspan="5">uniform reaction “through solution” production of damaging gel through water inhibition in pores and macrocracks
non-uniform “topochemical reaction” non-uniform gel pressure in microcracks and pores</td><td>“chemical equilibrium” asymptotic nature no supple-mentary gel supply</td></tr>
<tr><td rowspan="2">Pauta-tsananon & Saouma (2013)</td><td colspan="8">process</td><td rowspan="2">sigmoid
Jin et al. (2000) ASTM C1260 (waste glass)</td></tr>
<tr><td>reaction rim</td><td>gel diffusing into paste</td><td colspan="5">formation of gel pressure</td><td></td></tr>
<tr><td rowspan="2">Saouma & Perotti (2006)</td><td colspan="5">< T_L</td><td colspan="3">T_L <</td><td rowspan="2">sigmoid
concrete cylinder under compressive stresses</td></tr>
<tr><td colspan="5">retardation effect by gel absorption in concrete porosity (macrocracks by tensile cracking, diffused microcracks under high compression), low humidity</td><td colspan="2">reduction in crack opening displacement</td><td></td></tr>
<tr><td rowspan="2">Larive (1998)</td><td colspan="5">T_L</td><td colspan="3">T_L+2Tc</td><td rowspan="2">sigmoid
French limestone</td></tr>
<tr><td colspan="5">latency time = initiation/acceleration period</td><td colspan="2">characteristic time = decreasing reaction rate period</td><td></td></tr>
<tr><td rowspan="4">ISE (1992)</td><td colspan="8">expansion index of concrete structure</td><td rowspan="4">cracking depends on the humidity and state of reinforcement</td></tr>
<tr><td colspan="3">I</td><td colspan="2">II</td><td>III</td><td>IV</td><td>V</td></tr>
<tr><td colspan="3">expansion < 0.06%</td><td colspan="2">0.06-0.1%</td><td>0.1-0.15%</td><td>0.15-0.25%</td><td>0.25% <</td></tr>
<tr><td></td><td>(0.015-0.025% concrete cracking strain)</td><td>expan-sion of 0.04% is of no concern</td><td colspan="2">does not affect com-pressive strength significant-ly</td><td>reductions in bending and com-pression capacities</td><td>secondary reinforcement will not sufficiently reduce free expansion</td><td>(0.23% steel yield strain), internal delamination reduces shear and bond strength</td></tr>
<tr><td rowspan="2">Nielsen (2000)</td><td colspan="8">period</td><td rowspan="2">straight lines of k x T field concrete</td></tr>
<tr><td colspan="2">initiation</td><td colspan="5">propagation</td><td>rest</td></tr>
<tr><td rowspan="2">Grattan-Bellew (1981)</td><td colspan="8">phase</td><td rowspan="2">straight lines of k x square route T concrete prism</td></tr>
<tr><td colspan="2">induction</td><td colspan="5">main expansion</td><td>late-expansion</td></tr>
</table>

progress of ASR and expansion in terms of crack indices (Fig. D.12b) [1]. However, it should be kept in mind that although a small thin section around a reacted aggregate particle can be used for diagnosing ASR, it could overestimate crack index compared with examination of thin sections that cover a longer length of core specimen. In addition, concrete under constraint produces heterogeneous expansion in orientation and depth, hence petrographic examination (rating of petrographic stage, measurement of crack index) should ideally be made in three directions (Z, X, Y) at a certain range of depth of core (Fig. D.13).

Magnitude of cracking versus petrographic stage: petrographic stages of ASR, (i) to (vi), roughly correlate with (1) the magnitude of cracking and (2) the grade of maximum crack width, both measurable on-site (Table D.11) [1]. These parameters are ranked in arithmetic progression at regular intervals, proportional to the log of a wide range of on-site crack index and maximum crack width. According

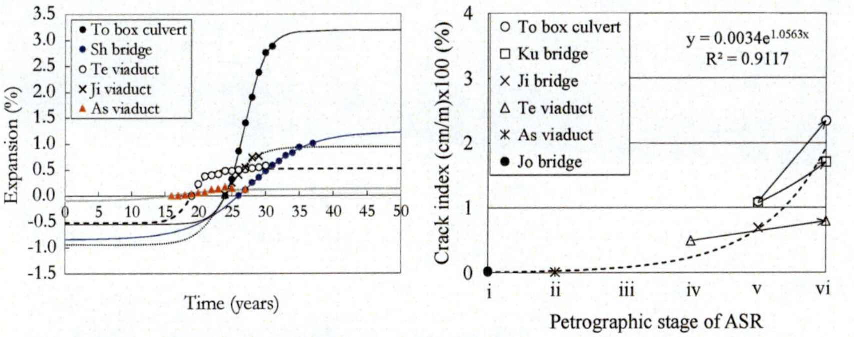

(a) Past expansion of field concretes before monitoring started (andesite and granitic rocks), drawn using data from [13] and fitted to Larive's sigmoidal curve

(b) Local crack index from visual inspection plus thin section petrography of DEF-free concrete (andesite and granitic rocks), including the same group data as (a)

Fig. D.12 Expansion from monitoring versus crack index of concrete [1]

to classical Weber-Fechner's law, this means that perceived stimuli by the human sense (magnitude, rank and stage) are proportional to the log of actual intensity of stimuli (crack index and maximum crack width).

Boundaries of each rank for ASR-affected concrete structures in Japan are tentatively suggested as (Eq. D.2–D.3, where C(%) = on-site crack index.

$$\text{M: magnitude of cracking} = \ln(C) + 5.0 \tag{D.2}$$

$$\text{Maximum crack width (mm)} = 0.2 \times \exp[1.5(M - 2)] \tag{D.3}$$

Table D.11 On-site crack index, maximum crack width, and petrographic stage of ASR, including RC and non-RC structures [1]

On-site crack index (1m x 1m)		Petrographic stage of ASR									
		i	ii	iii		iv		v		vi	
		On-site maximum crack width (mm)									
M	C (%)	no crack		< 0.1	0.1-0.2	0.2-0.4	0.4-0.9	0.9-1.9	1.9-4.0	4.0-8.5	8.5-18
1.0	<0.02	x	x								
1.5	0.02-0.05			x							
2.0	0.05-0.08				x						
2.5	0.08-0.14					x	x				
3.0	0.14-0.22						x				
3.5	0.22-0.37							x			
4.0	0.37-0.61								x		
4.5	0.61-1.00								x	x	x
5.0	1.00-1.66									x	x
5.5	1.66-2.73										x

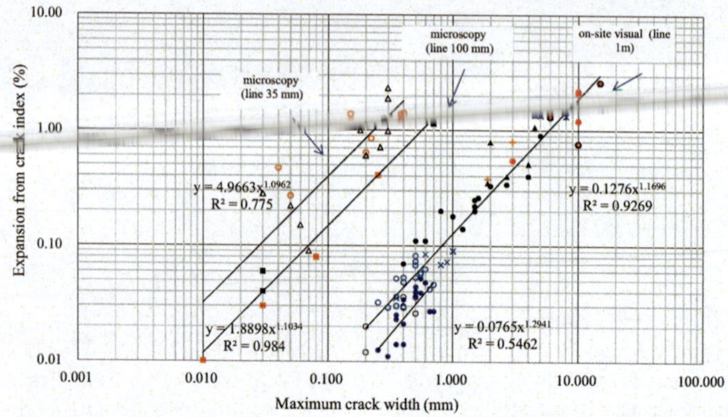

Fig. D.13 Crack indices versus maximum crack width of field concrete with several coverage lengths of examination. Adapted from [1] drawn using data partly from [21]

Results of petrographic examination should be interpreted with care, because the crack width and displacements observed on the concrete surface generally differ from those within the interior of the structure. Among severely cracked RC structures, variation of up to the magnitude 1 can be seen depending on the difference in the constraint and rupture of rebars of structure (Table D.11).

References

1. Katayama, M.: An attempt to estimate past expansion of concrete based on petrographic stage of alkali-silica reaction. In: Proceedings of the 39th International Conference on Cement Microscopy, pp. 217–236. Toronto, Canada (2017)
2. Katayama, T., Tagami, M., Sarai, Y., Izumi, S., Hira, T.: Alkali-aggregate reaction under the influence of deicing salts in the Hokuriku district. Japan. Mater. Char. **53**, 105–122 (2004)
3. Katayama, T., Oshiro, T., Sarai, Y., Zaha, K., Yamato, T.: Late-expansive ASR due to imported sand and local aggregates in Okinawa Island, southwestern Japan. In: Proceedings of the 13th International Conference on Alkali-Aggregate Reaction (ICAAR), pp. 862–873. Trondheim, Norway (2008)
4. CAN/CSA-A23.1-M90: Concrete materials and methods of concrete construction. In: Methods of Test for Concrete Canadian Standard Association (2009)
5. ASTM C227-10: Standard Test Method for Potential Alkali Reactivity of Cement-Aggregate Combinations (Mortar-Bar Method). Annual Book of ASTM Standards (2010). https://doi.org/10.1520/C0227-10
6. Katayama, T.: Petrography of alkali-aggregate reactions in concrete-reactive minerals and reaction products. East Asia Alkali-Aggregate Reaction Seminar. Tottori, pp. A45–A59. Tottori University, Supplementary paper, Tokyo (1997)

7. Katayama, T., Kaneshige, Y.: Diagenetic changes in potential alkali-aggregate reactivity of volcanic rocks in Japan-a geological interpretation. In: Grattan-Bellew, P.E. (ed.) Proceedings of the 7th International Conference on Alkali-Aggregate Reaction, pp. 489-493. Ottawa, Canada (1987)
8. Katayama, T., StJohn, D., Futagawa, T.: The petrographic comparison of some volcanic rocks from Japan and New Zealand—Potential reactivity related to interstitial glass and silica minerals. In: Proceedings of the 8th International Conference on Alkali-Aggregate Reaction (ICAAR), pp. 537–542. Kyoto, Japan (1989)
9. Katayama, T., Helgason, T., Olafsson, H.: Petrography and alkalireactivity of some volcanic aggregates from Iceland. In: Proceedings of the 10th International Conference on Alkali-Aggregate Reaction (ICAAR), pp. 377–384. Melbourne, Australia (1996)
10. Katayama, T.: Introduction to alkali-silica reaction, Part 3. How to diagnose ASR. Concrete J. (Japan Concrete Inst. JCI) **52**(12) (2014) (in Japanese)
11. Katayama, T.: Late-expansive ASR in a 30-year old PC structure in eastern Japan. In: Proceedings of the 14th International Conference on Alkali-Aggregate Reaction (ICAAR). paper 030411-KATA-05. Austin, Texas, USA (2012)
12. Katayama, T.: Petrographic study of alkali-aggregate reactions in Concrete. Ph.D. thesis. University of Tokyo (2012)
13. Langdon, S.: The effects of alkali-aggregate reactivity on the mechanical properties of concrete. M.S. thesis, University of Newfoundland. St. John's, Newfoundland, Canada (1998)
14. Katayama, T.: ASR gel in concrete subject to freeze-thaw cycles—Comparison between laboratory and field concretes from New found land, Canada. In: Proceedings of the 13th International Conference on Alkali-Aggregate Reaction (ICAAR), pp. 174–183. Trondheim, Norway (2008)
15. Saouma, V., Martin, R., Hariri-Ardebili, M.A., Katayama, T.: A mathematical model for the kinetics of the Alkali-Silica chemical reaction. Cement Concrete Res. **68**, 184–195 (2015)
16. Jensen, V.: Alkali aggregate reaction in southern Norway. Ph.D. thesis. The Norwegian Institute of Technology, University of Trondheim (1993)
17. Menéndez, E.: Evaluation and gradation of simultaneous damage in concrete affected by alkali-silica reaction and sulfate attack. Int. Rev. Civil Eng. (I.RE.C.E.) **2**(2) (2011)
18. Leemann, A., Griffa, M.: Diagnosis of alkali-aggregate reaction in dams. Tech. rep. Publication 290934. SFOE-Project SI/500863-01. EMPA, Laboratory for Concrete/Construction Chemistry (2013). http://www.bfe.admin.ch/dokumentation/energieforschung/index.html?lang=de&publication=11174
19. Grattan-Bellew, P., Katayama, T.: So-called alkali-carbonate reaction (ACR). In: Sims, I., Poole, A. (ed.) Alkali-Aggregate Reaction in Concrete: A World Review, Chapter 3, pp. 63–88. CRC Press (2017)
20. Mori, T., Kanehira, K.: X-ray energy spectrometry for electron-probe analysis. J. Geol. Soc. Japan **90**(4), 271–285 (1984). https://doi.org/10.5575/geosoc.90.271

21. H. E. P. Corporation.: Committee report on the investigation of alkali aggregate reaction. Tech. rep. See p. 17. Hanshin Expressway Public Corporation & Hanshin Expressway Management Technology Center (1986) (in Japanese)
22. Katayama, T., Sarai, Y., Sawaguchi, H.: Diagnosis of ASR in airport pavements in Japan; Early-expansive sand aggregates missed by conventional tests. In: Proceedings of the 15th International Conference on Alkali-Aggregate Reaction (ICAAR), pp. 1219–1228. Sao Paolo, Brazil 92016)
23. JCI-TC115FS §4.3: Method of measuring expansion ratio of core specimen immersed in alkali solution (immersion method using alkali solution). In: Committee Report on Diagnosis of ASR and its Ideal Future, pp. 316–319. Tokyo, (2014) (in Japanese)
24. Larive, C.: Apports Combinés de l'Experimentation et de la Modélisation à la Comprehension del'Alcali-Réaction et de ses Effets Mécaniques. Ph.D. thesis. Laboratoire Central des Ponts et Chaussées, Paris (1998). https://hal.inria.fr/docs/00/52/06/76/PDF/1997TH_LARIVE_C_NS20683.pdf

Appendix E
Sigmoidal Model for Expansion

E.1 Original Larive Model

Larive [1], on the basis obient and mechanical conditions, proposed an analytical model that governs concrete expansion. This thermodynamically-based, semi-analytical model was then calibrated using laboratory results in order to determine two key parameters: the latency time and characteristic times shown in Fig. E.1 for the normalized expansion.

$$\xi(t,T)=\frac{1-e^{-\frac{t}{\tau_c(T)}}}{1+e^{-\frac{(t-\tau_l(T))}{\tau_c(T)}}} \tag{E.1}$$

where τ_l and τ_c are the latency and characteristic times, respectively. The first corresponds to the inflection point, while the second is defined relative to the intersection of the tangent at τ_L with the asymptotic unit value of ξ. Like all chemical reactions, AAR is subject to Arrhenius Law [2], which relates the dependence of the rate constant k of a chemical reaction on absolute temperature (T expressed in degrees Kelvin, $T^oK = 273 + T^oC$) and activation energy E_a.

$$k = Ae^{-\frac{E_a}{RT}} \tag{E.2}$$

substituting k with τ_L and τ_C, Ulm et al. [3] has shown that these values at temperature T can be expressed in terms of the corresponding values at temperature T_0 through:

$$\begin{aligned}\tau_l(T) &= \tau_l(T_0)\exp\left[U_l\left(\frac{1}{T}-\frac{1}{T_0}\right)\right]\\ \tau_c(T) &= \tau_c(T_0)\exp\left[U_c\left(\frac{1}{T}-\frac{1}{T_0}\right)\right]\end{aligned} \tag{E.3}$$

© RILEM 2021
V. E. Saouma (ed.), *Diagnosis & Prognosis of AAR Affected Structures*,
RILEM State-of-the-Art Reports 31,
https://doi.org/10.1007/978-3-030-44014-5

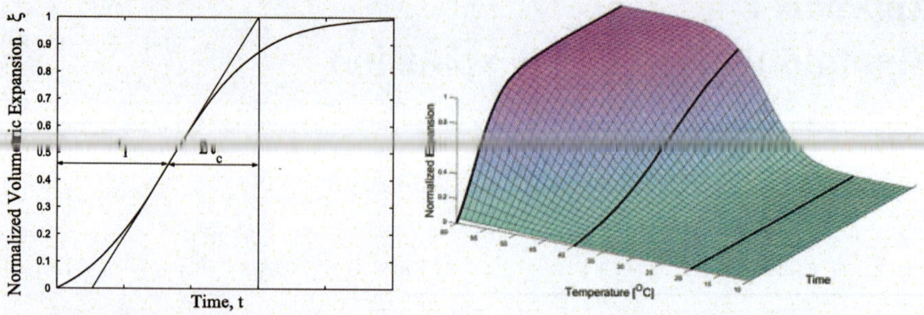

(a) Normalized Expansion Curve ($\xi(t) = \epsilon^{AAR}_{F,Vol}(t)/\epsilon(\infty)$)

(b) Effect of time and temeprature

Fig. E.1 Normalized sigmoid expansion curve [1]

Fig. E.2 Determination of activation energies

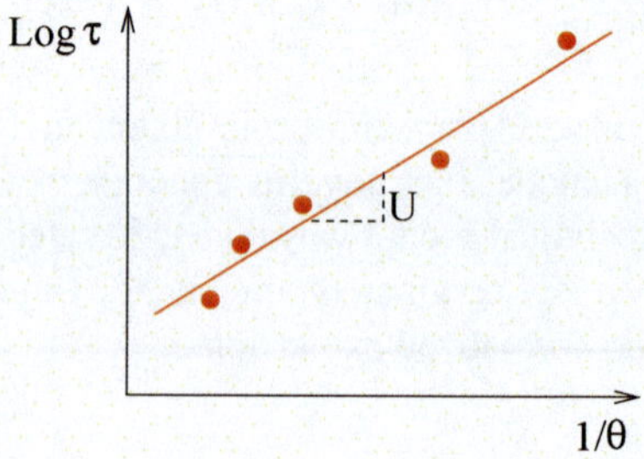

where U_l and U_c are the activation energies required to trigger the reaction for latency and characteristic times, respectively. Activation energies can be easily determined by rewriting Eq. E.3 in its non-exponential form:

$$\ln k = \ln\left(Ae^{-\frac{E_a}{RT}}\right) = \ln A - \frac{E_a}{RT} \tag{E.4}$$

which is the equation of a straight line with slope $-E_a/RT$ (Fig. E.2). We can thus determine the activation energy from values of k observed at different temperatures by simply plotting k as a function of $1/T$, Fig. E.2. Activation energies for Eq. E.3 were determined to be:

$$\begin{aligned} U_l &= 9,400 \pm 500K \\ U_c &= 5,400 \pm 500K \end{aligned} \tag{E.5}$$

Finally, further insight into the kinetics of the AAR reaction can be found in [4].

E.2 Brunetaud Extension

Brunetaud [5] extended the original Larive model for a continuously growing expansion over a long period while investigating delayed etringite formation (EDF). The original three parameters model (τ_L, τ_c, and ε^{∞}, Φ, δ) is now enriched with two correction terms (Φ and δ; $\Phi < \delta$))

$$\varepsilon(t) = \varepsilon^{\infty} \frac{1 - e^{-\frac{t}{\tau_c(T)}}}{1 + e^{-\frac{(t-\tau_l(T))}{\tau_c(T)}}} \left(1 + \frac{\tau_c}{\tau_l} \alpha_{W/C} \frac{\beta \tau_l}{\beta \tau_l + t - t_0} \right) \quad \text{From his Thesis} \tag{E.6}$$

where $\alpha_{w/c}$ impacts only the sigmoid curve, $\alpha_{0.48} = 0.0065$ and $\alpha_{0.35} = 0.0160$, and β dampens the linear portion of the expansion. It was determined that $\beta = 0.3$ would yield good results.

[6] have revisited this equation and adapted it to ASR as found it to provide a better model

$$\varepsilon(t) = \varepsilon^{\infty} \frac{1 - e^{-\frac{t}{\tau_c(T)}}}{1 + e^{-\frac{(t-\tau_l(T))}{\tau_c(T)}}} \left(1 - \frac{\Phi}{t + \delta} \right) \quad \text{Katayama} \tag{E.7}$$

The two parameters (Φ and δ) are only empirical and must satisfy $0 < \Phi < \delta$.

This formula, including a negative sign in the correction term, was used by Martin et al. [7] to describe long-term expanding behavior of delayed ettringite formation, and later by Kawabata et al. [6] for ASR expansion of alkali-wrapping concrete prisms that did not converge expansion in a short term.

E.3 Curve Fitting

When discrete expansion measurements are taken, one can extract the three parameters characterizing Eq. E.1 from a Matlab code (with the *curve fitting* package loaded), Fig. E.2. The following code could be further extended to determine activation energies, or the more suited Bruneteaud 5 parameters model (Fig. E.3).

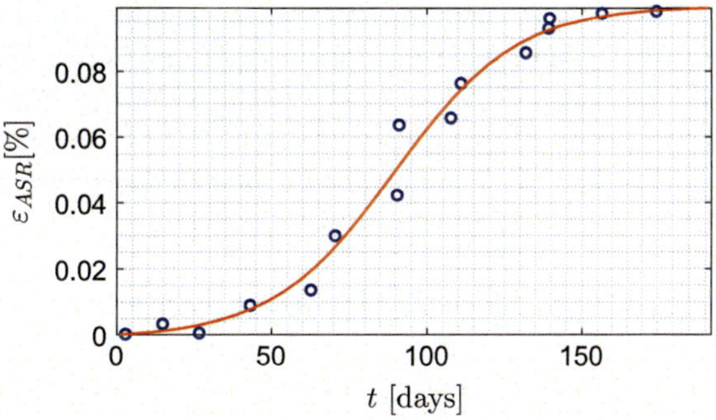

Fig. E.3 Fitted curve to discrete experimental data points

```
%==========================================================
clc; clear; close all;
%==========================================================
% Load data point from test results
[test_data, ~] = xlsread('Expansion_ASR.xlsx','A2:B16');
XX = test_data(:,1); YY = test_data(:,2);
%==========================================================
% Set up fittype and options for data fitting
ft = fittype('(1-exp(-t/tc))/(1+exp((tl-t)/tc))*einf', 'independent',...
't', 'dependent', 'y', 'coefficients',{'tc','tl','einf'});
opts = fitoptions('Method', 'NonlinearLeastSquares',...
'Algorithm','Trust-Region');
opts.DiffMinChange=1e-12;opts.Display='off';opts.TolX= 1e-09;
opts.Lower= [0 0 max(YY)]; % tc tl einf
opts.MaxFunEvals=1000;opts.MaxIter=1000;opts.TolFun=1e-09;
%==========================================================
% Fit Larive model to data points
[myfit, gof, fitout] = fit(XX, YY, ft, opts);
Optimized_coeff=coeffvalues(myfit);
tau_c_0=Optimized_coeff(1);
tau_l_0=Optimized_coeff(2);
epsi_inf=Optimized_coeff(3);
for j = 1:1:max(XX)*1.1
epsi(j)=epsi_inf*(1-exp(-j/tau_c_0))/(1+exp(-(j-tau_l_0)/tau_c_0));
time(j)=j;
end
%==========================================================
% Figure
h1 = plot(XX, 100*YY, 'bo', 'linewidth', 2.0);
hold on
h2 = plot(time, 100*epsi, 'r-', 'LineWidth', 1.5);
xlabel('$t$ [days]', 'Interpreter', 'latex', 'FontSize', 16, 'FontName', 'Arial');
ylabel('$\varepsilon_{ASR} [\%]$', 'Interpreter', 'latex', 'FontSize', 16, 'FontName', 'Arial');
grid minor;
xlim([0 max(XX)*1.1]);ylim([0 inf])
%==========================================================
```

References

1. Larive, C.: Apports Combinés de l'Experimentation et de la Modélisation à la Comprehension del'Alcali-Réaction et de ses Effets Mécaniques. Ph.D. thesis. Laboratoire Central des Ponts et Chaussées, Paris (1998). https://hal.inria.fr/docs/00/52/06/76/PDF/1997TH_LARIVE_C_NS20683.pdf
2. Arrhenius, S.: On the reaction velocity of the inversion of cane sugar by acids. Zeitschrift für physikalische Chemie **4**, 226 (1889)
3. Ulm, F., Coussy, O., Li, K., Larive, C.: Thermo-chemo-mechanics of ASR expansion in concrete structures. ASCE J. Eng. Mech. **126**(3), 233–242 (2000)
4. Saouma, V., Martin, R., Hariri-Ardebili, M.A., Katayama, T.: A mathematical model for the kinetics of the Alkali-Silica chemical reaction. Cement Concrete Res. **68**, 184–195 (2015)
5. Brunetaud, X.: Étude de l'influence de différents paramètres et de leurs interactions sur la cinétique de l'amplitude de la réaction sulfatique interne au béton. Ph.D. thesis. Châtenay-Malabry, Ecole centrale de Paris (2005)
6. Kawabata, Y., Yamada, K., Ogawa, S., Martin, R., Sagawa, Y., Seignol, J.F., Toutlemonde, F.: Correlation between laboratory expansion and field expansion of concrete: prediction based on modified concrete expansion test. In: Proceedings of the 15th International Conference on Alkali-Aggregate Reaction (ICAAR). Sao Paolo, Brasil (2016)
7. Martin,R., Renaud,J., Toutlemonde, F.: Experimental investigations concerning combined delayed ettringite formation and alkali aggregate reaction. In: 6th International Conference on Concrete Under Severe Conditions CONSEC10, p. 8. Mexico, June (2010)

Appendix F
Sample Conditioning and Core Cutting

Frederic Dunand, Tetsuya Katayama, Leandro Sanchez

F.1 Sample Conditioning

Fredric Dunand, Leandro Sanchez

It is well established that the specimens conditioning history (e.g. wetting/drying conditions) prior to testing changes the mechanical responses of concrete [1].

When AAR test procedures are carried out on concrete cylinders cast and tested at the same laboratory, a strict control of the specimen's moisture condition before testing is often applied/selected (e.g. storage in the moist curing room at a controlled temperature). However, when cores drilled from structures in service are evaluated or even laboratory samples should be transported before testing, variations in the moisture condition of the test specimens from coring (or shipping) to testing can occur which may significantly affect the final results. Although it has been shown that some laboratory samples dried than re-wetted resume their expansion behaviour despite increasing the variability among companion samples, it is frequently desirable to transport samples procured in the field in such a state that reflects as closely as possible the condition of their environment. More importantly, field experience is that there can be considerable variation between samples when drying occurs between field and laboratory [2].

F. Dunand
EPFL Lausanne, Lausanne, Switzerland
e-mail: cyrille.dunant@epfl.ch

T. Katayama
Taiheiyo Consultant Co., Ltd., Sakura, Japan
e-mail: katayama@catv296.ne.jp

L. Sanchez
University of Ottawa, Ottawa, Canada
e-mail: Leandro.Sanchez@uottawa.ca

© RILEM 2021

V. E. Saouma (ed.), *Diagnosis & Prognosis of AAR Affected Structures*,
RILEM State-of-the-Art Reports 31,
https://doi.org/10.1007/978-3-030-44014-5

For conventional procedures, ASTM C 42 requires that cores should be tested "in the same moisture condition than that they were in the field". However, a moisture gradient may be observed within concrete elements in contact or not with an external source of moisture thus causing possible variations in the test results. In order to minimize changes in moisture conditions, it is recommended to wrap and store the cores for at least 5 days before testing [3]. Canadian Standards suggest that the cores extracted from structures be rewetted for 48 hours in a moist curing room prior to testing for conventional mechanical procedures (e.g. compressive strength and modulus of elasticity), thus allowing to evaluate the samples with the same moisture's condition [4]. Otherwise, there is scant literature on the conditioning of ASR samples for transport, so most of these recommendations are based on the works of Lindgard et al. [5], complemented by the works on sample drying from Toutlemonde and Le Maou [6] and Sanchez et al. [7]. These works do not address directly the question of the sample's transport, but give guidance as to the effect or properties of various protocols.

F.1.1 Samples Extraction

While the extraction and transportation of concrete cores, ASTM C 42 requires that cores be tested "in the same moisture condition than that they were in the field". However, a moisture gradient is often observed within concrete elements in contact or not with an external source of moisture, which very likely increases the variations in the test results. In order to minimize changes in moisture conditions, it is recommended to wrap the cores in plastic film (or suitable materials—see section below) and store them for at least five days before testing ASTM, C42/C42M-18a [3].

F.1.2 Samples Sealing and Conditioning Prior to Testing

The materials which are typically used for sample conditioning are:

- Polyurethane sheets: these plastic sheets allow a tight binding of the sample, however, they are slightly porous and when used exclusively, a sample needs to be wrapped with 20 layers of the material for good isolation. A better method is to use sturdier sheets of plastic and vacuum seal the sample.
- Adhesive aluminium sheets: such sheets are used to control the humidity of samples in creep experiments, they are the most effective means to produce a seal. However, this cover can be easily damaged, and it is recommended that a plastic wrapping be added to protect the aluminium. Further, removing the aluminium is difficult if it is applied directly on the sample.
- Damp cotton cloth: damp cotton cloth can be used to avoid desiccation in conjunction with plastic wrapping. It is placed between the samples and a polyurethane

wrap. This keeps the sample for drying, but provides a significant source of water and can therefore induce leaching. As aluminium is known to play a role in the kinetics of ASR [8], it may be undesirable to have the aluminium for in direct contact with the sample, beyond considerations on the ease of unwrapping. It is however the most reliable method to preserve the humidity of the sample over long periods of time. Vacuum sealing samples in plastic provide good isolation for typical transportation times and may be easier to apply when on-site. In any case, using damp cloth to provide a source of moisture is not recommended as it can lead to leaching, does not preserve the sample in its pristine state, and is delicate to do correctly.

The procedure of samples conditioning before testing is still largely debatable. Although re-saturation was verified to be convenient, re-setting somehow the initial condition of ASR affected samples in a number of works [9, 10], some researchers found that this procedure might cause dramatic effects on AAR prognosis, making its further development unreliable [2]. Otherwise, drying effect is one of the main parameters affecting the reliability of AAR diagnosis and prognosis test procedures and thus it should be avoided. Moreover, it has been found that drying effects happen even in well-wrapped concrete specimens sealed in plastic sheets, but with a much lower degree compared to unsealed samples. Therefore, a maximum of a 48-hour rewetting period as by CSA23.2-14C (48 h "re-saturation" in the moist curing room (protected from running water) at 23 ± 2°C and 100% R.H.) was found largely suitable to contribute at reducing the deleterious effect of drying of AAR test specimens, restoring their initial conditions and thus enabling reliable condition assessment. Finally, it should be noted that re-wetting procedures tend to increase the variability of results of companion samples.

F.1.3 Procedure

In order to minimize the transportation and storage effects on reliability of AAR diagnosis and prognosis test procedures, 6 very important steps are needed to mitigate further problems due to transportation and/or storage of AAR specimens:

1. After casting or extraction, the samples should be weighted and well-wrapped.
2. In the case of testing drilled cores, the specimens should be stored in the laboratory for at least five days prior to testing, in order to homogenize their moisture degree;
3. Once wrapped, in the case of mechanical testing, the ends of the samples should be prepared (e.g. grinding or capping). The length to diameter ratio of the samples should be selected as a function of the test to be performed but shall not be lower than 2.
4. In the case of mid/long-term transportation and or storage, a temperature equal or slightly lower to 12 °C should be selected. This temperature was found to be able to avoid AAR further development.

5. After transportation and/or period of storage, the samples should be weighted again to assess potential drying. Then, in the case of significant difference from their original weight, samples shall be placed in the moisture curing room [4] at $23 \pm 2\,^{o}$C and 100% R.H. (protected from running water) until either the samples reach their initial measured weight or for a maximum of 48 hours. The period of re-saturation shall not exceed 48 hours;
6. Samples are ready for testing.

F.1.4 Shipment

During transportation and shipment, cores must be placed in a temperature-controlled box to ensure $T \leq 12$°C. It was found that

FedEx has temperature range that vary between 2 and 8 °C for up to 96 h (4 days), [11].

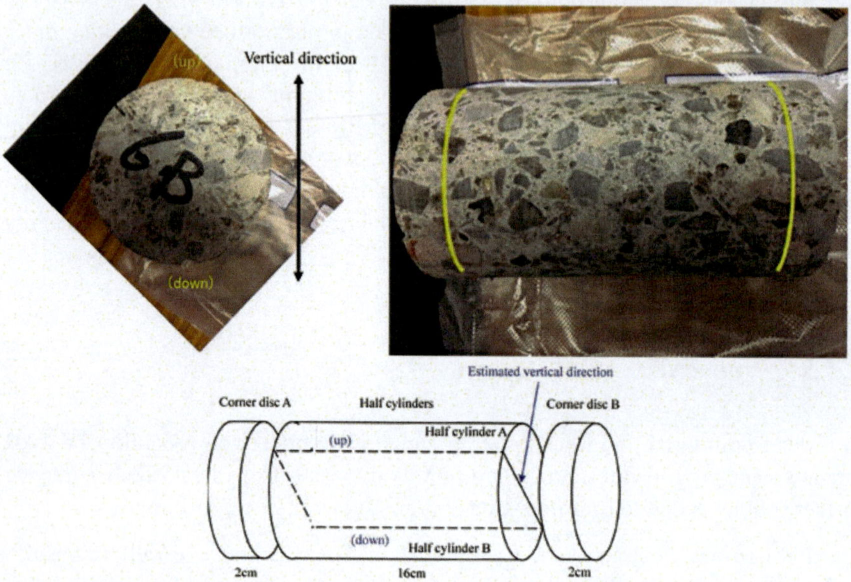

Fig. F.1 Recommended procedure for core cutting

F.2 Core Cutting

Recommended core cutting procedure for petrographic analysis is illustrated in Fig. F.1.

References

1. Mindess,S., Young, J.F., Darwin, D.: Concrete, 2nd edn. Pearson Education Ltd. (2003)
2. Merz, C.: Personal Communication to Sanchez L (2015)
3. ASTM, C42/C42M-18a: Standard test method for obtaining and testing drilled cores and sawed beams of concrete. Tech. rep. West Conshohocken, PA (2018)
4. CSA23.2-14C: Obtaining and Testing Drilled Cores For Compressive Strength Testing. In: Canadian Standards Association (2009)
5. Lindgard, J., Sellevold, E., Thomas, M., Pedersen, B., Justnes, H., Ronning, T.: Alkali-silica reaction (ASR)-performance testing: Influence of specimen pre-treatment, exposure conditions and prism size on concrete porosity, moisture state and transport properties. Cement Concrete Res. **53**, 145–167 (2013)
6. Toutlemonde, F., Le Maou, F.: Protection des éprouvettes de béton vis-àvis de la dessication-Le point sur quelques techniques de laboratoire. Tech. rep. 203. Bulletin des laboratoires des ponts et chaussées (1996)
7. Sanchez, L., Fournier, B., Jolin, M., Bastien, J.: Evaluation of the Stiffness Damage Test (SDT) as a tool for assessing damage in concrete due to alkalisilica reaction (ASR): input parameters and variability of the test responses. Construct. Build. Mater. **77**, 20–32 (2015)
8. Larive, C.: Apports Combinés de l'Experimentation et de la Modélisation à la Comprehension del'Alcali-Réaction et de ses Effets Mécaniques. Ph.D. thesis. Laboratoire Central des Ponts et Chaussées, Paris (1998). https://hal.inria.fr/docs/00/52/06/76/PDF/1997TH_LARIVE_C_NS20683.pdf
9. Chappex, T.: The role of aluminium from supplementary cementitious materials in controlling alkali-silica reaction. Ph.D. thesis. École Polytechnique Féddrale de Lausanne (2012)
10. Sanchez, L.: Contribution to the assessment of damage in aging concrete infrastructures affected by alkali-aggregate reaction. Ph.D. thesis. Laval University, Department of Geology and Geological Engineering (2014)
11. FedEx: Temperature Controlled Shipping Boxes (2017). http://www.fedex.com/us/healthcare/product-integrity/temperature-control/

Appendix G
Alkali Silica Reactor

G.1 Sample Conditioning

Some of the tests reported in this book require special storage of specimens at temperature higher than 38special test chamber (reactor) composed of a container capable of maintaining a nearly 100% relative humidity Fig G.1.

Nixon [1] characterize the reactor as:

> *Rigid containers for the test specimens made of plastic or other material resistant to corrosion by a solution of sodium hydroxide at a temperature of 80° C for a prolonged period of time. Each container must be of such dimension and shape to accommodate at least three specimens and must be provided with lids or other suitable means to prevent loss of moisture by leaking or evaporation. The prisms must be positioned and supported in such a way that the solution has access to the whole of the bar. It should further be ensured that the specimens do not*

(a) Large container (b) Small container

Fig. G.1 Containers and reacotr (Courtesy Dr. A. Leemann (Empa)

© RILEM 2021

V. E. Saouma (ed.), *Diagnosis & Prognosis of AAR Affected Structures*,
RILEM State-of-the-Art Reports 31,
https://doi.org/10.1007/978-3-030-44014-5

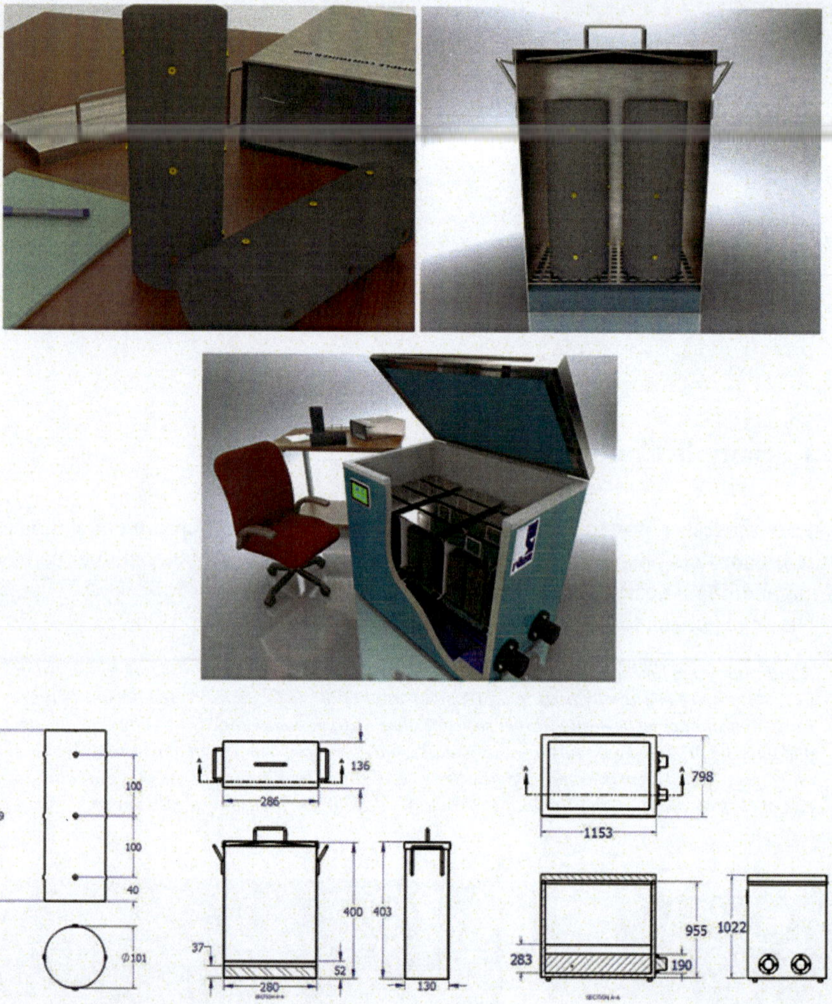

Fig. G.2 Specimen, container and reactor (all dimensions in mm); adapted from RILEM/TC 258-AAA-WP1-N022, courtesy of Dr. Jan Lingård

touch each other or the sides of the container. The specimens, if stood upright in the solution, shall not be supported by the steel pins.

[1]

As such an equipment is not commonly available in concrete laboratories, Fig. G.2 provide some guidance for in-house manufacturing.

Reference

1. Nixon, P.: Detection of potential alkali-reactivity of aggregates. The ultra accelerated mortar-bar-test. Mater. Struct. **33**(229), 283–289 (2000)

Subject Index

© RILEM 2021
V. E. Saouma (ed.), *Diagnosis & Prognosis of AAR Affected Structures*,
RILEM State-of-the-Art Reports 31,
https://doi.org/10.1007/978-3-030-44014-5

Author Index

© RILEM 2021

V. E. Saouma (ed.), *Diagnosis & Prognosis of AAR Affected Structures*,
RILEM State-of-the-Art Reports 31,
https://doi.org/10.1007/978-3-030-44014-5

M

N